THERMODYNAMICS OF IRREVERSIBLE PROCESSES

THERMODYNAMICS OF IRREVERSIBLE PROCESSES

Applications to Diffusion and Rheology

Gerard D. C. Kuiken
Delft University of Technology
Delft, The Netherlands

JOHN WILEY & SONS
Chichester · New York · Brisbane · Toronto · Singapore

Baffins Lane, Chichester
West Sussex PO19 1UD, England
Telephone: National Chichester (0243) 779777
International +44 243 779777

Other Wiley Editorial Offices

John Wiley & Sons, Inc., 605 Third Avenue,
New York, NY 10158-0012, USA

Jacaranda Wiley Ltd, 33 Park Road, Milton,
Queensland 4064, Australia

John Wiley & Sons (Canada) Ltd, 22 Worcester Road,
Rexdale, Ontario M9W 1L1, Canada

John Wiley & Sons (SEA) Pte Ltd, 37 Jalan Pemimpin #05-04,
Block B, Union Industrial Building, Singapore 2057

British Library Cataloguing in Publication Data

A catalogue record for this book is available from the British Library

ISBN 0 471 94844 6

From camera-ready copy supplied by the author using Tex.
Printed and bound in Great Britain by Bookcraft (Bath) Ltd, Midsomer-Norton, Avon.

Series Preface

Theoretical Chemistry is one of the most rapidly advancing and exciting fields in the natural sciences today. This Series is designed to show how the results of theoretical chemistry permeate and enlighten the whole of chemistry together with the multifarious applications of chemistry in modern technology. This is a series designed for those who are engaged in practical research, in teaching and for those who wish to learn about the role of theory in chemistry today. It will provide the foundation for all subjects which have their roots in the field of theoretical chemistry.

How does the materials scientist interpret the properties of a novel doped-fullerene superconductor or a solid-state semiconductor? How do we model a peptide and understand how it docks; and how does an astrophysicist explain the components of the interstellar medium? Where does the industrial chemist turn when he wants to understand the catalytic properties of a zeolite or a surface layer? What is the meaning of "far-from-equilibrium" and what is its significance in chemistry and in natural systems? How can we design the reaction pathway leading to the synthesis of a pharmaceutical compound? How does our modelling of intermolecular forces and potential energy surfaces yield a powerful understanding of natural systems at the molecular and ionic level? All these questions will be answered within our series which covers the broad range of endeavour referred to as 'theoretical chemistry'.

The aim of the series is to present the latest fundamental material for research chemists, lecturers and students across the breadth of the subject, reaching into the various applications of theoretical techniques and modelling. The series concentrates on teaching the fundamentals of chemical structure, symmetry, bonding, reactivity, reaction mechanism, solid state chemistry and applications in molecular modelling. It will emphasise the transfer of theoretical ideas and results to practical situations so as to demonstrate the role of theory in the solution of chemical problems in the laboratory and in industry.

D. Clary, A. Hinchliffe and D. S. Urch
June 1994

Preface

The macroscopic theory of the thermodynamics of irreversible processes, abbreviated by the acronym TIP, is a part of the continuum theory and attempts to describe within the framework of the continuum point of view the linear constitutive equations of state. The TIP aims for unification and generalization of the formulation of the equations describing systems with a linear constitution. The results apply often to quasi-linear systems. It unifies the formulations of the continuum theory used in many branches of physics and chemistry such as mechanics, fluid dynamics, magnetohydrodynamics, mass transport, and thermodynamics, and provides us with the equations of state needed to describe the transport processes discussed in interdisciplinary areas. The methodology of the thermodynamics of irreversible processes applies to all linear processes and processes that can be treated as linear within the bounds of the model representations of the relevant system.

This textbook has grown out of lectures at the Delft University of Technology in the Netherlands by the author and the late Professor H.J. Merk. The first chapter discusses the continuum point of view of matter. In the second chapter the basics are summarized of the classical thermodynamics needed in the subsequent discussion. Although classical, a derivation of the second law of thermodynamics is given in this chapter, which the author found illustrative and easy to understand even for students in a first year's undergraduate course. By introducing a vector representation for heat engines a clear picture is offered to distinguish between the irreversible, the reversible and the impossible processes. Chapter three provides an overview of the TIP axioms and some commonly encountered general axioms. The chapter is concluded with a diagram showing the methodology used in the thermodynamics of irreversible processes, and illustrated by some examples. In chapter four, the balance equations for mass, momentum, angular momentum, energy and entropy are formulated for multicomponent systems subjected to thermal and electromagnetic fields.

In chapter five the Onsager–Casimir reciprocal relations are derived. The derivation does not take Boltzmann's entropy identification for granted, but derives the probability distribution by solving the Fokker–Planck equation for aged systems. By doing so, it is shown that the Boltzmann distribution is one possible probability distribution that yields the Onsager symmetry relations. The derivation shows that not all Gaussian probability distributions that satisfy both the regression axiom and microscopic reversibility yield the Onsager reciprocal relations. The opinion in literature that microscopic reversibility brings about the Onsager symmetry relations is true only if one starts with the Boltzmann distribution.

The second half of the book discusses the applications of the general theory of the thermodynamics of irreversible processes to multicomponent diffusion in chapter six and to rheology in chapter seven. Diffusion is a vector process, and as such not coupled with scalar and tensor processes. Mass transfer has important applications, and has cross-effects with vector processes only. Diffusion due to heat transfer is an example.

Cross-effects in nondilute systems can be dealt with systematically only by the methods developed in the TIP. However, for dilute systems the kinetic theory of gases and fluids provides values of the constitutive coefficients and yields detailed results, not obtainable from the TIP. On the other hand the thermodynamics of irreversible processes generalizes several known dilute-gas results to arbitrary fluids, if these fluids do not show strong nonlinear effects or are not too complicated like polymer melts. The results of the kinetic theory of gases turn out to apply to dense gases and liquids as well, if the mass fractions or the molar fractions are systematically replaced by the corresponding chemical potentials. The results of the thermodynamics of the irreversible processes reduce to the results of the kinetic theory of gases for ideal systems. It is pointed out that a general definition of the heat flux in a multicomponent system is not possible, but that the definition is determined by the thermodynamic processes under which the heat transport takes place.

The Maxwell–Stefan equations for the description of diffusion are derived directly from the TIP and are used in chapter six as the starting point for describing diffusion. The conditions are derived for the possibility of negative Maxwell–Stefan diffusivities, which are experimentally found in electrolyte systems. It is shown that the validity of the Maxwell–Stefan equations is not restricted to dilute systems or to ideal gases. The relevant Maxwell–Stefan diffusivities are not always well known but can be calculated from the Fick diffusivities by using the relation between the two types of diffusivities. Estimates for the diffusivities are given. A discussion is given of why negative diffusivities are not likely to be found in non-electrolyte systems. In numerical calculations of mass transfer problems the Maxwell–Stefan diffusion equations can be a real advantage over other descriptions.

For binary systems the various diffusion descriptions can easily be transformed into each other. For multicomponent systems this is not so easy. The diffusivities for other diffusion descriptions can be calculated from the Maxwell–Stefan diffusivities and the relevant transformations are derived. A brief discussion of the method of de Groot and Mazur is followed by considering the validity of the idea of mechanical equilibrium used by de Groot and Mazur. Finally the simplifications for obtaining the most simple diffusion equations are explored.

In chapter seven the thermodynamic theory of the relaxation processes is treated. This theory is based on the thermodynamics of the irreversible processes combined with the assumption that the system has internal mechanisms. Systems with internal mechanisms are called incomplete systems, and such systems cannot be described completely by variables which are in principle measurable only macroscopically. The complete systems discussed in the previous chapters do not show relaxation. Relaxation phenomena occur only if the systems are incomplete. Internal processes found in such systems cannot be observed directly externally. A simple example is a dipole for which only the potential difference and the current between both poles are measured. These are the externally measurable quantities. The dipole itself could consist of a complicated network with many branches. The potential differences and currents through those branches cannot be measured externally and have to be considered as internal quantities in a macroscopic description. A system in which internal processes proceed, is incomplete only insofar as the external measurable quantities do not describe the system completely.

In general it might therefore happen that within the system, modifications and processes occur which are not directly measurable externally and that therefore the external variables describe the system incompletely. The internal mechanisms could be of microscopic nature. The mechanisms, which might proceed under certain circumstances on a microscopic level, can be observed directly only by using Röntgen rays or by other methods to investigate the micro structure. The microscopic mechanisms are modeled in this chapter by internal variables. Elimination of the unmeasurable internal variables results in relaxation equations, for which in the thermodynamic theory the conditions—adiabatic, isochoric, isotherm, isobaric and so on—under which the relaxation processes proceed can be taken into account consistently. For the scalar processes this is extensively discussed, and it is shown that one internal mechanism already gives rise to several processes with different relaxation times. For the vector processes the finite heat waves are discussed and for the tensor processes the linear viscoelastic bodies. The nine viscoelastic classes include the degeneracies possible for continuous spectra. These degeneracies yield fractional differential equations and are not strictly linear.

In Appendix A, the electromagnetic field equations for moving polarizable media, used in chapter four, are derived with a minimum of concepts. The basic concept is that there are charged particles, which produce an electric and a magnetic field depending on the state of motion of the charged particle. The Ampére, Boffi, and Minkowski formulations of the Maxwell field equations for moving media are then obtained. Appendix B reviews the tensor notation.

I am indebted to the late Professor Merk for his introductions in the theory of the thermodynamics of irreversible processes, and to Professor R.B. Bird for his stimulation to write this book, and his kindness to correct the English of the final version during his appointment as J.M. Burgers Professor in Delft.

Delft, The Netherlands G.D.C. Kuiken
April 1994

Contents

Nomenclature

Number at the end of the symbol name refers to the page of definition.

a	coefficient defined in (7.184) [Pa kg m^{-3}], 288
a	velocity of sound [m/s], 298
a_j	surface of volume b_j [m^2], 369
A	affinity for thermal relaxation [-], 295
A	constant used in (7.438) [various], 351
A	scalar affinity [various, J mol^{-1}], 259
A	zero pressure value of pV [N m], 17
A_α	affinity [J/mol], 129
$\mathcal{A}$	magnetic vector potential [N/A], 373
A_{ij}	empirical parameter [-], 212
A_k	the cross-section of the k-th conductor [m^2], 371
$[A]$	matrix defined by (6.136) [-], 221
$d\vec{A}$	surface element [m^2], 62
b	coefficient defined in (7.184) [Pa K kg J^{-1}], 288
b_α	phenomenological coefficients [various, J mol^{-1} K^{-1}], 261
b_e	phenomenological coefficient [Pa/K], 260
b_j	tiny volume around singularity at $\vec{r}_j$ [m^3], 369
$\vec{b}$	a vector field [various], 374
$\vec{b}$	constant vector specifying the origin O, 251
B	constant of integration [K^{-1}], 296
$\vec{B}$	magnetic flux density [kg A^{-1} s^{-2}], 114
c_0	speed of light in vacuum [299 792 458 m/s], 114
$c_{\alpha\beta}$	phenomenological coefficients [various, Pa], 261
c_{ee}	phenomenological coefficient [Pa], 259
$c'_{e\gamma}$	phenomenological coefficients [various, Pa], 259
c_K	partial molar density [mol m^{-3}], 107
c_p	specific heat at constant pressure [J K^{-1} kg^{-1}], 276
c_v	specific heat at constant volume [J K^{-1} kg^{-1}], 276

$\boldsymbol{c}$	Cauchy deformation tensor [-], 250
C	closed current loop [m], 374
C	the Euler constant [0.5772···], 355
C	heat capacity [J/K], 86
$\bar{C}_K$	defined in (7.324) [Pa^{-1}], 318
$[C_K]$	chemical symbol of element K [mol], 112
C_p	heat capacity at constant pressure [J/K], 43
C_V	heat capacity at constant volume [J/K], 43
$\boldsymbol{C}$	Green deformation tensor [-], 251
$đ$	inexact differential [-], 21
$\vec{d}_K$	diffusion vector [m^{-1}], 183
$\vec{d}'_K$	diffusion vector [m s^{-2}], 220
$[\vec{d}'_L]$	column matrix of the vectors $[\vec{d}'_L]$ [m s^{-2}], 221
$\vec{d}^{\star}_A$	molar diffusion vector [m^{-1}], 199
$\vec{d}^{\circ}_A$	barycentric diffusion vector [m^{-1}], 199
$\vec{D}$	electric displacement [A s m^{-2}], 114
D/Dt	material or substantial derivative [s^{-1}], 110
$D^{\star}/Dt$	molar time derivative [s^{-1}], 111
De	Deborah number, 311
$Đ$	Maxwell–Stefan diffusivity [$\text{m}^2\ \text{s}^{-1}$], 232
$Đ_{KL}$	Maxwell–Stefan diffusivities [$\text{m}^2\ \text{s}^{-1}$], 189
$Đ^{J\to 1}_{IJ}$	diffusivity at infinite dilution [$\text{m}^2\ \text{s}^{-1}$], 215
D'_{AB}	molar diffusivity relative to mass average velocity [$\text{m}^2\ \text{s}^{-1}$], 200
$\mathbb{D}_{AB}$	Fick diffusivity [$\text{m}^2\ \text{s}^{-1}$], 197
$\mathbb{D}^{(p)}_A$	binary pressure diffusivity [$\text{m}^2\ \text{s}^{-1}$], 197
$\mathbb{D}^{(p)}_A$	pressure diffusivity [$\text{m}^2\ \text{s}^{-1}$], 198
$D^{(T)}_K$	thermal diffusivity [$\text{m}^2\ \text{s}^{-1}$], 191
$\mathbb{D}_K$	diffusivity defined by (6.197) [$\text{m}^2\ \text{s}^{-1}$], 232
$\mathbb{D}^{(T)}_A$	binary thermal diffusivity [$\text{m}^2\ \text{s}^{-1}$], 197
$\boldsymbol{D}$	rate of deformation tensor defined in (4.117) [s^{-1}], 123
$\mathfrak{D}$	entropy dissipation $= T\,\pi_{(\text{s})}$ [W m^{-3}], 67
$\mathfrak{D}^{+}$	dissipation due to internal processes [W m^{-3}], 268
$\mathcal{D}$	operator equal to material time derivative D/Dt [s^{-1}], 273

e	elementary charge [1.602 177 33 10^{-19} C], 368
$e_{(\mathrm{tot})}$	total specific energy defined in (4.104) [J/kg], 120
$\vec{e}_\theta$	unit vector in θ-direction [-], 373
$\exp(-\mathbf{M}t)$	defined in (5.86) [various], 172
$\boldsymbol{e}$	spatial strain tensor [-], 251
$\mathbf{e}$	alternating tensor [-], 82
E	energy [J], 20
E_{kin}	kinetic energy [J], 20
E_{pot}	potential energy [J], 20
Ei	exponential integral function [-], 355
$E_{\bar{m}}(x)$	Mittag-Leffer function [-], 357
$\vec{E}$	electric field strength [kg m A^{-1} s^{-3}], 114
$\vec{E}'$	transformation of $\vec{E}$ to the rest system [N/C], 118
$\boldsymbol{E}$	material strain tensor [-], 251
$\mathfrak{E}$	elastic function $= \rho_\circ (f - f_\circ)$ [J m^{-3}], 300
$\mathring{\mathfrak{E}}$	defined in (7.326) [J m^{-3}], 319
f	functional defined in (2.29), 40
f	specific free energy [J/kg], 105
f	real valued function, 12
$\vec{f}^{(\mathrm{em})}$	electromagnetic force [m s^{-2}], 116
$\vec{f}'$	external force fields other than gravity [m s^{-2}], 121
F	free energy [J], 101
F_K	intensive variable defined in (5.20) [various], 154
$F(\tau)$	relaxation spectrum [Pa s^{-1}], 330
$\bar{F}(\tau)$	retardation spectrum [Pa^{-1} s^{-1}], 322
$\boldsymbol{F}$	deformation gradient [-], 250
g	specific Gibbs function [J/kg], 105
$g(t)$	relaxing shear modulus [Pa], 330
g_K	partial specific Gibbs function [J/kg], 107
g_{KM}	positive definite matrix [various], 157
$\vec{g}$	gravity [m s^{-2}], 117
$\mathbf{g}$	metric of O-space [various], 158
G	Gibbs function [J], 101

G	shear modulus of spring in mechanical Maxwell model [Pa], 303
$\bar{G}$	shear modulus of spring in mechanical Kelvin model [Pa], 302
$G(t)$	shear relaxation modulus [Pa], 327
G_∞	equilibrium shear modulus [Pa], 328
$G_\circ$	glasslike shear modulus [Pa], 329
$G^\star(z)$	complex shear modulus [Pa], 326
h	specific enthalpy [J/kg], 105
$\hbar$	Planck constant [1.054 572 66 J s], 377
$h(t)$	function defined in (7.335) [-], 321
H	enthalpy [J], 101
H	Hamiltonian [J], 161
$H(t)$	unit step function [-], 240
$\vec{H}$	magnetic field strength [A m^{-1}], 114
$\vec{H}$	vector affinity [various], 259
I	magnitude of electric current $\vec{I}$ [A], 372
I_ϵ	first invariant of $\boldsymbol{\epsilon}$ [-], 258
$\vec{I}$	electric current [A], 371
$\mathring{II}_D$	second invariant of $\mathring{\boldsymbol{D}}$ that is equal to $\mathring{\boldsymbol{D}} : \mathring{\boldsymbol{D}}$ [s^{-2}], 311
$\mathbf{I}$	unit tensor [-], 73
$\Im$	imaginary part of a complex function, 360
$j(t)$	retarded creep compliance [Pa^{-1}], 322
$j^\star(z)$	retarded complex shear compliance [Pa^{-1}], 325
$\vec{j}$	total electric current density [A m^{-2}], 116
J_∞	equilibrium shear compliance [Pa^{-1}], 328
$J_\circ$	instantaneous complianceglassy compliance [Pa^{-1}], 322
$J^\star(z)$	complex shear compliance [Pa^{-1}], 325
$J(t)$	shear creep compliance [Pa^{-1}], 322
$\vec{J}$	electric current density [A m^{-2}], 115
$\vec{J}$	Onsager flux [various], 160
$\vec{J}_\mathrm{d}$	displacement current [A m^2], 385
$\vec{J}_\mathrm{m}$	current density due to magnetization [A m^{-2}], 389
k	positive constant [-], 100
k	wave number [rad/m], 299

$k_A^{(T)}$	thermal diffusion ratio [-], 197
k_S	constant [W/K], 174
k_B	Boltzmann constant [$1.380,658 \times 10^{-23}$ N m/K], 8
$k_{\alpha\beta}$	phenomenological coefficients [various, J^2 mol^{-2}], 261
$k_{e\alpha}$	phenomenological coefficient [various, J/mol], 260
k_{ee}	phenomenological coefficient [Pa], 260
k_M	constant [J s K^{-1}], 174
K	spring constant [Pa], 291
K'	differential of K [Pa s^{-1}], 338
$\bar{K}'$	differential of $\bar{K}$ [Pa^{-1} s^{-1}], 329
$\bar{K}(\tau)$	function defined by (7.372) [Pa^{-1}], 329
Kn	the Knudsen number [-], 7
$\vec{K}(t)$	Onsager force [various], 158
K_s	isentropic bulk modulus [Pa], 276
K_T	isothermal bulk modulus [Pa], 276
$K(\tau)$	function defined by (7.390) [Pa], 338
$\vec{K}$	magnetic Lorentz force [N], 379
$\vec{K}_{ij}$	force between two charged particles i and j [C], 367
$\vec{K}_{ij}$	force between two charges i and j [N], 368
l	positive constant [-], 105
$l_{\alpha\beta}$	phenomenological coefficients [various], 261
$l^\star$	linear dimension of the physical volume element [m], 5
$d\vec{l}$	element of thin wire conductor [m], 373
L	phenomenological coefficient, 68
L	characteristic length [m], 5
L_D	osmotic diffusional coefficient [m^2 Pa^{-1} s^{-1}], 95
L_{DP}	ultrafiltration coefficient [m^2 Pa^{-1} s^{-1}], 95
$[L'']$	matrix of phenomenological coefficients L''_{KL} [kg s m^{-3}], 222
L_P	permeability coefficient [m^2 Pa^{-1} s^{-1}], 94
L_{PD}	osmotic volume flow coefficient [m^2 Pa^{-1} s^{-1}], 95
$L_{\alpha\beta}$	chemical coefficient [mol^2 m^{-3} J^{-1} s^{-1}], 134
$\mathcal{L}$	Laplace transform [-], 324

$\mathcal{L}$	linear operator [-], 148
m	molar mass [kg/mol], 78
m_e	mass of the electron [9.109 3897 10^{-31} kg], 377
m_K	molar mass of component K [kg/mol], 112
$\vec{m}$	magnetic dipole moment [A m^2], 376
Ma	Mach number [-], 135
M_K	mass of component K [kg], 23
$d_e M_K$	external exchange of mass [kg], 112
$d_i M_K$	internal exchange of mass [kg], 112
$\{M_K\}$	set of M_K [kg], 100
$\vec{M}$	total magnetization [A/m], 390
$\vec{M}_A$	magnetization [A/m], 389
$\vec{M}_{eq}$	equivalent magnetization [A/m], 390
$\mathbf{M}$	matrix defined in (5.70) [various], 168
$\vec{\mathcal{M}}$	angular momentum on a current loop C [J], 380
n	number of charges per unit volume [C m^{-3}], 371
n_o	Loschmidt constant [2.686 763 $\times 10^{25}$ m^{-3}], 55
n^s	number of current loops per unit volume [m^{-3}], 389
n_{SE}	Stokes–Einstein number [-], 216
$\vec{n}$	outwardly directed unit normal vector to a surface [-], 62
N	number of components, 99
N	number of internal variables in anelastic and viscous deformation [-], 313
N_A	Avogadro constant [6.022 1367$\times 10^{23}$ mol^{-1}] , 217
$\bar{N}(s)$	function defined in (7.461) [s/Pa] , 358
$N(s)$	function defined in (7.465) [Pa s^3] , 359
p	pressure [Pa], 17
$\vec{p}$	electric dipole moment [C m], 376
P	conserved quantity defined in (6.215) [-], 236
P_1	conditional probability [-], 145
P_2	compact notation for P_1 [-], 147
P_α	defined in (6.214) [-], 236
ΔP	pressure difference over a membrane [Pa], 94
P_K	intensive variable [various], 154

$[P_L]$	one mole of the product L [mol], 112
P_m	generalized conditional probability [-], 145
$\vec{P}_j$	a generalized force, 19
$\vec{P}$	polarization vector [C m], 387
$\vec{P}_{\text{eq}}$	equivalent polarization [C m^{-2}], 390
$\mathring{\boldsymbol{P}}$	tensor affinity [various, Pa], 259
$\mathbf{P}^{(\text{em})}$	Maxwell stress tensor, the pressure is defined as positive [Pa], 116
$\widetilde{\mathbf{P}}$	pressure stress tensor defined in (4.95) [Pa], 118
$\mathbf{P}_K$	partial pressure stress tensor [Pa], 118
$\mathcal{P}$	a material point [-], 71
$\mathcal{P}$	fixed point in space, 3
q	electric charge [C], 367
q	total electric charge per unit of mass [C/kg], 115
q_K	electric charge per unit of mass of component K [C/kg], 114
$\bar{q}$	molar electric charge [C/mol], 78
$\bar{q}_K$	electric charge per mole of component K [C/mol], 114
q_α	$= \mathring{\boldsymbol{\varepsilon}}_\alpha$ internal variable [s^{-1}] , 315
$\vec{\hat{q}}_P$	principal vectors, 316
$\vec{q}_K$	principal vector, 315
$đQ_{\text{C}}$	Clausius uncompensated heat [J], 49
Q	heat transferred to the system [J], 21
Q_K	component of $\vec{q}$, 317
$Q_{(\text{p})}$	reaction heat [J m^{-3} s^{-1}], 126
$\mathbf{Q}$	half of second moment [-], 150
$\mathbf{Q}$	symmetry operation tensor [-], 73
r	number of degrees of freedom [-], 155
r_K	rate of transformation of species K [mol m^{-3} s^{-1}], 109
$\vec{r}$	position vector [m], 367
(r, θ, z)	cylindrical coordinate system [various], 372
R	distance to wire conductor [m], 372
R	gas constant [J kg^{-1} K^{-1}], 42
R	global reservoir system [various], 155
R	radius of a particle [m], 216

$R_{(A)}$	molar gas constant [8.314 471 J mol^{-1} K^{-1}], 203
R_K	gas constant of component K [J kg^{-1} K^{-1}], 204
R_K	reservoir [various], 155
$[R_K]$	one mole of the reactant K [mol], 112
Ro	Rossby number [-], 238
$R(\vec{\zeta},)$	auxiliary function [-], 151
$\vec{R}_+$, $\vec{R}_-$	position from $\vec{r}$ to dipole positive and negative charge [m], 376
$\mathbf{R}$	correlation matrix [various], 166
$\Re$	real part of complex function [-], 298
s	specific entropy [J K^{-1} kg^{-1}], 104
s_α	$= \mathring{\boldsymbol{\sigma}}$ deviatoric stress [Pa], 315
$d\vec{s}_j$	displacement of a generalized force, 19
s_K	specific entropy of component K [J K^{-1} kg^{-1}], 102
$\delta\vec{s}$	infinitesimal displacement [m], 379
S	entropy (2.56) [J/K], 46
d_eS	entropy flux [J/K], 50
d_iS	entropy production [J/K], 50
Sc	Schmidt number [-], 237
Sr	Strouhal number [-], 136
$\mathcal{S}$	Stieltjes transformation [-], 346
t	time [s], 5
t	parameter [s], 256
$t_{(x)}$	Celsius temperature [°C], 16
$t_{(n)}^{(\alpha)}$	natural time of internal process α [s], 244
$t^{(m)}$	macroscopic time scale [s], 136
$t_{(n)}$	natural time [s^{-1}], 78
$\vec{t}^{\times}$	cross of an antisymmetric tensor, 78
$\Delta t^\star$	time resolution [s], 6
$t^{(o)}$	observation time [s], 244
T	thermodynamic temperature [K], 17
$T(\theta)$	defined on page, 41
Tr	truncation number [-], 311
$T^\star$	defined in (2.2) [K], 17

u	specific internal energy [J/kg], 104
u_K	partial internal energy [J/kg], 120
$\vec{u}$	displacement vector [m], 251
$\widetilde{u}$	internal energy defined in (4.107) [J/kg], 121
U	internal energy (2.5) [J], 20
U	potential energy of an electric field [J], 378
$U_{\rm m}$	potential energy of a magnetic field [J], 378
v	specific volume [$\rm m^3\ kg^{-1}$], 104
$v_{\rm g}$	magnitude of group velocity [m/s], 299
v_K	specific volume of component K [$\rm m^3\ kg^{-1}$] , 102
$v_{\rm s}$	magnitude of slip velocity [m/s], 8
$\vec{v}$	velocity [m/s], 62
$\vec{v}_K$	velocity of component K [m/s], 109
$\vec{v}^\star$	molar velocity [m/s], 200
$\vec{v}_j$	drift velocity of charge q_j [m/s], 371
$\vec{v}^\star$	molar-average velocity [m/s], 111
$\vec{v}_\star$	arbitrary velocity [m/s], 184
V	volume [$\rm m^3$], 99
dV	volume element [$\rm m^3$], 61
$\vec{V}_K$	barycentric diffusion velocity [m/s], 110
$(\Delta V)^-$	volume element < the physical volume element $(\Delta V)^\star$ [$\rm m^3$], 155
ΔV	arbitrary control volume around point $\mathcal{P}$ [$\rm m^3$], 3
$(\Delta V)^\star$	physical volume element [$\rm m^3$], 4
$\vec{V}_K^\diamond$	generalized diffusion velocity [m/s], 191
$\vec{V}_{KL}$	$\equiv \vec{V}_K - \vec{V}_L$ [m/s], 187
$\mathbf{V}$	symmetric matrix [various], 170
$\mathcal{V}$	electrostatic potential [V], 369
$\mathcal{V}_{\rm m}$	magnetic potential [N/A], 375
w_K	weight fraction of component K [-], 104
W	work exerted on the system (2.3) [J], 19
$W_1(\vec{\zeta}(t))$	probability that $\vec{\zeta} \in \left[\vec{\zeta}, \vec{\zeta} + d\vec{\zeta}\right]$ at time t [-], 144
W_i	probability function [-], 144
$W^\star$	work done by the system [J], 22

x_K	mole fraction [-], 108
$\vec{x}$	spatial coordinate [m], 5
$d\vec{x}$	spatial line element [m], 249
X'	specific extent of change [various, mol/kg], 276
X'	specific internal variable for thermal relaxation [J/kg], 295
$\vec{X}$	vector in material coordinate system [m], 249
$d\vec{X}$	material line element [m], 249
$\vec{X}_K$	process force [m s^{-2}], 221
dX'_α	normalized extent of reaction [mol kg^{-1}], 113
$[\vec{X}]$	column matrix with vectors $\vec{X}_K$ [m s^{-2}], 221
$y_{(k)}$	extensive property, 3
$y_{(V)}$	space average of extensive property, 3
ΔY	total value of the extensive property in ΔV, 3
Y_K	extensive variable [various], 154
z	state of a system, 12
z	complex number [s^{-1}], 324
z'	state of a system, 12
z_K	energy of component K due to radiation [J m^{-3}], 121
Z_K	extensive variable defined in (5.19) [various], 154
Z	set of states, 12

Greek Letters

α	cubic expansion coefficient [K^{-1}], 102
α	factor [-], 64
α	thermal expansion coefficient [K^{-1}], 276
$\alpha_A^{(p)}$	pressure diffusion factor [-], 198
$\alpha_A^{(T)}$	thermal diffusion factor [-], 197
α_K	activity [-], 207
$\alpha_{KL}^{(p)}$	multicomponent pressure factor [-], 234
$\alpha_{KL}^{(T)}$	multicomponent thermal diffusion factor [-], 232
β	factor [-], 64

γ	isentropic index [-], 135
γ_A	activity coefficient [-], 212
$\gamma_{\alpha K}$	weight fraction of element α in component K [-], 236
γ_K	activity coefficient [-], 208
Γ	the gamma function [-], 350
Γ	thermodynamic correction factor [-], 196
Γ_{KL}	thermodynamic correction factors [-], 232
δ	difference relative to the value in the reference state [-], 275
$\delta(t)$	delta function, 359
ϵ	component of the deformation tensor [-], 239
ϵ_0	permittivity of vacuum [8.854 187 817... 10^{-12} F/m], 114
ϵ_K	positive for even and negative for odd varibles [-], 70
$\boldsymbol{\epsilon}$	diagonal matrix [-], 173
$\boldsymbol{\epsilon}$	infinitesimal strain tensor [-], 252
$\boldsymbol{\epsilon}'$	sum of elastic and anelastic relative deformation [-], 257
$d\vec{\zeta}$	small volume in ζ space [-], 143
ζ_K	O-variable [various], 158
$\vec{\zeta}(t)$	stochastic process [-], 143
$\langle \vec{\zeta}', \tau \mid \vec{\zeta} \rangle$	conditional average [-], 151
$\boldsymbol{\zeta}_\gamma$	tensor extent of advancement [various, -], 259
η	constitutive coefficient [Pa s], 133
η	thermal efficiency [-], 25
η	viscosity coefficient [Pa s], 134
η	viscosity of dashpots in mechanical Maxwell model [Pa s], 303
$\bar{\eta}$	viscosity of dashpots in mechanical Kelvin model [Pa s], 302
η_{k}	cooling energy ratio [-], 29
$\eta_\circ^-$	stress relaxation viscosity [Pa s], 330
η_{w}	coefficient of performance [-], 29
$\vec{\eta}_\beta$	vector extent of advancement [various], 259
θ_i	temperature of reservoir i [K], 35
$\theta^\star$	arbitrary temperature [K], 41
κ	compressibility [m^2/N], 8
κ	constitutive coefficient [Pa s], 133

κ	dilatational viscosity coefficient [Pa s], 134
$\kappa_{\alpha\beta}$	phenomenological coefficient [various, mol^2 J^{-1} m^{-3} s^{-1}], 268
$\kappa_{\alpha \mathrm{v}}$	phenomenological coefficient [various, mol m^{-3}], 268
κ_K	partial compressibility coefficient of component K [Pa^{-1}], 209
κ_σ	dilatational viscosity [Pa s], 137
κ_s	isentropic compressibility [Pa^{-1}], 276
κ_T	isothermal compressibility [Pa^{-1}], 102
$\kappa_{\mathrm{v}\alpha}$	phenomenological coefficient [various, mol m^{-3}], 268
κ_{vv}	phenomenological coefficient [Pa s], 268
λ	constant defined in (7.451) [various], 356
λ	factor [-], 64
λ_α	chemical viscosity coefficient [mol m^{-3}], 134
$\lambda_{\beta\alpha}$	phenomenological coefficient [various], 269
$\lambda_{\beta \mathrm{q}}$	phenomenological coefficient [various], 269
λ	heat conductivity [W m^{-1} K^{-1}], 192
$\lambda_{\mathrm{q}\beta}$	phenomenological coefficient [various], 269
λ_{qq}	phenomenological coefficient [W m^{-1} K^{-1}], 269
λ'	heat conductivity defined by (6.43) [W m^{-1} K^{-1}], 193
Λ	mean free path [m], 7
$\boldsymbol{\lambda}$	conductivity tensor [W K^{-1} m^{-1}], 90
μ	atomic magnetic moment [A m^2], 376
μ	dashpot constant [s^{-1}], 291
μ_0	permeability of vacuum [4π 10^{-7} N A^{-2}], 114
μ_B	Bohr magneton [9.274 0154 10^{-24} J/T], 377
μ_K	chemical potential [J/kg], 100
$\tilde{\mu}_K$	chemical potential defined in (6.96) [J/kg], 206
μ_{KL}	chemical coefficients [J kg^{-2}], 102
μ_K^+	chemical potential defined in (6.89) [J/kg], 205
$\mu_K^\circ(T)$	chemical potential defined in (6.86) [J/kg], 204
ν	kinematic viscosity [m^2/s], 79
ν	number of internal variables causing anelastic deformation [-], 313
$\nu_{\alpha K}$	stoichiometric number of species C_K in reaction α [-], 112
ξ	potential energy [J/kg], 120

ξ_α	extent of reaction [mol], 112
ξ_α	scalar extent of advancement [various, mol m^{-3}], 259
$\vec{\xi}$	conditional average [-], 149
$\vec{\xi}$	regression of the fluctuations [-], 151
π_K	rate of production of mass of component K [kg m^{-3} s^{-1}], 109
$\Delta\pi$	osmotic pressure difference over a membrane [Pa], 94
$\pi_{(\mathrm{s})}$	rate of entropy production [J K^{-1} s^{-1}], 67
$\pi_{(\psi)}$	production of Ψ, 62
$\vec{\pi}$	arbitrary constant vector field [various], 374
$\vec{\pi}$	constant vector field [-], 65
$\Pi_{(\mathrm{S})}$	entropy production [J K^{-1} s^{-1}], 159
$\boldsymbol{\Pi}$	deviatoric stress tensor defined in (4.127) [Pa], 125
ρ	density [kg m^{-3}], 61
$\rho^{(\mathrm{e})}$	electric charge density [C m^{-3}], 388
$\rho_{\mathrm{f}}^{(\mathrm{e})}$	density of the free charged particles [C m^{-3}], 388
$\rho_{\mathrm{f}}^{(\mathrm{e})}$	charge density [C m^{-3}], 368
ρ_K	density of component K [kg m^{-3}], 109
σ	reflection coefficient [-], 96
σ_{AB}	mean diameter of molecules A and B [m], 194
$\sigma^{(\mathrm{e})}$	surface charge per unit of area [C m^{-2}], 387
σ_K	diameter of the molecule K [m], 194
$\vec{\sigma}$	angular momentum [kg m2 s^{-1}], 376
$\vec{\sigma}$	traction exerted on the material by the surroundings [Pa], 255
$\boldsymbol{\sigma}$	stress tensor, pressure is defined as negative [Pa], 256
τ	dimensionless time difference [-], 149
τ	relaxation time [s], 6
τ_{q}	relaxation time [s], 274
$\bar{\tau}_{\mathrm{q}}$	retardation time [s], 274
ϕ	volume fraction [-], 216
$\phi_{\alpha\beta}$	coefficient defined in (7.97) [various, Pa^{-1} s^{-1}], 270
$\phi'_{\gamma\beta}$	phenomenological coefficient [various, Pa^{-1} s^{-1}], 270
$\phi_{\gamma\mathrm{v}}$	coefficient defined in (7.97) [various, Pa^{-1} s^{-1}], 270
$\phi'_{\gamma\mathrm{v}}$	phenomenological coefficient [various, -], 270

$\phi_K(T)$	unspecified function of T [J/kg], 204
$\vec{\phi}_{(s)}$	rate of entropy flux [J K^{-1} s^{-1}], 66
$\phi(t)$	macroscopic quantity [various], 149
$\phi_{v\gamma}$	coefficient defined in (7.97) [various, Pa^{-1} s^{-1}] , 270
$\phi'_{v\gamma}$	phenomenological coefficient [various, -] , 270
ϕ_{vv}	coefficient defined in (7.97) [Pa^{-1} s^{-1}] , 270
ϕ'_{vv}	phenomenological coefficient [Pa s] , 270
$\phi_{(\psi)}$	general process flux , 66
ϕ	flux of the property Ψ , 63
$\vec{\Phi}_{(s)}$	entropy flux [W K^{-1} m^{-3}] , 89
$\vec{\Phi}$	volume flow over a membrane [m/s], 94
$\vec{\Phi}_{(e)}$	energy flux [J m^{-2} s^{-1}], 121
$\vec{\Phi}_K$	barycentric diffusion flux [kg m^{-2} s^{-1}], 110
$[\vec{\Phi}]$	matrix with diffusion fluxes $\vec{\Phi}_K$ [kg m^{-2} s^{-1}], 222
$\vec{\Phi}_{(q)}$	reduced heat flux defined in (4.130) [J m^{-2} s^{-1}], 126
$\vec{\Phi}_{(s)}$	barycentric entropy flux [J K^{-1} m^{-2} s^{-1}], 131
$\vec{\Phi}_{(u)K}$	partial energy flux vector [J m^{-2} s^{-1}], 121
$\vec{\tilde{\Phi}}_{(u)}$	internal energy flux [J m^{-2} s^{-1}], 122
φ	defined in figure 2.1 [-], 26
φ_K	fugacity [Pa], 206
φ_∞	fluidity for retardation, steady-state fluidity [Pa^{-1} s^{-1}], 318
$\vec{\chi}(\vec{X})$	deformed configuration [m], 250
ψ	specific value of Ψ, 61
ψ_K	partial specific quantity of component K, 105
$\psi(\vec{r},)$	scalar function [V], 369
$\Psi^{(e)}$	electric flux [V m], 369
Ψ	general extensive property, 61
ω	circle frequency [rad/s], 298
Ω	constant normalization factor [-], 170
Ω_B	defined by (5.102) [-], 175
Ω_M	defined by (5.110) [-], 177
Ω_S	defined by (5.106) [-] , 176

$\vec{\Omega}$	angular velocity [rad/s] , 92

Subscripts

A	referring to binary component, 195
B	referring to binary component, 195
ind	referring to induction, 381
I	referring to irreversible, 37
K	referring to component K property or parameter, 99
M	referring to all masses, 100
M'	referring to all masses except M_K, 100
N	referring to Nth component, 99
R	referring to reversible, 37
s	referring to solvent, 209
v'	referring to that I'_ϵ is constant, 262
α	referring to variable of state, 53
α	referring to internal process, 258
β	referring to variable of state, 53
β	referring to internal process, 258
γ	referring to internal process, 258
$\circ$	indicates that the value applies for $t \to 0$, 328
∞	indicates that the value applies for $t \to \infty$, 328

Superscripts

(a)	referring to anelastic behavior, 240
(a)	referring to internal coefficients, 280
A	referring to antisymmetric part of a tensor, 77
(e)	referring to elastic behavior, 240
(em)	referring to electromagnetic, 116
(eq)	referring to equilibrium value, 255
(id)	referring to ideal system, 205

(p)	referring to plastic behavior, 240
(o)	referring to constrained coefficients, 277
(s)	referring to scalar, 67
S	referring to symmetric part of a tensor, 77
(t)	referring to tensor quantity, 67
T	referring to transpose of a tensor, 77
(v)	referring to vector, 67
(v)	referring to viscous behavior, 240
(α)	referring to internal process α, 244

Miscellaneous

$\oint$	integration around a cyclic process [-], 22
$\hat{a}$	referring to complex amplitude [-], 298
$⨍$	principal value of integral [-], 369
$\circ$	referring to deviatoric part of a tensor defined in (3.44), 78
$\oint$	line integral, 48
$\cdot$	short notation for the substantial derivative D/Dt [s^{-1}], 255

1 The Continuum View of Matter

Summary: This short chapter discusses the range of applicability of a macroscopic theory.

In classical thermodynamics the theory of heat and the theory of mechanics are combined into one macroscopic theory. The macroscopic point of view of nature is sometimes called a phenomenological theory, but this phrasing should be considered as less appropriate, since phenomenology is concerned only with the description of the natural phenomena that precedes the understanding and explanation of the observed phenomena.

The explanation of physical phenomena needs a model for the system in which the phenomena arise. In addition, it needs a model for the matter contained within the system. In formulating these models the choices of proper space and time scales are very important. These choices determine also the validity and the range of applicability of a model. Classical physics models matter as it is seen by the naked eye, and as it is daily experienced. No use is made of amplifying instruments. In the classical continuum theory neither microscopic nor astronomic observations are done.

Matter seems to the naked eye to be connected and continuous in its structure. Everyday experience suggests therefore the *continuum* or *macroscopic* view of matter. However, it is well known that matter is made up of molecules*. Molecules execute fast and irregular motions of heat not seen by the naked eye.

From the macroscopic point of view the *continuum hypothesis* of matter assumes therefore that matter is perfectly continuous in structure and connected. The discrete nature of matter is ignored, and only the the relations between macroscopic phenomena are important. Similarly, classical thermodynamics is a macroscopic theory that disregards also the molecular structure of matter. In a continuum approach primarily the total is regarded, while the parts exist only in their relation to the whole.

At first sight the continuum view and the molecular view of matter appear contradictory. However, under certain circumstances both views converge to the same results if the macroscopic quantities can be identified with the statistical averages of the corresponding molecular quantities. In the calculation of the statistical averages, matter is postulated to have no structure as seen from the continuum point of view. No preferred

* All types of elementary particles are summarized here in the word *molecule*. A molecule is in this sense the smallest particle of matter considered having all the properties of that substance.

directions can be assigned. Identification of microscopic averages with the macroscopic quantities shows that then the average quantity of the molecular actions between the molecules is considered as a macroscopic field quantity. Calculations with molecular models are limited to simple representations of the forces between the molecules. Only the most simple materials or models can be treated in detail with a molecular theory. In general there is no cause for a contradiction between the macroscopic view and the microscopic view of matter in classical physics. No contradiction is found if the length and the time scales of the investigated physical phenomena—described by a macroscopic theory—do not interfere with the molecular or the astronomical length and time scales. So, on one hand the macroscopic length scales should be much larger than the molecular ones, while on the other hand these scales should not be astronomical, as no relativistic theories are considered. These restrictions are implied in the word 'classic' in classical physics. Within these limitations, the physical space used in the formulation of the classical macroscopic theories may be assumed to be Euclidean. In Euclidean space the theorem of Pythagoras applies, and a Cartesian coordinate system can be used. The discussion above is illustrated in more detail in Figure 1.1, where the space averages of the mass serve for illustration.

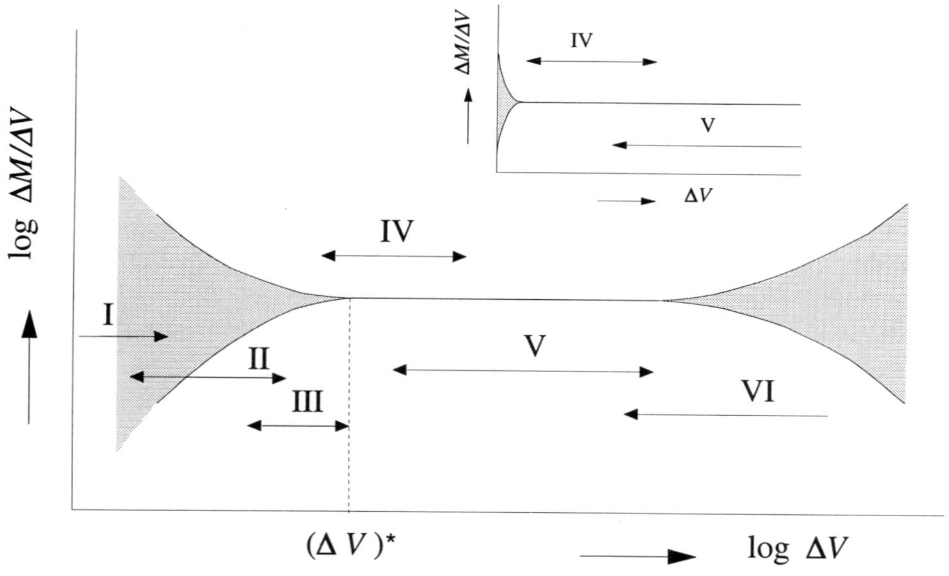

Figure 1.1. Space average of the mass in ΔV as a function of ΔV on a logarithmic scale. Macroscopic theories: region IV. High energy physics and quantum mechanics: region I. Molecular theories: region II. Fluctuation theories: region III. Global macroscopic theories: region V (the small figure on a linear scale is added for proper interpretation of the validity range of the macroscopic region). Astrophysics: region VI.

In figure 1.1 the space averages of the mass are depicted along the vertical axis. Along the horizontal axis the volume of an arbitrary volume ΔV around a point $\mathcal{P}$ is depicted. This arbitrary volume is used to determine the space averages and contains N molecules, N being very large. Suppose that $y_{(k)}$ is the *extensive property** of the kth 'molecule' considered. The total value ΔY of the extensive property $y_{(k)}$ in ΔV becomes

$$\Delta Y = \sum_{k=1}^{N} y_{(k)}$$

The space average of the extensive property $y_{(k)}$ is then given by

$$y_{(V)} = \Delta Y / \Delta V$$

If N is large enough, the stochastic parts of $y_{(k)}$ cancel each other with sufficient accuracy. Sufficient means in this context: within the error of measurement. Subsequently, $y_{(V)}$ becomes a macroscopic quantity, which may show spatial variations of the order of the macroscopic length scales. The compensation of the stochastic contributions in attaining the averages is in no way trivial. In the past the high molecular velocities and the much lower macroscopic velocities were considered as paradoxical. Starting from our global experience (region V in figure 1.1) the average value of $y_{(V)}$ will initially change systematically and approach a constant value. This happens if ΔV is small enough to ensure that within the error of measurement the averaged stochastic properties are almost equally distributed over ΔV, while further small shrinkages of ΔV do not result in a noticeable change in the quantity $y_{(V)}$. However, if ΔV becomes too small, the arbitrary volume contains an insufficient number of molecules to assure that the stochastic character of the molecular properties is compensated completely. The result is that $\Delta Y / \Delta V$ varies irregularly between certain limits as a function of time. These variations are called fluctuations, which corresponds to a coarse mapping of the molecular motions. Fluctuations are depicted in region III, sometimes called the *mesoscopic* region.

Region III is also important for the thermodynamics of irreversible processes for deriving the Onsager–Casimir reciprocal relations. In this mesoscopic region the macroscopic quantities are subjected to fluctuations. Reducing the dimensions of the arbitrary volume around the point $\mathcal{P}$ in region III increases the fluctuations of $y_{(V)}$. In a further shrinkage of the arbitrary volume, the *microscopic* region II is reached and the dimensions of the arbitrary volume have become so small that knowledge of the behaviour of a single molecule

---

* A property is called extensive if its value for the total system is equal to the sum of its values in the subsystems into which the system is divided. Mass and volume are examples.

and knowledge of the interaction between the molecules become important. The description of the dynamics of molecules is a subject of statistical physics. In the microscopic region II, the smallest length scale is of the order of the dimensions of the molecule. Such a length scale is much smaller than the length scale that could be assigned to a ΔV still containing many molecules. Usually, the transitions from one region to another are the most difficult regions to analyze. For example, in denser systems nonbinary collisions also occur, and the number of possible collisions, needed for the calculation of the statistical averages, is not completely known. Another difficulty—also encountered in dilute systems—is whether it is acceptable to identify an averaged quantity with the corresponding continuum property. If such an identification is justified, then the more detailed theory might provide a better understanding of the underlying physics than the more global theories do.

Quantum mechanics and high energy physics investigate the basic structure of matter, and in these fields the smallest length scales are considered. In the interactions of light and matter, large scale phenomena can be produced, which are not part of a macroscopic theory. An example is the blueness of the sky*, explained by quantum electrodynamics. Similarly, many other observable effects could be produced using lasers. The interactions of light with matter on the smallest scales may change the frequencies of the light, such that the effect of the interactions can be observed by the naked eye. Although visible by the naked eye, these effects do not belong to the macroscopic theories, since their roots are found in region I. Besides, there is no such thing as the localization of the 'matter' of light.

The length scales of the phenomenological descriptions correspond to those found in our earthly, everyday experiences. These length scales of the global phenomenological descriptions are depicted in region V of figure 1.1. Finally, the largest length scales are considered in astrophysics.

To summarize, the macroscopic level of description assumes arbitrary volumes large enough that fluctuations of a macroscopic quantity are not measurable in a macroscopic measurement. Suppose that $(\Delta V)^\star$ is the smallest arbitrary volume in which a fluctuation of the macroscopic property considered is just not measurable. Such an arbitrary volume is called a *physical volume element* to distinguish it from the mathematical volume elements, which can shrink to zero by definition.

Due to the limitations set by the finite physical volume element, it is obvious that the macroscopic value of the extensive property Y has to be defined as the spatial average

$$y^\star = \lim_{\Delta V \to (\Delta V)^\star} \Delta Y / \Delta V$$

---

* Feynman R.P. 1985. *QED, the Strange Theory of Light and Matter.* Princeton University Press, New Jersey, p. 100.

With this formal limit it is guaranteed that $y^\star$ does not experience measurable stochastic variations for arbitrary volumes of the size $(\Delta V)^\star$. However, accepting these smallest permissible volumes could also imply that spatial matter filled domains should be divided into connected areas with volumes of the order of $(\Delta V)^\star$. Such an approach would then result in a difficult analysis due to the extreme number of small regions to identify. Nevertheless, a finite discrete measurement and signal theory might be built on these grounds* in principle, although the description of the laws describing the interactions between these regions is not easy to give. Therefore, the continuum hypothesis simplifies theory considerably: first, by assuming that matter consists of contiguous connected wholes and second, by assuming that the discrete nature of matter can be neglected. Consequently, it is said that the linear dimension $l^\star$ of the physical volume element is much smaller than the distances L over which the macroscopic quantities vary observably, implying that from the macroscopic point of view the volume of a physical volume element is infinitesimal. Then the macroscopic properties y may be approximated by a point function, because in accepting the continuum hypothesis the following limit is assumed to hold

$$\lim_{\Delta V \to (\Delta V)^\star} \Delta Y/\Delta V = \lim_{\Delta V \to 0+} \Delta Y/\Delta V = y(\vec{x}, t)$$

The first equality denotes the transition to the continuum theory. In the application of the continuum hypothesis it is assumed further that the macroscopic quantities $y(\vec{x}, t)$ are continuous differentiable functions of the spatial coordinate $\vec{x}$ and time coordinate t. The above limit is acceptable by the assumption that the macroscopic quantities may vary considerably only over macroscopic distances L much larger than the linear dimension $l^\star$ of the physical volume element, or $l^\star/L \to 0$. From a mathematical point of view this limit means that the rational numbers used in physical experiments are extrapolated to the domain of real numbers. With this extrapolation, the description changes into a field description and a field theory results.

Besides the extensive properties there are also intensive properties. An intensive property is independent of the dimensions of the total system, and in contrast to an extensive property its global value is not equal to the sum of its values in subsystems of the total system. The intensive quantities are defined at equilibrium. Examples of intensive properties are the temperature and the pressure. In contrast with the extensive properties, the intensive properties do not change systematically if the arbitrary volume is contracted around point $\mathcal{P}$, and a different discussion is needed for a field theory of the intensive quantities in non-equilibrium systems. In non-equilibrium

* Van der Zanden J. 1984. Measurement theory and signal theory. *International Journal of General Systems*, 10: 15–46.

systems, such as moving fluids, intensive properties like the pressure and the temperature are accepted as locally defined quantities. However, to define the intensive quantities a thought experiment is necessary, in which during the measurement of the intensive quantity equilibrium is supposed to exist between this quantity and measuring sensor. This sensor should be tiny, so that its influence on the quantity to be measured can be neglected. Vstovsky* emphasizes similarly that the macroscopic description should be appropriate for a special class of experiments, namely those experiments done with measuring devices having a small but finite time resolution $\Delta t^\star$. This model implies that the physical volume element has to be large enough to be able to follow the macroscopic changes of the molecular mechanisms under specified circumstances. The relaxation times τ of the molecular mechanisms are in these circumstances much smaller than the finite time resolution $\Delta t^\star$ of the sensor. Molecular (microscopic) mechanisms and associated processes are not directly observable. As such, these processes and mechanisms are called *hidden* or *internal*. For applying the continuum theory it is therefore assumed that, in a thought experiment, the physical volume element reaches equilibrium in times of the order of the relaxation time τ. Next, this system is brought into contact with the tiny sensor of the measurement system. The combined system has to be an isolated system that reaches equilibrium with the tiny sensor within a time $\Delta t^\star \gg \tau$. Intensive properties, like pressure and temperature, become local and instantaneous defined physical quantities, if $\Delta t^\star$ is much smaller than the slow macroscopic changes in time of the pressure and the temperature. This discussion illustrates that for the application of the continuum view of matter the limit for $\Delta t^\star \to 0^+$ is assumed in addition to the limit for $(\Delta V)^\star \to 0^+$. In accepting both limits, it is assured that the physical volume element reaches local equilibrium within the finite time resolution, and pressure and temperature become locally and instantaneously defined quantities.

The continuum point of view presupposes both limiting procedures, and it is now somewhat paradoxical that this macroscopic view also uses the idea of 'particle'. In contrast with an elementary particle this 'particle' is a small quantity of matter with a volume of the order of a physical volume element. The matter in this tiny volume element is assumed to have been spread uniformly over that volume and to have a fixed identity. Macroscopically the volume element is infinitesimal. This *particle hypothesis* is a consequence of the continuum point of view and is based on the experience that under macroscopic circumstances it is possible to identify a small quantity of matter with color or radioactive tracers and that such a marked particle can be followed in space. Leonardo da Vinci† (1452–1519) pointed out the possibility

* Vstovsky V.P. 1976. On the dynamical basis of macroscopic theory. *Journal of Statistical Physics*, 15: 105–121.

† Leonardo da Vinci, *Del Moto e Misura dell'Acqua.* Edited by Carusi E. and Favaro A.

of coloring particles in a flowing continuum to make their path lines visible. The particle axiom of the continuum theory is again restricted to practical limits. After all, during long intervals a colored particle will spread out over space due to diffusion of the tracer. The result is that a small particle loses its fixed identity. In conclusion, the particle axiom can be applied only with enough accuracy within certain limited times of observation, yielding an upper time limit for the application of the continuum theory.

From the continuum point of view of matter a finite region of matter is divided into a large number of contiguous physical volume elements having the following characteristics:

(1) each volume element contains enough molecules, such that the result of averaging coincides with the gross macroscopic properties of interest,
(2) the linear dimensions of each physical volume element are very small with respect to the distances over which the macroscopic properties vary.

From condition two and the particle axiom it is deduced that in the continuum hypothesis a material body is considered as a set of 'material points', in which a material point is a particle in the sense as discussed above. In addition, it is supposed that all macroscopic quantities are single-valued and differentiable as many times as required for the theory. Exceptions are not allowed in spatial regions, but only in regions of zero measure, such as points, lines or surfaces. This is acceptable because in classical continuum theory singularities and discontinuities are confined to points, lines or surfaces.

Those conditions for the validity of the continuum theory are vague only because the questions of the order of magnitudes of the length and time scales have been considered. For specific materials, these conditions can be made more explicit. For example, for a dilute gas many results are known. A gas is called dilute if binary collisions between the gas molecules dominate. For such a gas the conditions for applying the continuum theory can be given in more detail. Suppose that Λ denotes the mean-free-path of the molecules. The mean free path is the average distance travelled by a molecule between two collisions. The Knudsen number is defined by

$$Kn = \Lambda / L$$

where L denotes a characteristic length over which the macroscopic variables change. For example, in a stationary tube flow L is of the order of the diameter of the tube. The macroscopic or continuum view of matter holds if

$$Kn \ll 1$$

1924. Bologna. See also: Rouse H. and Ince S. 1955. *History of Hydraulics*. Supplement to *La Houille Blanche*. Republicated as: Rouse H. and Ince S. 1957. *History of Hydraulics*. Iowa Institute of Hydraulic Research. State University of Iowa. See p. 47.

General observations offer only estimates of the order of magnitudes. Practical limits are specified by measurements. Experiments with gases have shown that within reasonable limits of the accuracy of the measurements the continuum view may already be applied for dilute gases for $Kn < 0.1$. For $Kn > 10$ the dimensions of the vessel containing the gas are small compared with the mean free path of the molecules. Such a gas is extremely rarefied and is known as a Knudsen gas*. Obviously for this extremely dilute gas the continuum hypothesis of a substance fails, and it fails also for $Kn \approx 1$, where the mean free path and the variation of the mean velocity of the molecules are of the same order. This region is called the slip region†, because the flow of such a diluted gas near a solid boundary is imagined to slip along the wall with a velocity v_s. In the first order approximation v_s is roughly Λ/L times a characteristic speed. Very dilute gases do not satisfy the *no-slip condition.* The no-slip condition is the condition that the velocity of a fluid close to a wall is equal to the velocity of that wall.

Example: The root mean square fluctuations of the pressure and the temperature are inversely proportional to the square root of the arbitrary volume. For the mean square fluctuations of the pressure, the following result can be derived‡

$$\overline{(\Delta p)^2} = -kT\left(\frac{\partial p}{\partial V}\right)_S = -\frac{k_B T}{\kappa V}$$

where κ denotes the compressibility and k_B the Boltzmann constant:

$$k_B = 1.380\,658 \times 10^{-23} \quad \mathrm{N\ m\ K^{-1}}$$

For water the compressibility is small, with

$$\kappa = -\frac{1}{V}\frac{\partial V}{\partial p} \simeq 500 \times 10^{-12} \quad \mathrm{m^2\ N^{-1}}$$

Substitution of this value for water under normal temperature and pressure conditions yields

$$\overline{(\Delta p)^2} \approx \frac{8 \times 10^{-12}}{V} \quad \mathrm{N\ m^{-2}}$$

* Knudsen M. 1934. *The Kinetic Theory of Gases. Some modern aspects.* Methuen's monographs on physical subjects. Methuen & Co, London.

† Chapman S. and Cowling T.G. 1970. *The Mathematical Theory of Non-Uniform Gases.* 3rd ed. Cambridge University Press, Cambridge. See p. 99; Maxwell J.C. 1879. On stresses in rarefied gases arising from inequalities of temperature. See p. 253 in appendix pp. 250–256: On the condition to be satisfied by a gas at the surface of a solid body. *Philosophical Transactions of the Royal Society of London*, 170: 231–256.

‡ Landau L.D. and Lifshitz E.M. 1959. *Statistical Physics, Course of Theoretical Physics*, Vol. **5**. Pergamon Press, London. (Eq. 111.11) on p. 353.

Suppose that a fluctuation of

$$\Delta p \sim 0.1 \text{ mm } H_2O \sim 1 \ \text{ N m}^{-2}$$

is just measurable with a pressure gauge. The linear dimensions of the physical volume element $l^\star$ for which no fluctuations are found, even if the time resolution of the pressure gauge is zero, become

$$l^\star \sim \left(\frac{8 \times 10^{-12}}{1}\right)^{\frac{1}{3}} \ \text{m} = 0.2 \text{ mm}$$

The large mean pressure fluctuations in water are effectively seen for the Brownian motion of suspended pollen and milk particles. It is left as an exercise to show that the linear dimensions of the physical volume element are much smaller than those associated with the pressure, even for almost neglectable mean fluctuations of the density and of the temperature.

1.1. EXERCISES

Exercise 1.1. Calculate $l^\star$ for a pressure fluctuation of 1 N m^{-2} in an ideal gas.

Exercise 1.2. Discuss why these pressure fluctuations are not measured. Hint: compare the period of the sound oscillations in a volume with linear dimensions of order of magnitude $l^\star$ with the finite time resolution of the sensor. See also exercise 2.1.

Exercise 1.3. Calculate the temperature fluctuations for the $l^\star$ in the first exercise.

Exercise 1.4. Under what conditions is the continuum approximation useful to model a physical process in a galaxy. Discuss the type of physical process you have in mind.

æ

2 Classical Thermodynamics

Summary: This chapter recapitulates the zeroth, first, second and third law of thermodynamics for global systems. Readers well educated in classical thermodynamics are familiar with these laws, but they still might want to read the discussion about equilibrium, the first law for open systems and a not previously published vector presentation for heat engines given in the section of the second law.

Classical thermodynamics is based on the everyday experience that one has with matter. For the formulation of these experiences, the concept of *systems* is basic in thermodynamics. This idea originated in thermodynamics and has been adopted by other sciences.

In the first place, the global physical system in its totality has to be described correctly. In the second place, a model of this system has to be constructed. Such a model of the physical reality is an idealization that is used to answer questions about the system. The model has to be chosen such that the relevant facets of the physical system and those in the derived model coincide. In other aspects the model and the physical system may differ. However, the desired coincidence has to be within the asserted limits of measurement errors.

The study of nature is concerned mostly with *ideal models.* In ideal models all kinds of disturbing or irrelevant factors are nelected. By putting forward models of nature, the complex reality is described more simply and is as such more approachable. The creation of an ideal model rests on the possibility of making a distinction between the essential and insignificant properties of the system, and its behavior, such that the concentration is directed completely to the essentials. Humans can distinguish between the essential and the non-essential. This skill is the root of modern sciences, and enables us to investigate the complex world with relatively simple idealizations. In this sense the notion *ideal* is understood as the art of limitation and simplification. A perfect model as such is to be seen as a simplified image of the reality that is too complicated to be studied in all its perspectives. Often, it happens that models and the results obtained with the models are used outside their previously accepted limits of validity. Experiments are in those cases imperative to test the predictions of the model. Experiments are also crucial for knowing the range of applicability of the model.

The use of perfect models reaches back to the ancient Greek philosophers, who apply this idea rather reversely: the complicated human experiences are a result of the dazzling of the humans by the gods. The human experiences are therefore considered as confusing images of the divine reality, which is in essence simple and harmonic but may be complicated in its consequences. In that sense, a model is an ideal based on a divine status. The ancient Greek philosophers were looking for the divine ideal, of which everyday reality is only an incomplete reflection. As a consequence, the Greek philosophers had little interest in the daily, earthly matters, like hair, mud, dirt, or anything else, which is vile and paltry*, in contrast with the Indian philosophers, who combined both facets in their studies†.

In thermodynamics, a system is a part of the space, whether or not occupied by a collection of material bodies, on which the attention of the researcher is concentrated. This idea is within the bounds of human capabilities, as we are unable to study the whole universe as one entity in all its details. Explaining the boundaries of a system is seen as essential, because a system has in principle a finite, spatial extent. The boundaries belong to the system. That part of the universe directly involved with the behavior of the system, although it does not belong to the system, is the *environment* of the system. In this approach the universe is divided into a system and its surroundings. It is pointed out that the surroundings may interact with the system. The type of interaction that the surroundings have with the system is specified by the boundary conditions.

In thermodynamics the interaction between the system and its surroundings is crucial. A first rough classification of the systems can therefore be made with respect to the nature of this interaction. The following types of systems may be envisaged.

(1) *Isolated systems.* An isolated system does not have any interaction with its surroundings. Such a system is considered as a closed universe, without any exchange of work or heat with its surroundings.
(2) *Adiabatic systems.* An adiabatic system cannot exchange heat with its surroundings, but may exchange work with its surroundings.
(3) *Diabatic systems.* A diabatic system may exchange work and heat with its surroundings.
(4) *Closed systems.* A closed system refers to a fixed quantity of matter, and it cannot exchange mass with or from its surroundings, but it may exchange work and heat.
(5) *Open systems.* An open system is a system that may exchange matter with its surroundings.

Systems are also classified according to their particular *state*. In thermodynamics the existence of states of a system is a basic assumption of the theory.

---

* Plato 428–348 B.C. Parmenides [130]. See also Aristotle 384–322 B.C. *Metaphysics*, Book VI.

† Seal B. 1915. *Positive Science of the Ancient Hindus.* Longmans and Co., London; Scott Blair G.W. 1949. *A Survey of General and Applied Rheology.* Sir Isaac Pitman & Sons, Ltd, London. See pp. 5–8.

For an abstract theory there is no need for a further explanation of the concept of state. (Compare this with the acceptance of the concept of a *point* in geometry.) Essential for the usefulness of the concept of state is that the various states can be discriminated, and that each state can be described unambiguously by a real set of state variables. This set of variables need not be finite. However, the state variables have to make up an independent and complete set of variables. This means that from the macroscopic point of view all properties of the system can be described completely. Variables of state are therefore macroscopic quantities that can describe completely all measurable properties of the system.

A system is said to have undergone a *thermodynamic process*, or in short a *process*, if at two different times the state of the system is different.

External variables of state specify the position and the velocity of the system as a whole (for example the position and the velocity of the center of mass of the body). *Internal variables of state* are all other variables. A *function of state* is a function that depends explicitly on the variables of state only.

In thermodynamics, the attention is concentrated especially on the internal state of the system, that is, on the determination of the internal state in terms of a few internal variables of state. If motions take place within the system, a thermodynamic formulation of the system may be impossible. For example, this occurs if in a finite region infinite discontinuities arise. Turbulent flow through a contraction is illustrative. In such a case it is not possible to talk of a thermodynamic state of the system!

According to Falk and Jung* state and state functions can be formulated formally and generally as follows: postulate that for every system considered there exist various states z. These states can be discriminated from each other, and together these states make up a set Z. This set need not be continuous. Between two states $z \in Z$ and $z' \in Z$ transitions are allowed. This means that alterations of state $z \to z'$ are possible. Since the set Z need not be continuous, it is not necessary that the transition $z \to z'$ be realized along a path belonging to the set Z. Now, a function of state f is defined as a one-to-one mapping of real numbers f to the states $z \in Z$.

Systems can also be distinguished from one another as regards the composition of the matter confined in the system, or whether or not the matter depends on the caloric state of the system. A few frequently used types of systems are:

(1) *Mechanical or hydrodynamic systems.* A hydrodynamic or mechanical system is a system in which the properties are assumed to be independent of the caloric state of the system.

---

* Falk G. and Jung H. 1959. Axiomatik der Thermodynamik, *Handbuch der Physik*, Band **3**, Teil 2, 1959, S. 119–175.

(2) *Thermodynamic systems.* A thermodynamic system is a system in which the properties do depend on heat and the caloric state. The formulation of these systems depends always on one extra variable of state more than is needed for the formulation of a purely mechanical system.

(3) *Homogeneous systems.* A homogeneous system is a system in which all parts of the system are equal from the macroscopic point of view. Nevertheless, a homogeneous system could be a mixture of chemical substances, called *components* of the system.

(4) *Simple systems.* A simple system is a homogeneous system with only one phase. A phase is a part of the system that is completely homogeneous. A single phase may be a multicomponent system.

(5) *Multiphase, heterogeneous* or *complex systems.* A multiphase system is a system in which more than one phase coexists in the system.

(6) *Discontinuous systems.* A discontinuous system can be considered as a heterogeneous system with a finite number of phases, since the properties are distributed uniformly only over the phases but might show a discontinuity at the boundaries of the phases. Dividing surfaces are found between the phases. These interfaces are homogeneous in a direction parallel to the dividing surface and very thin in the perpendicular direction. If it is necessary to account for all the properties of the dividing surfaces, the dividing surface has to be formulated as a phase. The phase of this surface of discontinuity is usually described by only one extra independent variable of state. Most often the size of its surface is chosen as this independent variable.

(7) *Continuous systems.* A continuous system is a heterogeneous system with an infinite number of infinitesimal phases. A well-known example is a column of gas in a gravitational field.

2.1. THE ZEROTH LAW OF THERMODYNAMICS

2.1.1. Equilibrium. Equilibrium is also a fundamental concept in the thermodynamic description of systems. The existence of an equilibrium state is based on the experience that under constant external conditions of the system, the state of the system remains constant in time, not counting an initial time that is needed for the transition from a nonequilibrium state to the equilibrium state. At first glance equilibrium could be defined as the state, in which *all* quantities describing the system do not change noticeably during the time of observation, while in addition within the system no flow of matter or energy is observed. But, the idea of equilibrium is relative and depends equally on the observer's time scale as well as on the accuracy of the measuring devices used. In conclusion, it is found that a closed system subjected to constant external conditions will reach equilibrium after a sufficiently long initial time. This is a macroscopic experience that applies within the space and time scales used. However, from a microscopic point of view equilibrium cannot be attained in principle, as the axiom of Poincaré–Bendixson* shows. The Poincaré–Bendixson axiom expresses that for *every* closed system the motions are quasi-periodic. For the thermodynamic macroscopic system the recurrence time related to the quasi-periodic motions is much longer than the time scales involved in the macroscopic changes of state†.

The neighbouring equilibrium states are influenced through the common walls. Two types of walls are essential for thermal interactions, namely, the adiabatic wall and the diabatic or diathermal wall. The systems enclosed by these walls are classified in accord with these walls either as an adiabatic system or as a diabatic system.

An adiabatic system is a system enclosed by a thermally insulated or *adiabatic wall.* Only through movements of (parts of) the enclosing adiabatic wall or through external electrodynamic and/or gravitational fields can the state of an adiabatic system be changed from outside.

A diabatic system is enclosed by a diathermal wall. A wall is *thermally conducting* or *diathermal* if the wall is rigid and impermeable for matter and has the following property. If two systems, each in complete internal equilibrium, are brought into contact with each other through a diathermal wall then those two systems are usually not in mutual equilibrium‡ or: if two systems in complete internal equilibrium are separated by a diathermal wall, and this wall is also impermeable for either mechanical or chemical contact

---

* Poincaré H. 1890. Sur le problème des trois corps et les équations de dynamic. *Acta Mathematica*, 13: 1–270, see pp. 67–73. For the proof of the theorem, see Cesari L. 1971. *Asymptotic Behaviour and Stability Problems in Ordinary Differential Equations*, 3rd ed., Academic Press, New York; Andronov A.A., Leontovich E.A., Gordon I.I. and Maier A.G. 1973. *Qualitative Theory of Second-order Dynamic Systems.* Wiley, New York.

† Loschmidt J. 1876. Über den Zustand des Wärmegleichgewichtes eines Systems von Körpern mit Rücksicht auf die Schwerkraft I.; II.; III. *Sitzungsberichte der Mathematisch-Naturwissenschaftlichen Klasse der Kaiserlischen Akademie der Wissenschaften*, 73: 128–142; 73: 366–372; 75: 287–298; Boltzmann L. 1877. Bemerkungen über einige Probleme der mechanische Wärmetheorie. *Sitzungsberichte der Mathematisch-Naturwissenschaftlichen Klasse der Kaiserlischen Akademie der Wissenschaften*, 75: 62–100, refers to Loschmidt on page 67; Zermelo E. 1896. Über einen Satz de Dynamik und die mechanische Wärmetheorie. *Annalen der Physik*, 57: 485–494; Zermelo E. 1896. Über die mechanische Erklärung irreversibler Vorgänge. Eine Antwort auf Hrn. Boltzmann's 'Entgegnung'. *Annalen der Physik*, 59: 793–801; Smoluchowski M. von. 1912. Experimentell nachweisbare, der üblichen Thermodynamik widersprechende Molekularphänomene, *Physikalisches Zeitschrift*, 13: 1069–1080; Smoluchowski M. von. 1913. Gültigkeitsgrenzen des zweiten Hauptsatzes der Wärmetheorie, *Physikalisches Zeitschrift*, 14: 261–262. See also Chandrasekhar S. 1943. Stochastic problems in physics and astronomy. *Reviews of Modern Physics*, 15: 1–89.

‡ Guggenheim E.A. 1949. *Thermodynamics, an Advanced Treatment for Chemists and Physicists.* North-Holland Publ. Co., Amsterdam; Haase R. 1969. *Thermodynamics of Irreversible Processes.* Addison-Wesley Publ. Co., Reading, Massachusetts.

between the systems, then the variables of state of one system are no longer independent of the variables of state of the other system*.

From experience it is found that if two separate systems each in complete internal equilibrium are brought into contact with each other through a diathermal wall, then the two systems will eventually reach a new state of mutual equilibrium called *thermal equilibrium*. Experience shows furthermore, that if two systems are each in complete thermal equilibrium with a third system, these two systems are also in thermal equilibrium with each other. Systems are in equilibrium with each other if no interactions exist between the systems and the two separate systems are in equilibrium. Moreover, experience teaches that thermal equilibrium systems are determined by a few macroscopic variables of mechanical, chemical or electrodynamic origin and only *one* extra typical thermodynamic variable. These experimental findings are generalized in the

Zeroth law of thermodynamics:

(1) Each isolated system will adjust itself, and reaches equilibrium in a finite time.
(2) This equilibrium is transitive, that is, if a system A is in equilibrium with the systems B and C, then the system B is in equilibrium with the system C.
(3) Humans do have a natural feeling of 'hot' and 'cold', like boiling water is hotter than melting ice. This intuitive experience for hotness and coldness is accounted for by *one* extra variable of state.

The zeroth law of thermodynamics has been postulated and added to the already known laws of thermodynamics due to a suggestion of Fowler† to emphasize that equilibrium cannot be taken for granted but is a basic postulate of the theory. The zeroth law of thermodynamics is also basic for the definition of the empirical temperature. The zeroth law precedes logically all other laws of thermodynamics.

An important perfect process in thermodynamics can be discussed after having defined equilibrium, namely, the process that is accomplished through equilibrium states. On first examination, one is interested in the equilibrium properties of the system and the relations between the accessible equilibrium states. To get these relations, it is necessary to vary the equilibrium states, such that every variation passes through a succession of equilibrium states. Processes, for which the system is all the time in equilibrium are called *quasi-static*. A quasi-static process is therefore a process for which at every moment the state can be formulated with the same variables of state as those in

---

* Zemansky M.W. 1968. *Heat and Thermodynamics*, 5th ed. McGraw-Hill, New York, 1968.
† Fowler R.H. and Guggenheim E.A. 1939. *Statistical Thermodynamics*. Cambridge University Press, Cambridge. See p. 56.

equilibrium. Because time is not a classical thermodynamic variable, it is possible to introduce the idea of a quasi-static process. Physically, a quasi-static process can be realized only if the changes of state progress sufficiently slowly. What sufficiently slowly is, depends on the nature of the process. Each process has a *natural time* t_n, which might be thought of as a measure of the time the system—not being in equilibrium—needs to reach equilibrium after it is isolated. If the time scales of the changes of state in some process are much larger than those natural times t_n, the changes of state can be assumed to be quasi-static. Of course, a quasi-static process is an idealization that can only be realized within a certain accuracy. However, for the experimental determination of the equilibrium properties of a system this idealization and its realization are prerequisites.

2.1.2. The empirical temperature. The empirical temperature is the mutual property of systems that are in thermal equilibrium with each other. The experience that a system can be warmer or colder than another system yields a sequence monotonically changing caloric states. Due to the zeroth law of thermodynamics this sequence of caloric states makes up a one-dimensional sequence. This sequence has to be mapped on the sequence of real numbers. Typical for the formulation of the temperature scale one needs to have at one's disposal

(1) Two well defined and easily reproducible caloric states called 'fixed points'. The temperature at these fixed points is recorded by a number that is arbitrary.
(2) A property having a one-to-one correspondence with the sequence of caloric states.
(3) A material that has the property mentioned in item (2).

Water is a material that satisfies the requirements of the list. Two fixed points were defined, one for melted ice, called the *ice point*, and one for boiling water at 76 cm mercury, called the *steam point*. Until 1990 the values zero and one hundred were assigned to these fixed points. In agreement with these points* the *Celsius temperature scale* is defined as

$$t_{(x)} = \frac{A_x - A_0}{A_{100} - A_0} 100 \tag{2.1}$$

in which the property A is by definition a linear function of the empirical temperature. The Celsius temperature $t_{(x)}$ is indicated by °C, the unit degree Celsius†.

* Celsius originally assigned to these fixed points the values 100 (melting) and 0 (boiling). Either Linneaus or Jean Pierre Christin from Lyon has reversed this order.

† Note: since 1990 the Celsius temperature scale is defined differently and is coupled directly to the Kelvin temperature scale.

Giauque* pointed out that for the definition of the degree Celsius it is not necessary to define the two fixed points, of which one is chosen at the freezing point and the other at the boiling point of water. To demonstrate this, consider the gas thermometer, and measure the pressure and the volume at a fixed point. Repeat the measurement for different values of p and extrapolate the product pV to the limit $p \downarrow 0$. By using careful experimental procedures, it is found for any real gas kept at a constant temperature that the product pV has the same finite value in the limit as the pressure tends to zero (Boyle). Suppose that the obtained limiting value of pV is A_0. Repeat the procedure at an arbitrary caloric state. Suppose that now the limit value of pV is equal to A. Experiments with the gas thermometer have proven that the ratio A/A_0 is independent of the mass and the nature of the gas. This result is an important property of the gas thermometer. Using this property the empirical temperature is defined by $T^\star/T_0^\star = A/A_0$. This empirical temperature defines the so-called *Avogadro temperature scale.* According to this temperature scale $A = \lim_{p\downarrow 0} pV = RT^\star$ for all gases. The Avogadro temperature scale is the international standard temperature scale. The fixed point is the triple point of water $T_3^\star = 273.16$ K, to which the value A_3 corresponds. The 'International Temperature Scale of 1990 (ITS–90)' has defined another 16 reproducible equilibrium points, also used as fixed points. However, the triple point of water is the only assigned point of the thermodynamic scale, and it defines the Kelvin temperature scale by

$$T^\star = 273.16 \frac{A}{A_3} \text{ K} \tag{2.2}$$

Nevertheless, the Kelvin scale is an empirical scale based on the properties of a particular substance, namely an ideal gas confined in a gas thermometer. Only with reference to the second law of thermodynamics will it be shown that the Kelvin scale is identical with the thermodynamic temperature scale, a substance-independent temperature scale. The kelvin, the unit of the thermodynamic temperature T denoted by the symbol K, is defined as the 1/273.16th part of the thermodynamic temperature of the triple point of water. With this definition all other temperature scales are now based on the Kelvin scale.

As already remarked the Celsius scale is derived from the Kelvin scale by defining 100 °C as the boiling point of water at the condensation point of water vapor at 76 cm mercury. The measurement of this point coincides with 373.15 K on the Kelvin scale. From this observation it follows that 1 °C equals 1 K. However, in the new ITS-90 the boiling point of water is not a fixed point any more, since modern measurement techniques have shown

---

* Giauque W.F. 1939. A proposal to redefine the thermodynamic temperature scale: with a parable of measures to improve weights. *Nature*, 143: 623–626.

Table 2.1. The defining fixed points of the ITS-90.

Number	Temperature		Substance[a]	State[b]
	T_{90}/K	t_{90}/°C		
1	3 tot 5[c]	-270.15 tot -268.15	He	V
2	13.8033	-259.3467	e-H_2	T
3	≈17[d]	≈-256.15	e-H_2 (or He)	V (or G)
4	≈20.3[d]	≈-252.85	e-H_2 (or He)	V (or G)
5	24.5561	-248.5939	Ne	T
6	54.3584	-218.7916	O_2	T
7	83.8058	-189.3442	Ar	T
8	234.3156	-38.8344	Hg	T
9	273.16	0.01	H_2O	T
10	301.9146	29.7646	Ga	M
11	429.7485	156.5985	In	F
12	505.078	231.928	Sn	F
13	692.677	419.527	Zn	F
14	933.473	660.323	Al	F
15	1234.93	961.78	Ag	F
16	1337.33	1064.18	Au	F
17	1357.77	1084.62	Cu	F

[a] All substances—except ^{3}He—are of natural isotopic composition, e-H_2 is hydrogen at the equilibrium concentration of the ortho- and para molecular forms.
[b] V: vapor pressure point, T: triple point, G: gas thermometer point, M: melting point, F: freezing point (temperature, at a pressure of 101 325 Pa, at which the solid and liquid phases are in equilibrium). The purity of the metals is better than 99.9999%.
[c] Between 0.65 K and 5.0 K the temperature is defined by equations relating the vapor pressures of ^{3}He and ^{4}He.
[d] H_2-boiling points are difficult to realize. Temperatures from 3.0 K to 24.5561 K are defined by the gas thermometer.

that the boiling point of water is not 100 °C, but 99.974 °C. Nowadays, a temperature difference of 1 °C is *defined* to be equal to 1 K in the new ITS-90. The relation between the thermodynamic temperature T_{90} in kelvin and the Celsius temperature t_{90} in °C is

$$t_{90} = (T_{90} - 273.15)\ °\mathrm{C}$$

The triple point of water is therefore fixed at 0.01 °C. This point is reliably and simply reproducible. Of all fixed points, it is the most accurate fixed point to determine (reproducibility better than 0.1 mK*, that is, better than 0.4 ppm). The expectation† is that the ITS-90 is the final International Temperature Scale, so that no discrepancy is made between a measured temperature and the

* Furukawa G.T. and Bigge W.R. 1982. Reproducibility of some triple point of water cells. *Temperature, its Measurement and Control in Science and Industry.* American Institute of Physics, 5: 291–297.
† Quinn T.J. 1989. News from the BIPM. *Metrologia*, 26(1): 69–74; *Techniques for*

thermodynamic temperature. The ITS-90 will agree with the thermodynamic temperature T at 0.65 K with an error of less than 1 mK and with an error of several millikelvins at 100 °C. Because recently uncertainties have been reported in the measurements of the gas thermometer (about 5 mK at 230 °C and about 30 mK at 660 °C), the accuracy above 200 °C is more difficult to assess. These uncertainties also affect the measurements done with the optical pyrometer, and the uncertainty at 1000 °C is estimated to be 50 mK at least. Very important is the improvement obtained with the new ITS-90, where the large variations in $dT^\star/dT$ existent in the IPTS-68 have disappeared in the new ITS-90.

2.2. THE FIRST LAW OF THERMODYNAMICS

2.2.1. Energy. As a macroscopic theory, classical thermodynamics is based completely on experimental facts. These facts need only to be brought into a general formulation and expressed as axioms to formulate classical thermodynamics. For closed systems the mechanical work W is defined by

$$W = \sum_j \int \vec{P}_j \cdot d\vec{s}_j \tag{2.3}$$

where $\vec{P}_j$ denotes a *generalized force* and $d\vec{s}_j$ denotes the displacement of the point of application of the force $\vec{P}_j$. The variables of state which determine the position of the point of application of the forces are called the *generalized coordinates*. Work is measurable and known by its definition (2.3).

Use next the fact that for every system a wall can be found with the following property: if a system enclosed by such a wall is in equilibrium, then the state of that system can only be changed by performing work. Changes are therefore exclusively possible by

(1) Movement of the wall or parts of the wall.
(2) Long range forces from gravitational fields or external electrodynamic fields.

Such a wall is called a *thermally insulated wall* or an *adiabatic wall.* A process that takes place in a system enclosed by an adiabatic wall is called an *adiabatic process.*

Experience has shown that the work W for *adiabatic* changes of state is determined only by the initial and the final thermodynamic states and not by

---

Approximating the International Temperature Scale of 1990, Bureau International des Poids et Mesures, Sèvres, 1990. *Supplementary Information for the International Temperature Scale of 1990*, Bureau International des Poids et Mesures, Sèvres, 1990, which includes a copy of Preston-Thomas H. 1990. The International Temperature Scale of 1990 (ITS-90). *Metrologia*, 27: 3–10.

the intermediate states through which the system passes nor by the path the process has followed from the initial state I to the final state II. Obviously, this observation suggests the existence of a function of state E, for which

$$\Delta E = E_{II} - E_I = W \tag{2.4}$$

for all *adiabatic* processes that connect the initial state I to the final state II. The function of state E is called the *energy* of the system. The kinetic and the potential energy contributions can be separated from the total energy. With this separation it is possible that a term U is left over that is defined by this separation

$$E = E_{\text{kin}} + E_{\text{pot}} + U \tag{2.5}$$

The kinetic energy can be calculated according to the basic principles of mechanics provided that the velocity of the system and the velocities of all parts in the system are known. If there are external force fields, the potential energy can be calculated by means of the position coordinates of the system and of its parts. The mechanically calculated kinetic energy and potential energy are denoted by E_{kin} and E_{pot} in (2.5). For a purely mechanical system we have

$$\Delta E_{\text{kin}} + \Delta E_{\text{pot}} = W \tag{2.6}$$

From experience, it is found that for a thermodynamic system the definitions (2.4) and (2.5) apply and *not* (2.6). This means that the energy has still an extra term, besides the well defined kinetic and potential energy that is called the *internal energy* and is denoted by U in (2.5). This internal energy is determined exclusively by the internal state of the system and as a consequence is explained as a thermodynamic function of state. If one assumes that the E_{kin} and the E_{pot} are calculated purely mechanically, then the internal energy is defined by (2.4) and (2.5).

For two arbitrary states it seems that ΔE is defined only if those two states can be changed into each other adiabatically. This raises the question whether it is always possible to connect two states adiabatically. Experience shows again that this question is answered affirmatively, meaning that into *one* direction at least it is always possible to transform two arbitrary states adiabatically into each other. This implies that for any two states, ΔE can always be defined by using an adiabatic process. For a non-adiabatic isolated system experience shows that changes of state may occur while the system is in internal equilibrium and no external work on the system is done. However, this happens only in the case that the system is not in thermal equilibrium with its surroundings. Then the temperature of the system differs from the temperature of its surroundings. On closed systems, the surroundings could perform actions that have nothing to do with performing work as defined by (2.3). These kinds of actions between system and surroundings are indicated by the name *heat exchange.* If a change of state of a closed system takes place, the quantity of heat Q transferred from the surroundings to the system is defined by

$$Q \equiv \Delta E - W \tag{2.7}$$

In this definition of heat, W is the work exerted on the system during a change of state, and this work can be measured or calculated, while ΔE is defined by (2.4) with reference to adiabatic changes. From mechanics, it follows that E_{kin} and E_{pot} are proportional to the mass of the system. From experiments it is concluded that the internal energy U is also proportional to the mass of the system, so that U is an *extensive quantity* (extensive quantities are proportional to the mass of the system). These experiences are summarized in

The first law of thermodynamics:

For every closed system, there exists an extensive function of state E, called the energy of the system, with the following properties:

(1) For adiabatic changes of states $I \rightarrow II$:

$$\Delta E = E_{II} - E_I = W \tag{2.8}$$

where W is the work performed on the system during the adiabatic change of state.

(2) For an arbitrary change of state $I \rightarrow II$:

$$\Delta E = W + Q \tag{2.9}$$

where Q is the heat transferred from the surroundings to the system during the change of state.

(3) For an arbitrary infinitesimal process in a closed system:

$$\boxed{dE = đW + đQ} \tag{2.10}$$

where dE is a exact differential, and $đW$ and $đQ$ stand for a diminutive of work and of heat respectively.

A diminutive is denoted by the symbol $đ$. The diminutives $đW$ and $đQ$ are inexact differentials. Sometimes, $đW$ and $đQ$ are called incomplete differentials.

The fact that $đW$ and $đQ$ are inexact differentials is closely related to the fact that W and Q are not functions of state. W and Q depend not only on the initial and the final state, but also on the path traversed from the initial state to the final state.

For $\Delta U = 0$ and $Q = 0$ (2.9) reduces to the energy axiom of mechanics:

$$\Delta E_{\text{kin}} + \Delta E_{\text{pot}} = W$$

For an isolated system one has by definition $Q = 0$, $W = 0$, so that

$$E = E_{\text{kin}} + E_{\text{pot}} + U = \text{ constant}$$

In this formulation, the first law expresses the conservation of energy.

A cycle is a process for which the final state and the initial state are the same. If the system executes a cycle, then $\oint dE = 0$, in which $\oint$ symbolizes integration around the cycle. From (2.10) it follows that:

$$\oint đW + \oint đQ = 0 \tag{2.11}$$

This formulation of the first law of thermodynamics expresses the equivalence of heat and work, that is, both are forms of energy. It is noticed that $\oint đQ$ is the quantity of heat transferred to the system by processes that make up a closed cycle. Furthermore, the quantity

$$W^{\star} = -\oint đW \tag{2.12}$$

is introduced as the work that is done *by* the system *on* its surroundings in a cycle. The substitution of $W^{\star}$ into (2.11) yields for the first law applied to a cycle

$$\oint đQ = W^{\star} \tag{2.13}$$

For $\oint đQ = 0$ it follows that $W^{\star} = 0$. This implies that work cannot be gained from 'nothing', that an engine cannot create its own energy or that *perpetual motion of the first kind* is impossible. In the past this empirical fact has been frequently used for the foundation of the first law of thermodynamics.

The first law of thermodynamics is an expression in the three quantities U, Q and W, and refers to a change of state of a macroscopic system in two different states of equilibrium. As a consequence, the first law is not concerned with the intermediate states. Since the internal energy is defined in terms of the initial and the final state of a process, the internal energy U is *not* defined during the process. In classical thermodynamics the variable of *time* is *not* a thermodynamic variable. This is in contrast with classical mechanics where for instance, the mechanical kinetic energy has a unique value in terms of the *instantaneous* velocities of the system.

2.2.2. The first law of thermodynamics for open systems. Generalization of the first law of thermodynamics for open systems is difficult. Suppose that for open systems the energy E is also a single-valued extensive function of state. Using a simple one-phase system, it is easy to demonstrate that the generalization is difficult. Such a simple one-phase system has no kinetic energy or potential energy and therefore this system has only internal energy. The state of the phase in this one-phase system is now completely described by the temperature T, the pressure p and the masses $M_1, M_2, \cdots, M_N$ of the N constituent components. For a closed phase we have

$$U = U(T, p, M_K) \tag{2.14}$$

For an infinitesimal change of state it therefore follows that

$$dU = \frac{\partial U}{\partial T}\, dT + \frac{\partial U}{\partial p}\, dp + \sum_{K=1}^{N} \frac{\partial U}{\partial M_K}\, dM_K \tag{2.15}$$

In a closed phase the M_K's might change due to chemical reactions. Obviously, (2.15) holds also for an open phase, in which the masses M_K change due to transport from the surroundings to the system. After all, (2.15) contains only variables of state and applies independently of the processes responsible for the change of the masses M_K.

This holds in general. The variables of state will always include the masses of the components, and therefore U is always a function of these masses. Since U (or E) is a function of state, it follows that U (or E) depends only on dM_K ($K = 1, 2, \cdots, N$) and not on the processes involved in the changes of the masses M_K.

This does not apply for the heat Q and the work W, since these quantities depend on the path traversed by the processes responsible for the change of state, and because heat and work are related to the interaction of the system with its environment. This has already been commented on by Defay*. As a consequence, it is concluded that heat and work can be indefinite for open systems. This is easily illustrated if an open phase is considered, for which the displacement coordinates are kept constant, while matter is pressed through a semi-permeable wall. In this process compression work is performed without a change in volume. Hence, the definition (2.3) of the work W does not apply any more! It is now no longer possible to define the expansion work for an open phase unambiguously, and as a result the definition of the heat by (2.9) becomes ambiguous.

From a microscopic point of view, this indefiniteness of the work is understood statistically as follows. The expansion work $p\,dV$ is the average work needed to overcome the molecular force fields. Yet the question remains how this work has to be calculated if molecules are leaving and entering the system, in which case the definition of work is no longer unambiguous.

The definitions of the heat Q and the work W applied to open systems become indefinite, which cannot be avoided in principle. For open systems,

* Defay M.R. 1929. Introduction à la thermodynamique des systèms ouverts. Académie Royal de Belgique. *Bulletin de la Classe des Sciences*, 15: 678–688.

for instance, one might proceed as follows. If W' denotes the work done by a change of the displacement coordinates of the open system, while the masses M_K $(K = 1, 2, \cdots N)$ are kept constant, then formally we have

$$\Delta E = Q + W' \tag{2.16}$$

In this expression, it is *not* permissible to interpret Q as heat, but it is always permissible to consider the quantity Q as a quantity that balances the energy equation. Sometimes, $đQ$, defined by

$$đQ = dE - đW' - \sum_{K=1}^{N} h_K \, (dM_K)_{\text{ext}} \tag{2.17}$$

is interpreted as 'heat'. In this interpretation, $(dM_K)_{\text{ext}}$ is the change of mass due to an external cause of transfer of mass to and from the system and h_K the specific enthalpy of component K. This interpretation is at best 'reasonable', but not imperative. Indeed, (2.17) reduces to the first law of thermodynamics for $(dM_K)_{\text{ext}} = 0$ for all K, because then $đW' = đW$. According to Tolhoek and De Groot*, (2.17) is also in agreement with a gas kinetic interpretation. In such an interpretation, the quantity of heat $đQ$ defined in (2.10), is the average value of the change of the microscopic energy caused by a transfer of the kinetic energy among the molecules. In contrast with the potential functions in thermodynamics this kinetic energy has a well-defined zero point. Note, that the definition of $đQ$ in (2.17) is independent of a change of the reference value of the internal energy of the molecules of component K transferred to and from the system, and similarly the heat $đQ$ defined by (2.10), the quantity $đQ$ in (2.17) can be interpreted as measuring the change of microscopic kinetic energy only. Haase† too, suggests a reduction of the type (2.17). From a purely thermodynamic point of view, (2.17) is arbitrary if $đQ$ is defined as 'heat'. For the time being, (2.16) is maintained, in which Q is assumed to be defined by an energetic balance of the energy equation.

2.3. THE SECOND LAW OF THERMODYNAMICS

2.3.1. The second law of thermodynamics according to Kelvin–Planck. Like the zeroth and the first law, the second law of thermodynamics is also based on experimental facts. A closed system operating in a cycle is considered for the axiomatization of these experimental results. For a cycle it follows from the first law of thermodynamics that

$$\oint đQ + \oint đW = 0 \tag{2.18}$$

* Tolhoek H.A. and De Groot S.R. 1952. A discussion of the first law of thermodynamics for open systems. *Physica*, 18: 780–790.

† Haase R. 1953. Zur Thermodynamik der irreversibler Prozesse III. *Zeitschrift für Naturforschung*, 8a: 729–740.

The work $W^\star$ transferred in one period of the cycle by the system to its surroundings is defined by (2.12). During such a cycle, the system absorbs the quantity of heat Q_1 and gives off the quantity of heat $|Q_2|$, or alternatively absorbs a quantity of heat $-Q_2$, yielding

$$W^\star = Q_1 + Q_2 \qquad \text{with} \qquad Q_1 > 0 \quad \text{and} \quad Q_2 < 0 \tag{2.19}$$

in which $Q > 0$ is defined as the quantity of heat absorbed by the system, and $Q < 0$ the quantity of heat discarded by the system.

If it is assumed that the objective of a cycle is the performance of work, for which a certain quantity of heat Q_1 is supplied to the system, then the discarded quantity of heat $|Q_2|$ has to be considered as a loss. The efficiency η of a cycle therefore becomes

$$\eta = \frac{W^\star}{Q_1} = 1 + \frac{Q_2}{Q_1} = 1 - \frac{|Q_2|}{|Q_1|} \tag{2.20}$$

which is called the *thermal efficiency.*

The mechanical apparatus that takes the system through a cycle exchanging heat with its surroundings is called a *heat engine*, while the system is called the *working substance.* According to the first law of thermodynamics a heat engine produces only work if $Q_1 + Q_2 > 0$. Examples of heat engines are power plants, gas turbines, and automobile engines.

Schematically, the working of a heat engine can be represented as follows

(1) There is a process or a series of processes during which the working substance absorbs heat from an external system at a high temperature. This external system is called the *hot reservoir.*
(2) There is a process or a series of processes during which the working substance discards heat to a second external system that is called the *cold reservoir.*
(3) All processes together make up a cycle.

This scheme is illustrated in figure 2.1. The working of a heat engine is completely described if Q_1 and $W^\star$ are given, since in a cycle $W^\star = Q_1 + Q_2$ also applies. With this simple relation, it is possible to represent a heat engine by a vector in the $Q_1, W^\star$ diagram*. In the vector diagram in figure 2.1, the heat engines are represented in the right half-plane, since heat engines receive the heat Q_1 from the hot reservoir.

---

* The diagram has been designed using the suggestions found in: Schenck H. Jr. 1961. A useful thermodynamic diagram. *American Journal of Physics*, 29: 703–704; Lufburrow R.A. 1963. Classes of Carnot cycles. *American Journal of Physics*, 31: 480–481; Bolder H. 1964. Een meetkundige beschouwing over de tweede hoofdwet van de thermodynamica. *Nederlands Tijdschrift voor Natuurkunde*, 30: 245–253 (In Dutch).

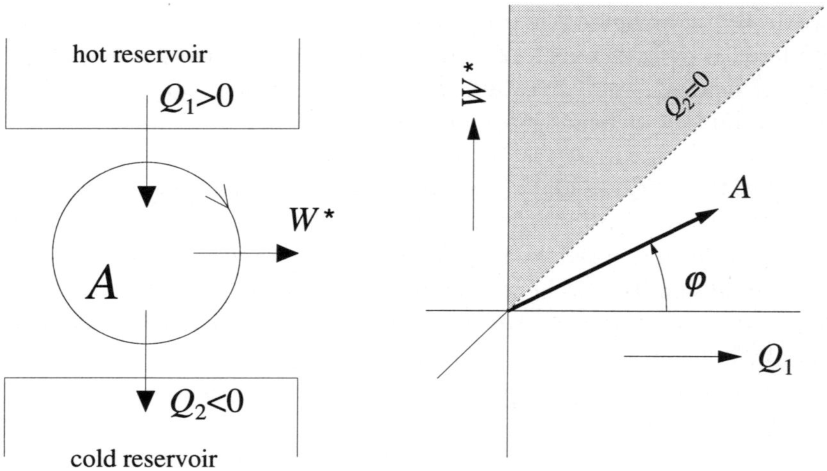

Figure 2.1. Schematic picture of a heat engine A and its vector representation in the $Q_1, W^\star$ diagram. Representations of heat engines with a positive efficiency are mapped in the domain $0° < \varphi < 45°$.

Using this vector representation for heat engines, the thermal efficiency η is immediately read from the diagram in figure 2.1 as $\eta = \tan\varphi$. Clearly, heat engines symbolized by points on the positive Q_1 axis have no thermal efficiency at all. Then the heat Q_1 taken from the hot reservoir is simply transferred to the cold reservoir with $Q_1 = -Q_2$. However, this process represents an important natural process in which heat from a hot body flows to a cold body. This process shows that it is possible to obtain thermal equilibrium between two bodies, initially at different temperatures. The heat engines mapped on the negative $W^\star$ axis convert the work received from the environment completely into heat that is transferred to the cold reservoir. This natural process is called friction. The heat engines symbolized by points in the fourth quadrant combine processes in which work is converted into heat discarded to the cold reservoir with processes in which the heat from the hot reservoir is transferred to the cold reservoir. These heat engines deliver no useful energy to the environment and these machines have a negative thermal efficiency.

Heat engines with a positive thermal efficiency are represented by points in the first quadrant. In the diagram in figure 2.1, the 45° line is the line for which $Q_2 = 0$. Processes represented by points on this line have a thermal efficiency of 100%. In such processes, heat is absorbed from the hot reservoir without a transfer of heat to the cold reservoir. On the other hand, experience has shown that it is impossible to construct a heat engine that converts heat absorbed from the hot reservoir into work without giving off heat to the cold reservoir. As a result, it is not possible to convert heat from a single reservoir completely into work. Should this be possible, it would be possible for example to use the ocean or the atmosphere for the heat reservoir and obtain useful work from an engine only by extracting heat from the ocean or the atmosphere.

These processes, which convert the heat absorbed from the cold reservoir completely into work, are represented by points on the positive $W^\star$ axis in the diagram of figure 2.1. The points in the region between the 45° line and the positive $W^\star$ axis represent heat engines that extract only heat from the cold reservoir as well as from the hot reservoir and convert the total amount of absorbed heat into work. The first law of thermodynamics does not exclude this possibility. Therefore, the exclusion of this possibility is a new experimental fact that has to be formulated as a postulate of the theory. An engine that performs work simply by extracting heat from a single heat reservoir represents a *perpetual motion of the second kind.*

The exclusion of a perpetual motion of the second kind goes back to Kelvin. Planck modified Kelvin's original statement. This modification can be formulated as follows.

The second law of thermodynamics according to Kelvin–Planck. It is impossible to construct a heat engine that drives a working substance through a cycle with the sole result that in each period of the cycle heat is absorbed from a reservoir and is converted into an equivalent quantity of work.

2.3.2. The second law of thermodynamics according to Clausius. The Clausius statement of the second law can be discussed as follows. First, it is noticed that in principle it is possible to reverse the cycle of a heat engine, if work is done on the working substance. In the reversed cycle, Q_1, Q_2, and $W^\star$ get an opposite sign. Now, the quantity of heat Q_2 is abstracted from the cold reservoir and the quantity of heat Q_1 is transferred to the hot reservoir. Such an engine is called a refrigerator or a heat pump, depending on the objective of the engine. If the objective is the absorption of heat from the cold reservoir it is called a refrigerator. If the objective is the discharge of heat to the hot reservoir it is called a heat pump.

A symbolic representation of a refrigerator (or a heat pump) is sketched in figure 2.2. In the vector representation in the $Q_1, W^\star$ diagram in figure 2.2, the refrigerators and heat pumps are mapped onto the left half-plane, because in both cases the heat $-Q_1$ is supplied to the hot reservoir. In those cases, one refers to a refrigerator or a heat pump because $|Q_1| > |Q_2|$, that is, more heat is transferred to the hot reservoir than heat is absorbed from the cold reservoir. This follows from $Q_1 + Q_2 = W^\star$ or $Q_1 = -Q_2 + W^\star$, in which Q_2 is positive and $W^\star$ negative, hence $|Q_1| > |Q_2|$. The objective of a refrigerator is the absorption of heat from the cold reservoir. To acquire this, work has to be provided on the system and the useful efficiency of a refrigerator is $Q_2/|W^\star|$. This efficiency is called the *coefficient of performance* or *cooling energy ratio.* The objective of a heat pump is the transfer of heat to the hot reservoir at the expenditure of as little work as possible. Therefore, the useful efficiency is given by $|Q_1|/|W^\star|$, which is also called the *coefficient of performance.*

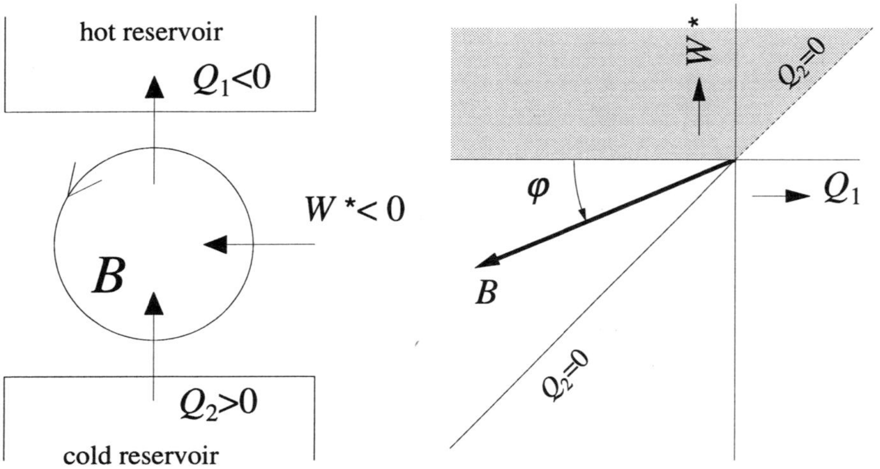

Figure 2.2. Schematic picture of a refrigerator (or a heat pump) B, and its vector representation in the $Q_1, W^\star$ diagram. Representations of refrigerators or heat pumps with useful coefficients of performance are mapped in the domain $0° < \varphi < 45°$.

Furthermore, experience has shown that work always has to be done on the working substance to absorb heat from the cold reservoir and to transfer the heat to the hot reservoir. Although, according the first law of thermodynamics, such a process is possible for $W^\star = 0$, it turns out that it is not possible in practice. Clausius has generalized this negative experience in the following statement of the second law.

The second law of thermodynamics according to Clausius. It is impossible to construct an engine that drives a working substance through a cycle with the sole result that in each period of the cycle heat is transferred from a cooler body to a hotter body.

The statement of Clausius in fact expresses an irreversibility principle: 'no heat can flow spontaneously from a body with a low temperature to a body with a high temperature'. The reverse process is possible and discussed before, since otherwise thermal equilibrium between two bodies with a different temperature would never be obtained. According to Clausius the transport of heat is therefore an irreversible process.

The images of the impossible processes of Clausius are given by points on the negative Q_1 axis in the $Q_1, W^\star$ vector diagram in figure 2.2. No images of refrigerators or heat pumps are found on this axis. Due to the principle of Kelvin–Planck no real cycles can be symbolized by the points on the positive $W^\star$ axis. As a consequence, the processes symbolized by the second quadrant represent the impossible processes that transport a part of the heat extracted from a colder body to a hotter body without an input of work and convert the remainder of the heat completely into work.

The points on the 45° line in the third quadrant represent the heat pumps for which all work performed on the working substance is converted into the quantity of heat $-Q_1$ supplied to the hot reservoir. Points in the domain between the 45° line and the negative $W^\star$ axis symbolize the dissipative (friction) processes, in which the work is converted into heat and discarded to the cold and the hot reservoirs. For the realization of the objectives of heat pumps and refrigerators the heat Q_2 has to be absorbed from the cold reservoir. For these engines it is imperative to perform work on the working substance. The vectors symbolizing these machines are therefore mapped onto the region between the 45° line ($Q_2 = 0$) and the negative Q_1 axis. If the vector B in the vector diagram of figure 2.2 represents a heat pump or a refrigerator, then the efficiency of the heat pump or its coefficient of performance is $\eta_{\mathrm{w}} = \cot\varphi$. Due to the Clausius principle, $\varphi > 0$ in this vector diagram. To summarize, the vectors symbolizing the heat pumps and the refrigerators are found in the area below the negative Q_1 axis limited by the 45° line, for which $Q_2 = 0$. The value of φ cannot be specified further until the heat reservoirs as well as the cycle are given in more detail. In contrast to the efficiency of the heat pump, the efficiency of the refrigerator—the cooling energy ratio η_{k}—cannot readily be seen from the vector diagram in figure 2.2.

2.3.3. Equivalence of Kelvin–Planck and Clausius statements. The statements of Kelvin–Planck and Clausius are equivalent. This means that the statement of Kelvin–Planck implies the statement of Clausius and vice versa. If the truth of the statement of Kelvin–Planck is denoted by K and the one of Clausius by C, the equivalence is written as

$$K \Longleftrightarrow C \tag{2.21}$$

where the symbol $\Longleftrightarrow$ indicates equivalence. The symbol $\Longrightarrow$ is used to indicate the implication. The meaning of the equivalence (2.21) is that

$$K \Longrightarrow C \qquad \text{and} \qquad C \Longrightarrow K \tag{2.22}$$

If $-K$ denotes the falsity of the Kelvin–Planck statement and $-C$ the falsity of the Clausius statement, then (2.22) also holds when

$$-K \Longrightarrow -C \qquad \text{and} \qquad -C \Longrightarrow -K \tag{2.23}$$

The equivalence of both statements is easier to prove by using this negative formulation rather than the positive formulation (2.22). It is easier because only one counter example is needed. Therefore, (2.23) will be proved.

If the statement of Clausius is violated then the statement of Kelvin–Planck is also violated. If C is violated, it is possible to construct a refrigerator that transfers the quantity of heat Q_2 from the cold reservoir to the hot reservoir *without* any input of work. Between both reservoirs a heat engine also operates. This engine does not violate any thermodynamic law.

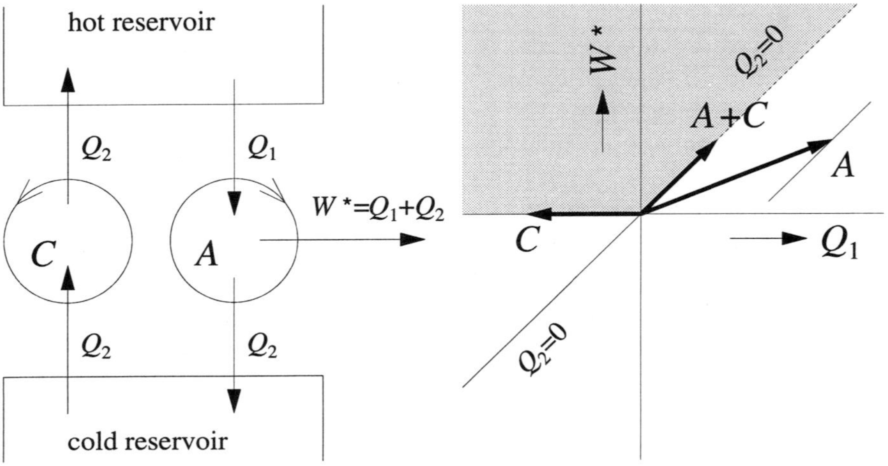

Figure 2.3. Schematic representation of an engine built up by combining a cooling engine and a heat engine. The combined engine is used for the proof that $-C \Longrightarrow -K$. The right-hand side shows its vector representation in the $Q_1, W^\star$ diagram.

The symbolic representation of the refrigerator C and of the heat engine A is sketched in figure 2.3 on the left-hand side, and on the right-hand side the vector representations. The heat engine absorbs the quantity of heat Q_1 from the hot reservoir, does the work $W^\star$ and loses the quantity of heat $-Q_2$ to the cold reservoir. The refrigerator and the heat engine are now combined to make up together a single engine. This combined engine does the work $W^\star$ and absorbs the quantity of heat $Q_1 + Q_2$ from the hot reservoir. Since no heat is delivered to or subtracted from the cold reservoir, this combined engine converts *all* heat into work extracted from a single reservoir. This is contrary to the statement of Kelvin–Planck, so it proves $-C \Longrightarrow -K$.

A simple direct graphical representation of the proof of $-C \Longrightarrow -K$ is given in the vector diagram in figure 2.3. Remember that for processes symbolized by the Q_1 axis it holds that $Q_1 = -Q_2$. The heat engine A has to be chosen such that the combined engine $A + C$ is symbolized by a vector along the line $W^\star = Q_1$, where $Q_2 = 0$, which is the representation of all impossible processes with an efficiency of 100%. All heat engines with images on the 45° line through A make up a combined engine $A + C$ symbolized by vectors along the line $Q_2 = 0$. Obviously, there are many heat engines available and appropriate for the proof of $-C \Longrightarrow -K$.

If the statement of Kelvin–Planck is violated then the statement of Clausius is also violated. Consider an engine K that violates the statement of Kelvin–Planck. This engine realizes the work $W^\star$ only by absorbing heat from the heat reservoir. The work $W^\star$ is used to drive a refrigerator B that takes out the quantity of heat Q_2 from the cold reservoir and transfers the quantity of heat $-(Q_1 + Q_2)$ to the hot reservoir.

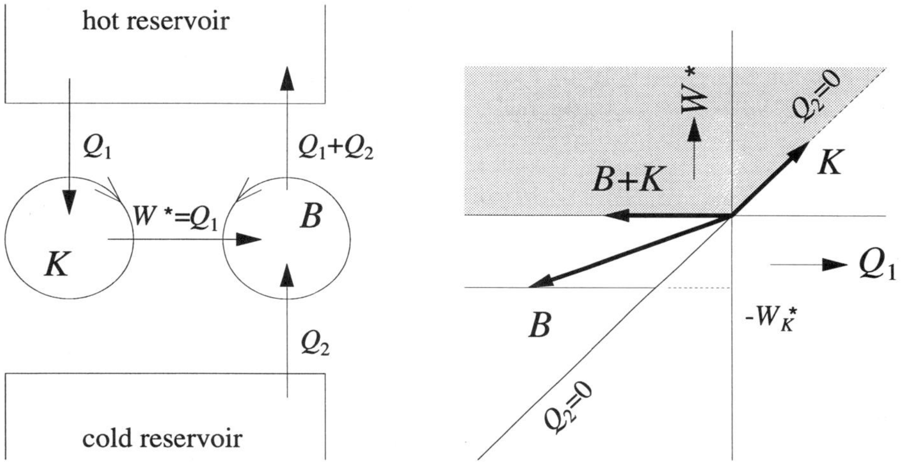

Figure 2.4. Schematic representation of an engine built up by combining a heat engine and a cooling engine. The combined engine is used for the proof that $-K \Longrightarrow -C$. Its vector representation is given in the $Q_1, W^\star$ diagram.

A schematic representation and the associated vector representations of the engine K with the cooling engine B are depicted in figure 2.4. Both engines are considered to work together. This combined engine does not perform work, and also no work is supplied to the engine. The result of the combined action of the refrigerator and the heat engine together is that the quantity of heat Q_2 is absorbed from the cold reservoir and is transferred to the hot reservoir. This is in contradiction with the statement of Clausius. Therefore, it proves $-K \Longrightarrow -C$.

The vector diagram in figure 2.4 gives a simple illustration of the proof. The representation of the engine K is found on the 45° line, for which $W^\star = Q_1$ or equivalently $Q_2 = 0$. Appropriate cooling engines B are represented by vectors lying on the line $-W_K^\star$, where $W_K^\star$ is the useful work delivered by the engine K. The representation of the combined engine is found by vector addition to lie on the negative Q_1 axis that symbolizes the impossible processes according to the Clausius statement of the second law of thermodynamics.

As a consequence, the conclusion is reached that both statements are equivalent, and that either one can be used for the further development of the theory.

2.3.4. The second law according to Carathéodory. Finally, a more local definition of the second law of thermodynamics has been given by Carathéodory. The Kelvin–Planck and the Clausius formulations of the second law of thermodynamics are global formulations. In an attempt at an axiomatization of thermodynamics, in particular the second law, Carathéodory* used a mathematical theorem. This theorem states that an inexact differential such as $X\,dx+Y\,dy+Z\,dz$, which is known as the Pfaffian form of the differential, has a factor such that it becomes a exact differential if, and only if, there exists at least one point in the neighborhood of a point P which cannot be reached from P along any curve on the surface determined by $X\,dx + Y\,dy + Z\,dz = 0$. Such a factor is called an integrating factor. The statement of Carathéodory of the second law of thermodynamics reads:

The second law of thermodynamics according to Carathéodory. Arbitrarily close to a given thermal equilibrium state there exist equilibrium states that cannot be reached by an adiabatic change.

The statement of Carathéodory formalizes the conclusion taken from the experiments carried out by Joule† with great care between 1843 and 1849, that the change of state cannot be reversed adiabatically. For instance, it is impossible to return the weight that drives a rotating stirrer heating the water by friction in an insulated container, to its original height by an adiabatic change of state. The original temperature of the water can only be restored by bringing the container in contact with a colder body.

The experiments of Joule are associated with points on the negative $W^\star$ axis in the vector diagram in figure 2.1. The points on the positive $W^\star$ axis are associated with vectors representing the impossible processes of Kelvin–Planck, and as a consequence symbolize states that cannot be reached. The statement of Carathéodory is closely related to the statement of Kelvin–Planck‡, and as such is equivalent to the statements of Kelvin–Planck

---

* Carathéodory C. 1909. Untersuchungen über die Grundlagen der Thermodynamik. *Mathematische Annalen*, 67: 355–386; Carathéodory C. 1925. Über die Bestimmung der Energie und der absoluten Temperatur mit Hilfe von reversiblen Prozessen. *Berliner Berichte*, 39–47. See also Born M. 1921. Kritische Betrachtungen zur Darstellung der Thermodynamik. *Physikalisches Zeitschrift*, 22: 218–224, 249–254 and 282–286; Kuiken G.D.C. 1979. Note sur la définition de l'entropie. *Entropie*, 85: 15–19; Whaples G. 1952. Carathéodory's temperature equations. *Journal for Rational Mechanical Analysis*, 1: 301–307.

† Joule J.P. 1843. On the caloric effects of magneto-electricity, and the mechanical value of heat. *Philosophical Magazine*, 23: 263–276, 347–355, 435–443; Joule J.P. 1845. On the changes of temperature produced by the rarefaction and condensation of Air. *Philosophical Magazine*, 26: 369–383; Joule J.P. 1847. On the mechanical equivalent of heat, as determined by the heat evolved by the friction of fluids. *Philosophical Magazine*, 31: 173–176; Joule J.P. 1849. On the mechanical equivalent of heat. *Philosophical Magazine*, 35: 533–534.

‡ 'Für das Axiom des zweiten Hauptsatzes habe ich eine Definition gewählt, die der

and Clausius. However, there is some difference, since the Carathéodory statement refers to equilibrium processes, while the Kelvin–Planck and Clausius statements do not. We will postpone that discussion until after the discussion of the reversible processes and the completion of the vector diagram.

2.3.5. Irreversible and reversible processes. As remarked before, with the Clausius statement of the second law of thermodynamics an irreversibility principle has already been formulated. The concept of irreversibility plays an important role in the thermodynamics and as such has to be illustrated further. It is no surprise that the idea of irreversibility is again based completely on experience.

For constant external conditions, the experimental evidence has already indicated that all processes in a system evolve spontaneously in one direction, and in particular, in the direction of the equilibrium state determined by the external conditions. Processes that evolve in one direction only are called *irreversible* processes. The conclusion is therefore: *in a system under constant external conditions, the processes take place irreversibly.*

The external conditions need not be constant. Suppose that for a given system A the external conditions are not constant. Now a secondary system can be brought into contact with the system A. The secondary system B can be chosen such that the system B performs all desired actions onto the system A. Furthermore, the system B can be chosen such that the external conditions of the combined system $A+B$ are constant. In the system $A+B$ all processes evolve in one direction, and therefore this holds also for the processes in the subsystem A of $A+B$. This reasoning yields the conclusion that *every process that proceeds with a finite velocity is irreversible.*

Planck* called a process that actually occurs in nature a *natural process.* Natural processes proceed irreversibly in a direction towards equilibrium. It never occurs in nature that a process spontaneously proceeds away from equilibrium. Such a process is called an *unnatural process.* The *reversible processes* are thought to form the limiting case between the natural and the unnatural processes. These processes can be recognized if one investigates under what conditions a reversible process might take place. The spontaneous processes in nature operate with finite velocity and are irreversible. Therefore, the reversible processes do *not* proceed with a finite velocity, and a reversible process is thought to proceed very slowly. Each system that is not in equilibrium will tends towards equilibrium in a one-way direction. As a consequence, non-equilibrium states cannot occur in a reversible process, and

---

Planckschen sehr verwandt ist; ...'. Quote from page 356 of Carathéodory C. 1909. Untersuchungen über die Grundlagen der Thermodynamik. *Mathematische Annalen*, 67: 355–386.

* Planck M. 1887. Ueber das Princip der Vermehrung der Entropie. *Annalen der Physik und Chemie, Leipzig*, 30: 562–582. See p. 563.

a reversible process must have a quasi-static character. All these observations concerning reversibility are formulated negatively, while we have to establish to what extent reversible processes are possible.

According to Gibbs*, all states of the system differing infinitesimally from the equilibrium state are reversible. To appreciate this, consider a system that is in equilibrium under constant external conditions. In this equilibrium system, small changes are considered that are possible without external actions and without breaking possible existing constraints in the system. These constraints are called passive forces by Gibbs. Any realizable infinitesimal alteration from the equilibrium state has to be a reversible one. If this is not true, then, the initial state was not an equilibrium state of the system. These infinitesimal changes of state are indeed a limiting case between the natural and the unnatural processes. After all, the infinitesimal changes cannot represent a natural process; since natural processes proceed towards a state of equilibrium and the system is already in a state of equilibrium. The conclusion is: *virtual infinitesimal changes in a system that is in a state of complete equilibrium, are reversible.*

Changes that are possible with respect to the external conditions and the existence of possible constraints, are called *virtual* in analogy to the nomenclature used in mechanics. This last conclusion is closely related to the statistical fluctuation theory. According to the statistical theory, a thermodynamic equilibrium system is not a changeless configuration, but an average of a large number of possible configurations and motions of the constituent molecules and atoms. From this point of view, a system—macroscopically in equilibrium—will follow all changes that are consistent with the conditions of the system. After such a change the system returns to the state of equilibrium. A temporary change of state, for example caused by a tiny impulse, is reversible.

A reversible process can be contemplated as follows: from a state of equilibrium an infinitesimal reversible change of state is produced. Now, the external conditions of the system have to be changed such that the system is also brought into equilibrium with its surroundings by an infinitesimal change. This procedure is continued. Such a process proceeds very slowly and conxists of a continuous series of equilibrium states. A reversible process is therefore quasi-static, but the reverse statement is not true in general. To appreciate the last remark, it will be shown that irreversibility is closely related to dissipation effects.

Consider a closed and adiabatically isolated system. On this system the

* Gibbs J.W. 1875–1878. On the equilibrium of heterogeneous substances. *Transactions of the Connecticut Academy*, 3: 108–248, 343–524. Reprinted in: *The Scientific Papers of J. Williard Gibbs*, Vol. One: *Thermodynamics*. Dover Publications, New York, 1961. See p. 61.

work W is done, and this work is completely converted into internal energy according the first law: $W = \Delta U$. Now restore the old situation. Remove temporarily the adiabatic isolation and withdraw the quantity of heat $\Delta Q = +\Delta U$ from the system. This quantity of heat has to be converted fully into work. Then the system *and* its direct environment are restored in the initial state. According to Kelvin–Planck it is impossible to convert $\Delta Q = \Delta U$ completely into work. Therefore, the process is irreversible. The conversion of work into internal energy is always irreversible and is called *dissipation*, because work is obviously dissipated and as a consequence cannot be recovered completely. It follows from experiments that in all motions dissipation always arises due to friction. Friction can be reduced to a larger extent, but in practice at normal temperatures it can never be totally eliminated. If effects of friction could be eliminated completely, a movement could persist for ever without external interference. This kind of impractical motion is called a *perpetual motion of the third kind*. The perpetual motion of the first and of the second kind are corollaries of the laws of thermodynamics. The perpetual motion of the third kind is a conclusion drawn from the fact that friction cannot be totally eliminated.

It is noted that quasi-static processes may be dissipative. *A quasi-static process is reversible only if no dissipation effects are involved in the process.* This is illustrated as follows: a quasi-static process passes through a continuous series of equilibrium states and, as a consequence, the quasi-static process is reversible. However, it has to be done such that all work that is performed by the environment on the system, can equally be performed by the system on the environment. This last requirement holds only if during the process no dissipation effects occur. In a quasi-static process, work could be done on a closed adiabatic system, but this process is not reversible!

2.3.6. Carnot processes and the absolute temperature. For the mathematical formulation of the second law, a reversible cycle is needed. Isothermal and adiabatic processes can be idealized as reversible processes. An ideal engine may be supposed to exist using those two processes. Before the proper foundation of the first law of thermodynamics, Carnot published in 1824 an article* in which he described a simple cycle operating between two heat reservoirs, and having maximum thermal efficiency. This engine is called the *Carnot engine*, and is made up in a cycle with the following four steps

(1) At a constant temperature θ_1 the working substance absorbs *reversibly* the quantity of heat Q_1 from the hot reservoir.

* Carnot N.L.S. 1824. *Réflexions sur la Puissance Motrice du Feu et sur les Machines propres à Développer cette Puissance.* Bachelier, Paris. Translation in the Dover publication: *Reflections on the Motive Power of Fire.* Dover, New York, 1960.

(2) The system performs work adiabatically in a *reversible* process, such that the temperature of the working substance is lowered from θ_1 to θ_2.
(3) At a constant temperature θ_2 the working substance rejects *reversibly* the quantity of heat Q_2 to the cold reservoir.
(4) The system undergoes a *reversible* adiabatic process, such that the temperature of the working substance rises from θ_2 to θ_1.

A Carnot process operates therefore between two heat reservoirs of constant temperature θ_1, respectively θ_2, with $\theta_1 > \theta_2$.

A heat reservoir is an idealized system that can be specified as follows. A *heat reservoir* is a system enclosed by an adiabatic wall and as such it can exchange energy only by heat transfer with $\Delta U = Q$. The mass of the system is extremely large, and the heat reservoir may reject or absorb an unlimited quantity of heat without an appreciable change in the internal energy or temperature.

In addition, the mass of a heat reservoir is supposed to be so large that during the absorption and the rejection of quantities of heat there are no detectable changes in any internal thermodynamic coordinate of the reservoir. The changes are very small and extremely slow—so slow and tiny that the changes are surely quasi-static ones and no dissipative effects are developed. In other words: if the heat reservoir absorbs or rejects heat, the changes *within* the reservoir are reversible. In future discussions, a heat reservoir has to be understood as described above, with all the properties mentioned. It is noted that heat reservoirs can also be approximated by using the phase transitions of material bodies.

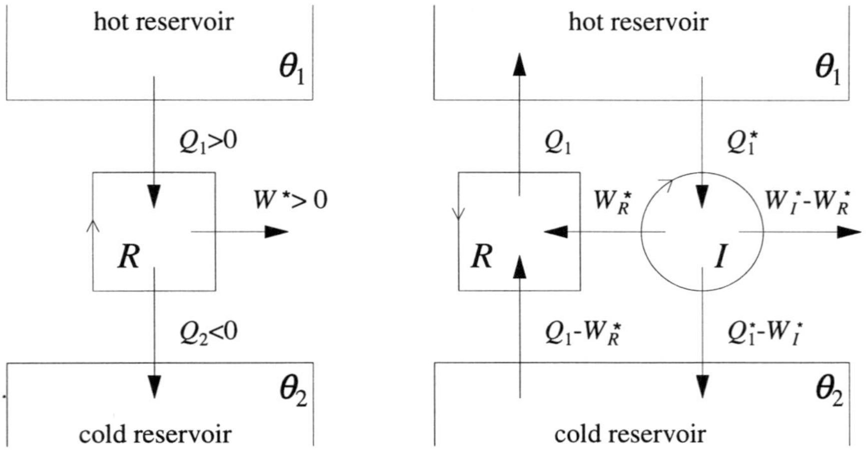

Figure 2.5. Symbolic representation of a reversible Carnot process R together with a symbolic representation of the reversible Carnot R with an arbitrary Carnot engine I.

A Carnot process operates between two heat reservoirs with constant temperatures θ_1 and θ_2, respectively. In principle, it is possible that reversible Carnot processes also exist, because for isothermal as well as for adiabatic processes the changes can be performed reversibly in principle.

In the left-hand side of figure 2.5 a reversible Carnot engine is symbolized. The reverse of a Carnot engine is a refrigerator or a Carnot heat pump. The proposition of Carnot can now be discussed. It reads:

Proposition of Carnot. No engine, which operates between two given heat reservoirs, has a higher thermal efficiency than a reversible Carnot engine operating between the same two heat reservoirs.

The thermal efficiency of the reversible Carnot power engine in the left-hand side of figure 2.5 is compared with the thermal efficiency of the irreversible power engine in the right-hand side of figure 2.5. In principle the reversible Carnot engine can be operated in the opposite direction as a heat pump or refrigerator. Consider for the proof the symbolic representation of a power engine I and a reversible Carnot power engine, but operating as a refrigerator R in the right-hand side of figure 2.5. Both engines operate between the two heat reservoirs at the given temperatures θ_1 and θ_2 respectively. Suppose that the thermal efficiency of the engine I is larger than the thermal efficiency of the reversible Carnot engine R, or

$$\eta_I > \eta_R \qquad \text{or} \qquad \frac{W_I^\star}{Q_1^\star} > \frac{W_R^\star}{Q_1} \tag{2.24}$$

Suppose, now, that the engine I drives the Carnot engine as a refrigerator and that both engines operate together as a single engine between the hot reservoir and the cold reservoir. This combined engine absorbs the heat

$$(Q_1 - Q_1^\star) + (W_I^\star - W_R^\star) \tag{2.25}$$

from the cold reservoir. The heat $Q_1^\star - Q_1$ is supplied to the hot reservoir, and $W_I^\star - W_R^\star$ is available to perform useful work. The incorrectness of the point of departure follows by either the Clausius principle or the Kelvin–Planck principle. Contradiction of the last principle results if $Q_1^\star = |Q_1|$, while (2.24) shows that $W_I^\star > W_R^\star$ and that the engine I drives the engine R. Since the quantity of heat $Q_1^\star$ needed to drive the engine I is delivered by the reversible refrigerator R the hot reservoir could be made small and be assimilated into the combined engine $I+R$. From (2.25) it follows that the net effect is that the quantity of heat $W_I^\star - W_R^\star > 0$ absorbed from the cold reservoir is completely converted into useful work. This is in contradiction with the Kelvin–Planck principle. Contradiction of the Clausius principle results if it is assumed that the engine I supplies just enough work to drive the reversible engine, that is, if $W_I^\star = W_R^\star$. Referring to (2.24) it follows that $|Q_1| > Q_1^\star$. The net effect is then that the quantity of heat $Q_1 - Q_1^\star$ is transferred from the cold reservoir to the hot reservoir without doing work. This is in contradiction to the principle of Clausius. Therefore, the relation

$$\eta_I \leq \eta_R \tag{2.26}$$

has to be satisfied, from which follows the

Corollary: All reversible Carnot engines operating between the same two thermal reservoirs have the same thermal efficiency.

This corollary is demonstrated by considering two reversible Carnot engines that operate between the same two thermal reservoirs. These Carnot engines are denoted by R_1 and R_2 respectively. If the engine R_1 drives the engine R_2 as a refrigerator, then it follows from (2.26) that $\eta_{R_1} \leq \eta_{R_2}$. If R_2 drives the engine R_1 as a refrigerator, then it follows from (2.26) that $\eta_{R_2} \leq \eta_{R_1}$. This proves the corollary that $\eta_{R_1} = \eta_{R_2}$, from which one arrives at the

Conclusion: the nature of the working substance used in a reversible Carnot cycle has no influence on the thermal efficiency of the engine.

The thermal efficiency of a reversible engine is therefore a property of the surroundings, namely, the heat reservoirs at the temperatures θ_1 and θ_2 and not of the design of the process or of the choice of the materials or gases used for the working substance. If an engine is reversible, it does not matter what the design is, because the heat supplied to the engine at the temperature θ_1 and the heat absorbed from the engine at the temperature θ_2 do not depend on the design of the engine itself.

The vector representation for cycles in a $Q_1, W^\star$ diagram can now be completed, since now the heat reservoirs are specified, while from the Carnot corollary it is known that there are no engines with a higher thermal efficiency than the reversible engines. The reversible engines are symbolized in figure 2.6 by the straight line through the origin of the $W^\star, Q_1$ plot. Evidently, $W^\star$ is proportional to Q_1 for reversible engines.

If the vector A represents a reversible Carnot process, the reverse process is symbolized by the vector $A^\star = -A$. Vectors lying in the half plane above the line RR do not represent real Carnot processes. Below the line RR only vectors representing irreversible processes can be found, since the reverse process is symbolized by a vector in the forbidden shaded area in figure 2.6. See for example the process E of which the vector $E^\star = -E$ represents an impossible process. The line RR of the reversible processes divides the domain of the irreversible processes from the area of the impossible processes. In the language of Planck: the line RR divides the domain of the natural processes from the domain of the unnatural processes, therefore, the line RR represents the border between the natural processes and the unnatural processes. As a consequence, the reversible processes are by definition symbolized by vectors on the line RR.

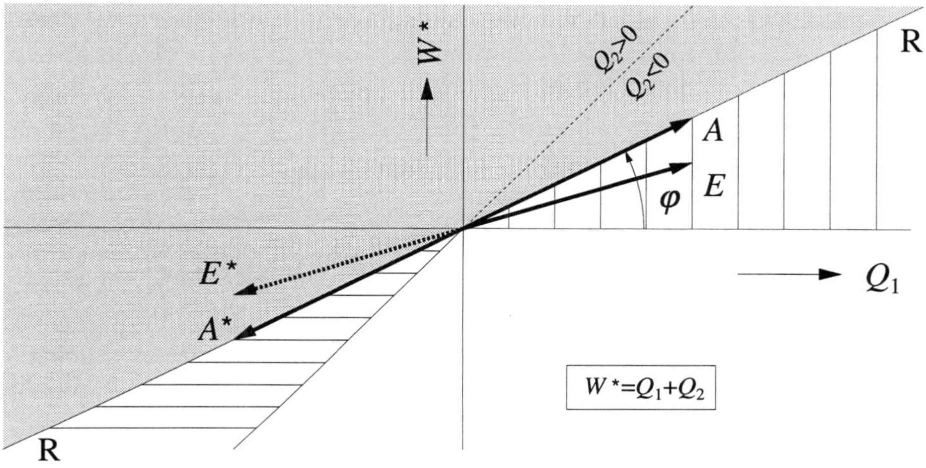

Figure 2.6. Q_1, W^* diagram for the representation of Carnot processes by given temperatures θ_1 and θ_2 of the hot and cold heat reservoir respectively. The line RR is the line of the reversible Carnot processes. In the shaded area above that line there are no possibilities for real processes. Below the line, vectors can be drawn symbolizing the irreversible processes. The angle φ is determined by the temperatures of the heat reservoirs. In the first quadrant, the vertically lined area is the region of the irreversible power engines that can be thought of as a combination of a reversible process and a process symbolized by the negative W^* axis (the negative W^* axis represents the friction processes in which all work is converted into heat). The horizontally lined area in the third quadrant is the region of the irreversible refrigerators and heat pumps that can be thought of as a combination of a reversible process with a process on the positive Q_1 axis (the positive Q_1 axis represents processes in which heat leaks to the cold reservoir).

The vector representation of a Carnot process is based only on the first law of thermodynamics, and as a consequence this representation applies to reversible as well as to irreversible cycles; therefore figure 2.6 can be regarded as a good illustration of the Planck definition of reversibility.

With the vector representation of Carnot processes, the reference to equilibrium processes in the Carathéodory statement of the second law of thermodynamics can be appreciated. The Carathéodory local definition of the second law can be represented by small circles with a radius r having their centers lying on the line of reversible processes. For all small values of r, one half of that circle lies in the half-plane of impossible processes that cannot be reached by any real neighboring process. These neighboring processes are represented by points in the other half of the circles, and these are mapped in the half-plane of the real irreversible processes. This is true only for circles with their centers lying on the line of reversible processes, for all other points in the half-plane of irreversible processes this conclusion cannot be drawn.

Finally, we have to calculate the dependence of the angle φ ($\tan\varphi$ gives the thermal efficiency of the Carnot engines) upon the temperatures of the heat reservoirs. This dependence can be expressed in terms of the absolute temperature, which is derived in the next section.

2.3.7. Derivation of the absolute temperature. For natural processes it follows from figure 2.6 that

$$\eta = \frac{W^\star}{Q_1} \leq \tan\varphi \tag{2.27}$$

in which φ is the angle between the positive Q_1 axis and the dividing line RR between the natural and the unnatural processes. The equality sign holds for reversible Carnot processes and the inequality sign holds for the irreversible processes. Therefore

$$\frac{Q_1 + Q_2}{Q_1} \leq \tan\varphi \qquad \text{or} \qquad \frac{Q_1}{-Q_2} \leq \frac{1}{1 - \tan\varphi} \tag{2.28}$$

The quantities of heat Q_1 and Q_2 taken from the heat reservoirs depend on the temperatures θ_1 and θ_2 of the heat reservoirs. Apparently, φ is determined only by the temperatures θ_1 and θ_2 of the two heat reservoirs, and the thermal efficiency of a reversible engine is thus only a function of the temperatures of the heat reservoirs. From the equality sign of (2.28) it follows for reversible Carnot processes that

$$\frac{|Q_1|}{|Q_2|} = f(\theta_1, \theta_2) \tag{2.29}$$

where f is an unknown functional. Consider first reversible Carnot processes only. Introduce three heat reservoirs with the temperatures θ_1, θ_2, and θ_3 respectively, and two reversible Carnot processes R_1 and R_2. Suppose that R_1 operates between the heat reservoirs with the temperatures θ_1 and θ_2; and that R_2 operates between the heat reservoirs with the temperatures θ_2 and θ_3. Relation (2.29) applies for the reversible processes R_1 and R_2 too, yielding

$$\frac{|Q_1|}{|Q_2|} = f(\theta_1, \theta_2) \qquad \text{and} \qquad \frac{|Q_2|}{|Q_3|} = f(\theta_2, \theta_3) \tag{2.30}$$

Consider both engines R_1 and R_2 to operate as a single reversible engine between the reservoirs with the temperatures θ_1 and θ_3. In that situation we have

$$\frac{|Q_1|}{|Q_3|} = f(\theta_1, \theta_3) \tag{2.31}$$

From the equations (2.30) and (2.31) it follows that:

$$\frac{|Q_1|}{|Q_2|} \equiv \frac{|Q_1|/|Q_3|}{|Q_2|/|Q_3|} = \frac{f(\theta_1, \theta_3)}{f(\theta_2, \theta_3)} = f(\theta_1, \theta_2) \tag{2.32}$$

The temperature θ_3 is an arbitrarily chosen temperature, which is not found in the last equation on the right-hand side of (2.32). The ratio on the left-hand side of this equation is therefore independent of θ_3, and the temperature θ_3 may be fixed at an arbitrary value $\theta^\star$. With the definition of $T(\theta) = f(\theta, \theta^\star)$ the equations in (2.32) reduce to

$$\frac{|Q_1|}{|Q_2|} = f(\theta_1, \theta_2) = \frac{T(\theta_1)}{T(\theta_2)} \tag{2.33}$$

The function $T(\theta)$ is not completely determined by (2.33)*. Suppose that analogously $f(\theta_1, \theta_2)$ is given by a function $\hat{T}(\theta)$. Then, we have for all values θ_1 and θ_2

$$\frac{T(\theta_1)}{T(\theta_2)} = \frac{\hat{T}(\theta_1)}{\hat{T}(\theta_2)} \qquad \text{or} \qquad \frac{T(\theta_1)}{\hat{T}(\theta_1)} = \frac{T(\theta_2)}{\hat{T}(\theta_2)} = \beta \tag{2.34}$$

from which follows that for all θ

$$T(\theta) = \beta \hat{T}(\theta) \tag{2.35}$$

The function $T(\theta)$ is therefore fixed up to a constant factor. If this constant factor has been chosen, then $T(\theta)$ is for each temperature θ a known quantity that is defined as the *thermodynamic temperature.* The statements of the second law limit the value of φ between 0 and $\pi/4$, and from (2.28) it is found that for (2.33) we have the inequalities

$$1 < f(\theta_1, \theta_2) = \frac{T_1}{T_2} < \infty \tag{2.36}$$

where T_i is written as shorthand for $T(\theta_i)$. The right-hand side inequality shows that T has the same sign for all values of θ, and that the factor β can be chosen such that $T > 0$. The left-hand side inequality shows that $T_1 > T_2$. These results are summarized by: *The thermodynamic temperature satisfies the condition that $T(\theta) > 0$ for all θ. The thermodynamic temperature is a monotonically increasing function of θ.* With this definition (2.33) becomes

$$\frac{|Q_1|}{|Q_2|} = \frac{T_1}{T_2} \tag{2.37}$$

---

* Bolder H. 1964. Een meetkundige beschouwing over de tweede hoofdwet van de thermodynamica. *Nederlands Tijdschrift voor Natuurkunde*, 30: 245–253 (In Dutch).

This result again shows clearly that the ratio of the energy flows to and from a reversible Carnot engine can be determined only by the temperatures of the heat reservoirs and not by the design of the engine. From (2.37) and (2.28) it is deduced that the angle φ in figure 2.6 is given by

$$\varphi = \arctan\left(1 - \frac{T_2}{T_1}\right) \tag{2.38}$$

The region of the natural processes in the vector representation for cycles is by this result completely known. For Carnot processes, it can be asserted that

$$\frac{|Q_1|}{|Q_2|} \leq \frac{T_1}{T_2} \tag{2.39}$$

where the equality holds for the reversible processes and the inequality for the irreversible processes. The ratio Q/T is sometimes called the *reduced quantity of heat**, although Prigogine uses this name for another quantity. In a reversible Carnot process, the sum of the reduced quantities of heat is zero for each period of the cycle, since $Q_1/T_1 = -Q_2/T_2$.

2.3.8. The thermodynamic temperature and the kelvin. The now introduced *thermodynamic temperature* or *absolute temperature* has to be related with the already introduced Kelvin temperature scale defined in (2.2). To obtain this relation, use can be made of the fact that for reversible Carnot processes the ratio Q_1/Q_2 is independent of the working substance. For the working substance, an ideal gas can be chosen. For an ideal gas the relation $A = \lim_{p\downarrow 0} pV = RT^\star$ applies by definition. For all pressures sufficiently low:

$$pV = RT^\star \tag{2.40}$$

By using the second law of thermodynamics it will be shown in section 2.3.11 that for reversible processes (2.40) implies that U depends only on $T^\star$. Therefore, for ideal gases we have

$$U = U(T^\star) \tag{2.41}$$

In the equilibrium state the internal energy of ideal gases depends only on the temperature. Since the second law is used to derive this property needed to relate the Kelvin scale with the thermodynamic scale, it is obvious also that the formalization of the experimental temperature is possible only by using the second law.

* Schottky W., Ulich H. and Wagner C. 1929. *Thermodynamik.* Julius Springer, Berlin.

The heat in a Carnot process, using an ideal gas for the working substance, can be calculated with the heat capacity. The definitions of the heat capacity at constant pressure and at constant volume are respectively:

$$C_p = \left(\frac{\partial Q}{\partial T^\star}\right)_p = \left[\frac{\partial(U + pV)}{\partial T^\star}\right]_p \tag{2.42}$$

$$C_V = \left(\frac{\partial Q}{\partial T^\star}\right)_V = \left(\frac{\partial U}{\partial T^\star}\right)_V \tag{2.43}$$

For an ideal gas $C_V = dU/dT^\star$ is a function of $T^\star$ only, yielding

$$đQ = C_V\, dT^\star + p\, dV \tag{2.44}$$

A reversible, and as such a quasi-static, Carnot process can be described using an ideal gas as the working substance. This process is sketched in a p, V diagram in figure 2.7. For isothermal absorption of heat, according to (2.44)

$$đQ = p\, dV = \frac{RT^\star}{V}\, dV \qquad \text{or} \qquad Q = \int \frac{RT^\star}{V}\, dV \tag{2.45}$$

Integration shows that for the isothermal processes $a \to b$ and $c \to d$

$$Q_1 = RT^\star \ln\frac{V_b}{V_a} \qquad \text{and} \qquad Q_2 = RT^\star \ln\frac{V_d}{V_c} \tag{2.46}$$

or

$$\frac{Q_1}{Q_2} = \frac{T_1^\star \ln(V_b/V_a)}{T_2^\star \ln(V_d/V_c)} \tag{2.47}$$

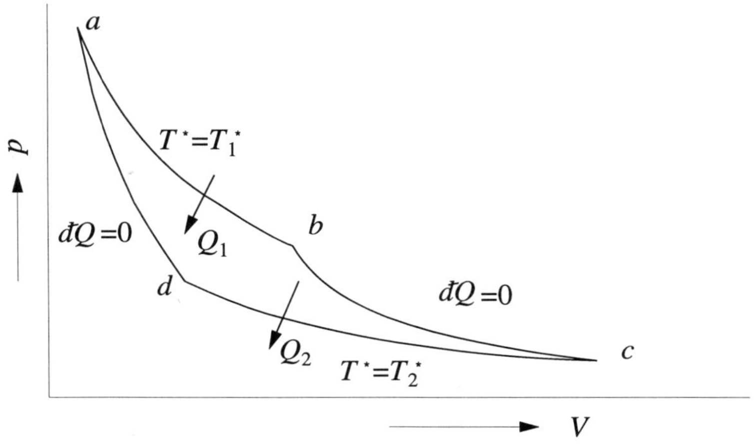

Figure 2.7. Pressure-volume diagram of a reversible Carnot process with an ideal gas.

For an adiabatic process we have

$$C_V\,dT^\star = -p\,dV = -\frac{RT^\star}{V}\,dV \qquad \text{or} \qquad \frac{C_V\,dT^\star}{T^\star} = -R\frac{dV}{V} \tag{2.48}$$

For the adiabatic processes $b \to c$ and $d \to a$ it follows that

$$\int_{T_1^\star}^{T_2^\star} \frac{C_V(T^\star)\,dT^\star}{T^\star} = R\ln\frac{V_c}{V_b} \quad \text{and} \quad \int_{T_1^\star}^{T_2^\star} \frac{C_V(T^\star)\,dT^\star}{T^\star} = R\ln\frac{V_d}{V_a} \tag{2.49}$$

From this result $\ln(V_c/V_b) = \ln(V_d/V_a)$ or $\ln(V_c/V_d) = \ln(V_b/V_a)$, so that finally (2.47) becomes

$$\frac{|Q_1|}{|Q_2|} = \frac{T_1^\star}{T_2^\star} \tag{2.50}$$

Comparison of (2.50) with (2.37) shows that the Kelvin scale can be taken to be equal to the absolute temperature scale. The fixed point of the absolute scale is the triple point T_3, which corresponds to $|Q| = |Q_3|$, thus

$$T = 273.16\frac{|Q|}{|Q_3|} \tag{2.51}$$

From the definition it follows that the absolute temperature is defined independently of the working substance. This was not quite so for the Kelvin scale, since for the empirical Kelvin scale it was only assumed—by extrapolating the measurements—that the Kelvin temperature does not depend on the type of *gas* used in the gas thermometer. With the above discussion it follows that this temperature is completely independent of the working substance, and in this sense absolute. (For measurements of temperatures below 1 K usually a kind of Carnot process is used that operates almost reversibly.)

2.3.9. The absolute zero. Suppose that a Carnot process operates between a heat reservoir with a temperature T_3 and one with a temperature $T < T_3$. Let Q_3 be positive. Then Q cannot also be positive. If Q and Q_3 were positive, then the only action in this process is the absorption of heat from some heat reservoirs in each period of the cycle and the production of an equivalent quantity of work. This is forbidden by the Kelvin–Planck principle of the second law . The lowest possible value of Q is obviously $Q = 0$. To this value corresponds $T = 0$. For all substances $T = 0$ is therefore the lowest temperature. As remarked, $Q = 0$ is in conflict with the Kelvin–Planck principle of the second law, and as a consequence the point $T = 0$ is unattainable, but $T = \epsilon$ may be attained if ϵ is positive and arbitrarily small. The point $T = 0$ is the *absolute zero.* Note that in this discussion, the declared 'unattainability' concerns unattainability by Carnot processes.

At the absolute zero temperature point all Carnot engines would have a thermal efficiency of 100%, since $\eta_R = 1 + (Q_2/Q_1) = 1 - (T_2/T_1)$ and $\eta_R = 1$ for $T_2 = 0$. This thermal efficiency is unattainable, but may be approached arbitrarily closely. The thermal efficiency of all heat engines is in principle always less than 100%. It is to be remarked that these conclusions follow from the first law and from the second law statements of Kelvin–Planck and Clausius respectively.

2.3.10. Mathematical formulation of the second law of thermodynamics for closed systems. For a Carnot process (2.39) or (2.52) applies

$$\frac{Q_1}{T_1} + \frac{Q_2}{T_2} \leq 0 \tag{2.52}$$

An arbitrary cycle can be considered as a series of infinitesimal Carnot processes, each of which absorbs such small quantities of heat at the temperature T_1 or rejects such small quantities of heat at temperature T_2 so that no changes of temperature occur in the environment of the system. By considering all infinitesimal Carnot processes as a discretization of an arbitrary process, then the quantities of heat Q_1 are absorbed at various temperatures T_1, and the quantities of heat Q_2 are rejected at various temperatures T_2. For an infinitesimal Carnot process

$$\frac{đQ_1}{T_1} + \frac{đQ_2}{T_2} \leq 0 \tag{2.53}$$

Summation over all infinitesimal Carnot processes yields

$$\oint \frac{đQ}{T} \leq 0 \tag{2.54}$$

where $đQ$ is the infinitesimal quantity of heat supply in the cycle studied. In (2.54), the equal sign applies to reversible processes and the unequal sign to irreversible processes. Equation (2.54) is known as the *Clausius proposition.*

Assume that the cycle is reversible and that I and II are two states in this cycle. Since the cycle is reversible states I and II are equilibrium states. Now the equality sign applies in (2.54)

$$\int_I^{II} \frac{đQ}{T} + \int_{II}^{I} \frac{đQ}{T} = 0 \tag{2.55}$$

and, as a consequence, the states I and II can be attained along all paths. For a reversible process, $\int_I^{II} (đQ/T)$ is independent of any path from state I to state II, if this path can be traversed reversibly. As a consequence, it follows that there has to exist a function of state S such that

$$\int_I^{II}{}_{\text{rev.}} \frac{đQ}{T} = S_{II} - S_I \tag{2.56}$$

The function of state S is defined by (2.56) up to an additive constant and is called *entropy* according to Clausius*. For an infinitesimal reversible process we have particularly

$$\frac{đQ}{T} = dS \qquad \text{(reversible)} \tag{2.57}$$

dS is a exact differential in this expression. Apparently, $1/T$ is the integrating factor for the inexact differential $đQ$, just as $1/p$ is the integrating factor for the inexact differential $đW$.

For irreversible processes, one may proceed as follows. Suppose that the transition between two equilibrium states I and II is irreversible. Suppose furthermore that it is possible that the transition $I \to II$ can also be carried out along a reversible path. Then the following cycle can be considered

$$\int_{I\ \text{irr.}}^{II} \frac{đQ}{T} + \int_{II\,\text{rev.}}^{I} \frac{đQ}{T} < 0 \tag{2.58}$$

Substitution of (2.56) yields:

$$S_{II} - S_I > \int_{I\ \text{irr.}}^{II} \frac{đQ}{T} \tag{2.59}$$

$\oint(đQ/T)$ is evidently a measure of the irreversibility of a cycle, and $S_{II} - S_I - \int_I^{II}(đQ/T)$ is a measure of the irreversibility of a transition $I \to II$. The above argument holds only for closed systems. For closed systems the second law of thermodynamics is summarized as follows:

The second law of thermodynamics for closed systems. For closed systems there exists an extensive function of state S that in a transition from an equilibrium state I to an equilibrium state II such that for reversible processes

$$S_{II} - S_I = \int_I^{II} \frac{đQ}{T} \tag{2.60}$$

---

* Clausius R.J.E. 1865. Ueber verschiedene für die Anwendung bequeme Formen der Hauptgleichungen der mechanischen Wärmetheorie. *Annalen der Physik. Leipzig*, (5)125: 353–400. See p. 390: 'Sucht man für S einen bezeichnenden Namen, so könnte man, ähnlich wie von der Grösse U gesagt ist, sie sey der *Wärme- und Werkinhalt* des Körpers, von der Grösse S sagen, sie sey der *Verwandlungsinhalt* des Körpers. Da ich es aber für besser halte, die Namen derartiger für die Wissenschaft wichtiger Grössen aus den alten Sprachen zu entnehmen, damit sie unverändert in allen neuen Sprachen angewandt werden können, so schlage ich vor, die Grösse S nach dem griechischen Worte ἡ τροπὴ [sic], die Verwandlung, die *Entropie* des Körpers zu nennen. Das Wort *Entropie* habe ich absichtlich dem Worte *Energie* möglichst ähnlich gebildet, denn die beiden Grössen, welche durch diese Worte benannt werden sollen, sind ihren physikalischen Bedeutungen nach einander so nahe verwandt, dass eine gewisse Gleichartigkeit in der Benennung mir zweckmässig zu seyn scheint.'

for reversible processes and for irreversible processes

$$S_{II} - S_I > \int_I^{II} \frac{đQ}{T} \tag{2.61}$$

In these expressions T is the absolute temperature of that part of the system that exchanges the quantity of heat $đQ$ with the surroundings. The extensive function of state S is called the *entropy** of the system.

The experiences that have led to the second law corollaries have now been formulated mathematically.

For infinitesimal processes the first and the second laws of thermodynamics yield

$$dS \geq \frac{đQ}{T} = \frac{dU - đW}{T} \tag{2.62}$$

if the kinetic and potential energy have been disregarded. For a single phase system and for quasi-static changes, $đW = -p\,dV$ applies, and (2.62) becomes for reversible processes

$$\boxed{dU = T\,dS - p\,dV} \tag{2.63}$$

This is the *fundamental equation of Gibbs*, which is a relation involving only state properties† and as a consequence is applicable to irreversible processes

---

* As already remarked on page 37, the proposition of Carnot that the thermal efficiency of an irreversible engine is always less than the thermal efficiency of a reversible Carnot engine, can readily be seen from figure 2.6. For the Carnot cycle using an ideal gas for the working substance $Q_1/T_1 = Q_2/T_2$ (2.50) follows, and from $\eta_I \leq \eta_R$ (2.26) by substitution of $W^\star$ the starting point follows for the mathematical derivation of the second law of thermodynamics, although achieved by using the Kelvin temperature. If no derivation is given showing that the Kelvin temperature is equal to the absolute temperature, then the proposition of Clausius and the mathematical expression of the second law as given in this section is obtained. In this way the use of partial differentials has been avoided, and the function of state entropy may be introduced in first year courses.

† The cycles of thermal engines are composed of processes in which one of the state properties T, S, p, or V is kept constant. The processes are called isothermal, adiabatic, isobaric, or isochoric, respectively. Of these, only the isothermal and the adiabatic process can be performed reversibly and, as a consequence, the importance of the Carnot cycle as a fundamental cycle becomes evident. The intensive functions of state p and T are the reciprocals of the integrating factors for of the work $đW$ and the heat $đQ$ respectively. The contribution $p\,dV$ to the energy U by varying V is usually thought to be understood, while the contribution $T\,dS$ to U by vaying S is found difficult. This does not seem correct. The volume of a gas cylinder can easily be estimated, but not the pressure. However, $dV = -đW/p$ shows that the effort (work) to change the volume by some amount is large if the pressure is high in the cylinder. Although both the entropy and the temperature cannot be estimated easily in the cylinder, $dS = đQ/T$ is an expression similar to $dV = -đW/p$, and the effort (heat) to change the entropy by some amount is large if the temperature is high in the cylinder. Just as it is difficult to change V for high pressures, it is difficult to change S for high temperatures,

as well, since these properties depend on the state alone irrespective of the type of process that produces the state. However, for the calculation of the entropy difference in an arbitrary irreversible process a reversible path is however needed in general. In the fundamental equation of Gibbs the first and the second laws of thermodynamics are combined into one equation and this equation contains the thermodynamic knowledge about the system.

2.3.11. The internal energy of an ideal gas. The assumption (2.41) that the internal energy of an ideal gas depends only on the temperature, can be derived using the equation (2.40) for an ideal gas. With $T = T^\star$, the ideal gas law (2.40) becomes

$$pV = RT \tag{2.64}$$

Dividing the fundamental equation of Gibbs (2.63) by T, the fundamental equation may be written as

$$dS = \frac{1}{T}\,dU + \frac{p}{T}\,dV \tag{2.65}$$

The internal energy U can be considered to depend on a mechanical coordinate, say V, and on a thermal coordinate, say T. With $U = U(T, V)$ the differential dS becomes

$$dS = \frac{1}{T}\left(\frac{\partial U}{\partial T}\right)_V dT + \frac{1}{T}\left\{p + \left(\frac{\partial U}{\partial V}\right)_T\right\} dV \tag{2.66}$$

Since dS is an exact differential, it follows that*

$$\frac{1}{T}\frac{\partial^2 U}{\partial T \partial V} = -\frac{1}{T^2}\left\{p + \left(\frac{\partial U}{\partial V}\right)_T\right\} + \frac{1}{T}\left\{\left(\frac{\partial p}{\partial T}\right)_V + \left(\frac{\partial^2 U}{\partial V \partial T}\right)\right\} \tag{2.67}$$

or

$$\left(\frac{\partial U}{\partial V}\right)_T = -p + T\left(\frac{\partial p}{\partial T}\right)_V = T^2\left(\frac{\partial (p/T)}{\partial T}\right)_V \tag{2.68}$$

The first equallity in (2.68) is the Helmholtz equation. With this result and with (2.43) both partial differentials of $U(V, T)$ are known and U can be

---

* This result follows directly from the Green theorem. Let G be a closed bounded region. If the functions $P(x, y)$, $Q(x, y)$, $\dfrac{\partial P}{\partial x}$ and $\dfrac{\partial Q}{\partial y}$ are continuous in an open region O containing G, then

$$\iint_G \left(\frac{\partial Q}{\partial x} - \frac{\partial P}{\partial y}\right) dx\,dy = \oint (P\,dx + Q\,dy)$$

in which the circle in the line integral on the right-hand side indicates that the integral is to be performed around the boundary K of G counterclockwise.

integrated. For an ideal gas it follows from (2.64) that $p/T = R/V$, and so from (2.68) we get

$$\left(\frac{\partial U}{\partial V}\right)_T = 0 \tag{2.69}$$

In addition, using (2.64) again,

$$\left(\frac{\partial U}{\partial p}\right)_T = \left(\frac{\partial U}{\partial V}\right)_T \left(\frac{\partial V}{\partial p}\right)_T = -\frac{V}{p}\left(\frac{\partial U}{\partial V}\right)_T \tag{2.70}$$

Furthermore, since $V/p \neq 0$, we may use (2.69) to obtain

$$\left(\frac{\partial U}{\partial p}\right)_T = 0 \tag{2.71}$$

so that U is evidently only a function of T

$$U = U(T) \tag{2.72}$$

Note that the above proof applies only to closed systems, but the result is more general.

2.3.12. Entropy production. Clausius has introduced for irreversible processes the *uncompensated entropy* defined by

$$dS = \frac{đQ}{T} + \frac{đQ_{\mathrm{C}}}{T} \geq \frac{đQ}{T} \tag{2.73}$$

The Clausius uncompensated entropy is $đQ_{\mathrm{C}}/T$ in this equation. Remember that $đQ$ was the heat the system exchanges with its environment and, as such, $đQ_{\mathrm{C}}$ is obviously the heat produced *within* the system by irreversible processes (for reversible processes we had $đQ_{\mathrm{C}}/T = 0$).

The Clausius* terminology for the 'uncompensated entropy' was not a very fortunate choice. Already Bertrand† had considered the entropy as a physical quantity that can be distributed spatially over a system and can be transported. Bertrand emphasized that this view could not be based

---

* Clausius R.J.E. 1865. Ueber verschiedene für die Anwendung bequeme Formen der Hauptgleichungen der mechanischen Wärmetheorie. *Annalen der Physik. Leipzig*, 125: 353–400. See p. 396: '*uncompensirte Verwandlung*'; Clausius R.J.E. 1856. Über die bewegende Kraft der Wärme, und die Gesetzte, welche sich daraus für die Wärmelehre selbst ableissen lassen. *Annalen der Physik und Chemie, Leipzig*, 19: 368–397, 500–524; Clausius R.J.E. 1897. *Theorie Mécanique de la Chaleur*. Monceaux, Bruxelles.

† Bertrand J.L.F. 1887. *Thermodynamique*. Gauthier-Villars, Paris.

on thermostatics*. Entropy is in this view a quantity that can flow and can be produced. Entropy is interpreted similarly in the thermodynamics of irreversible processes. For closed systems, it is therefore assumed that entropy can flow from the environment to the system and vice versa.

Suppose that for an infinitesimal process this flow of entropy is given by

$$d_{\mathrm{e}}S = \frac{đQ}{T} \tag{2.74}$$

Besides this flow of entropy there is a production of entropy by irreversible processes. Denote this entropy production by $d_{\mathrm{i}}S$ that is defined by

$$d_{\mathrm{i}}S = \frac{đQ_{\mathrm{C}}}{T} \geq 0 \tag{2.75}$$

From (2.73) it is seen that dS can be split into two parts

$$\boxed{dS = d_{\mathrm{e}}S + d_{\mathrm{i}}S \qquad \text{with} \qquad d_{\mathrm{e}}S = \frac{đQ}{T} \quad \text{and} \quad d_{\mathrm{i}}S = \frac{đQ_{\mathrm{C}}}{T} \geq 0} \tag{2.76}$$

For irreversible processes in a closed system the entropy production is always positive, while for reversible processes the entropy production is zero. The second law of thermodynamics is formulated with this observation as follows.

The second law of thermodynamics. In a closed thermodynamic system the entropy production is always positive for real (irreversible) processes, and for reversible processes zero.

The entropy can accordingly have only 'sources' and never 'sinks'. Of course, it is possible that dS is positive or negative, since this depends on the sign and magnitude of $d_{\mathrm{e}}S$ ($d_{\mathrm{e}}S$ can be positive as well as negative).

From the above statements it *cannot* be shown that for open systems the entropy production must necessarily be positive. If this is postulated, however, then the classical second law of thermodynamics is actually broadened.

---

* Bertrand J.L.F. 1887. See p. 266: Quand tous les points du corps que l'on étudie n'ont pas conservé, pendant la transformation des températures égales pour tous, que deviennent le rapport dQ/T de la quantité de chaleur fournie au corps à la température T, et l'intégrale de ce rapport?

Faut-il décomposer le corps en éléments infiniment petits et réunir les intégrales relatives á chacun d'eux? La démonstration des théorèmes n'y autorise pas.

Lorsque le corps, en changeant de volume n'exerce pas sur les corps environments une action égale à celle qu'il pourrait vaincre; lorsque, en le comprimant, on exerce sur lui une action supérieure à la résistance dont il est capable, les molécules, dans le cas d'un gaz particulièrement, prennent des vitesses finies, inégales en grandeur, differentes en direction, et, pendant la durée de cette perturbation, le mot *pression*, appliqué aux corps, cesse d'avoir un sens defini.

For closed systems therefore

$$S_{II} - S_I \geq 0 \tag{2.77}$$

If in addition the universe is assumed to be a closed system, the well-known statement of Clausius* results:

The entropy of the universe tends towards a maximum.

Just like the temperature, the entropy is strictly defined only in equilibrium. For non-equilibrium states the entropy may be generalized analogously to the temperature by using the kinetic theory of matter. For the classical thermodynamics, this solution is not available, since the theory is macroscopic, at least at first glance. The question arises if (2.77) is meaningful. It is to be expected that in a closed system the equilibrium state is determined unambiguously, and one cannot make such a statement about two states of true equilibrium in a closed system. This difficulty can be avoided if assumptions about the constraints in the system are introduced. So, if frozen† equilibrium is possible, S_{II} is for example the entropy of the unconstrained equilibrium state, and S_I is the entropy of the frozen equilibrium.

If irreversible processes take place between the two equilibrium states I and II, the entropy—especially the entropy difference (2.77)—can be defined only if state II can be reached from state I also along a reversible path. This is formulated as follows. If after removing a constraint the processes take place irreversibly, a reversible process must be possible such that, using only external actions, the system is brought from the final state back into the initial state. Only then the thermodynamic definition and determination of the entropy difference is possible. As a rule, it is possible to examine whether a reversible

---

* Clausius R.J.E. 1865. Ueber verschiedene für die Anwendung bequeme Formen der Hauptgleichungen der mechanischen Wärmetheorie. *Annalen der Physik, Leipzig*, 125: 353–400. See the last two sentences p. 400: **1**) 'Die Energie der Welt ist constant. **2**) Die Entropie der Welt strebt einem Maximum zu'. Although Clausius has emphasized that the previous statements are a summary of the first and the second law only with respect to all assumptions used, his statement was nevertheless often misused, and has even been used for a proof of the existence of God by Pope Pius XII. Die Gottesbeweise im Lichte der modernen Naturwissenschaft. *Ansprache Papst Pius' XII an die Mitglieder der Päpstlichen Akademie der Wissenschaften am 22. November 1951*, Morus-Verlag, Berlin. See also: Fast J.D. 1969. Entropie, kosmos en geloof. *Acta Technica Belgica, EPE*, 5: 28–35, and Boltzmann L. 1897. Zu Hrn. Zermolo's Abhandlung 'Über die mechanisch Erklärung irreversibler Vorgange', *Annalen der Physik*, 60: 392–398. Boltzmann discusses in that article that it is highly probably due to the necessary fluctuations in thermal equilibrium that even in an enormous large dead universe tiny parts, like our earth, are living for times that are small only with respect to the age of the universe.

† The equilibrium is metastable and exist solely due to an internal constraint. By removing the constraint natural processes will occur. For example, opening of a valve enables water to flow; with the addition of a suitable catalyst chemical reactions result.

inverse path is feasible, and furthermore, it turns out that the constraints can be assumed to play the role of a finite number of *possible* extensive quantities.

Sometimes it is not necessary to require a reversible inverse path. This is the case for instance if the entropy difference between two states can be calculated theoretically, as for ideal gas mixtures. In general, however, a reversible inverse path is needed!

2.4. THE THIRD LAW OF THERMODYNAMICS

The third law of thermodynamics is concerned with the unattainability of the absolute zero of the thermodynamic temperature scale by Carnot processes. With the remark that a thermal efficiency of 100% of any Carnot engine is inaccessible, this unattainability has already been suggested (but not more than that), and is now introduced as a postulate:

The third law of thermodynamics. It is impossible by any procedure, irrespective of the level of the idealization of the procedure, to cool the temperature of a system to the absolute zero in a finite number of operations.

Again, this third law of thermodynamics is based on experimental evidence. Nernst* called this law the *Principle of Unattainability of the Absolute Zero*. Similarly, Fowler and Guggenheim† called this principle the *unattainability statement of the third law of thermodynamics*. There exist of course several alternative formulations of the third law‡. Among these, the formulation of Nernst-Simon is important

Nernst-Simon heat theorem. For any isothermal process involving only phases in internal equilibrium or, alternatively, in frozen metastable equilibrium not disturbed by the process, the following statement applies

$$\lim_{T\to 0} \Delta S = 0 \tag{2.78}$$

In this theorem ΔS denotes the entropy difference between two states of the isothermal process.

The third law of thermodynamics and the Nernst-Simon heat theorem are equivalent. There has been considerable confusion about the equivalence of the two statements. About thirty years after the 1906 publication of Nernst§

* Nernst W. 1926. *The New Heat Theorem.* Methuen, London. Reprinted by Dover, New York, 1969. See p. 87.

† Fowler R.H. and Guggenheim E.A. 1939. *Statistical Thermodynamics.* Cambridge University Press, Cambridge, p. 224.

‡ Wilks J. 1961. *The Third Law of Thermodynamics.* Oxford University Press, London, p. 142.

§ Nernst W. 1906. Über die Berechnung chemischer Gleichgewichte aus thermischen Messungen. *Nachrichten der Königlichen Gesellschaft der Wissenschaften, Göttingen, Mathematisch-Physische Klasse*, Heft I.

agreement had been obtained about the final formulation of the third law of thermodynamics. Particularly the experiments and the arguments of Simon* in the period from 1927 until 1937 contributed a great deal to the proper understanding of the third law.

Following Fowler and Guggenheim it will be shown that the third law and the heat theorem of Nernst-Simon are equivalent. Consider any adiabatic process such as a change in volume, a change of an external field, a chemical reaction and the like. This process is formally denoted by $\alpha \to \beta$, where α and β represent the number of the variables of state of the system. The subscripts α and β will be used to indicate the properties of the system in the states α and β respectively. Define

$$C_\alpha = \left(\frac{\partial Q}{\partial T}\right)_\alpha = T\left(\frac{\partial S}{\partial T}\right)_\alpha \tag{2.79}$$

and analogously for C_β. Evidently, C_α and C_β are the heat capacities of the system in the state α and β respectively. Then the entropy in the state α is

$$S_\alpha = S_\alpha^{(0)} + \int_0^T \frac{C_\alpha}{T}\, dT \tag{2.80}$$

and the entropy in the state β

$$S_\beta = S_\beta^{(0)} + \int_0^T \frac{C_\beta}{T}\, dT \tag{2.81}$$

$S_\alpha^{(0)}$ and $S_\beta^{(0)}$ are the limiting values as $T \to 0$ of S_α, and S_β respectively. Suppose that the integrals in (2.80) and (2.81) converge. This means that it is assumed that for $T \to 0$ the quantities C_α and C_β tend to zero sufficiently fast for the entropies to remain finite. This assumption is supported by the quantum theory.

Suppose now that the process considered is such that an adiabatic change of the state α with the temperature T' to the state β with the temperature T'' can take place. From the second law of thermodynamics, it follows that the increase of entropy by a change of state for an adiabatic and closed system is either positive or zero. The system is of course assumed to be closed, to ensure an unambiguous definition of the quantity of heat dQ. Evidently the lowest possible temperature will be attained if the process operates reversibly, since the dissipative effects in any irreversible process result in an increase of temperature. Propose now, that the process takes place reversibly and

---

* Sir Francis Simon. 1956. The Third Law of Thermodynamics. An Historical Survey, 40th Guthrie Lecture: delivered before the Physical Society on 13th March 1956. *Year Book of the Physical Society*, London, pp. 1–29.

adiabatically, meaning that the entropy is constant, that is, the process operates isentropically. In that case $S_\alpha = S_\beta$, and hence

$$S_\alpha^{(0)} + \int_0^{T'} \frac{C_\alpha}{T} \, dT = S_\beta^{(0)} + \int_0^{T''} \frac{C_\beta}{T} \, dT \tag{2.82}$$

If the change of state $\alpha \to \beta$ results in a state at zero absolute temperature, then T'' would be zero, or

$$S_\beta^{(0)} - S_\alpha^{(0)} = \int_0^{T'} \frac{C_\alpha}{T} \, dT \tag{2.83}$$

The condition of intrinsic stability of the phases in the system being considered results in the conclusion that $C_\alpha > 0$ is, so that according to (2.83)

$$S_\beta^{(0)} - S_\alpha^{(0)} > 0 \tag{2.84}$$

If $S_\beta^{(0)} - S_\alpha^{(0)} > 0$, then it is possible to choose a T' such that (2.83) is satisfied and the absolute zero of the thermodynamic temperature can be reached. According to the conjecture of the third law of thermodynamics this is impossible, hence

$$S_\beta^{(0)} \leq S_\alpha^{(0)} \tag{2.85}$$

If the reverse process is considered, then from similar reasoning the conclusion is reached that

$$S_\beta^{(0)} \geq S_\alpha^{(0)} \tag{2.86}$$

Then from the third-law premise that $T = 0$ cannot be reached, it is deduced that $S_\beta^{(0)} = S_\alpha^{(0)}$, which is equivalent to the Nernst-Simon heat theorem

$$\lim_{T \to 0} \Delta S = 0 \tag{2.87}$$

Conversely, the heat theorem (2.87) implies that the absolute zero point is unattainable. If (2.87) applies, then according to (2.82) it follows for an adiabatic reversible process $\alpha \to \beta$ that

$$\int_0^{T'} \frac{C_\alpha}{T} \, dT = \int_0^{T''} \frac{C_\beta}{T} \, dT \tag{2.88}$$

and $T'' = 0$ would be attained if $\int_0^{T'} (C_\alpha / T) \, dT = 0$. Since $C_\alpha > 0$ and $T' \geq 0$ this condition is not satisfied, and $T'' = 0$ is not reachable. For the reverse process $\beta \to \alpha$ the proof of the unattainability of the absolute zero is analogous.

The above proof shows that the third law of thermodynamics and the Nernst-Simon heat theorem in the formulation (2.87) are equivalent if and only if it is possible to connect the states α and β by a reversible path. Furthermore, it is important that $C_\alpha > 0$, and $\beta > 0$. This signifies that during the process all phases are in stable equilibrium. No phase transitions are allowed, and if the phase is in metastable equilibrium, the metastable character of this equilibrium is completely frozen. Only under these conditions are the third law of thermodynamics and the Nernst-Simon heat theorem equivalent. The heat theorem emphasizes the condition that the process is isothermal, because ΔS is in this theorem the difference in entropy of two states at the same temperature. For a statistical interpretation of the third law, see for instance Fowler and Guggenheim*.

2.5. EXERCISES

Exercise 2.1. The mean-recurrence time Θ_n of a fluctuation in the number of molecules n with mass m in a spherical δV with radius r is approximately†

$$\Theta_n \approx \tfrac{1}{3}\pi r \left(\frac{m}{\langle n\rangle k_B T}\right)^{\frac{1}{2}} \exp\left[(n - \langle n\rangle)^2 / 2\langle n\rangle\right]$$

The average number of molecules $\langle n\rangle$ is given by $n_o \Delta V$, where n_o is the Loschmidt constant ($n_o = 2.686\,763 \times 10^{25}\ \mathrm{m}^{-3}$). The mean-free path of oxygen is 9.93×10^{-8} m, approximately one tenth of a μm. Calculate that the mean-recurrence time for a 1 percent fluctuation of the average number of molecules at room temperature is about one second for a radius of 2.5 times the mean-free path of the oxygen molecules and increases to 10^6 s for a δV with a radius of 0.3 μm and to 10^{68} s for a δV with a radius of 0.5 μm. Discuss the fact that, due to the long mean-recurrence times, the necessary fluctuations in $(\Delta V)^\star$ are not noticeable and that the continuum approximation applies to much smaller $(\Delta V)^\star$ than the fluctuations of a macroscopic property suggest.

Exercise 2.2. Discuss whether the zeroth law is essential for defining the intensive variables pressure and temperature.

Exercise 2.3. A function of state implies equilibrium. In nonequilibrium systems the assumption of local equilibrium defines the functions of state locally. Due to the small but finite resolution time of the tiny sensor of the thermometer, the temperature can be defined locally, and by using the

---

* Fowler R.H. and Guggenheim E.A. 1939. *Statistical Thermodynamics.* Cambridge University Press, Cambridge, §539.

† Chandrasekhar S. 1943. Stochastic Problems in Physics and Astronomy. *Reviews of Modern Physics*, 15: 1–89; Bartlett M.S. 1950. Recurrence Times. *Nature*, 165: 727–728.

Carathéodory formulation of the second law, the entropy can also be defined locally*. A space- and time-dependent 'nonequilibrium' temperature field $T = \partial U/\partial S$ is then independent of the temperature gradients. Discuss whether large temperature gradients are acceptable in a continuum description.

Exercise 2.4. Calculate φ in figure 2.6 for the early heat engines that use steam at one bar pressure for the hot reservoir and water for the cold reservoir. Calculate the second law efficiency $\epsilon = W_E^\star/W_A^\star$ (see figure 2.6) if the thermal efficiency for the real irreversible process is 2%. Reflect on the smallness of the area available for useful heat engines for the early heat engines.

Exercise 2.5. A standard Otto cycle consists of two adiabatic and two isochoric processes. Assume that during the isochoric processes the heating and cooling proceed according to constant average temperature rates r_h and r_c. Discuss that the adiabatic processes can be considered as instantaneous. Show that the cycle period t_c and the reversible work are given by

$$t_\mathrm{c} = \frac{\Delta T_\mathrm{h}}{r_\mathrm{h}} + \frac{\Delta T_\mathrm{c}}{r_\mathrm{c}} \qquad \text{and} \qquad W^\star = C_{V\mathrm{h}}\Delta T_\mathrm{h} - C_{V\mathrm{c}}\Delta T_\mathrm{c}$$

where ΔT_h, ΔT_c denote the absolute value of the temperature difference during heating and cooling, and $C_{V\mathrm{h}}$, $C_{V\mathrm{c}}$ are the mean heat capacities at constant volume during heating and cooling respectively. Show that the reversible thermal efficiency η and reversible power $P_\mathrm{R} = W^\star/t_\mathrm{c}$ are given by

$$\eta = 1 - \frac{C_{V\mathrm{c}}}{C_{V\mathrm{h}}}\frac{\Delta T_\mathrm{c}}{\Delta T_\mathrm{h}} = 1 - \frac{C_{V\mathrm{c}}}{C_{V\mathrm{h}}}\frac{1}{r^{\gamma-1}} \quad \text{and} \quad P_\mathrm{R} = \frac{C_{V\mathrm{c}} - C_{V\mathrm{h}}r^{1-\gamma}}{(1/r_\mathrm{c}) + (1/r_\mathrm{h})r^{1-\gamma}}$$

where r is the compression ratio (maximum volume divided by minimum volume), and γ is the isentropic index. Assume that the thermal and friction losses can be lumped in a global term that is linear in the piston velocity and show that the loss power is $b(r-1)^2$ where b a factor that depends on the piston position at the minimum volume, the friction coefficient that accounts for the global losses and the time spent in a stroke. Show that the thermal efficiency becomes†

$$\eta(r) = 1 - \frac{C_{V\mathrm{c}}}{C_{V\mathrm{h}}}\frac{1}{r^{\gamma-1}} - \frac{b(r-1)^2}{C_{V\mathrm{h}}}\left((1/r_\mathrm{c}) + (1/r_\mathrm{h})r^{1-\gamma}\right)$$

and that for a real Otto engine with $C_{V\mathrm{c}} = 0.2988$ J/K, $C_{V\mathrm{h}} = 0.4372$ J/K, $r_\mathrm{c} = 12.3 \times 10^4$ K/s, $r_\mathrm{h} = 6.0 \times 10^4$ K/s, $b = 32.5$ W, and $\gamma = 1.4$ a maximum thermal efficiency of 0.28 is found for a compression ratio of 9.3.

---

* Kuiken G.D.C. 1979. Note sur la définition de l'entropie. *Entropie*, 85: 15–19.

† Angulo-Brown F., Fernández-Betanzos J. and Díaz-Pico C.A. 1994. Compression ratio of an optimized air standard Otto-cycle model. *European Journal of Physics*, 15: 38–42.

3 Basic Axioms of the TIP

Summary: Six general axioms are discussed and four supplementary axioms. These basic axioms are used to derive the balance equations and the constitutive equations. The structure of the theory is given in a diagram, and illustrated for anisotropic heat conduction and for membrane permeability.

3.1. GENERAL AXIOMS OF THE TIP

3.1.1. TIP-axiom I: validity of the classical thermodynamics. A basic presumption for the application of the classical thermodynamics is that every system reaches equilibrium after isolation and a certain time lapse. This equilibrium is described by variables of state that are independent of time and space coordinates. The classical thermodynamics is valid in this equilibrium state, for which it has to be remarked that a complete isolation is possible only if there are no external force fields interacting with the system. If such force fields are present, a system can be in a state of equilibrium in which the phases are inhomogeneous. In a gravitational field, the atmosphere can be in equilibrium with a layered structure, in which the thermodynamic state depends on the height. Since gravitation is always present on earth, its influence can be ignored only if the length scales of the system are small enough with respect to the length scales inherent in this layered structure. Elastic bodies can be in equilibrium with the applied loads, in which the variables of states need not be homogeneous. Then, the system is not isolated, since the load is caused by the contacts between the system and its environment.

Summarizing: *if* a system can be isolated, then the phases are homogeneous in equilibrium. If by the presence of external forces, long range actions between the system and its environment occur, equilibrium can still be realized, but the phases are not in general homogeneous in this equilibrium. Therefore, it is often customary to assume that the length scales of the system are such that the influence of long range forces can be neglected.

3.1.2. TIP-axiom II: local and instantaneous equilibrium. If the system is not in equilibrium, the thermodynamic quantities will in general be functions of space and time.

For sufficiently small deviations from equilibrium, the system can be divided into tiny (physical) volume elements, each of which can be regarded as a small homogeneous equilibrium system. The length and time scales of these subsystems are infinitesimally small from a macroscopic point of view, but from the molecular point of view still large, such that the subsystem contains enough molecules so that the averages taken over the number of molecules have deterministic significance. A result of the existence of such physical volume elements is that the matter is assumed to behave as a continuum, such that the thermodynamics of irreversible processes—akin to classical thermodynamics—is a continuum theory, in which all quantities are continuously differentiable functions of space and time. Exceptions are allowed solely on isolated points, on curves and on surfaces; that is, in regions having a zero measure.

The thermodynamics of irreversible processes imposes in fact more strict requirements than a continuum theory does, since it requires in addition that the infinitesimal subsystems are in equilibrium at every moment. If there is no equilibrium, the concepts of temperature and pressure would be meaningless. However, practice shows that, even away from equilibrium, the equilibrium concepts of temperature and pressure can be used intelligibly. Consider the temperature as an example. Measurement of the temperature demands that in principle there is equilibrium between the thermometer and the system. For this to be realized it is necessary that the system itself is in equilibrium. If the temperature is measured in practice and there is no equilibrium, the measurement is meaningful* if it is assumed that a small system exists around the thermometer that is in instantaneous equilibrium with the thermometer. Here, the significance of a 'small' surrounding system is that the dimensions of the system must be small with respect to the macroscopic distances over which the temperature varies noticeably (= measurably). On the other hand the system has to be large with respect to the dimensions of the sensor of the thermometer, to ensure that the sensor does not influence noticeably the state of the local system. The significance of 'instantaneous' is that the macroscopic time scale, over which the temperature as a function of time varies, is large with respect to the time the thermometer needs to restore its equilibrium if the external conditions of the thermometer change. This is the *principle of local and instantaneous equilibrium*, referred to as TIP-axiom

* Kuiken G.D.C. 1979. Note sur la définition de l'entropie. *Entropie*, 85: 15–19. In the small volume element containing the tiny sensor of the temperature measuring device, the temperature is fluctuating around the measured equilibrium value. Due to finite resolution time of the thermometer, those fluctuations cannot be recorded by the sensor. As a consequence, the temperature fluctuations within the physical volume element are unattainable in a temperature measurement. Using a reasoning similar to the one used in the Carathéodory formulation of the second law of thermodynamics, the temperature and the entropy can be defined locally. From this reasoning it is found that large temperature gradients are acceptable. Besides, it is not necessary to introduce the local temperature of the working substance—the physical volume element—by presuming that the temperatures of the hot and cold reservoir differ infinitesimally. Therefore, a temperature of the working substance itself could be assumed, which is not defined in classical thermodynamics. Note that all definitions of temperature assume the existence of (local) equilibrium, since the definitions involve functions of state, for example $T = \partial U/\partial S$.

II. This principle limits the area of applications of the thermodynamics of the irreversible processes to sufficiently small deviations from the equilibrium state*.

Local and instantaneous equilibrium implies that temperature and pressure still have a useful meaning. Then, all classical thermodynamic functions can be given a sensible meaning, such as the internal energy and the entropy. Furthermore, all thermodynamic relations can be applied as far as a local and instantaneous content can be assigned to those relations. As a result, the validity of classical thermodynamics is a basic assumption of the thermodynamics of irreversible processes. Applications of this axiom II demand, among other conditions, that the extensive quantities of the classical thermodynamics can be replaced by definite analogons that can be considered as extensive point functions. By use of local and instantaneous equilibrium, the classical thermodynamic knowledge of the system is summarized in a similar *specific fundamental equation of Gibbs*. The validity of the principle of local and instantaneous equilibrium means that the differences from equilibrium are not too large, since this principle implies that the system can be divided into subsystems that are tiny with respect to the macroscopic length scales, but large with respect to the dimensions of the measuring instruments.

If in non-equilibrium situations the functions of state differ from subsystem to subsystem, then the functions of state become continuous differentiable functions of space and time. In those cases the system is called a *continuous system* or an *inhomogeneous system*, in which gradients of the quantities of state give rise to *irreversible processes of the second kind* or *transport processes*. Classical examples of gradient processes are:

- transport of heat by temperature gradients (Fourier's law),
- transport of matter by concentration gradients (Fick's law),
- transport of electrical charges by gradients of the electric potential (Ohm's law).

If the transport processes are very fast, then the phases remain in practice homogeneous. In that case the *irreversible processes of the first kind* or *relaxation processes* can arise, such as for example:

- chemical reactions,
- dielectric relaxation,
- magnetic relaxation,
- elastic after-effect (viscoelasticity).

The 'very fast progress' of transport phenomena implies in this regard that

* "Sufficiently" small is a vague term, implying that the limitations expressed by the continuum theory and the validity of the Gibbs–Duhem relation have also to be considered. The limitations may be less serious than expected, as shown by the applications of the Navier–Stokes equations.

the transport processes are fast with respect to the relaxation processes. If this is not so, irreversible processes of the first and the second kind can act simultaneously.

If the principle of the local and instantaneous equilibrium applies, then in systems not at equilibrium all thermodynamic quantities become definable. Additionally, for the application of the continuum theory, the thermodynamic quantities are assumed to be continuous differentiable functions of the position vector $\vec{x}$ and the time t ($\vec{x}$ is defined with respect to a point O rigidly attached to the observer). The basic formulas and equations of the classical thermodynamics may be applied locally, and in the continuum limit the differentials can be replaced by 'grad', $\partial/\partial t$, or $\mathrm{D}/\mathrm{D}t$.

Discontinuous systems can be considered as forming a limiting case of the continuous systems. These systems consist of homogeneous phases, separated by dividing walls. The transport processes take place now only through these surfaces of discontinuity (membrane processes). Membrane processes are dominant if these processes are slow with respect to the transport processes *within* the phases and the possible homogeneous relaxation processes.

3.1.3. TIP-axiom III: balance equations. This axiom implies that for the system a set of balance equations applies, these being based on generalizations from empirical facts. As such, these facts make up by themselves one or more axioms. The balance equations are derived as balances of quantities, like mass, momentum, angular momentum, energy, electrical charge, referring to the matter in an arbitrary control volume $V_{(c)}$ around a point $\mathcal{P}$, bounded by a control surface $A_{(c)}$ and located completely within the body concerned. This volume serves to 'control' (in the sense of 'account for') the quantities being described in the balance equations.

For these derivations a *reference system* has to be constructed. A reference system is a physical concept, and in principle it is composed of a rigid arrangement of rigid bodies, such that during the observation time the arrangement, the shape and the size of the relevant bodies do not change within the accuracy of the measurement involved. In thought, an observer is rigidly tied to the reference system, so that concepts like 'observer' and 'reference system' are equivalent and can be used interchangeably.

In one reference system an observer can install infinitely many *coordinate systems*. Coordinate systems are mathematical constructions that are not involved with the physical phenomena to be studied. As a consequence, the mathematical formulation of the physical laws and equations must not depend on the arbitrarily chosen coordinate system. This requirement suggests the use of coordinate-independent objects, like scalars, vectors and tensors. Otherwise, these objects need not be invariant with respect to moving frames.

In the continuum theory, one uses for the formulation of the balance equations—paradoxically—the notion of *particles*. A particle is in this context a tiny quantity of matter enclosed in a physical volume element that as a consequence contains a large number of molecules. From the macroscopic point of view, a particle is infinitesimally small and in that sense one considers the particles as *material points*. In relation to a reference system, a particle has at each moment a position specified by the position vector $\vec{x}$. In addition, the particle has with respect to the chosen reference system a velocity $\vec{v}$, which may be a function of $\vec{v}$ and t.

Suppose now that the balance of the extensive property Ψ has to be described. The tensor character of this property is given, but not yet specified. Denote the specific value of Ψ by ψ. According to international recommendations* the extensive quantities per unit of mass are called *specific*. The amount Ψ in a 'fixed' control volume—i.e., fixed in relation to the reference system—becomes

$$\Psi_{(c)} = \iiint_{V_{(c)}} \rho\psi \, dV \tag{3.1}$$

where ρ denotes the density. (One can also use a control volume moving with the fluid for deriving the balance equations, but we do not do that here. The transformation of the control volume moving with the fluid to a volume fixed in space can formally be obtained using the transport theorem of Reynolds†). The rate of change of Ψ can be obtained using the following arguments. For a stationary $V_{(c)}$ the change of this property as a function of time is

$$\frac{d}{dt}\Psi_{(c)} = \iiint_{V_{(c)}} \left(\frac{\partial \rho\psi}{\partial t}\right) dV \tag{3.2}$$

This change is caused by

(1) Flow (convection) of matter through $A_{(c)}$ along with the property Ψ.
(2) Exchange of the property Ψ between the matter in $V_{(c)}$ and the matter outside $V_{(c)}$.
(3) Production of Ψ in $V_{(c)}$.

---

* Cohen E.R. and Giacomo P. eds., 1987. Symbols, Units, Nomenclature and Fundamental constants in Physics (1987 revision), document IUPAP-25 (SUNAMCO 87–1) SUN-Commission. International Union of Pure and Applied Physics, *Physica*, 146A: 1–67.

† The rate of change of the total Ψ over a material volume $V_{(c)}$ equals the rate of change of the total Ψ over the fixed volume that is the instantaneous configuration of $V_{(c)}$, plus the flux of $\vec{v}\Psi$ out of the bounding surface. Truesdell C. and Toupin R. 1960. *The Classical Field Theories. Encyclopedia of Physics*, Vol. III/1. Springer-Verlag, Berlin. See p. 347. Reynolds (1903) has extended the Leibniz formula for differentiating an integral. See Bird R.B., Stewart W.E. and Lightfoot E.N. 1960. *Transport Phenomena*, John Wiley & Sons, New York, 1960. Appendix A.5.

If the velocity of the matter is $\vec{v}$, then the first contribution is given by

$$-\iint_{A_{(c)}} (\vec{n}\cdot\vec{v})\,\rho\psi\,dA = -\iiint_{V_{(c)}} \operatorname{div}(\rho\vec{v}\psi)\,dV \tag{3.3}$$

In this expression the convention is used that the outward direction of the unit normal to the surface element $d\vec{A}$ of $A_{(c)}$ is considered positive. This explains the minus sign in (3.3). In addition the well-known divergence theorem of Gauss is applied.

The second contribution to (3.2) is more difficult to calculate. For this case the exchange of Ψ is divided into

(i) Exchange between the matter in $V_{(c)}$ and the matter outside $V_{(c)}$ not part of the volume V of the total system.
(ii) Exchange between the matter in $V_{(c)}$ and the matter outside the chosen control volume $V_{(c)}$ but within the volume V of the total system.

The first item bears on the *external actions* performed by external fields. These fields are caused by bodies not part of the system, of which the resulting external actions are incorporated in the production term (the notion 'production' used here has a broad meaning). The second item bears on the *internal actions within* the system considered. In the continuum theory, the internal actions have to be supplemented by axioms. If it is assumed that in the continuous phase the range of the intermolecular forces is of the order of the dimensions of the physical volume elements discussed in Chapter 1, and such that the distribution function of these forces is approximated by a delta function, the following axiom can be adopted

Axiom of the internal actions. The actions exerted by the matter of the body outside $V_{(c)}$ on the matter inside $V_{(c)}$ are equivalent to surface actions applied at points on the outside of $A_{(c)}$. The surface actions are distributed (absolute) continuously over $A_{(c)}$. These statements hold for all control volumes into which the body is divided.

To account for the third item, denote the 'production' of the property Ψ per unit of time and volume by $\pi_{(\psi)}$. Using (3.1) to (3.3) and the axiom of the internal actions, it follows that the integral (or global) balance equation of Ψ can be written as

$$\iiint_{V_{(c)}} \frac{\partial(\rho\psi}{\partial t}\,dV = \qquad)$$

$$-\iiint_{V_{(c)}} \operatorname{div}(\rho\vec{v}\psi)\,dV + \iint_{A_{(c)}} \Psi[\vec{x},t,d\vec{A}] + \iiint_{V_{(c)}} \pi_{(\psi)}\,dV \tag{3.4}$$

where $\Psi[\vec{x},t,d\vec{A}]$ is an operator that assigns to $d\vec{A}$ the surface action Ψ. It will be shown that the operator is a linear operator of $d\vec{A}$.

The global balance equation (3.4) applies to all possible control volumes, so it also applies in the limit $V_{(c)} \to 0^+$. If in this limit the integrands of the volume integrals will be finite, it follows from (3.4)

$$\iint_{A_{(c)}} \Psi[\vec{x}, t, d\vec{A}] = \mathcal{O}\left(V_{(c)}\right) \qquad \text{for} \qquad V_{(c)} \to 0^+ \tag{3.5}$$

If the surface actions are continuous, this condition results in the very useful flux theorem of the internal actions for the continuum theory and thermodynamics of irreversible processes

Flux theorem for internal actions. If the internal actions can be pictured as surface actions, for which:

(1) The surface actions are (absolutely) continuously distributed over every control surface.
(2) The limit transition (3.5) is valid.

then

$$\boxed{\Psi[\vec{x}, t, d\vec{A}] = -d\vec{A} \cdot \phi(\vec{x}, t) = -\vec{n} \cdot \phi(\vec{x}, t)\, dA} \tag{3.6}$$

where ϕ is the flux of the property Ψ, and the tensor ϕ is one order higher than the tensor Ψ (the minus sign is due to a historical convention).

The usual demonstrations of the flux theorem start with the introduction of a coordinate system in which an elementary cube or tetrahedron is sketched, whereupon the balance of linear momentum is applied to derive the components of the stress tensor of Cauchy. Such derivations are less elegant, since for obtaining a coordinate-independent representation of the internal surface actions, one starts with deriving the components of the stress tensor in some coordinate system. An illustrative coordinate-free proof has been proposed by Merk and Bolder in the 1960s, although not published.

Consider a small tetrahedron* ABCD split up by a surface ABE into two tetrahedrons (see figure 3.1). Assume

$$\Delta ABC = d\vec{A}_D \qquad \Delta BCD = d\vec{A}_A \qquad \Delta CDA = d\vec{A}_B \qquad \Delta DAB = d\vec{A}_C$$

and

$$\Delta ABE = d\vec{A} = \vec{n}\, dA$$

Assume further

$$CE = \alpha(CD) \qquad \text{and} \qquad ED = \beta(CD) \qquad \text{with} \qquad \alpha + \beta = 1$$

* A spatial region can always be divided into tiny tetrahedrons that can be considered as the elementary spatial volume. The $d\vec{A}$ space as well as the Ψ space are Euclidean spaces in which for instance addition and multiplication are defined. The mapping $d\vec{A} \to \Psi$ is linear if $\Psi[d\vec{A}_1 + d\vec{A}_2] = \Psi[d\vec{A}_1] + \Psi[d\vec{A}_2]$ and $\Psi[\lambda\, d\vec{A}] = \lambda\Psi[d\vec{A}]$.

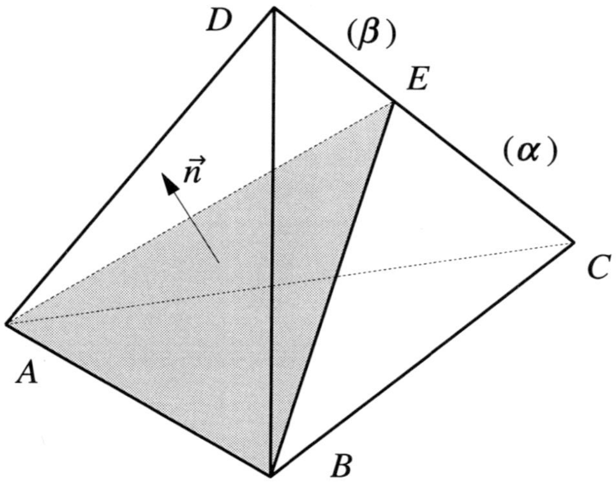

Figure 3.1. Elementary tetrahedrons to prove the flux theorem for internal actions.

With these definitions we have

$$\begin{aligned} \Delta BCE &= \alpha\, d\vec{A}_A \qquad & \Delta BED &= \beta\, d\vec{A}_A \\ \Delta ACE &= \alpha\, d\vec{A}_B \qquad & \Delta AED &= \beta\, d\vec{A}_B \end{aligned} \tag{3.7}$$

If $\Psi[\vec{x}, t, d\vec{A}]$ is distributed absolutely continuously over every control surface, then for a sufficiently small dA and finite $\lambda \geq 0$

$$\Psi[\lambda\, d\vec{A}] = \lambda \Psi[d\vec{A}] \tag{3.8}$$

where for brevity $\Psi[d\vec{A}]$ is denoted for $\Psi[\vec{x}, t, d\vec{A}]$. Application of (3.5) to the sufficiently small tetrahedron ABCE and use of (3.7) and (3.8) give

$$\alpha\Psi[d\vec{A}_A] + \alpha\Psi[d\vec{A}_B] + \Psi[d\vec{A}] + \Psi[d\vec{A}_D] = \mathcal{O}(V) \tag{3.9}$$

For $V \to 0$ the left-hand side of (3.9) with bounded values of Ψ is a function of the surface of the tetrahedron, whereas the right-hand side depends on its volume. So the right-hand side converges faster to zero than the left-hand side, meaning that for sufficiently small volumes the right-hand contribution can be neglected.

Application of (3.5) to the sufficiently small tetrahedron ABCD results analogously in

$$\Psi[d\vec{A}_A] + \Psi[d\vec{A}_B] + \Psi[d\vec{A}_C] + \Psi[d\vec{A}_D] = 0 \tag{3.10}$$

When this is used to eliminate the sum $\Psi[d\vec{A}_A]+\Psi[d\vec{A}_B]$ from (3.9), one gets

$$\Psi[d\vec{A}] = \alpha\Psi[d\vec{A}_C] - \beta\Psi[d\vec{A}_D] \tag{3.11}$$

where $\beta = 1 - \alpha$.

By a simple application of the Gauss divergence theorem a similar result is obtained for the surface element $d\vec{A}$. For a constant vector field $\vec{\pi}$

$$\iint_A \vec{n}\cdot\vec{\pi}\, dA = \vec{\pi}\cdot\iint_A \vec{n}\, dA = \iiint_V \operatorname{div}\vec{\pi}\, dV = 0$$

This applies to all vector fields $\vec{\pi}$, so that

$$\iint_A \vec{n}\, dA = \iint_A d\vec{A} = \vec{0} \tag{3.12}$$

Condition (3.5) changes into (3.12) with the substitution $\Psi = d\vec{A}$. The analysis for $\Psi[d\vec{A}]$ could be repeated for $d\vec{A}$ and the relation (3.11) with the substitution $\Psi[d\vec{A}] = d\vec{A}$ will then found. So

$$d\vec{A} = \alpha\, d\vec{A}_C - \beta\, d\vec{A}_D$$

Substitution of this result into (3.11) it follows for arbitrary α ($\beta = 1 - \alpha$) that

$$\Psi[\alpha\, d\vec{A}_C - \beta\, d\vec{A}_D] = \alpha\Psi[d\vec{A}_C] - \beta\Psi[d\vec{A}_D]$$

In combination with (3.8) this means that $\Psi[d\vec{A}]$ is a linear mapping or equivalently a tensor operator of $d\vec{A}$, so that (3.6) must be valid.

The flux theorem was derived by Cauchy in 1829 for the mechanical stresses, and is known in the theory of stresses as the *fundamental stress theorem of Cauchy*. In mechanics, the tractions at a control surface yield a flux tensor, which is understood to be the well-known stress tensor. The flux theorem shows that $\Psi[d\vec{A}]$ is a linear function of $d\vec{A}$. From the flux theorem also the Cauchy lemma follows

$$\Psi[-d\vec{A}] = -\Psi[d\vec{A}] \tag{3.13}$$

The Cauchy lemma expresses that actions performed by the matter outside V upon the matter inside V are equal in magnitude and opposite in direction to the actions performed by the matter inside V upon the matter outside V. In short: 'action equals reaction'. Surface actions are sometimes called *contact actions*, since these actions are also found to represent the forces between two and more bodies if these bodies are brought into contact with each other.

With the help of the flux theorem and the divergence theorem, the contribution of the contact actions is written as a volume integral

$$\iint_{A_{(c)}} \Psi[d\vec{A}] = -\iint_{A_{(c)}} \vec{n}\cdot\phi\,(\vec{x},t)\; dA = -\iiint_{V_{(c)}} (\operatorname{div}\phi)\; dV$$

Applied to the global balance (3.4), this equation is now completely expressed in volume integrals. Since this global balance equation is valid for any sufficiently regular control volume, the integrands yield the *differential* or *local balance equation.*

$$\boxed{\frac{\partial}{\partial t}(\rho\psi) = -\mathrm{div}\,\left(\rho\vec{v}\psi + \phi_{(\psi)}\right) + \pi_{(\psi)}} \tag{3.14}$$

The flux $\phi_{(\psi)}$ is introduced in the local balance equation to represent the transport of Ψ by molecular actions. The flux $\phi_{(\psi)}$ is therefore interpreted as a *process flux*, which determines the rate of change of the irreversible transport of Ψ. The same sign convention is used for $\rho\vec{v}\psi$ and $\phi_{(\psi)}$.

The production term $\pi_{(\psi)}$ is determined by physical axioms like the conservation of mass, the equations of motion of Newton and Euler. If $\pi_{(\psi)} = 0$, (3.14) expresses the conservation of the property Ψ. The balance equation then becomes a *conservation law* for Ψ.

The formulation of the balance equations leads to a definition of the process fluxes and as such is very important in the thermodynamics of irreversible processes. It will be shown that the process fluxes can be scalars, vectors or tensors.

3.1.4. TIP-axiom IV: entropy balance. As already remarked on page 49 entropy is considered to be a quantity that can be transported and produced. Besides, irreversible processes *within* the system generate a positive entropy production. These global experiences are now formulated locally in this entropy balance axiom. From the statistical foundation of the Onsager Casimir Reciprocal Relations it can be shown* that local stability of the local equilibrium system is guaranteed if the entropy production is non-negative. Summarizing the axiom we may write

Entropy balance. The entropy is a physical quantity that is distributed over space as a function of time and satisfies a balance equation, in which one term accounts for the rate of entropy production. This rate of entropy production is non-negative.

The local entropy balance is for a continuous phase analogous to (3.14)

$$\boxed{\frac{\partial}{\partial t}(\rho s) = -\mathrm{div}\,\left(\rho\vec{v}s + \vec{\phi}_{(\mathrm{s})}\right) + \pi_{(\mathrm{s})}} \tag{3.15}$$

with

$$\boxed{\pi_{(\mathrm{s})} \geq 0} \tag{3.16}$$

* Kuiken G.D.C. 1977. The derivation of the Onsager Casimir reciprocity relations without using Boltzmann's postulate, *Journal of Non-Equilibrium Thermodynamics*, 2: 153–168. See also Chapter 5.

In this equation s is identified as the specific entropy, $\vec{\phi}_{(s)}$ as the entropy flux and $\pi_{(s)}$ as the rate of entropy production.

The entropy flux and the rate of entropy production follow from

(1) The specific equation of Gibbs.
(2) The principle of local and instantaneous equilibrium.
(3) The balance equations of TIP-axiom III.

The rate of entropy production or entropy dissipation is of especial interest, and consists of a sum of products, of which one factor is a scalar, vector or tensor process flux. Symbolically, this is expressed as follows

$$\mathfrak{D} = T\,\pi_{(s)} = \Phi^{(s)}K^{(s)} + \vec{\Phi}^{(v)} \cdot \vec{K}^{(v)} + \Phi^{(t)} : \mathbf{K}^{(t)} \geq 0 \tag{3.17}$$

The superscript (s) denotes a scalar process, the superscript (v) a vector process and the superscript (t) a tensor process. Each term on the right-hand side of (3.17) can in turn be given as a sum of terms, since obviously more than one scalar, vector and tensor process may take place in a system.

The process fluxes are the quantities that are defined by the divergence term of the balance equations. The conjugate process forces, associated with these fluxes, may be interpreted as the driving forces of the process fluxes. The words 'flux' and 'force' are however more or less arbitrary and may be interchanged, if needed.

3.1.5. TIP-axiom V: phenomenological equations. The objective of the thermodynamics of irreversible processes is to obtain a mathematical formulation of the irreversible processes based on generally accepted axioms. This description of the irreversible processes found in a body is determined by the nature of the substance of the body. The nature of the matter is described as the *constitution* or *structure* of the matter. In a constitutive theory, we attempt to model the processes occurring in the matter by *phenomenological equations* or *constitutive equations.* The thermodynamics of the irreversible processes is thus to be considered as a constitutive theory, in which the deviations from the equilibrium state are small. Irreversible processes arise as a result of deviations from the equilibrium state. These irreversible processes are described by process forces and process fluxes that are zero in unconstrained equilibrium. In frozen equilibrium the process fluxes are zero, but one or more of the process forces may be unequal to zero.

For sufficiently small deviations from the equilibrium state it is assumed that the relations between the process fluxes and process forces are linear. The thermodynamics of the irreversible processes is a theory of the linear constitutive equations and is to be considered as a first approximation of non-equilibrium processes. As a consequence, the theory is a logical extrapolation from classical thermodynamics. Furthermore, the thermodynamics of the irreversible processes is the only unifying theory in which consistently the complete linear constitutive equations of the matter can be treated. A major obstacle for the extrapolation of the theory to nonlinear constitutive equations is the principle of local and instantaneous equilibrium, since this principle is needed to define the thermodynamic quantities; it holds with certainty only for small deviations of the equilibrium, such that the constitutive equations can be approximated by linear equations.

To ensure that none of the variables used in the set of variables encountered in the constitutive equations is forgotten, it is assumed that in principle every process flux can depend linearly on *all* process forces that are possible within the system. This is called the *principle of equipresence*, which holds where a process force is not forbidden by another generally accepted principle. The 'principle of equipresence' is much more a statement about the procedure to be followed than a principle enforced by empirical facts*. Other generally accepted principles, like the principle of material isomorphism, always reduce the number of variables.

The linear and homogeneous relations between the process forces and the process fluxes are based on experience and are called the (linear) *phenomenological equations*, while the coefficients in these equations are called the *phenomenological coefficients.*

According to (3.17) the process fluxes can be given by

$$\Phi^{(\mathrm{s})} \qquad \vec{\Phi}^{(\mathrm{v})} \qquad \mathbf{\Phi}^{(\mathrm{t})}$$

and the conjugate process forces by

$$K^{(\mathrm{s})} \qquad \vec{K}^{(\mathrm{v})} \qquad \mathbf{K}^{(\mathrm{t})}$$

With the index notation and the Einstein summation convention applied to Cartesian coordinates the phenomenological equations read

$$\begin{aligned}
\Phi^{(\mathrm{s})} &= L^{(\mathrm{ss})} K^{(\mathrm{s})} + L_k^{(\mathrm{sv})} K_k^{(\mathrm{v})} + L_{km}^{(\mathrm{st})} K_{mk}^{(\mathrm{t})} \\
\Phi_i^{(\mathrm{v})} &= L_i^{(\mathrm{vs})} K^{(\mathrm{s})} + L_{ik}^{(\mathrm{vv})} K_k^{(\mathrm{v})} + L_{ikm}^{(\mathrm{vt})} K_{mk}^{(\mathrm{t})} \\
\Phi_{ij}^{(\mathrm{t})} &= L_{ij}^{(\mathrm{ts})} K^{(\mathrm{s})} + L_{ijk}^{(\mathrm{tv})} K_k^{(\mathrm{v})} + L_{ijkm}^{(\mathrm{tt})} K_{mk}^{(\mathrm{t})}
\end{aligned} \tag{3.18}$$

As always the lowercase Latin subscripts indicate the Cartesian coordinates. The phenomenological coefficients are denoted by the symbols L.

Historical development has given rise to distinguishing the response of a material into direct and indirect responses or effects. The *direct effects* are described by phenomenological equations, in which a flux depends only on its conjugate force, such that in (3.18) $L_{\dots}^{(\mathrm{qr})} = 0$ for q $\neq$ r. In the

* Woods L.C. 1981. The bogus axioms of continuum mechanics. *Bulletin of the Institute of Mathematics and its Applications*, 17: 98–102.

thermodynamics of irreversible processes, the principle of equipresence is used to account for the—experimentally verified—*indirect effects* or *cross-effects*, in which $L^{(\mathrm{qr})}_{\cdots} \neq 0$ for q $\neq$ r. A well-known example is thermal diffusion: mass transfer by diffusion is caused not only by concentration gradients—'ordinary' diffusion—but also by temperature gradients—thermal diffusion.

It is to be noted that in the phenomenological equations no *memory-effects* occur. Memory effects can turn up in the mathematical description if not all variables can be described instantaneously and internal or hidden variables have to be used. Examples are given in the discussion of the relaxation phenomena in Chapter 7.

3.1.6. TIP-axiom VI: O.C.R.R. In this axiom the symmetry properties of the phenomenological coefficients are expressed that will be discussed extensively in Chapter 5. These symmetry properties of the phenomenological coefficients are known as the Onsager Casimir reciprocal relations. These reciprocal relations are not easily presented in a general formulation, since the tensorial character of the phenomenological coefficients has to be considered.

The original objective of Onsager* was to prove the symmetry of the thermal conductivity tensor. However, the demonstration was related only to homogeneous systems, in which no transport processes occur. Casimir† drew attention to this restriction, and extended the demonstration of Onsager to one for inhomogeneous systems. De Groot and Mazur‡ have taken into account the effect of external magnetic fields on the thermal conductivity, and those authors finally worked out the correct formulation of the reciprocal relations of the thermodynamics of the irreversible processes that are usually credited to Onsager and Casimir.

For example, consider a system in which only vector transport processes can occur. Then for N such processes the dissipation is given by

$$\mathfrak{D} = T\,\pi_{(\mathrm{s})} = \sum_{K=1}^{N} \vec{\Phi}_K \cdot \vec{K}_K \tag{3.19}$$

in which the uppercase Latin subscripts indicate the processes occurring. The Einstein summation convention does not hold for these subscripts,

---

* Onsager L. 1931. Reciprocal relations in irreversible processes I. *Physical Review*, 37: 405–426; Onsager L. 1931. Reciprocal relations in irreversible processes II. *Physical Review*, 38: 2265–2279.

† Casimir G. 1945. On Onsager's principle of microscopic reversibility. *Review of Modern Physics*, 17: 343–350.

‡ De Groot S.R. and Mazur P. 1954. Extension of Onsager's theory of reciprocal relations. I, II. *Physical Review*, 94: 218–224, 224–226. See also: De Groot S.R. and Mazur P. 1962. *Non-Equilibrium Thermodynamics*, North Holland Publ. Co., Amsterdam; and Meixner J. and Reik H.G. 1959. *Thermodynamik der irreversiblen Prozesse. Encyclopedia of Physics*, Vol. III/2. Springer-Verlag, Berlin. See p. 513.

since this convention applies only to subscripts denoting coordinates. The phenomenological equations read

$$\vec{\Phi}_K = \sum_{M=1}^{N} \mathbf{L}_{KM} \cdot \vec{K}_M \tag{3.20}$$

The Onsager Casimir reciprocal relations are with respect to an inertial system and in the presence of an external magnetic field $\vec{B}_0$

$$\boxed{\mathbf{L}_{KM}\left(\vec{B}_0\right) = \epsilon_K \epsilon_M \mathbf{L}^{\mathrm{T}}_{KM}\left(-\vec{B}_0\right)} \tag{3.21}$$

in which $\epsilon_K = 1$, if $\vec{K}_K$ is an even function of time and $\epsilon_K = -1$, if $\vec{K}_K$ is an odd function of time.

The Onsager Casimir reciprocal relations (O.C.R.R.) are very important in the thermodynamics of irreversible processes, since these relations reduce the number of independent phenomenological coefficients, and relate the cross-effects found in materials that can be described by constitutive equations for which the tensor phenomenological coefficients have off-diagonal components. From a physical point of view the relations are of interest, since the Onsager Casimir reciprocal relations can be derived only in general terms with reference to statistical mechanics and fluctuation theory, supplemented by the axiom of microscopic reversibility and the axiom of the regression of the fluctuations.

3.2. SUPPLEMENTARY CONSTITUTIVE AXIOMS

In the previous section six axioms have been discussed that are needed for the construction of a thermodynamic theory of irreversible processes. The objective of this theory is to develop a unifying treatment of the linear constitutive equations of matter. To accomplish this, the six TIP axioms are not sufficient, and they have to be supplemented with axioms that are needed in every classical constitutive theory. Four frequently used supplementary axioms are discussed in the following subsections. These axioms specify the invariance of the final equations with respect to Galilean transformations, coordinate transformations, choice of dimensions and symmetry operations.

3.2.1. Axiom of Galilean-invariance. Galilean transformations describe the change of the reference system, in which the reference systems transform into one another by means of a uniform translation. In the classical theory, the physical laws and equations have to be invariant with respect to reference systems that are in relative translation at constant velocity. They are said to be invariant under Galilean transformations. This applies also to the constitutive equations that describe the response of a body to loadings and deformations.

Usually, Galilean invariance is combined with the classical axiom of time invariance: the physical laws and equations have to be invariant with respect to an arbitrary choice of the zero point of the time scale. Time invariance is also applied to the constitutive equations, both in the linear and in the nonlinear constitutive theories. A consequence of applying time invariance is that aging of a substance is eliminated from consideration. In the linear theory aging is unimportant and in the nonlinear theories aging is accounted for separately. In the classical theory intervals of time are invariant, that is, independent of the state of motion of the observer, and every observer is allowed to choose the zero point of his chronology arbitrarily. A quantity that is invariant under Galilean transformations and the choice of the zero point of the time scale is called a *Galilean invariant.*

The axiom of Galilean invariance leads quite naturally to the use of Galilean invariants. It is often not possible to prove that certain quantities are Galilean invariant. Frequently, this has to be postulated by experience. Only for the kinematical quantities can Galilean invariance be proved. For example the position vector $\vec{x}$ of a material point is not a Galilean invariant, but the operator 'grad' is a Galilean invariant.

Galilean invariance can have far reaching consequences. If the thermodynamic quantities are Galilean invariants, then these quantities cannot depend explicitly on the position vector $\vec{x}$ and the time t. To illustrate: suppose that the specific entropy s of the material point $\mathcal{P}$ in a non-equilibrium system depends explicitly on the position vector and on time

$$s(\mathcal{P}) = s(\vec{x}, t, T, \operatorname{grad} T, \cdots, \mathcal{P})$$

in which the specific entropy s, the temperature T and all the $\cdots$ variables are Galilean invariants. After applying a Galilean transformation, s transforms into s', T into T', $\vec{x}$ into $\vec{x}'$, t into t', and so on, with

$$\vec{x}' = \vec{b} + \vec{x} \qquad \text{and} \qquad t' = t_0 + t$$

and for which in a Galilean transformation $d\vec{b}/dt$ is a constant. After a Galilean transformation the specific entropy becomes

$$s'(\mathcal{P}) = s'(\vec{x}', t', T', \operatorname{grad} T', (\cdots)', \mathcal{P})$$

If s, T and the $\cdots$ variables are Galilean invariants, then

$$s(\mathcal{P}) = s(\vec{b} + \vec{x}, t_0 + t, T, \operatorname{grad} T, \cdots, \mathcal{P})$$

This last equation applies to all $\vec{b}$ and t_0, so that $s(\mathcal{P})$ cannot depend explicitly on $\vec{x}$ and t. Therefore

$$s(\mathcal{P}) = s(T, \operatorname{grad} T, \cdots, \mathcal{P}) \tag{3.22}$$

If in the neighborhood of stable equilibrium s is a maximum and has the value s^+, then for sufficiently small deviations of this equilibrium

$$s = s^+(T) + \mathcal{O}\left(|\mathrm{grad}\, T|^2\right) \tag{3.23}$$

in which the $\cdots$ variables are dropped for brevity. This result implies that for variations to and including the first order the principle of local and instantaneous equilibrium holds. In the transition of (3.22) into (3.23), it is assumed that close to the equilibrium state, s depends analytically on $\mathrm{grad}\, T$, such that s can be expanded with respect to $\mathrm{grad}\, T$ if $|\mathrm{grad}\, T|$ is small enough.

Similarly, the phenomenological equations may be made plausible. Suppose that the flux Φ depends on the process forces $\{K_1, \cdots, K_N\}$. The process fluxes and process forces are zero in an unconstrained equilibrium. If no equilibrium is found, then

$$\Phi = \Phi\left(\vec{x}, t, K_1, K_2, \cdots, K_N\right) \tag{3.24}$$

If Φ and $\{K_1, \cdots, K_N\}$ are Galilean invariants, then Φ can not depend explicitly on $\vec{x}$ and t if the phenomenological equations are Galilean invariant, so that (3.24) becomes

$$\Phi = \Phi\left(K_1, K_2, \cdots, K_N\right) \tag{3.25}$$

For sufficiently small variations of the state of equilibrium (3.25) can be expanded into

$$\Phi = \sum_{K=1}^{N} L_K K_K = \mathcal{O}\left(\|K_K\|\right) \tag{3.26}$$

Modern statistical mechanical research* has shown that for denser gases non-local quantities may become important. For the calculation of the statistical averages to obtain the macroscopic transport coefficients, the expansions with respect to the density are not known for dense gases. It is conjectured that a non-local integral operator is expanded with respect to a local differential operator that leads to divergences. Extrapolations to large variations of the equilibrium state or nonlinear and non-local descriptions are not yet clarified, and are insufficiently founded or experimentally justified to yield a basis for a nonlinear thermodynamic theory.

In conclusion, the axiom of Galilean invariance is a useful axiom for the appreciation of several TIP axioms.

---

* Cohen E.G.D. 1979. Kinetic theory, hydrodynamics and fluctuations, Notes of guest lectures at the Lorentz Institute Leiden, The Netherlands; Cohen E.G.D. 1984. The kinetic theory of fluids—An introduction. *Physics Today*, 37: 64–73.

3.2.2. Axiom of coordinate invariance. An arbitrary coordinate system can be set up in a reference system, which is needed for the mathematical description of the physical phenomena. Since the physical phenomena must not depend on the choice of the coordinate system, the description of the physical phenomena has to be coordinate invariant. This requirement suggests the use of coordinate-independent objects, like scalars, vectors, and tensors.

3.2.3. Axiom of dimensional invariance. The basic physical units are also chosen arbitrarily and therefore, the description of the physical phenomena may not depend on the chosen fundamental physical units. This requirement recommends the use of fundamental units that are consistent with each other, such as the SI units. The linear constitutive equations are then usually automatically dimensionally invariant, that is, independent of the physical dimensions of the quantities occurring in those equations.

3.2.4. Axiom of material isomorphism. This axiom might be formulated as follows: the constitutive equations have to be invariant with respect to symmetry operations that converts the considered material body into itself. If a symmetry operation is applied, the body cannot be distinguished physically from the situation before the application of the symmetry operation, so that is spoken of as *isomorphism*. The required invariance is accordingly trivial.

A symmetry operation can be seen as a point transformation, which transforms the considered body into itself. If $\vec{x}$ is a point of the body, then after the application of the symmetry operation this point has the position $\vec{x}'$ for example. If $\vec{x}$ and $\vec{x}'$ are measured from the same origin O, we have

$$\vec{x}' = \mathbf{Q} \cdot \vec{x} = \vec{x} \cdot \mathbf{Q}^{\mathrm{T}} \tag{3.27}$$

In a symmetry operation the length of a vector does not change, so $\vec{x}' \cdot \vec{x}' = \vec{x} \cdot \vec{x}$. From (3.27) it follows that

$$\vec{x}' \cdot \vec{x}' = \vec{x} \cdot \mathbf{Q}^{\mathrm{T}} \cdot \mathbf{Q} \cdot \vec{x} = \vec{x} \cdot \vec{x}$$

or with the introduction of the unit tensor $\mathbf{I}$

$$\vec{x} \cdot \left(\mathbf{Q}^{\mathrm{T}} \cdot \mathbf{Q} - \mathbf{I}\right) \cdot \vec{x} = 0$$

which applies to all $\vec{x}$, so that

$$\mathbf{Q}^{\mathrm{T}} \cdot \mathbf{Q} = \mathbf{I} \quad \rightarrow \quad \mathbf{Q}^{-1} = \mathbf{Q}^{\mathrm{T}} \tag{3.28}$$

This leads to

$$[\det \mathbf{Q}]^2 = \det\left(\mathbf{Q}^{\mathrm{T}} \cdot \mathbf{Q}\right) = \det(\mathbf{I}) = 1$$

and

$$\det(\mathbf{Q}) = \pm 1 \tag{3.29}$$

The plus sign corresponds to the proper symmetry operations: the rigid rotations. The minus sign corresponds to the improper symmetry operations, which involve not only the rigid rotations but also the reflections. The symmetry of a body is determined by a group of orthogonal transformations (3.27) that satisfy (3.28).

Material isomorphism can now be formulated as follows. Suppose that the flux Φ depends on the force K

$$\Phi = f(K) \tag{3.30}$$

After performing a symmetry operation, Φ transforms into Φ' and K into K', and

$$\Phi' = f'(K') \tag{3.31}$$

is valid for this operation. In a symmetry operation the matter after the application of the operation cannot be distinguished physically from the matter before the operation. This means that the relation between Φ and K has to be the same relation as between Φ' and K', so that

$$\Phi = f(K) \quad \rightarrow \quad \Phi' = f(K') \tag{3.32}$$

For the linear phenomenological relations (3.18) this means that

$$L = L' \qquad \vec{L} = \vec{L}' \qquad \mathbf{L} = \mathbf{L}' \qquad \cdots \tag{3.33}$$

for the symmetry group of the body considered. For the application of (3.33) the transformation rules of scalars, vectors, and tensors have to be known.

A scalar L has to be invariant with respect to symmetry operations. Although according to (3.33) this is true also for vectors and tensors, one has to consider the transformations to which those quantities are subjected. In the Euclidean space, a vector transforms as a line element $\vec{x}_{(1)} - \vec{x}_{(2)}$. From this and from (3.27) it follows that

$$\vec{L}' = \mathbf{Q} \cdot \vec{L} \qquad \text{or} \qquad L'_i = Q_{ij} L_j \tag{3.34}$$

A tensor of the second order can be defined as a homogeneous, linear mapping of two vector spaces onto each other

$$\vec{a} = \mathbf{L} \cdot \vec{b}$$

After a transformation this becomes

$$\vec{a}' = \mathbf{L}' \cdot \vec{b}' \qquad \text{with} \qquad \vec{a}' = \mathbf{Q} \cdot \vec{a} \quad \text{and} \quad \vec{b}' = \mathbf{Q} \cdot \vec{b}$$

such that

$$\mathbf{Q}\cdot\vec{a} = \mathbf{L}'\cdot\mathbf{Q}\cdot\vec{b}$$

or with (3.28)

$$\vec{a} = \mathbf{Q}^{\mathrm{T}}\cdot\mathbf{L}'\cdot\mathbf{Q}\,\vec{b} = \mathbf{L}\cdot\vec{b}$$

which has to hold for all vectors $\vec{a}$ and $\vec{b}$, yielding

$$\mathbf{L} = \mathbf{Q}^{\mathrm{T}}\cdot\mathbf{L}'\cdot\mathbf{Q} \qquad \text{or} \qquad \mathbf{L}' = \mathbf{Q}\cdot\mathbf{L}\cdot\mathbf{Q}^{\mathrm{T}} \tag{3.35}$$

When the summation convention is used, this is written in the index notation as

$$L'_{ij} = Q_{ip}Q_{jq}L_{pq} \tag{3.36}$$

Similarly, the transformation for higher order tensors can be inferred. For example for a third-order tensor it follows that

$$L'_{ijk} = Q_{ip}Q_{jq}Q_{kr}L_{pqr} \tag{3.37}$$

Symmetry principle for bodies having a symmetry center. If a substance has a symmetry center, then all the phenomenological tensors of odd order are zero.

PROOF: If a body has a symmetry center, then

$$\|\mathbf{Q}\| = \begin{pmatrix} -1 & 0 & 0 \\ 0 & -1 & 0 \\ 0 & 0 & -1 \end{pmatrix} \tag{3.38}$$

is a symmetry operation. With respect to this operation the phenomenological coefficients have to be invariant.

Substitution of (3.38) into (3.34) yields $L'_i = -L_i$, and because of (3.33) it follows that

$$L_i = -L_i \qquad \rightarrow \qquad L_i = 0$$

Similarly, in (3.37) we have $L'_{ijk} = -L_{ijk}$ if $\mathbf{Q}$ is given by (3.38), so that

$$L_{ijk} = -L_{ijk} \qquad \rightarrow \qquad L_{ijk} = 0$$

and the proof is analogous for higher order tensors.

For central symmetry, material invariance requires that all odd order phenomenological tensors vanish, and the phenomenological equations (3.18) reduce to

$$\begin{aligned} \Phi^{(\mathrm{s})} &= L^{(\mathrm{ss})}K^{(\mathrm{s})} + L^{(\mathrm{st})}_{km}K^{(\mathrm{t})}_{mk} \\ \Phi^{(\mathrm{v})}_i &= L^{(\mathrm{vv})}_{ik}K^{(\mathrm{v})}_k \\ \Phi^{(\mathrm{t})}_{ij} &= L^{(\mathrm{ts})}_{ij}K^{(\mathrm{s})} + L^{(\mathrm{tt})}_{ijkm}K^{(\mathrm{t})}_{mk} \end{aligned} \tag{3.39}$$

For a material possessing a symmetry center, the fluxes of even (odd) order depend in the linear constitutive equations only on the forces of even (odd) order. This corollary has been attributed to P. Curie*, although this author seemed to have thought more about that the fact that the types of symmetry of the various constitutive equations for the same isotropic materials† are related.

Symmetry principle for isotropic materials. For isotropic material bodies, the phenomenological tensors of even order are to be constructed using the unit tensor.

PROOF: Isotropic materials definitely have a symmetry center. Therefore, the symmetry principle for materials having a symmetry center applies, as well as the phenomenological equations (3.39). The condition $\mathbf{L}' = \mathbf{L}$ holds for all $\mathbf{Q}$ satisfying (3.28), and from (3.35) it is seen that

$$\mathbf{L} = \mathbf{Q} \cdot \mathbf{L} \cdot \mathbf{Q}^{\mathrm{T}}$$

and that

$$\vec{a} \cdot \mathbf{L} \cdot \vec{b}$$

is therefore a bilinear scalar form of $\vec{a}$ and $\vec{b}$ that is invariant with respect to all orthogonal transformations. The only bilinear scalar invariant form of $\vec{a}$ and $\vec{b}$ is $\vec{a} \cdot \vec{b}$, so that

$$\vec{a} \cdot \mathbf{L} \cdot \vec{b} = L\, \vec{a} \cdot \vec{b}$$

in which L an invariant scalar. If

$$\vec{a} \cdot (\mathbf{L} - L\,\mathbf{I}) \cdot \vec{b} = 0$$

for all $\vec{a}$ and $\vec{b}$, then the following relation holds

$$\mathbf{L} = L\,\mathbf{I} \qquad \text{or} \qquad L_{ij} = L\delta_{ij} \tag{3.40}$$

A fourth-order tensor requires that

$$a_{ji} L_{ijkm} b_{mk}$$

is a bilinear scalar invariant form of the second-order tensors $\mathbf{a}$ and $\mathbf{b}$. The irreducible terms of such a form are

$$(\mathbf{I} : \mathbf{a})(\mathbf{I} : \mathbf{b}) \qquad (\mathbf{a}^{\mathrm{S}} : \mathbf{b}^{\mathrm{S}}) \qquad (\mathbf{a}^{\mathrm{A}} : \mathbf{b}^{\mathrm{A}})$$

* Curie P. 1894. Sur la symmétrie dans les phénomènes physiques, symmétrie d'un champ électrique et d'un champ magnétique. *Journal de Physique et le Radium*, 3(3): 393–415.
† Truesdell C. 1984. *Rational Thermodynamics*, 2nd ed. Springer-Verlag, New York. See p. 390.

in which $\mathbf{a}^{\mathrm{S}} = \frac{1}{2}(\mathbf{a}+\mathbf{a}^{\mathrm{T}})$ is the symmetric part of $\mathbf{a}$ (with $\mathbf{a}^{\mathrm{T}}$ the transposed tensor of $\mathbf{a}$) and $\mathbf{a}^{\mathrm{A}} = \frac{1}{2}(\mathbf{a}-\mathbf{a}^{\mathrm{T}})$ the antisymmetric part of $\mathbf{a}$. Then

$$a_{ji}L_{ijkm}b_{mk} = L_{(1)}(\mathbf{I}:\mathbf{a})(\mathbf{I}:\mathbf{b}) + L_{(2)}(\mathbf{a}^{\mathrm{S}}:\mathbf{b}^{\mathrm{S}}) + L_{(3)}(\mathbf{a}^{\mathrm{A}}:\mathbf{b}^{\mathrm{A}}) \qquad (3.41)$$

where $L_{(1)}$, $L_{(2)}$ and $L_{(3)}$ are invariant scalars. For the symmetric, and antisymmetric second-order tensors, we have respectively

$$\mathbf{a}^{\mathrm{S}}:\mathbf{b}^{\mathrm{S}} = \tfrac{1}{4}\left(\mathbf{a}+\mathbf{a}^{\mathrm{T}}\right):\left(\mathbf{b}+\mathbf{b}^{\mathrm{T}}\right) = \tfrac{1}{2}\left(\mathbf{a}:\mathbf{b}+\mathbf{a}:\mathbf{b}^{\mathrm{T}}\right)$$
$$\mathbf{a}^{\mathrm{A}}:\mathbf{b}^{\mathrm{A}} = \tfrac{1}{4}\left(\mathbf{a}-\mathbf{a}^{\mathrm{T}}\right):\left(\mathbf{b}-\mathbf{b}^{\mathrm{T}}\right) = \tfrac{1}{2}\left(\mathbf{a}:\mathbf{b}-\mathbf{a}:\mathbf{b}^{\mathrm{T}}\right)$$

or written in Cartesian index notation using the summation convention

$$\mathbf{a}^{\mathrm{S}}:\mathbf{b}^{\mathrm{S}} = \tfrac{1}{2}a_{ij}\left(\delta_{ik}\delta_{jm}+\delta_{im}\delta_{jk}\right)b_{mk}$$
$$\mathbf{a}^{\mathrm{A}}:\mathbf{b}^{\mathrm{A}} = \tfrac{1}{2}a_{ij}\left(\delta_{ik}\delta_{jm}-\delta_{im}\delta_{jk}\right)b_{mk}$$

The bilinear scalar invariant (3.41) can be written with these results as

$$a_{ji}\left[L_{ijkm} - L_{(1)}\delta_{ij}\delta_{km} - \tfrac{1}{2}L_{(2)}\left(\delta_{ik}\delta_{jm}+\delta_{im}\delta_{jk}\right) - \tfrac{1}{2}\,L_{(3)}\left(\delta_{ik}\delta_{jm}-\delta_{im}\delta_{jk}\right)\right]b_{mk} = 0$$

This equation has to be satisfied for all a_{ij} and b_{mk}, so that

$$L_{ijkm} = L_{(1)}\delta_{ij}\delta_{km} + \tfrac{1}{2}L_{(2)}\left(\delta_{ik}\delta_{jm}+\delta_{im}\delta_{jk}\right) + \tfrac{1}{2}L_{(3)}\left(\delta_{ik}\delta_{jm}-\delta_{im}\delta_{jk}\right) \qquad (3.42)$$

The principle that for isotropic media the phenomenological coefficients have to be irreducible combinations of the unit tensor is in fact trivial, since in an isotropic medium no preferred directions exist. The unit tensor is the only invariant object for isotropic materials that can be used to construct phenomenological tensors.

Application of (3.40) and (3.42) to the phenomenological equations (3.39) yields

$$\boxed{\begin{aligned} \Phi^{(\mathrm{s})} &= L^{(\mathrm{ss})}K^{(\mathrm{s})} + L^{(\mathrm{st})}\left(\mathbf{I}:\mathbf{K}^{(\mathrm{t})}\right) \\ \vec{\Phi}^{(\mathrm{v})} &= L^{(\mathrm{vv})}\vec{K}^{(\mathrm{v})} \\ \Phi^{(\mathrm{t})} &= L^{(\mathrm{ts})}K^{(\mathrm{s})}\mathbf{I} + L^{(\mathrm{tt})}_{(1)}\left(\mathbf{I}:\mathbf{K}^{(\mathrm{t})}\right)\mathbf{I} + L^{(\mathrm{tt})}_{(2)}\mathbf{K}^{(\mathrm{t})\mathrm{S}} + L^{(\mathrm{tt})}_{(3)}\mathbf{K}^{(\mathrm{t})\mathrm{A}} \end{aligned}} \qquad (3.43)$$

The tensorial phenomenological equation can be regrouped as a sum of three terms. A second-order tensor can be split into a symmetric and an antisymmetric part, while the symmetric part can be split further into a deviator and a spherically symmetric part. The deviator of a tensor $\mathbf{t}$ is defined by

$$\overset{\circ}{\mathbf{t}} = \mathbf{t} - \tfrac{1}{3}\,(\mathbf{I} : \mathbf{t})\,\mathbf{I} \qquad \text{with} \qquad \mathbf{I} : \overset{\circ}{\mathbf{t}} = 0 \tag{3.44}$$

This means that $(3.43)_3$ can be rewritten as a sum of three terms

$$\begin{aligned} \Phi^{(t)\mathrm{A}} &= L^{(\mathrm{tt})}_{(3)} \mathbf{K}^{(t)\mathrm{A}} \\ \overset{\circ}{\Phi}{}^{(\mathrm{t})} &= L^{(\mathrm{tt})}_{(2)} \overset{\circ}{\mathbf{K}}{}^{(\mathrm{t})} \\ \tfrac{1}{3}\left(\mathbf{I} : \Phi^{(\mathrm{t})}\right) &= L^{(\mathrm{ts})} K^{(\mathrm{s})} + L^{(\mathrm{tt})}_{(1)} \left(\mathbf{I} : \mathbf{K}^{(\mathrm{t})}\right) + \tfrac{1}{3} L^{(\mathrm{tt})}_{(2)} \left(\mathbf{I} : \mathbf{K}^{(\mathrm{t})}\right) \end{aligned} \tag{3.45}$$

In addition, to the antisymmetric part of a second-order tensor a vector—the 'cross' of the tensor—can be assigned, such that

$$\mathbf{t}^{\mathrm{A}} \cdot \vec{a} = \vec{t}^{\,\mathrm{X}} \times \vec{a} \tag{3.46}$$

Isotropic materials are sometimes called amorphous. From the previous discussion, it follows that these materials have the simplest constitutive equations.

Materials having one preferred direction. It is not necessary for the material isomorphism of continuous bodies to satisfy the requirements of the group theory of crystal classes. For example, the material may be laminated or have a structure that consists of fibers aligned in one direction. It is also possible that in a continuous body one preferred direction is produced by mechanical processes, like stamping, stretching or rolling, or that an external magnetic field $\vec{B}_\circ$ introduces a preferred direction. In planes perpendicular to the preferred direction the material is then isotropic. This type of symmetry is called *transverse isotropy*.

An externally applied magnetic field does not always cause transverse isotropy. To examine to what extent this is the case, a natural time $t_{(n)}$ is assigned to the material. An electrically charged particle with the molar mass m and the molar electrical charge $\bar{q}$, has the tendency to execute a rotating motion around $\vec{B}_\circ$ with a time of rotation which is of the order of 1/(the cyclotron frequency)* or $\left((\bar{q}/m)B_\circ\right)^{-1}$. It is to be expected that $\vec{B}_\circ$ does *not* influence the behavior of the material if during the natural time the charged

---

* Cyclotron or Larmor-frequency $\omega_L = (\bar{q}/m)B$. See for instance: Brown S.C. and Ingraham J.C. 1967. Conduction of electricity in gases, in *Handbook of Physics*, 2nd ed., eds. Condon E.U. and Odishaw H. McGraw-Hill Book Co., New York, p. **4**–174.

particles can not even execute a fraction of the revolution around $\vec{B}_\circ$, that is if

$$t_{(n)} \ll \left(\frac{\bar{q}}{m} B_\circ\right)^{-1}$$

or

$$t_{(n)} \frac{\bar{q}}{m} B_\circ \ll 1 \tag{3.47}$$

For viscous fluids the natural time* can be estimated by

$$t_{(n)} \approx \frac{\nu}{a^2} \tag{3.48}$$

in which ν denotes the kinematic viscosity and a the velocity of sound. Combination of (3.47) and (3.48) results in

$$\nu \frac{\bar{q} B_\circ}{m a^2} \ll 1 \tag{3.49}$$

(Discuss whether this means that for incompressible fluids ($a = \infty$) this condition is always satisfied.) If (3.49) is *not* satisfied, then the applied magnetic field influences the behavior of the material and the principle attributed to Curie does not apply.

Symmetry principle for materials having one preferred direction. For bodies with a preferred direction the phenomenological tensors of even order depend not only on the unit tensor, the permutation tensor (or alternating tensor) as found from the symmetry principle for materials having a symmetry center, but in addition on a vector denoting the preferred direction.

PROOF: Suppose that the preferred direction is produced by an external magnetic field $\vec{B}_\circ$. The phenomenological coefficients has to remain invariant

---

* For a dilute gas $t_{(n)}$ is assumed to be equal to the mean collision time t_{coll}

$$t_{(n)} = t_{\text{coll}} \approx \frac{\Lambda}{\langle C_{\text{mol}} \rangle} = \Lambda \sqrt{\frac{\pi}{8} \frac{m}{k_B T}}$$

in which Λ is the mean-free-path length, and $\langle C_{\text{mol}} \rangle$ the mean peculiar velocity of the molecules. According to the kinetic theory we have

$$\nu \approx \tfrac{1}{2} \Lambda \langle C_{\text{mol}} \rangle \qquad \text{and} \qquad a^2 = \gamma \frac{p}{\rho} = \frac{\pi}{8} \gamma \langle C_{\text{mol}} \rangle^2$$

where γ denotes the ratio of specific heats, and a the velocity of sound. This yields for the natural time

$$t_{(n)} \approx \frac{\pi}{4} \gamma \frac{\nu}{a^2} = \mathcal{O}\left(\frac{\nu}{a^2}\right)$$

by performing the symmetry-operation (3.27), while the direction of $\vec{B}_\circ$ is also unchanged with respect to the material, so that

$$\vec{B}'_\circ = \mathbf{Q} \cdot \vec{B}_\circ = \vec{B}_\circ \cdot \mathbf{Q}^{\mathrm{T}} \tag{3.50}$$

First suppose that the phenomenological coefficient is a scalar L, then

$$L'\left(\vec{B}'_\circ\right) = L\left(\vec{B}_\circ\right)$$

or

$$L'\left(\vec{B}'_\circ\right) = L\left(\mathbf{Q} \cdot \vec{B}_\circ\right) = L\left(\vec{B}_\circ\right)$$

for all $\mathbf{Q}$ of the transverse-isotropy group. This means that L can depend only on the scalar invariant $B_\circ^2 \equiv \vec{B}_\circ \cdot \vec{B}_\circ$.

Suppose now that the phenomenological coefficient is a vector $\vec{L}$. Then for every symmetry operation

$$\vec{L}'\left(\vec{B}'_\circ\right) = \mathbf{Q} \cdot \vec{L}\left(\mathbf{Q} \cdot \vec{B}_\circ\right) = \vec{L}\left(\vec{B}_\circ\right) \tag{3.51}$$

To discuss the consequences, align the x_3-axis with the vector $\vec{B}$, such that $B_1 = 0$, $B_2 = 0$ and $B_3 = B$ (the subscript $\circ$ is temporarily dropped for brevity). For all rotations around the x_3-axis we have

$$\mathbf{Q} \cdot \vec{B} = \vec{B}$$

so that for this set of rotations

$$\mathbf{Q} \cdot \vec{L}\left(\vec{B}\right) = \vec{L}\left(\vec{B}\right) \tag{3.52}$$

For a rotation around the x_3-axis over an angle φ we have

$$\|\mathbf{Q}\| = \begin{pmatrix} \cos\varphi & -\sin\varphi & 0 \\ \sin\varphi & \cos\varphi & 0 \\ 0 & 0 & 1 \end{pmatrix} \tag{3.53}$$

yielding for (3.52)

$$\begin{aligned} L_1 &= L_1 \cos\varphi - L_2 \sin\varphi, \\ L_2 &= L_1 \sin\varphi + L_2 \cos\varphi, \\ L_3 &= L_3 \end{aligned}$$

Since this applies to all φ, it follows that $L_1 = 0$ and $L_2 = 0$. For a rotation of 180° around the x_1-axis

$$\|\mathbf{Q}\| = \begin{pmatrix} 1 & 0 & 0 \\ 0 & -1 & 0 \\ 0 & 0 & -1 \end{pmatrix} \tag{3.54}$$

Using this rotation, the condition (3.51) for $\vec{L} = L_3\vec{e}_3$, and $\vec{B} = B\vec{e}_3$ results in

$$L_3\left(\vec{B}\right) = -L_3\left(-\vec{B}\right)$$

showing that $\vec{L}$ is an odd function of $\vec{B}$. For a vectorial phenomenological coefficient it is therefore found that

$$\vec{L}\left(\vec{B}\right) = L\left(B^2\right)\vec{B} \tag{3.55}$$

For $\vec{B} = \vec{0}$, the principle attributed to Curie is again satisfied.

Finally, suppose that the phenomenological coefficient is a second-order tensor, say $\mathbf{L}$. For a second-order tensor a symmetry operation requires

$$\mathbf{L}'\left(\vec{B}'\right) = \mathbf{Q}\cdot\mathbf{L}\left(\mathbf{Q}\cdot\vec{B}\right)\cdot\mathbf{Q}^{\mathrm{T}} = \mathbf{L}\left(\vec{B}\right) \tag{3.56}$$

Align the x_3-axis again along the vector $\vec{B} = B\vec{e}_3$. For each rotation about the x_3-axis it is now found that

$$\mathbf{L}\left(\vec{B}\right) = \mathbf{Q}\cdot\mathbf{L}\left(\vec{B}\right)\cdot\mathbf{Q}^{\mathrm{T}} \tag{3.57}$$

For a 90° rotation about the x_3-axis, we have $x_1 \to -x_2$, $x_2 \to x_1$, $x_3 \to x_3$, so that (3.56) results in

$$\begin{pmatrix} L_{11} & L_{12} & L_{13} \\ L_{21} & L_{22} & L_{23} \\ L_{31} & L_{32} & L_{33} \end{pmatrix} = \begin{pmatrix} L_{22} & -L_{21} & -L_{23} \\ -L_{12} & L_{11} & L_{13} \\ -L_{32} & L_{31} & L_{33} \end{pmatrix}$$

from which it is deduced that

$$\begin{array}{lll} L_{11} = L_{22} & L_{12} = -L_{21} & L_{13} = -L_{23} \\ L_{23} = L_{13} & L_{31} = -L_{32} & L_{32} = L_{31} \end{array}$$

These relations are satisfied only if

$$L_{13} = L_{23} = 0 \qquad \text{and} \qquad L_{31} = L_{32} = 0$$

so that

$$\mathbf{L} = L_{11}\mathbf{I} + (L_{33} - L_{11})\, \vec{e}_3\vec{e}_3 + L_{12}\mathbf{e} \cdot \vec{e}_3, \tag{3.58}$$

where $\mathbf{e}$ denotes the third-order alternating tensor.

$\mathbf{Q}$ is given by (3.54) for a rotation of 180° about the x_1-axis. Condition (3.56) leads for this rotation to the conclusion that

$$\mathbf{L}\left(\vec{B}\right) = \mathbf{Q} \cdot \mathbf{L}\left(-\vec{B}\right) \cdot \mathbf{Q}^{\mathrm{T}}$$

It now follows that L_{11}, L_{22}, and L_{33} in (3.58) are even functions of $\vec{B}$, while L_{12} is an odd function of $\vec{B}$. This means that (3.58) can be written as (note that $\vec{B} = B\vec{e}_3$)

$$\mathbf{L} = L_{\circ}\left(B^2\right)\mathbf{I} + L_{\pi}\left(B^2\right)\vec{B}\vec{B} - L_{\tau}\left(B^2\right)\mathbf{e} \cdot \vec{B} \tag{3.59}$$

The last result shows clearly that $\mathbf{L}$ is constructed using the elements $\mathbf{I}$, the permutation tensor $\mathbf{e}$, and $\vec{B}$. These elements are then the only available elements to construct the phenomenological coefficients with the condition that for $\vec{B} = \vec{0}$ these coefficients reduce to the isotropic coefficients. The phenomenological coefficients are therefore expressed in irreducible combinations of

$$\mathbf{I} \qquad \mathbf{e} \qquad \vec{B} \tag{3.60}$$

such that for $\vec{B} = \vec{0}$ the isotropic coefficients result. Analogously, for the third-order phenomenological tensor it is found that

$$\begin{aligned} L_{ijk} = L_{(1)}\left(B^2\right)\delta_{ij}B_k + L_{(2)}\left(B^2\right)\delta_{ik}B_j + L_{(3)}\left(B^2\right)\delta_{jk}B_i \\ + L_{(4)}\left(B^2\right)B^2 e_{ijk} + L_{(5)}\left(B^2\right)B_iB_jB_k \end{aligned} \tag{3.61}$$

Note that $\mathbf{e} : (\vec{B}\vec{B}) = \vec{0}$, hence, $\mathbf{e}$ combines only with $\vec{B}$.

Finally, a fourth-order phenomenological tensor with the components L_{ijkm} may be constructed from

$$\left.\begin{array}{lllll} \delta_{ij}\,\delta_{km} & e_{ijk}\,B_m & e_{ijp}\,B_p\,\delta_{km} & \delta_{ij}\,B_kB_m & B_i\,B_j\,B_k\,B_m \\ \delta_{ik}\,\delta_{jm} & e_{ijm}\,B_k & e_{ikp}\,B_p\,\delta_{jm} & \delta_{ik}\,B_jB_m & \\ \delta_{im}\,\delta_{jk} & e_{imk}\,B_j & e_{imp}\,B_p\,\delta_{kj} & \delta_{im}\,B_jB_k & \\ & e_{mjk}\,B_i & e_{jkp}\,B_p\,\delta_{im} & \delta_{jk}\,B_iB_m & \\ & & e_{jmp}\,B_p\,\delta_{ik} & \delta_{jm}\,B_iB_k & \\ & & e_{kmp}\,B_p\,\delta_{ij} & \delta_{km}\,B_iB_j & \end{array}\right\} \tag{3.62}$$

In the thermodynamics of irreversible processes phenomenological tensors of higher order are rarely needed. Tensors of order higher than the fourth-order become rapidly more complicated. Simplifications can be found if one realizes that the thermodynamics of irreversible processes involves only small deviations from the equilibrium state. In a system with charged particles, equilibrium can be found if an external electro-magnetic field is absent (a system is difficult to isolate in the presence of an external electro-magnetic field!). Small deviations of the equilibrium arise therefore only if $|\vec{B}_\circ|$ is small enough that quadratic terms in $|\vec{B}_\circ|$ can be neglected. On first examination, the expansions can then be restricted to the first order in $|\vec{B}_\circ|$, and considerable simplifications in (3.59), (3.61), and (3.62) are achieved!

3.3. SUMMARY OF THE STRUCTURE OF THE TIP

In table 3.1 a summary of the structure of the thermodynamics of irreversible processes and its applications is given in a scheme similar to the Nassi–Shneiderman* scheme for structured computer programming (the 'if A then B', and 'do A while B' pictures are used). Every linear phenomenon can be analyzed with this scheme. This means that by using the techniques of thermodynamics of irreversible processes a uniform method is possible for all natural phenomena that can be linearly modeled. Obviously, this can be achieved only if the contents of each box of the scheme are well known and well understood. With the application of the principle of equipresence effects which are not well known can be calculated. This is particularly true for the cross-effects. . Of course, the results have to be confirmed by empirical facts.

Since the range of validity of the generalized axioms, on which the thermodynamics of the irreversible processes is based, is not always well known, it is necessary to verify new-found effects experimentally. These experiments may be called fundamental measurements if the experimental arrangement is sufficiently simple that the responses also can be calculated; then the experimental results can unambiguously be compared with the theoretical calculations. Also the phenomenological coefficients appearing in the theoretical results cannot be calculated theoretically in a macroscopic theory, and have to be measured by fundamental experiments as well.

If the results of the fundamental experiments confirm the theoretical calculations, the theory can be applied to practical problems that often involve complicated geometrical configurations. Due to the complications encountered in practice, and the influence of the historical developments, the continuum theory has been segmented into a variety of special fields, such as elasticity, fluid mechanics, heat transfer, mechanics, and others. In these individual disciplines, the attention is concentrated on a single aspect of the behavior of materials. More detailed investigations often show that a phenomenon cannot be explained based on a single discipline, but that more disciplines have to be considered. Particularly for the multi-disciplinary approach, the

* Nassi I. and Shneiderman B. 1973. Flow chart techniques for structured programming, *Sigplan Notices*, 8(8): 12–26.

Table 3.1. Structure of the thermodynamics of irreversible processes

System		
Propose an idealized model of the system.		
Repeat	Improve the model.	
	TIP-Axioma I: Classical thermodynamics applies.	
	TIP-Axiom II: Local and instanteneous equilibrium applies. Gibbs' specific fundamental equation ($du = T\,ds - p\,dv$) holds.	
		TIP-Axiom III: Balance equation for the mass, momentum, moment of momentum, energy and electric charge. Maxwell's e.m.-equations.
		Axiom of internal actions. **Flux theorem**.
		Local differential balance equations. Define the **fluxes**.
		TIP-Axiom IV: Calculate the local **entropy** balance equation using the Gibbs equation and the energy balance. Define the **entropy flux**.
		Calculate the **entropy production** ($\pi_{(s)} \geq 0$) and define the **forces**.
	TIP-Axiom V: Linear **phenomenological** equation between the process forces and the process fluxes. **Equipresence**.	
	TIP-Axiom VI: Onsager **C**asimir **R**eciprocal **R**elations (**O.C.R.R.**).	
	General Axioms: Galilean-invariance, Coordinate invariance, Dimensional invariance, Material Isomorphism.	
	Linear constitutive equations for diffusion, heat conduction, relaxation, electrical conduction, elastic after-effects, membrane processes, and so on.	
	Theoretical calculations of the processes.	
	Compare the calculations with **fundamental experiments**.	
Calculation with the proposed model equivalent to the results of the fundamental experiments?		
Applications: elasto mechanics, fluid dynamics, diffusion, linear rheology, and so on.		

thermodynamics of irreversible processes is a very appropriate starting point.

Finally, it is again remarked that the thermodynamics of irreversible processes is a linear theory, in which the phenomenological coefficients may be functions of the local and instantaneous thermodynamic state. This is already a quasi-linear extrapolation of the thermodynamics of the irreversible processes, which must be verified experimentally, since in a nonlinear extrapolation of a linear theory besides quasi-linear effects other essential nonlinear effects may also be important.

3.3.1. Anisotropic heat conduction in crystals. The development of the thermodynamics of irreversible processes probably started with the pioneering work of De Donder[*] in the field of chemical reactions and with Onsager[†], who wanted to prove the symmetry of the heat conduction tensor. Soret[‡], P. Curie[§] and Voigt[¶] did experiments with gypsum, erythrite, dolomite, and apatite, and their experiments did not contradict the assumption that the heat conduction tensor is symmetric. The symmetry of the heat conduction tensor does not follow from the symmetry properties of the crystal considered. Truesdell[‖] remarked that these experiments have not been repeated, and the symmetry of the heat conduction tensor has not been substantiated by experiments for the other eleven types of crystals or for transverse isotropic materials. Also a theoretical proof of the symmetry property of the heat conduction tensor could not be given. Only Onsager is thought to have given the proof based on fluctuation theory, the microscopic reversibility and a regression axiom that can be traced back to the theory of Einstein[#] on Brownian motion.

The heat conduction in a solid is not only of great historical importance,

---

[*] De Donder Th. 1920. Transformations physiques et chimiques des systèmes de Gibbs. Académie Royal de Belgique. *Bulletins de la Classe des Sciences*, 5(6): 315–328.

[†] Onsager L. 1931. Reciprocal relations in irreversible processes I. *Physical Review*, 37: 405–426; Onsager L. 1931. Reciprocal relations in irreversible processes II. *Physical Review*, 38: 2265–2279.

[‡] Soret Ch. 1893. De la conductibilité calorifique dans les cristaux. *Journal de Physique et le Radium*, 2(3): 241–259; Soret Ch. 1893. Sur l'étude expérimentale des coefficients rationnels de conductibilité thermique. *Journal de Physique et le Radium*, 2(3): 355–357; Soret Ch. 1893. *Société Physique et d'Histoire Naturelle de Genève, Procès-verbaux*, 29: 322–323; Soret Ch. 1894. *Société Physique et d'Histoire Naturelle de Genève, Procès-verbaux*, 32: 631–633.

[§] Curie P. 1893. À propos des éléments de cristallographie de M. Ch. Soret. *Société de Physique et d'Histoire Naturelle de Genève, Archives*, 29: 237–354.

[¶] Voigt W. 1903. Fragen der Kristallphysik, I. *Nachrichten der Gesellschaft der Wissenschaften zu Göttingen*, 3: 87–89.

[‖] Truesdell C. 1984. The Onsager relations, in *Rational Thermodynamics*, Springer-Verlag, New York, p. 369.

[#] Einstein A. 1906. Zur Theorie der Brownsche Bewegung. *Annalen der Physik*, 19: 289–306.

but it is also a relatively simple example of the general method used in the thermodynamics of irreversible processes.

System and model. The system consists of a piece of solid material, for which heat conduction is the main point of interest. This piece of matter is therefore idealized as follows:

(1) The body is a solid, such that no convection takes place in the material.
(2) The material of the body is a chemically simple material, such that no diffusion is possible.
(3) The body is a closed system.
(4) The external pressure on the body is constant.
(5) The thermal expansion of the body can be neglected.
(6) The body cannot be polarized, so that there is no electrical and/or magnetic polarization of the material of the body in the presence of external electromagnetic fields.
(7) No internal processes occur in the body, so that no relaxation phenomena take place in the body.
(8) The equilibrium state of the body is thermodynamically stable, so that the body returns to its previous equilibrium state as soon as any imposed disturbance is removed; furthermore the body has no tendency to change phases in the neighborhood of the equilibrium state.
(9) In the body a transport of heat can occur because of inhomogeneous temperature distributions.

The assumptions for the model are summarized in this list. Based on these starting points the theory of heat conduction has to be developed.

Classical thermodynamics of the model. In the first place the equilibrium properties of the system have to be known. For this model the equilibrium properties can be determined by quasistatic calorimetry. To this end, the body is brought into a thermostat and is left in the thermostat until thermal equilibrium between the body and the thermostat has been achieved. If this is realized with sufficient accuracy, then the temperature is almost homogeneously distributed over the body, so that we may speak of the temperature T of the body. Subsequently, a small quantity of heat $đQ$ is supplied quasistatically to the body. If the equilibrium is thermodynamically stable, then a small temperature increase dT is accompanied by the small supply of heat. The heat capacity C of the body is therefore measurable

$$C(T) = đQ/dT \tag{3.63}$$

Since the cubic expansion coefficient is assumed to be zero, the heat capacity depends only on the temperature, the size of the body and the nature of the material, but not on the pressure, so that there is no difference between the heat capacities at constant pressure or at constant volume.

As the body is a closed system, the function of state

$$U(T) = \int_{T_0}^{T} C(T)\, dT \tag{3.64}$$

can be defined, in which T_0 is a conveniently chosen reference temperature. The function of state U is a measure of the energy that is stored by the supply of heat in the body and is called the *internal energy*.

From the second law of classical thermodynamics, it follows that there must exist a function of state S, such that for a quasistatic supply of heat

$$S = \int_{T_0}^{T} dQ/T \tag{3.65}$$

This defines the entropy of the body, and the summary of (3.63)–(3.65) leads to

$$dU = C\, dT = T\, dS \tag{3.66}$$

showing that the thermodynamic fundamental equation of Gibbs is now $dU = T\, dS$.

Under the given circumstances the body has one independent variable, for example the temperature, while furthermore the volume V and the mass M of the body can vary. The quantities of state U, S, C, V and M are *extensive*. The temperature T is *intensive*.

The extensive quantities have to be defined per unit of mass (or per unit of volume, or per mol) for the application to a non-equilibrium situation, in which the thermodynamic quantities are not homogeneous any more. The following specific quantities therefore correspond to the previously mentioned extensive quantities

$$u = U/M \qquad s = S/M \qquad c = C/M \qquad v = V/M = 1/\rho \tag{3.67}$$

The extensive quantities related to the whole system are symbolized by capitals and the corresponding specific quantity by lowercase letters.

The mass M of the system is a constant for a closed system, so that (3.66) divided by M becomes

$$du = c\, dT = T\, ds \tag{3.68}$$

In this 'specific' formulation of the Gibbs fundamental equation, the complete thermodynamic knowledge of the system is summarized.

Suppose now that the temperature is not distributed homogeneously over the body. An example is a bar-shaped body, with the temperatures at the two end surfaces different, and for which the surface of the bar is isolated. Then, a one dimensional heat transport is to be expected, for which the temperature is no longer distributed homogeneously over the bar.

If the principle of local and instantaneous equilibrium holds, then the formulas and equations of classical thermodynamics apply, so that (3.68) remains valid, in which the differentials may be replaced by grad or by $\partial/\partial t$. Mathematically, the principle of local and instantaneous equilibrium can consequently be expressed as follows. In (3.68) it may be assumed that

$$d(\cdots) \quad \rightarrow \quad \text{grad}(\cdots) \tag{3.69a}$$

and the principle of instantaneous equilibrium means analogously

$$d(\cdots) \quad \rightarrow \quad \frac{\partial}{\partial t}(\cdots) \tag{3.69b}$$

These transitions from classical thermodynamics to the thermodynamics of the irreversible processes are essential.

Balance equation of energy. The ensuing step is to formulate the relevant non-equilibrium balance equations. The model here studied concerns only the balance of energy, in which the control volume is fixed with respect to the observer. The energy consists now only of internal energy, so that the total energy of the material in the control volume is

$$U_{(c)} = \iiint_{V_{(c)}} \rho u \, dV$$

From the assumptions of the model, it follows that the density ρ is a constant, and the differential of the internal energy is given by

$$\frac{d}{dt} U_{(c)} = \iiint_{V_{(c)}} \rho \frac{\partial u}{\partial t} \, dV \tag{3.70}$$

The matter outside the control volume $V_{(c)}$ exchanges heat with the matter inside $V_{(c)}$. To describe this internal heat exchange, the macroscopic theory has to postulate the 'axiom of the internal actions'. This axiom assigns to the heat passing per unit of time the surface element $d\vec{A}$ an operator that links the surface element $d\vec{A}$ to the surface action $q[d\vec{A}]$. There is no production of energy in this restricted model. The law of conservation of energy now becomes

$$\iiint_{V_{(c)}} \rho \frac{\partial u}{\partial t} \, dV = \iint_{A_{(c)}} q[d\vec{A}] \tag{3.71}$$

With the application of the 'flux action for internal actions' it follows that

$$q[d\vec{A}] = -\vec{n} \cdot \vec{\Phi}_{(q)} \, dA \tag{3.72}$$

in which $\vec{\Phi}_{(q)}$ is the heat flux vector. With the substitution of (3.72) and with the application of the Gauss integral transformation yield

$$\iint_{A_{(c)}} q[d\vec{A}] = -\iint_{A_{(c)}} \vec{n}\cdot\vec{\Phi}_{(q)}\,dA = -\iiint_{V_{(c)}} \mathrm{div}\,\vec{\Phi}_{(q)}\,dV \tag{3.73}$$

Substitution of (3.73) into (3.71) results in an integral balance equation that applies to all—also infinitesimally small—control volumes. Therefore, the following differential balance equation is deduced

$$\rho\frac{\partial u}{\partial t} = -\mathrm{div}\,\vec{\Phi}_{(q)} \tag{3.74}$$

The local balance equation is obtained with this equation. This balance equation contains the new quantity $\vec{\Phi}_{(q)}$. This is not a thermodynamic quantity of state, but a thermodynamic *flux quantity* that is called the *heat flux.*

Balance equation for the entropy. From the Gibbs fundamental equation (3.68), and by using the principle of local and instantaneous equilibrium, it follows that

$$\rho\frac{\partial u}{\partial t} = \rho T\frac{\partial s}{\partial t}$$

or with the substitution of the balance equation of energy (3.74)

$$\rho\frac{\partial s}{\partial t} = -\frac{1}{T}\mathrm{div}\,\vec{\Phi}_{(q)} = -\mathrm{div}\left(\frac{\vec{\Phi}_{(q)}}{T}\right) + \vec{\Phi}_{(q)}\cdot\mathrm{grad}\,T^{-1} \tag{3.75}$$

The left-hand side member and the right-hand side member of these last equations together can be considered as a balance of entropy

$$\frac{\partial}{\partial t}(\rho s) = -\mathrm{div}\,\vec{\Phi}_{(s)} + \pi_{(s)} \tag{3.76}$$

in which $\vec{\Phi}_{(s)}$ is the entropy flux, $\pi_{(s)}$ the entropy produced per unit of time and volume. The density ρ is now a constant.

The second law of classical thermodynamics is interpreted in the condition

$$\pi_{(s)} \geq 0 \tag{3.77}$$

in which the equality sign applies for equilibrium and the inequality sign for non-equilibrium.

The entropy flux and the rate of entropy production are specified according to (3.75) by

$$\vec{\Phi}_{(s)} = \vec{\Phi}_{(q)}/T \qquad \text{and} \qquad \pi_{(s)} = -T^{-2}\vec{\Phi}_{(q)}\cdot\mathrm{grad}\,T \geq 0 \tag{3.78}$$

The dissipation per unit of time and volume is therefore

$$\mathfrak{D} = T\pi_{(\mathrm{s})} = -T^{-1}\vec{\Phi}_{(\mathrm{q})} \cdot \operatorname{grad} T \geq 0 \tag{3.79}$$

The product $T\pi_{(\mathrm{s})}$ is called the dissipation function and is equivalent to the quantity of heat that is supplied irreversibly to the system. For a pure viscous flow the dissipation function is identical to the dissipation of the mechanical energy produced by the viscosity of the fluid.

In addition to the process flux $\vec{\Phi}_{(\mathrm{q})}$ just found, we now also obtain a non-equilibrium quantity $\operatorname{grad} T$. By analogy with the mechanical energy that exists of a product of force and velocity, the new process quantity $\operatorname{grad} T$ is regarded as the driving force of the process flux. Hence, the *process force* is specified by $-T^{-1}\operatorname{grad} T$.

Summarizing, the process fluxes are suggested by the balance equations, and the corresponding process forces from the bilinear form of the entropy production (or the dissipation).

Phenomenological equations. At equilibrium it is known that the entropy production $\pi_{(\mathrm{s})}$ is zero. From this fact and from $(3.78)_2$ it follows that in unconstrained equilibrium

$$\vec{\Phi}_{(\mathrm{q})} = \vec{0} \qquad \text{and} \qquad \operatorname{grad} T = \vec{0} \tag{3.80}$$

For small deviations of equilibrium state it is to be expected that the process fluxes depend linearly on the process forces. According to (3.80) this linear approximation results according to the thermodynamics of the irreversible processes in the following phenomenological equation

$$\vec{\Phi}_{(\mathrm{q})} = -T^{-1}\mathbf{L} \cdot \operatorname{grad} T \tag{3.81}$$

where $\mathbf{L}$ is the phenomenological tensor for the heat conduction that is usually related to the thermal conductivity tensor by

$$\boldsymbol{\lambda} = T^{-1}\mathbf{L} \tag{3.82}$$

so that the phenomenological equation becomes

$$\vec{\Phi}_{(\mathrm{q})} = -\boldsymbol{\lambda} \cdot \operatorname{grad} T \tag{3.83}$$

This might be called the *heat conduction law of Fourier*, which is conceived as a linear constitutive equation. From (3.79) another important property of the thermal conductivity tensor follows

$$(\operatorname{grad} T) \cdot \big(\boldsymbol{\lambda} \cdot (\operatorname{grad} T)\big) \equiv \boldsymbol{\lambda} : [(\operatorname{grad} T)(\operatorname{grad} T)] \geq 0 \tag{3.84}$$

To illustrate the consequences of this inequality, the conductivity tensor is split into a symmetrical and an antisymmetrical part, such that $\boldsymbol{\lambda}$ can be written as

$$\boldsymbol{\lambda} = \boldsymbol{\lambda}^{\mathrm{S}} + \boldsymbol{\lambda}^{\mathrm{A}} \tag{3.85}$$

The components of the antisymmetric tensor $\boldsymbol{\lambda}^{\mathrm{A}}$ satisfy $\lambda_{ij}^{\mathrm{A}} = -\lambda_{ji}^{\mathrm{A}}$. Now for every real valued vector $\vec{b}$

$$\boldsymbol{\lambda}^{\mathrm{A}} : (\vec{b}\,\vec{b}) \equiv \lambda_{ji}^{\mathrm{A}} b_i b_j = \lambda_{ji}^{\mathrm{A}} b_j b_i = -\lambda_{ij}^{\mathrm{A}} b_j b_i \equiv -\boldsymbol{\lambda}^{\mathrm{A}} : (\vec{b}\,\vec{b})$$

so that

$$\boldsymbol{\lambda}^{\mathrm{A}} : (\vec{b}\,\vec{b}) = 0 \tag{3.86}$$

The inequality (3.84) is therefore concerned only with the symmetrical part of the conductivity tensor

$$\boldsymbol{\lambda}^{\mathrm{S}} : (\vec{b}\,\vec{b}) \geq 0 \tag{3.87}$$

with $\vec{b} = \operatorname{grad} T$. If the equilibrium is stable, then the equality sign holds only for $\vec{b} = \vec{0}$. Hence, the symmetrical part of the conductivity tensor is positive definite. A necessary and sufficient condition for the positive definiteness is that all determinants of all submatrices of $||\boldsymbol{\lambda}||$ are positive.

Especially in applications it is assumed that $\boldsymbol{\lambda}$ is still a function of the temperature. In a linear theory, like the TIP, this is strickly speaking not allowed. However, (3.83) might be interpreted as the result of an expansion of a functional relation between $\vec{\Phi}_{(\mathrm{q})}$ and $\operatorname{grad} T$

$$\vec{\Phi}_{(\mathrm{q})} = \vec{\Phi}_{(\mathrm{q})}(T, \operatorname{grad} T) \tag{3.88}$$

In true equilibrium $T = T^{+}$ and $\operatorname{grad} T = \vec{0}$. Expansion of (3.88) around the equilibrium state leads to

$$\vec{\Phi}_{(\mathrm{q})} = -\boldsymbol{\lambda}(T^{+}) \cdot \operatorname{grad} T \tag{3.89}$$

By assuming that $\boldsymbol{\lambda}$ now depends on T, the quasilinear extrapolation of this expansion becomes

$$\vec{\Phi}_{(\mathrm{q})} = -\boldsymbol{\lambda}(T) \cdot \operatorname{grad} T \tag{3.90}$$

It is open to discussion to what extent this is acceptable, since in addition to the nonlinear effects in (3.90) other nonlinear effects may be important. The quasi-linear extrapolation displayed in (3.90) is to be seen as a useful inductive extension of the TIP, but it needs experimental verification.

The Fourier law (3.83) implies that the relation between $\vec{\Phi}_{(\mathrm{q})}$ and $\operatorname{grad} T$ is instantaneous without any memory-effects (if there are memory effects, then this relation is a functional). Possible memory effects are introduced in the thermodynamics of irreversible processes by internal relaxation processes (incomplete systems) that are discussed in Chapter 7. The phenomenological equations are on first examination instantaneous in the TIP, although for incomplete systems with internal variables this results—by elimination of the internal variables—in constitutive equations that involve memory effects.

Onsager Casimir reciprocal relations. Suppose that a heat conduction problem is described with respect to a relative frame of reference that rotates with respect to an inertial system with the angular velocity $\vec{\Omega}$. An external magnetic field $\vec{B}_\circ$ exists in this reference system (through the electromagnetic field equations of Maxwell an electrical field might be connected with the magnetic field). It is to be expected that $\boldsymbol{\lambda}$ is a function of $\vec{\Omega}$ and $\vec{B}_\circ$. Then the Onsager Casimir reciprocal relations (O.C.R.R.) are

$$\boldsymbol{\lambda}(T, \vec{\Omega}, \vec{B}_\circ) = \boldsymbol{\lambda}^{\mathrm{T}}(T, -\vec{\Omega}, -\vec{B}_\circ) \tag{3.91}$$

The demonstration of these essential relations is given by using the statistical fluctuation theory. The reciprocal relations (3.91) are with respect to an inertial system

$$\boldsymbol{\lambda}(T, \vec{B}_\circ) = \boldsymbol{\lambda}^{\mathrm{T}}(T, -\vec{B}_\circ) \tag{3.92}$$

If (3.85) is used, the Fourier law can be written as

$$\vec{\Phi}_{(\mathrm{q})} = -\boldsymbol{\lambda}^{\mathrm{S}}(T, \vec{B}_\circ) \cdot \operatorname{grad} T - \boldsymbol{\lambda}^{\mathrm{A}}(T, \vec{B}_\circ) \cdot \operatorname{grad} T \tag{3.93}$$

To an antisymmetric tensor $\boldsymbol{\lambda}^{\mathrm{A}}$ an axial vector $\vec{\lambda}^{\mathrm{X}}$ may be assigned, such that

$$\boldsymbol{\lambda}^{\mathrm{A}} \cdot \vec{b} = \vec{\lambda}^{\mathrm{X}} \times \vec{b} \tag{3.94}$$

and the substitution of this definition into (3.93) yields

$$\vec{\Phi}_{(\mathrm{q})} = -\boldsymbol{\lambda}^{\mathrm{S}}(T, \vec{B}_\circ) \cdot \operatorname{grad} T - \vec{\lambda}^{\mathrm{X}}(T, \vec{B}_\circ) \times \operatorname{grad} T \tag{3.95}$$

The Onsager Casimir reciprocal relations (3.92) yield for the symmetric and antisymmetric part of the conductivity tensor

$$\begin{aligned} \boldsymbol{\lambda}^{\mathrm{S}}(T, \vec{B}_\circ) &= \boldsymbol{\lambda}^{\mathrm{S}}(T, -\vec{B}_\circ) \\ \vec{\lambda}^{\mathrm{X}}(T, \vec{B}_\circ) &= -\vec{\lambda}^{\mathrm{X}}(T, -\vec{B}_\circ) \end{aligned} \tag{3.96}$$

$\boldsymbol{\lambda}^{\mathrm{S}}$ is therefore an even function of $\vec{B}_\circ$, while $\vec{\lambda}^{\mathrm{X}}$ is an odd function of $\vec{B}_\circ$.

It follows for $\vec{B}_\circ = 0$ that $\vec{\lambda}^{\mathrm{X}} = \vec{0}$, confirming that the thermal conductivity tensor is symmetric, which has not been contradicted by the experiments of

Soret and Voigt*. Apparently, in the absence of an external magnetic field,

$$\vec{\lambda}^{\mathrm{X}} = b(T, \vec{B}_\circ)\, \vec{B}_\circ \tag{3.97}$$

in which b is an even function of $\vec{B}_\circ$.

Substitution of (3.97) into (3.95) yields

$$\vec{\Phi}_{(\mathrm{q})} = -\boldsymbol{\lambda}^{\mathrm{S}} \cdot \operatorname{grad} T - b\, \vec{B}_\circ \times \operatorname{grad} T \tag{3.98}$$

The magnetic field produces a component of the heat flux that is perpendicular to grad T. This is called the *effect of Righi-Leduc* (1887), and $\vec{\lambda}^{\mathrm{X}}$ is the vector of Righi-Leduc. This effect is to be expected if heat is entirely or partly transported by free electrons or ions. The electrically charged particles circulate around the magnetic field lines and this motion produces the cross-effect of Righi-Leduc†.

Thermal energy equation of Fourier. Substitution of (3.83) into (3.74) gives

$$\rho \frac{\partial u}{\partial t} = \operatorname{div}(\boldsymbol{\lambda} \cdot \operatorname{grad} T)$$

or with the substitution of $(3.68)_1$

$$\rho c \frac{\partial T}{\partial t} = \operatorname{div}(\boldsymbol{\lambda} \cdot \operatorname{grad} T) \tag{3.99}$$

This equation can also be traced back to Fourier.

The input of the TIP considerations about heat conduction in crystals has been completed with these results. Subsequent simplifications concern aspects found in every constitutive theory, for instance material isomorphism. If it is assumed that the material is isotropic, then $\boldsymbol{\lambda} = \lambda \mathbf{1}$ and (3.99) reduces to the standard heat conduction equation $\rho c \partial T / \partial t = \lambda \nabla^2 T$.

3.3.2. Membrane permeability. The selective permeability of solvents through membranes is an illustrative example of the application of the

* Soret Ch. 1893. De la conductibilité calorifique dans les cristaux. *Journal de Physique et le Radium*, 2(3): 241–259; Soret Ch. 1893. Sur l'étude expérimentale des coefficients rationnels de conductibilité thermique. *Journal de Physique et le Radium*, 2(3): 355–357; Soret Ch. 1893. *Société Physique et d'Histoire Naturelle de Genève, Procès-verbaux.* 29: 322–323; Soret Ch. 1894. *Société Physique et d'Histoire Naturelle de Genève, Procès-verbaux*, 32: 631–633; Voigt W. 1903. Fragen der Kristallphysik, I. *Nachrichten der Gesellschaft der Wissenschaften zu Göttingen*, 3: 87–89.

† Hall E.H. 1938. The four magnetic transverse effects in copper and their changes with temperature: new measurements. *Proceedings of the American Academy of Arts and Sciences*, 72: 301–325, shows for copper at 24.9 °C a cross-effect of 2.698 $10^{-3}\,\lambda$. The effect is small and is difficult to measure, but demonstrable.

phenomenological equations and of the application of the Onsager Casimir reciprocal relations. Membrane permeability is important in the life of cells and tissues, and determines the effectiveness of the operation of artificial filters, like those used in haemodiafiltration and in desalination of sea water by electrolysis. Membrane processes can be characterized as discontinuous systems. On both sides of the membrane there may be different concentrations of substances to which the membrane may or may not be permeable. In both cases, the solutes may contribute to the differences in osmotic pressure* $\Delta\pi$ across the membrane. The flux is now the volume flow† per unit area $\vec{\Phi}_D$, resulting from the velocity difference of the solute relative to that of the solvent, and consequently is similar to a diffusion flux. In addition, a constant pressure difference ΔP can be maintained over the membrane, to which a volume flow $\vec{\Phi}_P$ is related. Using the thermodynamics of irreversible processes, the membrane transport processes will now be discussed.

With the introduction of these fluxes and forces the dissipation function (3.19) becomes

$$\mathfrak{D} = \vec{\Phi}_P\,\Delta P + \vec{\Phi}_D\,\Delta\pi \tag{3.100}$$

and consequently the phenomenological equations between the forces and the fluxes become

$$\begin{aligned}\vec{\Phi}_P &= L_P\,\Delta P + L_{PD}\,\Delta\pi\\ \vec{\Phi}_D &= L_{DP}\,\Delta P + L_D\,\Delta\pi\end{aligned} \tag{3.101}$$

for which the Onsager Casimir reciprocal relations assert that

$$L_{PD} = L_{DP} \tag{3.102}$$

The phenomenological equations (3.101) explain several phenomena found in membrane systems. For a semipermeable membrane, having on both sides solutions with the same concentration of the solute, $\Delta\pi = 0$. If a pressure difference $\Delta P \neq 0$ is maintained over the membrane, equation $(3.101)_1$ shows that there is a volume flow $\vec{\Phi}_P$ proportional to ΔP with the proportionality coefficient L_P. This coefficient is called the mechanical filtration coefficient of the membrane or the hydraulic membrane permeability. In addition to the volume flow it also follows from $(3.101)_2$ that a diffusional flow

$$\left(\vec{\Phi}_D\right)_{\Delta\pi=0} = L_{DP}\,\Delta P \tag{3.103}$$

* The osmotic pressure is the pressure difference that is needed to prevent the flow through a semipermeable membrane that separates the solvent and its solution, or solutions with different concentrations of solutes.

† Katchalsky A. and Curran P.F. 1975. *Nonequilibrium Thermodynamics in Biophysics.* Harvard University Press, Cambridge, MA, Chapter 10.

is produced, although $\Delta\pi = 0$. This phenomenon is called ultrafiltration, and the coefficient L_{DP} is a measure of the ultrafiltration characteristics of the membrane.

In a complementary experiment the pressure on both sides of the membrane is kept equal and the concentration of the solutes different, so that now $\Delta P = 0$ and $\Delta\pi \neq 0$. The osmotic pressure difference produces a diffusional flow, determined by the osmotic diffusional coefficient L_D and the osmotic pressure

$$\left(\vec{\Phi}_D\right)_{\Delta P=0} = L_D \, \Delta\pi \tag{3.104}$$

Again, the phenomenological equations (3.101) show that there is also a volume flow

$$\left(\vec{\Phi}_P\right)_{\Delta P=0} = L_{PD} \, \Delta\pi \tag{3.105}$$

associated with the osmotic pressure, although the hydrostatic pressure difference is kept at zero. This flow is called the osmotic flow, and L_{PD} is called the coefficient of the osmotic volume flow.

The Onsager Casimir reciprocal relations connect the phenomena of osmotic flow and ultrafiltration. From equations (3.102)–(3.105) it follows that

$$\left(\frac{\vec{\Phi}_P}{\Delta\pi}\right)_{\Delta P=0} = L_{PD} = L_{DP} = \left(\frac{\vec{\Phi}_D}{\Delta P}\right)_{\Delta\pi=0} \tag{3.106}$$

The volume flow per unit of osmotic pressure difference is found equal to the diffusional flow per unit of hydrostatic pressure difference over the membrane.

The introduction of the cross coefficients is necessary for the description of the transport phenomena occurring in membranes. Without the cross coefficients the selectivity of the membranes can not be modeled properly. A membrane is called a semipermeable membrane if it is permeable for some solutes and impermeable to other solutes. The selectivity of these membranes is reflected in the coefficient of osmotic flow L_{PD}, while the permeability coefficient L_P characterizes only the mechanical permeability of the membrane for a particular solute.

Living cells are surrounded by semipermeable membranes. By means of the phenomenon of osmosis, selective diffusion can take place to regulate the water balance in the tissues. Very high values may be obtained for the osmotic pressure. Pressures of about 10 MPa are found in some desert plants, with the record-holder being the saltbush *Atriplex Confertifolia*; the cells of the leaves of this plant can build up an osmotic pressure of 21.53 MPa. Human blood has an osmotic pressure of about 0.7 MPa. This osmotic pressure prevents the contents of the red blood cells from diffusing into the plasma. Physiological saline solutions has to be used with the same osmotic pressure as blood, if intravenous injections are administered, to avoid disturbing the water balance of the red blood cells by osmosis.

It is usually supposed that the osmotic pressure equals the hydrostatic pressure difference if the hydrostatic pressure difference over the semipermeable membrane stops the flow of either solute through the membrane. This is correct only if the membrane is ideal. Suppose that equilibrium is obtained after some time and no volume flow through the membrane is found, so that $\vec{\Phi}_P = \vec{0}$. With this condition (3.101) yields

$$(\Delta P)_{\vec{\Phi}_P=\vec{0}} = -\frac{L_{PD}}{L_P}\,\Delta\pi = \sigma\Delta\pi \tag{3.107}$$

in which σ is the *reflection coefficient**. It is readily deduced from (3.107) that $\Delta P = \Delta\pi$ only if $\sigma = 1$. Substitution of $\sigma = 1$ into (3.101) shows that no diffusional flow will be found if in addition $L_P = L_D$, and nothing will leak through the membrane under these conditions. The solute is blocked from passage through the membrane; it is 'reflected' from the membrane. Such a membrane is called an ideal membrane. The values of σ depend on the membrane as well as on the solution. For urea in human red blood cells σ is approximately 0.6. A porous glass filter is an example of a nonselective membrane, for which $\sigma = 0$. Negative values of σ are also possible. The phenomenon is then is called negative anomalous osmosis. See for further details and numerical values Katchalsky and Curran†.

3.4. EXERCISES

Exercise 3.1. If instead of the osmotic pressure heat transport is considered across a membrane, discuss the mass transport in an isobaric system and the heat transport in an isothermal system.

Exercise 3.2. Formulate the thermal energy equation using the heat flux (3.98).

Exercise 3.3. For variations to and including the first order the principle of local and instantaneous equilibrium holds. Does that imply that $(\Delta T/T)^2 \ll 1$ instead of $\Delta T/T \ll 1$ for the application of local and instantaneous equilibrium? Explain your answer.

---

* Staverman A.J. 1951. The theory of measurement of osmotic pressure. *Recueil des Travaux Chimiques des Pays-Bas*, 70: 344–352.

† Katchalsky A. and Curran P.F. 1975. *Nonequilibrium Thermodynamics in Biophysics.* Harvard University Press, Cambridge, MA, Chapter 10.

4 Multicomponent Simple Fluids

Summary: After an introduction, the balance equations and the entropy equation for multicomponent fluids are derived. The energy equation is also formulated as a temperature equation. The process fluxes and the process forces are defined by these balance equations and by the entropy equation.

4.1. COMPLETE AND INCOMPLETE SYSTEMS

If also for the non-equilibrium states of a system the thermodynamic state can be characterized completely by external variables that in principle are measurable macroscopically, then such a system is called a *complete system.*

However, it is possible that internal processes occur in a system that may be detected externally. A well-known example is the polarization in an electric dipole formed by placing a dielectric between two oppositely charged plates. Only the current and the electrical potential between the two poles are measurable externally. These two are the external variables. *Within* the dipole, which is viewed externally as a 'black box', a complicated electrical network may be present. The currents and electrical potentials in the components of this network are not externally measurable and may be thought of as internal quantities. The internal structure of the system is then not described completely by only the current and the potential between the plates. The internal variables needed to describe the internal structure of the dielectric are 'forgotten'. Such a system is called *incomplete.* The incompleteness of the system is removed if the variables describing the internal structure are incorporated in the description of the system. These variables are called *internal variables* or *hidden variables.* The instantaneous characterization demands a description by means of *all* variables, i.e. by the external and internal variables.

The general equations for complete systems are derived in this chapter. For the complete systems multicomponent non-polar fluids with chemical reactions are considered. These general equations obtained are then used as the starting points for examining the various scalar, vector and tensor processes.

4.2. SYSTEM AND IDEAL MODEL

The system consists of one stable fluid phase, with may be composed of N chemical components (or species), so that the phase is a multicomponent one. A fluid is defined as a medium that in equilibrium can resist only pressures. This definition is in close agreement with the experimental observations of Pascal (1647), who was the first to show that in water at rest the pressure on a small surface in its interior is independent of the orientation of that small surface. An equivalent definition is also given by Lamb (1879) in the opening chapter of his famous book *Hydrodynamics**. To this definition of a fluid (gas or liquid) may be added that away from equilibrium and in the absence of external electromagnetic fields the fluid is isotropic with respect to an inertial system. It is noted that isotropy excludes, for example, liquid crystals. By the above two requirements a fluid is defined, which means that it is is composed of relatively small molecules, like those of water and air. The possibility is not excluded that among the chemical components one or more chemical reactions can be found.

As always, a fluid is regarded as a continuum in the thermodynamics of irreversible processes. If there are N components in the fluid, then it is supposed, since the work of Fick† and Stefan‡, that the medium is composed of a superposition of N continua, in which each component is a continuum that executes its own individual motions. This means that at a point $\vec{x}$ in space, at time t, a particle of each component may be found, such that at one point of the fluid there may be N components. It is essential to assume that each particle of each component can be distinguished and can be followed on its trajectory in space. A velocity $\vec{v}_K$ can therefore be assigned to a particle of component K (chemical components are denoted by Latin capital subscripts). The ability to discriminate the particles of the various components and to follow their paths is a major important assumption of multicomponent continuous media. In addition it can be remarked, that it is senseless to think about components in a continuum theory if those components cannot be distinguished from each other in a continuum.

Because the thermodynamics of irreversible processes is a first order theory, the velocities of the components may not differ very much from each other, so that it must be required that

$$|\vec{v}_K - \vec{v}_L| \ll |\vec{v}|$$

where $\vec{v}$ is a characteristic velocity of the flow of the multicomponent fluid, often $\vec{v}$ is a hypothetical velocity.

If the above condition is satisfied, then the overall motion of a fluid can be characterized by one velocity $\vec{v}$ that is determined by one equation of motion. This model is called the *one-component model* description of a

---

* Sir Horace Lamb. 1945. "The fundamental property of a fluid is that it cannot be in equilibrium in a state of stress such that the mutual action between two adjacent parts is oblique to the common surface." *Hydrodynamics*. 6th ed. Dover Publications, New York, p. 1.

† Fick A. 1855. Ueber Diffusion. *Annalen der Physik und Chemie (Poggendorffer Annalen)*, 94: 59–86.

‡ Stefan J. 1871. Ueber das Gleichgewicht und die Bewegung, insbesondere die Diffusion von Gasmengen. *Sitzungsberichten Akademie Wissenschaften. Wien*, 63(2): 63–124.

multicomponent system, but in which the motions of the various individual components are described by diffusion equations.

If the motions of the N components are described by N velocities with their corresponding N equations of motion then this is called an N *component model* description of a multicomponent system. An N component model description might be needed if the differences between the velocities of the components are too large, as might be the case for charged particles in a plasma, in which fast electrons and slow ions are found. In addition, the N component model description is usually used in describing multiphase flow. In the subsequent discussion it is assumed that the one-component model is appropriate.

4.3. THERMODYNAMICS OF MULTICOMPONENT FLUIDS

4.3.1. The fundamental equation of Gibbs. The work done by a quasi-static change of volume on a homogeneous fluid is, according to Pascal, given by

$$dW = -p\,dV \tag{4.1}$$

with p the thermodynamic pressure and V the volume. For a closed system, the first and second laws of classical thermodynamics are summarized in

$$dU = T\,dS - p\,dV \tag{4.2}$$

It may be inferred from this equation that the internal energy depends on the extensive variables S and V

$$U = U(S, V) \tag{4.3}$$

which conclusion applies for closed systems.

An open system can exchange mass with its surroundings, so that the internal energy U depends also on the possible changes in the masses M_K of each component K. This implies that (4.3) has to be replaced by

$$U = U\left(S, V, M_1, M_2, \cdots, M_N\right) \tag{4.4}$$

where N is the number of chemical components. Equation (4.4) is the integrated form of the Gibbs fundamental equation for an open homogeneous multicomponent system. In differential form this equation becomes

$$dU = T\,dS - p\,dV + \sum_{K=1}^{N} \mu_K\,dM_K \tag{4.5}$$

where

$$\begin{aligned} T &= \left(\frac{\partial U}{\partial S}\right)_{V,M} \\ -p &= \left(\frac{\partial U}{\partial V}\right)_{S,M} \\ \mu_K &= \left(\frac{\partial U}{\partial M_K}\right)_{S,V,M'} \qquad \text{for} \quad K = 1, 2, \cdots, N \end{aligned} \tag{4.6}$$

The subscript M' stands for $M_1, \cdots, M_{K-1}, M_{K+1}, \cdots, M_N$, and the subscript M stands for $M_1, M_2, \cdots, M_N$. The temperature T and the pressure p are therefore still defined for closed systems (M is a constant for all K), but additionally the new intensive variables μ_K are introduced that stand for the *specific chemical* or *thermodynamic potential* of component K. These chemical potentials refer to one unit of mass of the component concerned.

The quantities U, S, V, $\{M_K\}$ are extensive ($\{M_K\}$ stand for the set of all M_K). The additive property of the extensive quantities implies that these quantities are multiplied by an arbitrarily chosen positive constant k if all the masses M_K at constant T, p, $\{\mu_K\}$ are multiplied by k too (keep in mind that this is possible only for open systems). So it is presumed therefore that

$$U' = kU, \qquad S' = kS, \qquad V' = kV, \qquad \{M'_K = kM_K\} \tag{4.7}$$

Substitution into (4.4) yields

$$U' = U'(S', V', \{M'_K\})$$

Differentiate this equation with respect to k, using (4.7) again to get

$$U = \frac{\partial U'}{\partial S'} S + \frac{\partial U'}{\partial V'} V + \sum_{K=1}^{N} \frac{\partial U'}{\partial M'_K} M_K$$

Let k approach unity, and then substitution of (4.6) gives

$$U = TS - pV + \sum_{K=1}^{N} \mu_K M_K \tag{4.8}$$

This is the Euler formula for homogeneous functions of the first degree, termed here the *Euler relation*, which is a special case of the Euler theorem for homogeneous functions of the nth degree. Differentiation of (4.8) results in

$$dU = T\,dS + S\,dT - p\,dV - V\,dp + \sum_{K=1}^{N} (\mu_K\,dM_K + M_K\,d\mu_K)$$

According to (4.5) this becomes

$$0 = S\,dT - V\,dp + \sum_{K=1}^{N} M_K\,d\mu_K \tag{4.9}$$

which is the *Gibbs–Duhem relation**.

* Mazur has remarked that it can be shown that if nonlinear relations between the forces and fluxes exist, the Gibbs–Duhem relation no longer holds. Personal communication from D. Bedeaux.

The independent variables in (4.5) are S, V, $\{M_K\}$. Sometimes, it is more advantageous to consider other independent variables. The transformation to other independent variables has to be done so that all information contained in (4.5) is maintained. This can be realized by *Legendre transformations** or *contact transformations*. In classical thermodynamics this comes down to the introduction of the *thermodynamic potential functions* (or the Massieu functions). The commonest potential functions are

$$\begin{aligned} H &= U + pV && \textit{enthalpy} \\ F &= U - TS && \textit{free energy} \text{ (Helmholz potential)} \\ G &= U - TS + pV && \textit{Gibbs function} \text{ (free enthalpy)} \end{aligned} \tag{4.10}$$

Using (4.5) the following fundamental equations are obtained

$$\begin{aligned} dH &= T\,dS + V\,dp + \sum_{K=1}^{N} \mu_K\,dM_K \\ dF &= -S\,dT - p\,dV + \sum_{K=1}^{N} \mu_K\,dM_K \\ dG &= -S\,dT + V\,dp + \sum_{K=1}^{N} \mu_K\,dM_K \end{aligned} \tag{4.11}$$

These along with (4.4) lead to four different *representations* for the complete knowledge of a system in the sense of the classical thermodynamics result from these equations, namely

$$\begin{aligned} U &= U\,(S, V, \{M_K\}) && \textit{internal energy representation} \\ H &= H\,(S, p, \{M_K\}) && \textit{enthalpy representation} \\ F &= F\,(T, V, \{M_K\}) && \textit{free energy representation} \\ G &= G\,(T, p, \{M_K\}) && \textit{Gibbs representation} \end{aligned} \tag{4.12}$$

These representations are equivalent provided that each potential function is used with the appropriate independent variables (the representation $H = H(T, p, \{M_K\})$ is therefore not an equivalent representation!).

The fundamental equation of Gibbs in (4.5) can be expressed by four equivalent formulations, as is illustrated in (4.12). *Thermodynamic equations of state* correspond to each representation. For example, the following thermodynamic equations of state correspond to the Gibbs representation $(4.12)_4$

$$\begin{aligned} S &= S\,(T, p, \{M_K\}) \\ V &= V\,(T, p, \{M_K\}) \\ \mu_K &= \mu_K\,(T, p, \{M_K\}) \end{aligned} \tag{4.13}$$

---

* Callen H.B. 1960. *Thermodynamics*, 1st ed. John Wiley & Sons, New York. See pp. 90–98.

These equations of state are related to each other by means of the *Maxwell relations*. For (4.13) for instance, it is found that

$$\left(\frac{\partial S}{\partial p}\right)_{T,M} = -\left(\frac{\partial V}{\partial T}\right)_{p,M}; \qquad \left(\frac{\partial S}{\partial M_K}\right)_{T,p,M'} = -\left(\frac{\partial \mu_K}{\partial T}\right)_{p,M}$$
$$\left(\frac{\partial V}{\partial M_K}\right)_{T,p,M'} = \left(\frac{\partial \mu_K}{\partial p}\right)_{T,M}; \qquad \left(\frac{\partial \mu_K}{\partial M_L}\right)_{T,p,M'} = \left(\frac{\partial \mu_L}{\partial M_K}\right)_{T,p,M'} \tag{4.14}$$

These Maxwell relations result from the fact that in $(4.11)_3$ all differentials are *exact* differentials.

The thermodynamic quantities of state cannot be measured directly. To obtain measurable quantities the thermodynamic equations of state in (4.13) have to be written in differential form, thus

$$\begin{aligned} dS &= \left(\frac{\partial S}{\partial T}\right)_{p,M} dT + \left(\frac{\partial S}{\partial p}\right)_{T,M} dp + \left(\frac{\partial S}{\partial M_K}\right)_{T,p,M'} dM_K \\ dV &= \left(\frac{\partial V}{\partial T}\right)_{p,M} dT + \left(\frac{\partial V}{\partial p}\right)_{T,M} dp + \left(\frac{\partial V}{\partial M_K}\right)_{T,p,M'} dM_K \\ d\mu_L &= \left(\frac{\partial \mu_L}{\partial T}\right)_{p,M} dT + \left(\frac{\partial \mu_L}{\partial p}\right)_{T,M} dp + \left(\frac{\partial \mu_L}{\partial M_K}\right)_{T,p,M'} dM_K \end{aligned} \tag{4.15}$$

The differential quotients in (4.15) are the *thermodynamic coefficients*, which are—possibly by means of the Maxwell relations—measurable. For fluids in the Gibbs representation, the following can be defined

$$\begin{aligned} C_p &= T\left(\frac{\partial S}{\partial T}\right)_{p,M} \qquad \alpha = \frac{1}{V}\left(\frac{\partial V}{\partial T}\right)_{p,M} \qquad \kappa_T = -\frac{1}{V}\left(\frac{\partial V}{\partial p}\right)_{T,M} \\ s_K &= \left(\frac{\partial S}{\partial M_K}\right)_{T,p,M'} \qquad v_K = \left(\frac{\partial V}{\partial M_K}\right)_{T,p,M'} \\ \mu_{KL} &= \left(\frac{\partial \mu_K}{\partial M_L}\right)_{T,p,M'} = \mu_{LK} \end{aligned} \tag{4.16}$$

where C_p is the heat capacity at constant pressure, α the cubic expansion coefficient, κ_T the isothermal compressibility, s_K the partial entropy of component K, v_K the partial volume of component K, and μ_{KL} the chemical coefficient.

The equations of state in differential form (4.15) become with (4.16) and the Maxwell relations (4.14)

$$
\begin{aligned}
dS &= \frac{1}{T}C_p\,dT - V\,\alpha\,dp + \sum_{K=1}^{N} s_K\,dM_K \\
dV &= V\,\alpha\,dT - V\,\kappa_T\,dp + \sum_{K=1}^{N} v_K\,dM_K \\
d\mu_L &= -s_L\,dT + v_L\,dp + \sum_{K=1}^{N} \mu_{LK}\,dM_K
\end{aligned}
\tag{4.17}
$$

Therefore, for a fluid with N chemical components $3+\frac{1}{2}N(N+5)$ independent thermodynamic coefficients have to be determined. With the substitution of these coefficients into (4.17), the thermodynamic equations (4.13) may be obtained by integration; these in turn may be substituted into $(4.11)_3$. Integration of this latter yields the fundamental equation $(4.12)_4$ in the Gibbs-representation.

Finally, it is remarked that adding subscripts to the partial differential coefficients (as in (4.6), (4.14), (4.15), (4.16)) is an unusual practice in mathematics. Since in classical thermodynamics, the mixed use of several representations simultaneously can be confusing, the subscripts are added to indicate which independent variables are being held constant to avoid misinterpretation.

4.3.2. Specific thermodynamic quantities. In the application of the principle of local equilibrium all extensive quantities are computed per unit mass (or per unit volume, or per mole). An extensive, physical quantity divided by the mass of the system in the bulk is called *specific*, according to the recommendations of the International Union of Pure and Applied Physics*. The system may consist of more than one component and/or more than one phase. Specific quantities are denoted preferably by lowercase Latin subscripts.

If M_K is the mass of component K in the system, then the total mass of the system is

$$
M = \sum_{K=1}^{N} M_K \tag{4.18}
$$

The specific quantities corresponding to the quantities in (4.4) are therefore

$$
u = \frac{U}{M} \qquad s = \frac{S}{M} \qquad v = \frac{V}{M} \qquad w_K = \frac{M_K}{M} \tag{4.19}
$$

* Cohen E.R. and Giacomo P. eds. 1987. Symbols, Units, Nomenclature and Fundamental constants in Physics (1987 revision), document IUPAP-25 (SUNAMCO 87–1) SUN-Commission. International Union of Pure and Applied Physics, *Physica*, 146A: 1–67.

in which w_K is the weight fraction of component K. From the latter definition it follows that

$$\sum_{K=1}^{N} w_K = 1 \tag{4.20}$$

so that weight fractions are linearly dependent.

Substitution of (4.19) into (4.5) yields

$$u\,dM + M\,du = \left(Ts - pv + \sum_{K=1}^{N} \mu_K w_K\right) dM + \left(T\,ds - p\,dv + \sum_{K=1}^{N} \mu_K\,dw_K\right) M \tag{4.21}$$

Dividing the equation of Euler (4.8) by M yields its specific form

$$u = Ts - pv + \sum_{K=1}^{N} \mu_K w_K \tag{4.22}$$

and substitution into (4.21) reduces this equation to

$$\boxed{du = T\,ds - p\,dv + \sum_{K=1}^{N} \mu_K\,dw_K} \tag{4.23}$$

which is the specific formulation of the Gibbs equation. This fundamental equation will be used to derive the entropy equation.

Differentiating (4.22) and using (4.23) thus gives the specific formulation of the equation of Gibbs–Duhem (4.9)

$$\boxed{-s\,dT + v\,dp = \sum_{K=1}^{N} w_K\,d\mu_K} \tag{4.24}$$

From (4.23) it follows, according to (4.20), that

$$u = u\,(s, v, w_1, w_2, \cdots, w_{N-1}) \tag{4.25}$$

so that u depends on the $N+1$ variables $s, v, w_1, w_2, \cdots, w_{N-1}$. This is not in contradiction with the fact that from (4.5) it follows that U depends on $N+2$ independent variables, since $U = Mu$, so that M is the missing independent variable in (4.25). The changeover to specific variables in the internal energy representation means therefore a transformation of the variables $S, V, M_1, M_2, \cdots, M_N$ to the variables $s, v, w_1, w_2, \cdots, w_{N-1}, M$.

The specific thermodynamic potentials (4.10) become

$$\begin{aligned} h &= H/M = u + pv && \textit{specific enthalpy} \\ f &= F/M = u - Ts && \textit{specific free energy} \\ g &= G/M = h - Ts && \textit{specific Gibbs function} \end{aligned} \tag{4.26}$$

From (4.23) and (4.26) we have

$$\begin{aligned} dh &= \quad T\,ds + v\,dp + \sum_{K=1}^{N} \mu_K\,dw_K \\ df &= -s\,dT - p\,dv + \sum_{K=1}^{N} \mu_K\,dw_K \\ dg &= -s\,dT + v\,dp + \sum_{K=1}^{N} \mu_K\,dw_K \end{aligned} \tag{4.27}$$

Instead of the specific quantities molar quantities can also be introduced. Molar is the name given if an extensive physical quantity is divided by the number of moles. If needed, the molar quantities will be introduced as appropriate.

4.3.3. Partial specific thermodynamic quantities. If Ψ denotes an arbitrary extensive quantity, then the corresponding partial specific quantity of component K is

$$\psi_K = \left(\frac{\partial \Psi}{\partial M_K}\right)_{T,p,M'} \tag{4.28}$$

Here, Ψ is supposed to depend on $T, p, \{M_K\}$

$$\Psi = \Psi\,(T, p, \{M_K\}) \tag{4.29}$$

For constant $T, p, \{\mu_K\}$ multiply all masses by l and put

$$\Psi' = l\,\Psi \qquad \text{and} \qquad M'_K = l\,M_K$$

yielding for (4.29)

$$\Psi' = \Psi'\,(T, p, \{M'_K\})$$

Differentiate this relation to l, then let l approach unity and apply (4.28). This results in

$$\Psi = \sum_{K=1}^{N} M_K \psi_K \qquad \text{or} \qquad \psi = \sum_{K=1}^{N} w_K \psi_K \tag{4.30}$$

This is again the Euler relation for homogeneous functions of the first degree. Relation (4.30) does not represent superposition in the strict sense, since ψ_K is defined by (4.28) as a differential quotient, which may depend on T, p and all w_K. The relation (4.30) would represent a superposition if ψ_K were equal to the value of ψ for pure component K not influenced by the presence of other components, and only for ideal systems is it assumed that one component is not influenced by another.

The differential of (4.29) is after using (4.28)

$$d\Psi = \left(\frac{\partial \Psi}{\partial T}\right)_{p,M} dT + \left(\frac{\partial \Psi}{\partial p}\right)_{T,M} dp + \sum_{K=1}^{N} \psi_K \, dM_K$$

or

$$\psi \, dM + M \, d\psi = M \left[\left(\frac{\partial \psi}{\partial T}\right)_{p,w} dT + \left(\frac{\partial \psi}{\partial p}\right)_{T,w} dp\right] + \sum_{K=1}^{N} \left[\psi_K w_K \, dM + \psi_K M \, dw_K\right]$$

or with the substitution of $(4.30)_2$

$$d\psi = \left(\frac{\partial \psi}{\partial T}\right)_{p,w} dT + \left(\frac{\partial \psi}{\partial p}\right)_{T,w} dp + \sum_{K=1}^{N} \psi_K \, dw_K \tag{4.31}$$

Again with the application of $(4.30)_2$, the following relation results

$$\sum_{K=1}^{N} w_K \, d\psi_K = \left(\frac{\partial \psi}{\partial T}\right)_{p,w} dT + \left(\frac{\partial \psi}{\partial p}\right)_{T,w} dp \tag{4.32}$$

This can be seen to be a generalization of the Gibbs–Duhem relation (4.24).

Suppose in (4.31) that $\psi = g$

$$dg = \left(\frac{\partial g}{\partial T}\right)_{p,w} dT + \left(\frac{\partial g}{\partial p}\right)_{T,w} dp + \sum_{K=1}^{N} g_K \, dw_K$$

Comparison of this result with $(4.27)_3$ shows that the chemical potential $\mu_K = g_K$ is a partial Gibbs function. This is already known from $(4.11)_3$

$$\mu_K = \left(\frac{\partial G}{\partial M_K}\right)_{T,p,M'} = g_K$$

For future applications dh is calculated, where h is assumed to depend on $T, p, \{w_K\}$. Use (4.31) with the substitution $\psi = h$

$$dh = \left(\frac{\partial h}{\partial T}\right)_{p,w} dT + \left(\frac{\partial h}{\partial p}\right)_{T,w} dp + \sum_{K=1}^{N} h_K \, dw_K \tag{4.33}$$

From $(4.27)_1$ it follows with the substitution of the heat capacity $(4.16)_1$

$$\left(\frac{\partial h}{\partial T}\right)_{p,w} = T \left(\frac{\partial s}{\partial T}\right)_{p,w} = c_p \equiv C_p/M \tag{4.34}$$

And once more from $(4.27)_1$

$$\left(\frac{\partial h}{\partial p}\right)_{T,w} = T \left(\frac{\partial s}{\partial p}\right)_{T,w} + v \tag{4.35}$$

The fundamental equation $(4.27)_3$ implies the following Maxwell relation

$$\left(\frac{\partial s}{\partial p}\right)_{T,w} = - \left(\frac{\partial v}{\partial T}\right)_{p,w} = -v\,\alpha \tag{4.36}$$

where use is made of $(4.16)_2$. From (4.35) and (4.36) we obtain

$$\left(\frac{\partial h}{\partial p}\right)_{T,w} = v\,(1 - \alpha\,T) \tag{4.37}$$

Then (4.33) becomes with (4.34) and (4.37)

$$dh = c_p \, dT + v\,(1 - \alpha\,T)\; dp + \sum_{K=1}^{N} h_K \, dw_K \tag{4.38}$$

This equation will be needed in the discussion of the energy balance.

Finally, some remarks about the idea of 'concentration' are raised. The partial mass density ρ_K of component K is its mass per unit volume. Let the total mass density be denoted by ρ, then

$$\rho = \sum_{K=1}^{N} \rho_K \qquad \text{and} \qquad w_K = \frac{\rho_K}{\rho} \qquad \text{with} \qquad \sum_{K=1}^{N} w_K = 1 \tag{4.39}$$

The partial molar density c_K is also used, and it is frequently called 'the' concentration of component K. The partial molar density is the number of moles of component K per unit volume that corresponds to the mole fraction x_K

$$c = \sum_{K=1}^{N} c_K \qquad \text{and} \qquad x_K = \frac{c_K}{c} \qquad \text{with} \qquad \sum_{K=1}^{N} x_K = 1 \tag{4.40}$$

If m_K denotes the molar mass (the molecular weight) of component K, then

$$\rho_K = c_K m_K \tag{4.41}$$

The effective molar mass m of the system is defined analogously to (4.41)

$$\rho = c\, m \tag{4.42}$$

From these definitions follow two useful relations

$$\begin{aligned} m &= \frac{\rho}{c} = \sum_{K=1}^{N} x_K\, m_K \\ \frac{1}{m} &= \frac{c}{\rho} = \sum_{K=1}^{N} \frac{w_K}{m_K} \end{aligned} \tag{4.43}$$

The molar mass of a system therefore depends on the composition of the system. For ideal systems and in problems involving chemical reactions one has a preference for mole fractions, while in fluid dynamics there is a preference for the mass fractions. As a consequence, both mass and mole fractions are used in practice.

4.4. MASS BALANCE AND DIFFUSION EQUATIONS

The mass balances are based on the following axioms

(1) Mass is permanent, that is, it cannot be created or destroyed.
(2) Mass is invariant with respect to the state of motion of the body concerned.
(3) In case of changes due to chemical reactions the total of mass is conserved.

As a consequence, the mass M_K of component K in a spatially fixed control volume $V_{(c)}$ can be changed only by the flow of component K through the control surface $A_{(c)} \equiv \partial V_{(c)}$ and by the production of component K by (internal) scalar reactions. Mathematically formulated

$$\frac{d}{dt} M_K = \iiint_{V_{(c)}} \frac{\partial \rho_K}{\partial t}\, dV = -\iint_{A_{(c)}} \vec{n} \cdot (\rho_K \vec{v}_K)\, dA + \iiint_{V_{(c)}} \pi_K\, dV \tag{4.44}$$

in which ρ_K is the density of component K, $\vec{v}_K$ the velocity of component K, $\vec{n}$ the outwardly directed unit normal vector to the surface $A_{(c)}$, and π_K the rate of production of mass of component K by internal processes per unit volume and per unit time. With the application of the Gauss divergence theorem (and since (4.44) applies for all $V_{(c)}$ and therefore also for very small $V_{(c)}$) (4.44) yields the N *partial mass balances** in differential form

$$\frac{\partial \rho_K}{\partial t} + \operatorname{div}(\rho_K \vec{v}_K) = \pi_K = m_K r_K \tag{4.45}$$

where r_K is the rate of transformation (i.e. the formation or the consumption) of species K in mol per volume per second.

The law of the *conservation of mass* now becomes

$$\sum_{K=1}^{N} \pi_K = 0 \tag{4.46}$$

Summation over K of (4.45) and the substitution of this conservation of mass relation yield

$$\boxed{\frac{\partial \rho}{\partial t} = -\operatorname{div} \rho \vec{v}} \tag{4.47}$$

The conservation of mass naturally suggests the barycentric velocity $\vec{v}$ as the reference velocity for the description of diffusion

$$\rho \vec{v} = \sum_{K=1}^{N} \rho_K \vec{v}_K \tag{4.48}$$

---

* Formally, the balance equations can also be derived using the Reynolds transport, stated first by Reynolds (1903) and proved by Spielrein (1916). It is assumed that the component K is enclosed by a control volume that has the same velocity as component K. If no chemical processes occur in this convected volume, then this volume can be seen to be 'material' in the sense that this volume consists permanently of the same particles K. If chemical processes take place in the volume, then component K is produced or changed into other components, and the convected volume cannot be a material volume. It is therefore better to speak of a K-convected volume, in which the particles K cannot leave this volume. It is remarked that in formulating the balance of mass, the mass of component K in a K-convected material volume can change only by chemical reactions if no mass is added from or removed to the exterior of the convected material volume. The law of conservation of mass of component K confined in a K-convected volume then becomes

$$\frac{\mathrm{D}_K}{\mathrm{D}t} \iiint_V \rho_K \, dV = \iiint_V \pi_K \, dV$$

This integral (or global) balance equation of mass can be transformed into a mass balance equation for component K. Furthermore, it is noted that the exclusion of the supply of mass from or the removal of mass to the outside of the K-convected volume is not a serious restriction. The external exchange of mass has to be done through the surfaces of the material volume, so that these exchanges can be accounted for in the boundary conditions of the local balance equations of mass.

with

$$\rho = \sum_{K=1}^{N} \rho_K \tag{4.49}$$

A reference velocity characterizes an average velocity of the fluid. In the barycentric description of the diffusion $\vec{v}$ is a reference velocity that defines the *mass average velocity*. The *molar average velocity* and the *volume average velocity* are other possible reference velocities. A time derivative can be introduced to each of the reference velocities. With the mass average velocity the well-known *barycentric time derivative* or *substantial derivative* is defined by

$$\frac{\mathrm{D}}{\mathrm{D}t} \equiv \frac{\partial}{\partial t} + (\vec{v} \cdot \mathrm{grad}\,) \tag{4.50}$$

With the barycentric time derivative, (4.47) can also be written as

$$\boxed{\frac{\mathrm{D}\rho}{\mathrm{D}t} = -\rho\,\mathrm{div}\,\vec{v}} \tag{4.51}$$

Equations (4.47) and (4.51) are two well-established formulations of the *continuity equation* of fluid dynamics and are known as the barycentric equation of continuity*. In using this one-component description with the barycentric velocity $\vec{v}$, N individual mass balances are not needed, but instead there are $N-1$ diffusion equations.

4.4.1. Diffusion equations. Diffusion can be considered as the motion of components with respect to each other. A reference velocity is needed to derive diffusion equations. The description of diffusion therefore depends on the choice of the reference velocity. When the barycentric velocity $\vec{v}$ is used, the barycentric diffusion velocity is introduced as

$$\vec{V}_K = \vec{v}_K - \vec{v} \tag{4.52}$$

along with the barycentric diffusion flux

$$\vec{\Phi}_K = \rho_K \vec{V}_K \tag{4.53}$$

With the substitution of this flux into (4.45) this equation can be written as

$$\frac{\partial \rho_K}{\partial t} + \mathrm{div}\,(\rho_K \vec{v}) = -\mathrm{div}\,\left(\rho_K \vec{V}_K\right) + \pi_K \tag{4.54}$$

or

$$\frac{\mathrm{D}\rho_K}{\mathrm{D}t} + \rho_K \mathrm{div}\,\vec{v} = -\mathrm{div}\,\vec{\Phi}_K + \pi_K \tag{4.55}$$

* The name barycentric originates from the fact that this equation guarantees that the hypothetical fluid moving with the velocity $\vec{v}$ is continuously connected.

Table 4.1. Diffusion Fluxes and Equations

Diffusion Flux	**Diffusion Equation**
$\vec{\Phi}_K = \rho_K(\vec{v}_K - \vec{v}) = \rho_K \vec{V}_K$	$\rho \frac{\mathrm{D}w_K}{\mathrm{D}t} = -\mathrm{div}\,\vec{\Phi}_K + \pi_K$
$\vec{\Phi}'_K = c_K(\vec{v}_K - \vec{v}) = c_K \vec{V}_K$	$\frac{\mathrm{D}c_K}{\mathrm{D}t} = -\mathrm{div}\,\vec{\Phi}'_K - c_K \mathrm{div}\,\vec{v} + r_K$
$\vec{\Phi}^\star_K = c_K(\vec{v}_K - \vec{v}^\star) = c_K \vec{V}^\star_K$	$c\frac{\mathrm{D}^\star x_K}{\mathrm{D}t} = -\mathrm{div}\,\vec{\Phi}^\star_K + \left(r_K - x_K \sum_{K=1}^N r_K\right)$

$\vec{\Phi}_K$ is the mass diffusion flux relative to the mass-average velocity $\vec{v}$.
$\vec{\Phi}'_K$ is the molar diffusion flux relative to the mass-average velocity $\vec{v}$.
$\vec{\Phi}^\star_K$ is the molar diffusion flux relative to the molar-average velocity $\vec{v}^\star$.

Combination with (4.51) gives

$$\boxed{\rho \frac{\mathrm{D}w_K}{\mathrm{D}t} = -\mathrm{div}\,\vec{\Phi}_K + \pi_K} \tag{4.56}$$

where $w_K = \rho_K/\rho$ is the mass fraction.

Of these N equations only $N-1$ equations are linearly independent due to $\sum_{K=1}^N w_K = 1$, $\sum_{K=1}^N \vec{\Phi}_K = \vec{0}$ and $\sum_{K=1}^N \pi_K = 0$. An equation (4.56) can be considered as the simplest form of a diffusion equation with the barycentric diffusion flux $\vec{\Phi}_K = \rho_K \vec{V}_K$. In the one-component model, $\vec{v}_K$ is not explicitly calculated, and a phenomenological equation has to be found for $\vec{\Phi}_K$. The flux $\vec{\Phi}_K$ can be expressed for example in terms of w_K by pursuing the methods of the TIP, so that the set of equations (4.56) can be solved in principle.

The tendency from theoretical considerations to choose the barycentric description comes up against the practical objection that in systems with chemical reactions and those involving ideal gas mixtures there is preference for working with the molar concentration c_K. This preference suggests the use of the *molar-average velocity*, defined by

$$c\,\vec{v}^\star = \sum_{K=1}^{N} c_K \vec{v}_K \tag{4.57}$$

The various descriptions used are summarized in the table, where $\mathrm{D}^\star/\mathrm{D}t$ is the molar time derivative defined by

$$\frac{\mathrm{D}^\star}{\mathrm{D}t} \equiv \frac{\partial}{\partial t} + (\vec{v}^\star \cdot \mathrm{grad}\,) \tag{4.58}$$

4.4.2. Chemical reactions. If chemical reactions occur in the system, the masses M_K can change by virtue of exchange of mass with the environment—mass transfer between the system and its surroundings—and also by reactions occurring within the system. The changes of M_K by the first mechanism are denoted by an external change $d_e M_K$, and the changes by the second mechanism are denoted by $d_i M_K$, so that

$$dM_K = d_e M_K + d_i M_K \tag{4.59}$$

The αth reaction between chemical components can be represented by

$$\sum_K \nu_{\alpha K}[R_K] \rightarrow \sum_L \nu'_{\alpha L}[P_L] \tag{4.60}$$

where $\nu_{\alpha K}$ is the molar or stoichiometric number of constituent K in the reaction α, $[R_K]$ the chemical symbol of one mole of the reactant K, and $[P_L]$ the chemical symbol of one mole of the product L.

Suppose for reactants that $\nu'_{\alpha K} = -\nu_{\alpha K}$ and for products that $\nu'_{\alpha L} = +\nu_{\alpha L}$, so that components that are formed as the reaction proceeds to the right are taken positively, and components that are consumed are counted negatively. If $[C_K]$ is the chemical symbol of one mole of constituent K ($[C_K] \in [R_K] \cup [P_K]$), then the reaction equation (4.60) of reaction α becomes

$$\sum_K \nu_{\alpha K}[C_K] \rightarrow 0 \tag{4.61}$$

Denoting by m_K the molar mass of component K, (4.61) implies that

$$\sum_K \nu_{\alpha K} m_K = 0 \tag{4.62}$$

This is the *stoichiometric equation*, which expresses the conservation of mass during a chemical reaction α. The stoichiometric numbers $\nu_{\alpha K}$ are the smallest whole numbers satisfying (4.62).

The advancement of reaction α is described by the *extent of reaction* or *reaction coordinate* ξ_α. The mass of component K formed by an advancement of $d\xi_\alpha$ mole of the reaction α is

$$d_\alpha M_K = \nu_{\alpha K} m_K \, d\xi_\alpha \tag{4.63}$$

If in the system n scalar reactions take place, then the internal increase of the mass M_K is given by

$$d_i M_K = \sum_{\alpha=1}^{n} \nu_{\alpha K} m_K \, d\xi_\alpha \tag{4.64}$$

for which, also with (4.62), it follows that

$$d_i M = \sum_{K=1}^{N} d_i M_K = \sum_{K=1}^{N} \sum_{\alpha=1}^{n} \nu_{\alpha K} m_K \, d\xi_\alpha = 0 \tag{4.65}$$

It is seen that the mass is conserved, but as a rule not the number of moles. As a consequence, the equations are most often simpler by using the masses rather than using the moles of the constituents. With (4.64) the increase of the mass fraction becomes

$$d_i w_K = d_i M_K / M = \sum_{\alpha=1}^{n} \nu_{\alpha K} m_K \, dX'_\alpha \tag{4.66}$$

where dX'_α denotes the normalized extent of reaction*, which is the extent of reaction per unit of mass that is $d\xi_\alpha / M$ (chemical engineers use also the extent of reaction per unit of volume, so that the change of the partial molar densities is considered or to use the *fractional conversion* for which the fractional change of a reactant is considered).

The production term π_K in the diffusion equation (4.56) can be expressed in the *rate of reaction* or *reaction velocity*

$$\pi_K = \rho \sum_{\alpha=1}^{n} \nu_{\alpha K} m_K \frac{\mathrm{D} X'_\alpha}{\mathrm{D}t} \tag{4.67}$$

For $d_e w_K = d_e (M_K / M)$ applies

$$\rho \frac{\mathrm{D_e} w_K}{\mathrm{D}t} = -\mathrm{div}\, \vec{\Phi}_K \tag{4.68}$$

From dimensional considerations it is seen that $\dim[DX'_\alpha / Dt] = \text{mol kg}^{-1}\ \text{s}^{-1}$. It is under discussion to what extent the rate of reaction has to be expressed in terms of the barycentric time derivative. A teleological argument is that in that way the simplest formulation of rate of reaction is obtained in connection with the already introduced barycentric velocity. To put it differently: in the barycentric description it is assumed that the reactions occur with respect to the centers of mass of the physical volume elements. Strictly, (4.67) is an extrapolation of classical thermodynamics, since for quasistatic processes $dX'_\alpha = 0$ for all α.

* Villermaux J. 1993. Kinetics of composite reactions in closed and open flow systems (IUPAP Recommendations 1993). *Pure and Applied Chemistry*, 65: 2641–2656.

4.5. ELECTROMAGNETIC BALANCE EQUATIONS

According to the equations of Maxwell for the electromagnetic field, the electromagnetic balances are independent of the mechanical balances. The electromagnetic field equations of Maxwell* read in usual notation

$$\begin{aligned} \operatorname{div} \vec{D} &= \rho q & \operatorname{div} \vec{B} &= 0 \\ \operatorname{curl} \vec{E} &= -\frac{\partial \vec{B}}{\partial t} & \operatorname{curl} \vec{H} &= \frac{\partial \vec{D}}{\partial t} + \vec{J} + \rho q \vec{v} \end{aligned} \tag{4.69}$$

where q is the electric charge per unit of mass. The physical dimensions of the electromagnetic quantities follow from the electromagnetic field equations, and it is found for example that

$$\begin{aligned} \dim[E] &= \text{kg m A}^{-1}\ \text{s}^{-3} & \dim[B] &= \text{kg A}^{-1}\ \text{s}^{-2} \\ \dim[D] &= \text{A s m}^{-2} & \dim[H] &= \text{A m}^{-1} \end{aligned} \tag{4.70}$$

where $\dim[\cdots]$ means: the variable $\cdots$ has the dimension of.

The Maxwell equations (4.69) are covariant (form invariant) for relativistic and semi-relativistic transformations. In the semi-relativistic transformations terms of the order of $(v/c_0)^2$ are neglected, where c_0 denotes the speed of light in vacuum. The *semi-relativistic Maxwell–Lorentz transformations* for electromagnetic fields are

$$\begin{aligned} \vec{E}' &= \vec{E} + \vec{v} \times \vec{B} & \vec{D}' &= \vec{D} + \frac{\vec{v} \times \vec{H}}{c_0^2} \\ \vec{H}' &= \vec{H} - \vec{v} \times \vec{D} & \vec{B}' &= \vec{B} - \frac{\vec{v} \times \vec{E}}{c_0^2} \end{aligned} \tag{4.71}$$

The *electromagnetic constitutive equations* for non-polar media are given by

$$\vec{D} = \epsilon_0 \vec{E} \qquad \text{and} \qquad \vec{B} = \mu_0 \vec{H} \tag{4.72}$$

with $c_0^2 = 1/(\epsilon_0 \mu_0)$, where ϵ_0 denotes the permittivity of vacuum and μ_0 the permeability of vacuum. Note that $(4.71)_2$ with (4.72) follows from $(4.71)_1$, and that $(4.71)_4$ follows from $(4.71)_3$. The quantities with an accent refer to the 'barycentric rest state': $\vec{v} = \vec{0}$.

If $\bar{q}_K$ is the electric charge of one mole of component K, then $\bar{q}_K / m_K = q_K$ is the electric charge of component K per unit of mass of component K

$$\rho_K\, q_K = c_K\, \bar{q}_K \tag{4.73}$$

* In Appendix A, these equations are derived, using a minimum of concepts. Readers might skip this section if they are interested in non-polar media only. The section is added to discuss the electromagnetic momentum, moment of momentum and energy.

For the complete system the total electric charge per unit of volume is

$$\rho q = \sum_{K=1}^{N} \rho_K q_K = \sum_{K=1}^{N} c_K\, \bar{q}_K \tag{4.74}$$

The barycentric electric current density is

$$\vec{J} = \sum_{K=1}^{N} \rho_K q_K \vec{V}_K \tag{4.75}$$

From these definitions and also the mass balances (4.45), the balance of the electric current is obtained. Multiply (4.45) with q_K, sum over K and use the definitions (4.74) and (4.75) to get

$$\frac{\partial \rho q}{\partial t} = -\mathrm{div}\left(\vec{J} + \rho q \vec{v}\right) + \sum_{K=1}^{N} q_K \pi_K \tag{4.76}$$

or with the material time derivative (4.50) and the continuity equation (4.51)

$$\boxed{\rho \frac{\mathrm{D}q}{\mathrm{D}t} = -\mathrm{div}\, \vec{J} + \sum_{K=1}^{N} q_K\, \pi_K} \tag{4.77}$$

Either (4.76) or (4.77) is the *electric current balance equation*, which reduces to the law of conservation of electric charge if $\sum_{K=1}^{N} q_K \pi_K = 0$.

The *electromagnetic momentum per volume* is given by

$$\vec{D} \times \vec{B} = \frac{1}{c_0^2}\left(\vec{E} \times \vec{H}\right) \tag{4.78}$$

Indeed, from (4.70) it follows that $\dim[\vec{D} \times \vec{B}] = \dim\,[\rho \vec{v}] = \mathrm{kg\ m^{-2}\ s^{-1}}$. Substitution of the Maxwell equations (4.69) into the time derivative of (4.78) gives

$$\begin{aligned}
\frac{\partial}{\partial t}\vec{D} \times \vec{B} &= \frac{\partial \vec{D}}{\partial t} \times \vec{B} + \vec{D} \times \frac{\partial \vec{B}}{\partial t} = \\
&= \left(\mathrm{curl}\ \vec{H} - \vec{J} - \rho\, q \vec{v}\right) \times \vec{B} - \vec{D} \times \mathrm{curl}\ \vec{E} = \\
&= \mathrm{div}\left[\epsilon_0 \vec{E}\vec{E} + \mu_0 \vec{H}\vec{H} - \frac{1}{2}\left(\epsilon_0 E^2 + \mu_0 H^2\right)\mathbf{I}\right] \\
&\quad - \vec{J} \times \vec{B} - \rho q\left(\vec{E} + \vec{v} \times \vec{B}\right)
\end{aligned} \tag{4.79}$$

so that

$$\boxed{\frac{\partial}{\partial t}\left(\frac{1}{c_0^2}\vec{E}\times\vec{H}\right) = -\mathrm{div}\,\mathbf{P}^{(\mathrm{em})} - \rho\vec{f}^{(\mathrm{em})}} \tag{4.80}$$

which can be considered as an electromagnetic balance of momentum, where the electromagnetic Maxwell stress tensor is given by

$$\mathbf{P}^{(\mathrm{em})} = -\left[\epsilon_0\vec{E}\vec{E} + \mu_0\vec{H}\vec{H} - \frac{1}{2}\left(\epsilon_0 E^2 + \mu_0 H^2\right)\mathbf{I}\right] \tag{4.81}$$

and the electromagnetic force is given by

$$\rho\vec{f}^{(\mathrm{em})} = \rho q\left(\vec{E} + \vec{v}\times\vec{B}\right) + \vec{J}\times\vec{B} \tag{4.82}$$

From (4.81) it follows that the Maxwell stress tensor is symmetric or

$$\boxed{\mathbf{P}^{(\mathrm{em})} = \mathbf{P}^{(\mathrm{em})^{\mathrm{T}}}} \tag{4.83}$$

where the superscript T indicates the transpose of a second order tensor. With the use of the Maxwell electromagnetic field equations and the constitutive equations (4.72) one derives

$$\begin{aligned}
\mathrm{div}\left(\vec{E}\times\vec{H}\right) &= -\vec{E}\cdot\mathrm{curl}\,\vec{H} + \vec{H}\cdot\mathrm{curl}\,\vec{E} = \\
&= -\vec{E}\cdot\left[\frac{\partial\vec{D}}{\partial t} + \vec{J} + \rho q\vec{v}\right] - \vec{H}\cdot\frac{\partial\vec{B}}{\partial t} = \\
&= -\frac{\partial}{\partial t}\frac{1}{2}\left(\epsilon_0 E^2 + \mu_0 H^2\right) - \vec{E}\cdot\left(\vec{J} + \rho q\vec{v}\right)
\end{aligned} \tag{4.84}$$

which results, after a rearrangement of terms, in the Poynting equation

$$\boxed{\frac{\partial}{\partial t}\frac{1}{2}\left(\epsilon_0 E^2 + \mu_0 H^2\right) = -\mathrm{div}\left(\vec{E}\times\vec{H}\right) - \vec{E}\cdot\left(\vec{J} + \rho q\vec{v}\right)} \tag{4.85}$$

The Poynting equation is to be seen as an electromagnetic energy balance equation, where $(\partial/\partial t)\frac{1}{2}(\epsilon_0 E^2 + \mu_0 H^2)$ is the electromagnetic energy per volume, $\vec{E}\times\vec{H}$ the electromagnetic flux vector (known as the Poynting vector) and $\vec{E}\cdot(\vec{J} + \rho q\vec{v}) = \vec{E}\cdot\vec{j}$ is the Joule heat that represents the loss of energy by the total electric current density $\vec{j} = \vec{J} + \rho\, q\vec{v}$.

Note, that in the derivation of (4.80) and (4.85) it was not necessary to use the Cauchy flux theorem for internal actions. Nevertheless, the divergence term is identified with the flux quantities introduced by the Cauchy axiom of internal actions. For example, in the momentum equation (4.80) the flux is in the divergence identified by (4.79) with a stress tensor, although the last quantity is a locally defined quantity, while the range of the electromagnetic stress is not local.

The electromagnetic balance equations (4.80) and (4.85) are formulated elegantly in the spatial description. For obtaining the material description use

$$\rho \frac{\mathrm{D}\Psi/\rho}{\mathrm{D}t} = \frac{\partial \Psi}{\partial t} + \mathrm{div}\ (\vec{v}\Psi) \tag{4.86}$$

which follows by applying the continuity equation (4.47). Ψ is in (4.86) an (extensive) property per volume. Using this result the material description of (4.80), for example, can be written as

$$\rho \frac{\mathrm{D}}{\mathrm{D}t}\left(\frac{1}{\rho c_0^2}\vec{E} \times \vec{H}\right) = -\mathrm{div}\left[\mathbf{P}^{(\mathrm{em})} - \vec{v}\,\frac{\vec{E} \times \vec{H}}{c_0^2}\right] - \rho \vec{f}^{(\mathrm{em})} \tag{4.87}$$

Clearly, the spatial descriptions are simpler than the material descriptions of the electromagnetic fields. Note also the minus sign of the convective transport of electromagnetic momentum!

4.6. MOMENTUM BALANCES

The actions that change the momentum of the matter in a control volume are divided into external and internal actions. For the external actions only gravitational actions are considered, because the electromagnetic actions are introduced through the complete momentum balance. For gravitational forces we have

$$\vec{f}_K = \vec{g} \qquad \Rightarrow \qquad \vec{f} = \sum_{K=1}^{N} w_K \vec{f}_K = \vec{g} \tag{4.88}$$

For constant $\vec{g}$

$$\vec{f} = \vec{g} = \mathrm{grad}\ (\vec{g} \cdot \vec{x}) \tag{4.89}$$

To assign a scalar potential for $\rho\vec{g}$ if $\vec{g}$ is not constant is not as easy as for constant $\vec{g}$.

The total momentum per unit of volume is the sum of the mechanical momentum and the electromagnetic momentum per unit volume

$$\sum_{K=1}^{N} \rho_K \vec{v}_K + \frac{1}{c_0^2}\vec{E} \times \vec{H} \tag{4.90}$$

It is obvious that one should introduce here the barycentric velocity $\vec{v}$

$$\rho\vec{v} = \sum_{K=1}^{N} \rho_K \vec{v}_K \qquad \Rightarrow \qquad \sum_{K=1}^{N} \rho_K \vec{V}_K = \vec{0} \tag{4.91}$$

The total momentum then becomes

$$\rho\vec{v} + \frac{1}{c_0^2}\vec{E} \times \vec{H} \tag{4.92}$$

The flux of momentum consists of the convective transport of momentum $\sum_{K=1}^{N} \rho_K \vec{v}_K \vec{v}_K$, the sum of the partial pressure stress tensors $\mathbf{P}_K$, and the electromagnetic flux of momentum (= the Maxwell stress tensor)

$$\sum_{K=1}^{N} (\rho_K \vec{v}_K \vec{v}_K + \mathbf{P}_K) + \mathbf{P}^{(\mathrm{em})} \tag{4.93}$$

Substitution of $(4.91)_2$ into (4.93) results in

$$\rho\vec{v}\vec{v} + \widetilde{\mathbf{P}} + \mathbf{P}^{(\mathrm{em})} \tag{4.94}$$

where

$$\widetilde{\mathbf{P}} = \sum_{K=1}^{N} \left(\mathbf{P}_K + \rho_K \vec{V}_K \vec{V}_K\right) \tag{4.95}$$

The differential balance equation for the total momentum becomes

$$\frac{\partial}{\partial t}\left(\rho\vec{v} + \frac{1}{c_0^2}\vec{E} \times \vec{H}\right) = -\mathrm{div}\left(\rho\vec{v}\vec{v} + \widetilde{\mathbf{P}} + \mathbf{P}^{(\mathrm{em})}\right) + \rho\vec{g} \tag{4.96}$$

The mechanical pressure stress tensor is introduced with the help of the Cauchy stress theorem that is derived as TIP axiom III. In the formulation of (4.96) it is assumed that no net barycentric momentum is produced by exchange of momentum between the chemical components caused by chemical reactions. This conclusion is based on the fact that a beaker, containing a reactant fluid, does not jiggle around, and by using the principle of local and instantaneous equilibrium.

Subtracting the electromagnetic balance of momentum (4.80) from (4.96) yields

$$\frac{\partial}{\partial t}(\rho\vec{v}) = -\mathrm{div}\left(\rho\vec{v}\vec{v} + \widetilde{\mathbf{P}}\right) + \rho q\vec{E}' + \vec{J} \times \vec{B} + \rho\vec{g} \tag{4.97}$$

where $\vec{E}' = \vec{E} + \vec{v} \times \vec{B}$ is the already defined transformation of $\vec{E}$ to the rest system in a semi-relativistic Lorentz transformation. With the continuity equation the first equation of motion (4.97) can be rewritten as

$$\boxed{\rho\frac{\mathrm{D}\vec{v}}{\mathrm{D}t} = -\mathrm{div}\,\widetilde{\mathbf{P}} + \rho q\vec{E}' + \vec{J} \times \vec{B} + \rho\vec{g} = -\mathrm{div}\,\widetilde{\mathbf{P}} + \rho\left(\vec{f}^{(\mathrm{em})} + \vec{g}\right)} \tag{4.98}$$

In a neutral fluid $\rho q = 0$, and the resulting equation is often used in magnetohydrodynamics (MHD flow).

Finally, it is pointed out that the balance equation (4.96) reduces for $\rho\vec{g} = \vec{0}$ to a law of conservation of momentum.

4.7. MOMENT OF MOMENTUM BALANCE

The proof of the symmetry of $\mathbf{P}^{(\mathrm{em})}$ has already been given in (4.83). The symmetry for the mechanical part of the stress pressure tensor for non-polar media can be derived from the total balance of the moment of momentum

$$\frac{\partial}{\partial t}\left[\vec{x}\times\left(\rho\vec{v}+\frac{1}{c_0^2}\vec{E}\times\vec{H}\right)\right]= -\operatorname{div}\left[\rho\vec{v}\,(\vec{x}\times\vec{v})+\vec{x}\times\widetilde{\mathbf{P}}+\vec{x}\times\mathbf{P}^{(\mathrm{em})}\right]+\rho\,\vec{x}\times\vec{g} \quad (4.99)$$

Taking the vector product of (4.80) with $\vec{x}$ gives

$$\frac{\partial}{\partial t}\vec{x}\times\left(\frac{1}{c_0^2}\vec{E}\times\vec{H}\right)=-\vec{x}\times\operatorname{div}\mathbf{P}^{(\mathrm{em})}-\rho\,\vec{x}\times\vec{f}^{(\mathrm{em})} \quad (4.100)$$

Because of the symmetry* of $\mathbf{P}^{(\mathrm{em})}$, this becomes

$$\frac{\partial}{\partial t}\vec{x}\times\left(\frac{1}{c_0^2}\vec{E}\times\vec{H}\right)=-\operatorname{div}\left(\vec{x}\times\mathbf{P}^{(\mathrm{em})}\right)-\rho\,\vec{x}\times\vec{f}^{(\mathrm{em})} \quad (4.101)$$

Subtracting (4.101) from (4.99) gives

$$\frac{\partial}{\partial t}\vec{x}\times\rho\vec{v}=-\operatorname{div}\left[\rho\vec{v}\,(\vec{x}\times\vec{v})+\vec{x}\times\widetilde{\mathbf{P}}\right]+\rho\,\vec{x}\times\vec{f}^{(\mathrm{em})}+\rho\,\vec{x}\times\vec{g}$$

or

$$\rho\frac{\mathrm{D}}{\mathrm{D}t}\,(\vec{x}\times\vec{v})=-\operatorname{div}\left(\vec{x}\times\widetilde{\mathbf{P}}\right)+\rho\,\vec{x}\times\vec{f}^{(\mathrm{em})}+\rho\,\vec{x}\times\vec{g} \quad (4.102)$$

The symmetry of the mechanical part of the stress pressure tensor follows if the vector product of (4.98) with $\vec{x}$ is taken and the final product is subtracted from (4.102), resulting in

$$\boxed{\widetilde{\mathbf{P}}=\widetilde{\mathbf{P}}^{\mathrm{T}}} \quad (4.103)$$

To obtain the symmetry of the stress tensor it is again assumed that there is no net production of moment of momentum by exchange of moment of

---

* Here use $\left(\vec{x}\times\operatorname{div}\mathbf{P}^{(\mathrm{em})}\right)=\operatorname{div}\left(\vec{x}\times\mathbf{P}^{(\mathrm{em})}\right)+\left(\mathbf{P}^{(\mathrm{em})\mathrm{T}}\cdot\operatorname{grad}\right)\times\vec{x}$. In the latter the last term is zero, because of the antisymmetric properties of the permutation tensor $\mathbf{e}$ and the symmetry of the tensor $\mathbf{P}^{(\mathrm{em})}$

$$\left(\mathbf{P}^{(\mathrm{em})\mathrm{T}}\cdot\operatorname{grad}\right)\times\vec{x}=-P_{ij}^{(\mathrm{em})}\frac{\partial}{\partial x_i}e_{jlm}x_m\vec{e}_l=-P_{ij}^{(\mathrm{em})}e_{jlm}\delta_{im}\vec{e}_l$$
$$=e_{ljm}P_{mj}^{(\mathrm{em})}\vec{e}_l=\mathbf{e}:\mathbf{P}^{(\mathrm{em})}=\vec{0}$$

momentum between the chemical constituents and/or by production due to chemical reactions. A possible exchange among the constituents of momentum, moment of momentum or energy, might cause only a redistribution of those quantities over the components, but does not give rise to a net production (at least this is assumed). A net production would result in a couple that is not compensated, and as a consequence would cause an infinitely fast rotation of a volume element around its center of mass. If there are intrinsic angular momenta, then such a fluid is called a *polar fluid*, in which the intrinsic angular momentum is balanced by couples, and then (4.103) need no longer hold*.

Under the given circumstances, the balance of angular momentum leads to the conclusion that the stress tensor is symmetric. It has to be pointed out that it cannot be proved that each partial stress tensor $\mathbf{P}_K$ is symmetric! From the derivation given, it is hardly necessary to emphasize that the symmetry relation (4.103) is not an intrinsic property of the stress tensor, as many text books assume, but is in fact an equation of motion, known as the (barycentric) *Cauchy second equation of motion.*

4.8. ENERGY BALANCE

Total energy is obtained by summing internal energy, kinetic energy, potential energy and electromagnetic energy. There is no net production of total energy by interconversion of energy between the constituents or by production of energy by chemical reactions. On a per-unit-volume basis this amounts to

$$\rho e_{(\mathrm{tot})} = \sum_{K=1}^{N} \rho_K u_K + \tfrac{1}{2} \sum_{K=1}^{N} \rho_K v_K^2 + \rho\xi + \tfrac{1}{2}\left(\epsilon_0 E^2 + \mu_0 H^2\right) \tag{4.104}$$

in which it is supposed that it is possible to calculate the partial internal energy u_K of component K per unit of mass, and where ξ is the potential energy defined by $\vec{g} = -\operatorname{grad}\xi$ with $\xi = -\vec{g}\cdot\vec{x}$. With the substitution of

$$\vec{v}_K = \vec{v} + \vec{V}_K \tag{4.105}$$

and $(4.91)_2$ the expression for total energy per unit volume becomes (note that v^2 represents the square of the velocity $\vec{v}$, and not the square of the specific volume v)

$$\rho e_{(\mathrm{tot})} = \rho\widetilde{u} + \tfrac{1}{2}\rho v^2 + \rho\xi + \tfrac{1}{2}\left(\epsilon_0 E^2 + \mu_0 H^2\right) \tag{4.106}$$

in which $\widetilde{u}$ is defined by

$$\rho\widetilde{u} = \sum_{K=1}^{N} \left(\rho_K u_K + \tfrac{1}{2}\rho_K V_K^2\right) \tag{4.107}$$

* Dahler J.S. and Scriven L.E. 1961. Angular momentum of continua..*Nature*, 192: 36–37.

For obtaining a balance equation for the change of the internal energy one can use the substitution

$$\psi \to e_{(\text{tot})}$$

in the general balance equation (3.14). On the right-hand side of the general balance equation, the convective transport of electromagnetic energy is given by the Poynting vector $\vec{E} \times \vec{H}$. The mutual internal surface actions performed by each component K are made up of the work exerted by the internal forces and the internal supply of energy associated with internal energy transport. With use of the flux theorem for internal actions, the flux of the internal energy supply is determined by the energy flux vector $\vec{\Phi}_{(\text{u})K}$, so that in the general balance equation the substitution

$$\vec{\phi}_{(\psi)} \to \sum_{K=1}^{N} \mathbf{P}_K \cdot \vec{v}_K + \sum_{K=1}^{N} \vec{\Phi}_{(\text{u})K}$$

can be made. The production term $\pi_{(\psi)}$ in the general balance equation now contains a contribution of external force fields $\vec{f}'$ acting on each mass particle other than the gravitational forces, which are accounted for in the potential energy, so that $\pi_{(\psi)} \to \sum_{K=1}^{N} \rho\vec{v}_K \cdot \vec{f}'_K = \rho\vec{v} \cdot \vec{f}' + \sum_{K=1}^{N} \vec{\Phi}_K \cdot \vec{f}'_K$. Finally, the external energy $\pi_{(\psi)} \to \sum_{K=1}^{N} \rho_K z_K$ can be supplied by radiation, where z_K is the energy per unit volume of component K due to radiation. This contribution will be neglected from now on. Summarizing, the law of *conservation of energy* reads

$$\frac{\partial}{\partial t}\rho e_{(\text{tot})} = -\text{div}\,\vec{\Phi}_{(\text{e})} + \rho\vec{v} \cdot \vec{f}' + \sum_{K=1}^{N} \vec{\Phi}_K \cdot \vec{f}'_K \tag{4.108}$$

in which

$$\vec{\Phi}_{(\text{e})} = \sum_{K=1}^{N} \left[\rho_K \vec{v}_K u_K + \rho_K \vec{v}_K \tfrac{1}{2} v_K^2 + \mathbf{P}_K \cdot \vec{v}_K\right] + \rho\,\vec{v}\,\xi + \vec{E} \times \vec{H} + \sum_{K=1}^{N} \vec{\Phi}_{(\text{u})K} \tag{4.109}$$

With the substitution of (4.105) and $\sum_{K=1}^{N} \rho_K \vec{V}_K = \vec{0}$, the energy flux $\vec{\Phi}_{(\text{e})}$ can be written as

$$\vec{\Phi}_{(\text{e})} = \rho\vec{v}\left(\widetilde{u} + \tfrac{1}{2}v^2 + \xi\right) + \widetilde{\mathbf{P}} \cdot \vec{v} + \vec{E} \times \vec{H} + \vec{\widetilde{\Phi}}_{(\text{u})} \tag{4.110}$$

with

$$\vec{\widetilde{\Phi}}_{(\text{u})} = \sum_{K=1}^{N} \left[\vec{\Phi}_{(\text{u})K} + \rho_K \vec{V}_K \left(u_K + \tfrac{1}{2}V_K^2\right) + \mathbf{P}_K \cdot \vec{V}_K\right] \tag{4.111}$$

After substitution of (4.104) and (4.110), the energy equation (4.108) becomes

$$\frac{\partial}{\partial t}\left[\rho\tilde{u} + \tfrac{1}{2}\rho v^2 + \rho\xi + \tfrac{1}{2}\left(\epsilon_0 E^2 + \mu_0 H^2\right)\right] = \\ -\operatorname{div}\left[\rho\vec{v}\left(\tilde{u} + \tfrac{1}{2}v^2 + \xi\right) + \widetilde{\mathbf{P}}\cdot\vec{v} + \vec{E}\times\vec{H} + \vec{\widetilde{\Phi}}_{(\mathrm{u})}\right] \\ + \rho\vec{v}\cdot\vec{f}' + \sum_{K=1}^{N}\vec{\Phi}_K\cdot\vec{f}'_K \qquad (4.112)$$

This is under the given conditions the *law of conservation of energy.* Subtraction of (4.85) from the above equation gives then

$$\frac{\partial}{\partial t}\left(\rho\tilde{u} + \tfrac{1}{2}\rho v^2 + \rho\xi\right) = -\operatorname{div}\left[\rho\vec{v}\left(\tilde{u} + \tfrac{1}{2}v^2 + \xi\right) + \widetilde{\mathbf{P}}\cdot\vec{v} + \vec{\widetilde{\Phi}}_{(\mathrm{u})}\right] \\ + \rho\vec{v}\cdot\vec{f}' + \sum_{K=1}^{N}\vec{\Phi}_K\cdot\vec{f}'_K + \vec{E}\cdot\left(\vec{J} + \rho q\vec{v}\right)$$

or by use of the continuity equation

$$\rho\frac{\mathrm{D}}{\mathrm{D}t}\left(\tilde{u} + \tfrac{1}{2}v^2 + \xi\right) = -\operatorname{div}\left(\widetilde{\mathbf{P}}\cdot\vec{v} + \vec{\widetilde{\Phi}}_{(\mathrm{u})}\right) \\ + \rho\vec{v}\cdot\vec{f}' + \sum_{K=1}^{N}\vec{\Phi}_K\cdot\vec{f}'_K + \vec{E}\cdot\left(\vec{J} + \rho q\vec{v}\right) \qquad (4.113)$$

Furthermore, the potential energy in a gravitational field satisfies the equation

$$\rho\frac{\mathrm{D}\xi}{\mathrm{D}t} = \rho\vec{v}\cdot\operatorname{grad}\xi = -\rho\,\vec{v}\cdot\vec{g}$$

so that (4.113), with $\vec{f} = \vec{f}' + \vec{g}$, becomes

$$\rho\frac{\mathrm{D}}{\mathrm{D}t}\left(\tilde{u} + \tfrac{1}{2}v^2\right) = -\operatorname{div}\left(\widetilde{\mathbf{P}}\cdot\vec{v} + \vec{\widetilde{\Phi}}_{(\mathrm{u})}\right) \\ + \vec{E}\cdot\left(\vec{J} + \rho q\vec{v}\right) + \rho\vec{v}\cdot\vec{f} + \sum_{K=1}^{N}\vec{\Phi}_K\cdot\vec{f}'_K \qquad (4.114)$$

By taking the scalar product of the momentum balance equation (4.98) with $\vec{v}$, the term $\vec{v}\times\vec{B}$ in $\vec{E}'$ does not contribute, since $\vec{v}\cdot(\vec{v}\times\vec{B}\,) = 0$. This results in the *balance equation for the mechanical energy*

$$\boxed{\rho\frac{\mathrm{D}}{\mathrm{D}t}\tfrac{1}{2}v^2 = -\vec{v}\cdot\operatorname{div}\widetilde{\mathbf{P}} + \vec{v}\cdot\left(\rho q\vec{E} + \vec{J}\times\vec{B}\right) + \rho\vec{v}\cdot\vec{f}} \qquad (4.115)$$

Subtract this balance from (4.114)

$$\rho \frac{\mathrm{D}\widetilde{u}}{\mathrm{D}t} = -\mathrm{div}\,\vec{\widetilde{\Phi}}_{(\mathrm{u})} + \vec{v}\cdot \mathrm{div}\,\widetilde{\mathbf{P}} - \mathrm{div}\left(\widetilde{\mathbf{P}}\cdot\vec{v}\right) + \vec{E}'\cdot\vec{J} + \sum_{K=1}^{N}\vec{\Phi}_K\cdot\vec{f}'_K \qquad (4.116)$$

Because of the symmetry of $\widetilde{\mathbf{P}}$, it follows that*

$$\vec{v}\cdot\mathrm{div}\,\widetilde{\mathbf{P}} - \mathrm{div}\left(\widetilde{\mathbf{P}}\cdot\vec{v}\right) = -\widetilde{\mathbf{P}}^{\mathrm{T}} : (\mathrm{grad}\,\vec{v}\,) = -\widetilde{\mathbf{P}} : \boldsymbol{D}$$

in which the barycentric *rate of deformation tensor* of Euler or *stretching tensor* is defined by

$$\boldsymbol{D} = \tfrac{1}{2}\left[(\mathrm{grad}\,\vec{v})^{\mathrm{T}} + \mathrm{grad}\,\vec{v}\right] \qquad (4.117)$$

The equation of the modified 'internal' energy becomes

$$\rho \frac{\mathrm{D}\widetilde{u}}{\mathrm{D}t} = -\mathrm{div}\,\vec{\widetilde{\Phi}}_{(\mathrm{u})} - \widetilde{\mathbf{P}} : \boldsymbol{D} + \vec{E}'\cdot\vec{J} + \sum_{K=1}^{N}\vec{\Phi}_K\cdot\vec{f}'_K \qquad (4.118)$$

This is the *thermal energy equation*, which is also called the *internal energy equation*, since this equation no longer contains the barycentric kinetic energy or the potential energy

According to (4.107), the modified 'internal' energy is given by

$$\widetilde{u} = u + \sum_{K=1}^{N} \tfrac{1}{2} w_K V_K^2$$

with

$$u = \sum_{K=1}^{N} w_K u_K \qquad (4.119)$$

In the thermodynamics of irreversible processes V_K is of first order in smallness. The terms $\rho_K V_K^2$, which can be interpreted as the kinetic energy of diffusion of component K, are of the second order and need not be considered as a contribution to the thermodynamic internal energy. From (4.111) it is seen that the kinetic energy of diffusion is counted differently than the internal energy u. However, for a consistent theory the second order terms are neglected and thus it may be asserted that

$$\widetilde{u} \approx u \qquad (4.120)$$

---

* Use is made of $\widetilde{\mathbf{P}}^{\mathrm{T}} : (\mathrm{grad}\,\vec{v}\,) = \widetilde{\mathbf{P}} : (\mathrm{grad}\,\vec{v}\,) = \widetilde{\mathbf{P}} : (\mathrm{grad}\,\vec{v}\,)^{\mathrm{T}}$.

The quantity u is interpreted as the thermodynamic internal energy in this equation, and the known thermodynamic formulas can be applied to u. Again, by neglecting second order terms in (4.95), $\rho_K \vec{V}_K \vec{V}_K$ can be discarded with respect to $\mathbf{P}_K$. Since the stress tensor $\mathbf{P}_K$ of component K is under normal flow conditions small to first order, second order terms can be neglected with respect to the first order terms

$$\widetilde{\mathbf{P}} \approx \mathbf{P} = \sum_{K=1}^{N} \mathbf{P}_K . \tag{4.121}$$

Analogously, by neglecting all* second order terms, the energy flux (4.111) reduces to

$$\widetilde{\vec{\Phi}}_{(\mathrm{u})} \approx \vec{\Phi}_{(\mathrm{u})} = \sum_{K=1}^{N} \vec{\Phi}_{(\mathrm{u})K} \tag{4.122}$$

so that the thermal energy equation (4.118) with neglect of all second order terms becomes

$$\boxed{\rho \frac{\mathrm{D}u}{\mathrm{D}t} = -\mathrm{div}\, \vec{\Phi}_{(\mathrm{u})} - \mathbf{P} : \boldsymbol{D} + \sum_{K=1}^{N} \vec{\Phi}_K \cdot \vec{f}'_K + \vec{E}' \cdot \vec{J}} \tag{4.123}$$

in which the last term on the right-hand side represents the Joule heat.

In applications, the left side of the energy equation is usually expressed as $\rho c_p \, \mathrm{D}T/\mathrm{D}t$. From the experimental point of view, the intensive variables T and p are well suited to characterize the state of fluids and solids, while for gases sometimes the variables T and V are simpler to use. To transform the thermal energy equation into an equation for $\mathrm{D}T/\mathrm{D}t$ first the specific enthalpy is introduced. If $\widetilde{u} \approx u$ and the internal energy u is considered as a thermodynamic quantity, then it follows from $(4.26)_1$

$$dh = du + \frac{1}{\rho}\, dp + p\, d\left(\frac{1}{\rho}\right) \tag{4.124}$$

By applying the principle of local and instantaneous equilibrium, and using the continuity equation, the material time derivative of h is found to be

$$\rho \frac{\mathrm{D}h}{\mathrm{D}t} = \rho \frac{\mathrm{D}u}{\mathrm{D}t} + \frac{\mathrm{D}p}{\mathrm{D}t} + p\, \mathrm{div}\, \vec{v} \tag{4.125}$$

* Kirkwood J.C. and Crawford B.L. Jr. 1952. The macroscopic equations of transport. *Journal of Physics and Chemistry*, 56: 1048–1051. Kirkwood and Crawford argue that the second order term $\rho_K \vec{V}_K u_K = \vec{\Phi}_K u_K$ shows the energy flux by diffusion produced volume increment, and should not be neglected. However, for a consistent first order theory this term is discarded, but might be considered as a nonlinear extrapolation if there is experimental evidence for such an extrapolation. The term $\mathbf{P}_K \cdot \vec{V}_K$ is also not taken into account, but similarly, this term can be considered as dissipation due to the diffusion velocities. However, this seems to be a small second order contribution under normal flow conditions, where a one-component description of diffusion is acceptable.

From this *enthalpy equation* it follows by the elimination of $\rho\,\mathrm{D}u/\mathrm{D}t$ by using (4.123), and again neglecting terms of the second order

$$\rho\,\frac{\mathrm{D}h}{\mathrm{D}t} = -\mathrm{div}\,\vec{\Phi}_{(\mathrm{u})} - \boldsymbol{\Pi} : \boldsymbol{D} + \sum_{K=1}^{N} \vec{\Phi}_K \cdot \vec{f}'_K + \vec{E}' \cdot \vec{J} + \frac{\mathrm{D}p}{\mathrm{D}t} \tag{4.126}$$

where the viscous stress tensor (also called the deviatoric stress tensor or the extra stress tensor) is defined by

$$\boldsymbol{\Pi} = \mathbf{P} - p\,\mathbf{I} \tag{4.127}$$

From thermodynamics—see (4.38)—it follows that

$$dh = c_p\,dT + \frac{1}{\rho}(1 - \alpha T)\,dp + \sum_{K=1}^{N} h_K\,dw_K$$

or by the application of the local and instantaneous equilibrium

$$\rho\frac{\mathrm{D}h}{\mathrm{D}t} = \rho c_p\,\frac{\mathrm{D}T}{\mathrm{D}t} + (1 - \alpha T)\,\frac{\mathrm{D}p}{\mathrm{D}t} + \sum_{K=1}^{N} h_K\,\rho\,\frac{\mathrm{D}w_K}{\mathrm{D}t}$$

Substitution of (4.126) and (4.56) into the above equation yields

$$\rho c_p\frac{\mathrm{D}T}{\mathrm{D}t} = -\mathrm{div}\,\vec{\Phi}_{(\mathrm{u})} - \boldsymbol{\Pi} : \boldsymbol{D} + \vec{E}' \cdot \vec{J} + \sum_{K=1}^{N} \vec{\Phi}_K \cdot \vec{f}'_K + \alpha T\,\frac{\mathrm{D}p}{\mathrm{D}t} + \sum_{K=1}^{N} h_K\left(\mathrm{div}\,\rho_K\vec{V}_K - \pi_K\right)$$

or

$$\rho c_p\,\frac{\mathrm{D}T}{\mathrm{D}t} + \sum_{K=1}^{N} \rho_K\vec{V}_K \cdot \mathrm{grad}\,h_K = -\mathrm{div}\,\vec{\Phi}_{(\mathrm{q})} - \boldsymbol{\Pi} : \boldsymbol{D} + \vec{E}' \cdot \vec{J} + \sum_{K=1}^{N} \vec{\Phi}_K \cdot \vec{f}'_K + \alpha T\,\frac{\mathrm{D}p}{\mathrm{D}t} + Q_{(\mathrm{p})} \tag{4.128}$$

with

$$\vec{\Phi}_{(\mathrm{q})} = \vec{\Phi}_{(\mathrm{u})} - \sum_{K=1}^{N} h_K\vec{\Phi}_K \tag{4.129}$$

$$Q_{(\mathrm{p})} = -\sum_{K=1}^{N} \pi_K h_K = \rho\sum_{K=1}^{N} \nu_{\alpha K}(m_K h_K)\,\frac{\mathrm{D}X'_\alpha}{\mathrm{D}t} \tag{4.130}$$

where in the second equality of (4.130) the expression (4.67) for π_K has been substituted. In (4.130) $m_K h_K$ is the partial molar enthalpy of component K and $Q_{(\mathrm{p})}$ is the reaction heat at constant pressure.

The flux $\vec{\Phi}_{(\mathrm{q})}$ represents in the thermal energy equation (4.128) the reduced heat flux of Prigogine, and $\sum_{K=1}^{N} \rho_K \vec{V}_K \cdot \mathrm{grad}\, h_K$ is the 'effect of diffusing heat capacities' of Ackermann*. The derivation† of Ackermann was not exact‡. The effect of diffusing heat capacities is neglected by most authors.

The thermal energy equation (4.128) can be regarded as the *temperature equation*. In the derivation of this equation the second order terms have been neglected. If all second order terms are consistently neglected, then

$$\rho_K \vec{V}_K \cdot \mathrm{grad}\, h_K \qquad \text{and} \qquad \boldsymbol{\Pi} : \boldsymbol{D}$$

are second order terms in the balance equations and should as such be neglected. Then (4.128) simplifies to

$$\rho c_p \frac{\mathrm{D}T}{\mathrm{D}t} = -\mathrm{div}\, \vec{\Phi}_{(\mathrm{q})} + \alpha T \frac{\mathrm{D}p}{\mathrm{D}t} + Q_{(\mathrm{p})} + \sum_{K=1}^{N} \vec{\Phi}_K \cdot \vec{f}'_K + \vec{E}' \cdot \vec{J} \qquad (4.131)$$

From the temperature equation it follows that the reduced heat flux of Prigogine is the correct definition of the heat flux in a multicomponent system. Meixner and Reik§ showed that with this choice of the reduced heat flux, the energy equation is invariant for an arbitrary change of the undetermined zero level of the energy, which is a requirement, since only changes in energy are important and chemical reactions do not produce zero-point energy. The term $\sum_{K=1}^{N} h_K \vec{\Phi}_K$ represents a heat flow due to diffusion that is irreversible.

From the applications and the comparison with experiments it follows that in the right-hand side of (4.131) the second order term $-\boldsymbol{\Pi} : \boldsymbol{D}$ should

---

* Ackermann G. 1937. Wärmeübertragung und moleculare Stoffübertragung in gleichen Feld bei groszen Temperatur- und Partialdruckdifferenzen. *V.D.I.-Forschungsheft*, 382: 1–16.

† It is easy to see that for ideal systems the partial heat capacities determine these extra terms, since for ideal systems $\mathrm{grad}\, h_K = c_{p,K} \mathrm{grad}\, T$. If furthermore it is assumed that $\rho c_p = \sum_K \rho_K c_{p,K}$ then the first term of the thermal energy equation becomes

$$\sum_{K=1}^{N} \rho_K c_{p,K} \left(\frac{\partial}{\partial t} + \vec{v}_K \cdot \mathrm{grad} \right) T$$

This suggests a summation of partial energy equations. The effect of the diffusing heat capacities was also found by Ackermann in this way.

‡ Merk H.J. 1959. The macroscopic equations for simultaneous heat and mass transfer in isotropic, continuous and closed systems. *Applied Scientific Research*, A8: 73–99.

§ Meixner J. and Reik H.G. 1959. Thermodynamik der irreversiblen Prozesse. In the *Encyclopedia of Physics*. Springer-Verlag, Berlin, p. 438.

be included, but the effect of the diffusing heat capacities has never been well tested, so that this contribution is usually omitted*. Finally, the most commonly used temperature equation for multicomponent, non-polar fluids is given by

$$\rho c_p \frac{\mathrm{D}T}{\mathrm{D}t} = -\mathrm{div}\, \vec{\Phi}_{(\mathrm{q})} + \alpha T \frac{\mathrm{D}p}{\mathrm{D}t} - \boldsymbol{\Pi} : \boldsymbol{D} + Q_{(\mathrm{p})} + \sum_{K=1}^{N} \vec{\Phi}_K \cdot \vec{f}'_K + \vec{E}' \cdot \vec{J} \tag{4.132}$$

For a single-component system, this equation was first derived by Fourier[†] based on the theory of the heat substance, in which only the first term in the right-hand side of (4.132) is counted. Kirchhoff[‡] derived a thermal energy equation for infinitesimal motions of a gas with applications to wave propagation of gases in pipes. A rather general derivation of the thermal energy equation for a single-component system[§] was not formulated until C. Neumann[¶].

---

* Gal-Or B. 1981. *Cosmology, Physics, and Philosophy.* Springer-Verlag, New York, remarks on pp. 157–159 that the complete energy equation in terms of the temperature field T, has been formulated for the first time by Merk H.J. 1959. The macroscopic equations for simultaneous heat and mass transfer in isotropic, continuous and closed systems. *Applied Scientific Research,* A8: 73–99., and that the neglect of the term $\sum_K \vec{\Phi}_K \cdot \mathrm{grad}\, h_K$ is a result of the incorrect derivations in most textbooks.

† Fourier J.B.J. 1833. *Théorie Analytique de la Chaleur.* Paris (1922); Sur le mouvement de la chaleur dans les fluides. *Memoires de l'Académie de Science Institute de France,* 12(2): 507–530.

‡ Kirchhoff G. 1868. Ueber die Einfluss der Wärmeleitung in einem Gase auf die Schallbewegung. *Annalen der Physik und Chemie (Poggendorffer Annalen),* 134: 177–193.

§ For a pure component $\mathrm{D}p/\mathrm{D}t$ can be eliminated by use of $(4.17)_2$. Dividing this equation by M, with the application of the principle of local and instantaneous equilibrium, yields $\rho\kappa_T \mathrm{D}p/\mathrm{D}t = \mathrm{D}\rho/\mathrm{D}t + \rho\alpha \mathrm{D}T/\mathrm{D}t$. By substitution of this result into the temperature equation (4.132) for a single pure component and by using the relation between the heat capacities at constant pressure and constant volume: $c_p - c_v = \alpha^2 T/(\rho\kappa_T)$, the thermal energy equation for a pure component becomes

$$\rho c_v \frac{\mathrm{D}T}{\mathrm{D}t} = -\mathrm{div}\, \vec{\Phi}_{(u)} + \frac{c_p - c_v}{\alpha} \frac{\mathrm{D}\rho}{\mathrm{D}t} - \boldsymbol{\Pi} : \boldsymbol{D} + \vec{E}' \cdot \vec{J}$$

If the fluid velocities are small compared with the velocity of sound, the pressure variations due to the fluid motions are too small to produce appreciable density variations, and the fluid is assumed to be incompressible. In L.D. Landau and E.M. Lifshitz, *Fluid Mechanics.* Pergamon, 1959, §50, it is pointed out that for non-uniformly heated fluids the density varies with temperature, and this variation cannot be neglected even at small fluid velocities. Therefore, it is necessary in the 'incompressible approximation' for non-uniformly heated fluids to assume that the pressure is constant, and not the density, for the determination of the derivatives of the thermodynamic quantities. In this incompressible approximation the temperature differences in the fluid also have to be small.

¶ Neumann C. 1894. Ueber die Bewegung der Wärme in compressiblen oder auch incompressiblen Flüssigkeiten. *Bericht über die Verhandlungen der (Koniglich) Sachsischen Gesell-*

4.9. ENTROPY BALANCE EQUATION

The entropy production term in the balance equation for the entropy (3.15) has to be calculated from the specific fundamental equation of Gibbs and the balance equations. The Gibbs fundamental equation (4.23) reads

$$du = T\,ds - p\,dv + \sum_{K=1}^{N} \mu_K\,dw_K \tag{4.133}$$

Again, by applying the principle of local and instantaneous equilibrium and by substituting $v = 1/\rho$ this equation can be written as

$$\rho T\,\frac{\mathrm{D}s}{\mathrm{D}t} = \rho\frac{\mathrm{D}u}{\mathrm{D}t} - \frac{p}{\rho}\frac{\mathrm{D}\rho}{\mathrm{D}t} - \sum_{K=1}^{N} \mu_K\rho\,\frac{\mathrm{D}w_K}{\mathrm{D}t}$$

Substitution of the thermal equation of energy (4.123), in which is assumed that $\widetilde{u} = u$ (this condition is essential!), and use of the continuity equation (4.51) lead to

$$\rho T\,\frac{\mathrm{D}s}{\mathrm{D}t} = -\mathrm{div}\,\vec{\Phi}_{(\mathrm{u})} - \boldsymbol{\Pi} : \boldsymbol{D} + \sum_{K=1}^{N} \vec{\Phi}_K \cdot \vec{f}'_K + \vec{E}' \cdot \vec{J} - \sum_{K=1}^{N} \mu_K\rho\,\frac{\mathrm{D}w_K}{\mathrm{D}t} \tag{4.134}$$

The last term on the right-hand side can be rewritten, using the diffusion equation (4.56), as

$$\sum_{K=1}^{N} \mu_K\rho\,\frac{\mathrm{D}w_K}{\mathrm{D}t} = -\sum_{K=1}^{N} \mu_K\left[\mathrm{div}\,(\rho_K\vec{V}_K) - \pi_K\right] \tag{4.135}$$

in which, with the substitution of (4.67), the term $\sum_{K=1}^{N}\mu_K\pi_K$ becomes

$$\sum_{K=1}^{N} \mu_K\pi_K = -\rho\sum_{\alpha=1}^{\kappa} A_\alpha\dot{X}'_\alpha$$

with $\dot{X}'_\alpha \equiv \mathrm{D}X'_\alpha/\mathrm{D}t$ and the introduction of the affinity A_α of the reaction α

$$A_\alpha = -\sum_{K=1}^{N} \nu_{\alpha K}m_K\mu_K \tag{4.136}$$

---

schaft (Akademie) der Wissenschaften zu Leipzig. (Mathematisch-Physische Classe), 46: 1–24.

The introduction of the affinity can be traced back to De Donder* and turns out to be an extrapolation of classical thermodynamics. The affinity acts as the driving force of the progressing reaction and turns out to be in accordance with the experimental evidence. With the introduction of the affinity it becomes possible to predict the progress of chemical processes that would otherwise be impossible† without the introduction of these irreversible changes $A_\alpha \dot{X}'_\alpha$. Similarly, the introduction of the affinities is equally important in the discussion of systems having internal degrees of freedom.

For a system that is locally in complete true equilibrium $A_\alpha = 0$. The affinities are found only if also $dX'_\alpha \neq 0$, and adding the term $A_\alpha \dot{X}'_\alpha$ can, like the term $\boldsymbol{\Pi} : \boldsymbol{D}$, be considered as a contribution of the second order. These second order contributions have to be validated by experimental evidence if physico-chemical processes occur that give cause for irreversible processes of the first and the second kind.

The entropy balance equation (4.134) therefore becomes

$$\rho T \frac{\mathrm{D}s}{\mathrm{D}t} = -\mathrm{div}\, \vec{\Phi}_{(\mathrm{u})} - \boldsymbol{\Pi} : \boldsymbol{D} + \sum_{K=1}^{N} \vec{\Phi}_K \cdot \vec{f}'_K + \vec{E}' \cdot \vec{J} + \sum_{K=1}^{N} \mu_K \mathrm{div}\,(\rho_K \vec{V}_K) + \rho \sum_{\alpha=1}^{\kappa} A_\alpha \dot{X}'_\alpha$$

This equation has not yet been cast into the form of a barycentric entropy balance equation: $\rho\,\mathrm{D}s/\mathrm{D}t = -\mathrm{div}\,\vec{\Phi}_{(\mathrm{s})} + \pi_{(\mathrm{s})}$. To aim for this goal, divide the equation by T, to get

$$\begin{aligned} \rho \frac{\mathrm{D}s}{\mathrm{D}t} &= -\mathrm{div}\left[\frac{1}{T}\left(\vec{\Phi}_{(\mathrm{u})} - \sum_{K=1}^{N} \mu_K \rho_K \vec{V}_K\right)\right] \\ &+ \left(\vec{\Phi}_{(\mathrm{u})} - \sum_{K=1}^{N} \mu_K \rho_K \vec{V}_K\right) \cdot \mathrm{grad}\, T^{-1} - \frac{1}{T}\sum_{K=1}^{N} \rho_K \vec{V}_K \cdot \mathrm{grad}\, \mu_K \\ &- \frac{1}{T}\boldsymbol{\Pi} : \boldsymbol{D} + \frac{1}{T}\sum_{K=1}^{N} \vec{\Phi}_K \cdot \vec{f}'_K + \frac{1}{T}\vec{E}' \cdot \vec{J} + \frac{\rho}{T}\sum_{\alpha=1}^{\kappa} A_\alpha \dot{X}'_\alpha \end{aligned} \quad (4.137)$$

In this equation $\vec{\Phi}_{(\mathrm{u})}$ is still found, although for multicomponent fluids $\vec{\Phi}_{(\mathrm{q})}$ is a better choice for the heat flux. To get an equation in $\vec{\Phi}_{(\mathrm{q})}$, differentiate (4.10)

---

* De Donder Th. 1920. Transformations physiques et chimiques des systèmes de Gibbs. Académie Royal de Belgique. *Bulletins de la Classe des Sciences*, 5(6): 315–328; De Donder Th. and Van Rysselberghe P. 1936. *Thermodynamic Theory of Affinity.* Stanford University Press, California.

† Prigogine I. and Defay R. 1954. *Chemical Thermodynamics.* Longmans Green and Co., London. See p. 43.

partially with respect to M_K at constant T, p, M' and apply the definition (4.28) for partial quantities, giving

$$h_K = u_K + pv_K \qquad \text{and} \qquad g_K \equiv \mu_K = u_K - Ts_K + pv_K \tag{4.138}$$

so that

$$\mu_K = h_K - Ts_K \tag{4.139}$$

where $s_K = -(\mu_K - h_K)/T$ denotes the partial specific entropy. This is used to rewrite the first two terms in the right-hand side of (4.137)

$$\vec{\Phi}_{(\mathrm{u})} - \sum_{K=1}^{N} \mu_K \rho_K \vec{V}_K = \vec{\Phi}_{(\mathrm{q})} + T \sum_{K=1}^{N} \rho_K \vec{V}_K \, s_K \tag{4.140}$$

where $\vec{\Phi}_{(\mathrm{q})}$ is again the reduced heat of Prigogine. Substitution yields for (4.137)

$$\begin{aligned} \rho \frac{\mathrm{D}s}{\mathrm{D}t} &= -\mathrm{div}\left[\left(\vec{\Phi}_{(\mathrm{q})}/T\right) + \sum_{K=1}^{N} \rho_K \vec{V}_K \, s_K\right] \\ &+ \left(\vec{\Phi}_{(\mathrm{q})} + T\sum_{K=1}^{N} \rho_K \vec{V}_K s_K\right) \cdot \mathrm{grad}\, T^{-1} - \frac{1}{T}\sum_{K=1}^{N} \rho_K \vec{V}_K \cdot \mathrm{grad}\, \mu_K \\ &- \frac{1}{T}\boldsymbol{\Pi} : \boldsymbol{D} + \frac{1}{T}\sum_{K=1}^{N} \vec{\Phi}_K \cdot \vec{f}'_K + \frac{1}{T}\vec{E}' \cdot \vec{J} + \frac{\rho}{T}\sum_{\alpha=1}^{\kappa} A_\alpha \dot{X}'_\alpha \end{aligned} \tag{4.141}$$

In addition, the contributions of the diffusion flux $\vec{\Phi}_K = \rho_K \vec{V}_K$ to the entropy production can be assembled. According to $(4.17)_3$

$$d\mu_K = -s_K \, dT + (d\mu_K)_T$$

The application of the principle of local and instantaneous equilibrium implies that

$$\begin{aligned} \mathrm{grad}\, \mu_K &= -s_K \mathrm{grad}\, T + (\mathrm{grad}\, \mu_K)_T \\ &= T^2 s_K \mathrm{grad}\, T^{-1} + (\mathrm{grad}\, \mu_K)_T \end{aligned}$$

Substitution into (4.141) finally yields

$$\begin{aligned} \rho \frac{\mathrm{D}s}{\mathrm{D}t} &= -\mathrm{div}\left[\left(\vec{\Phi}_{(\mathrm{q})}/T\right) + \sum_{K=1}^{N} \vec{\Phi}_K s_K\right] + \vec{\Phi}_{(\mathrm{q})} \cdot \mathrm{grad}\, T^{-1} - \frac{1}{T}\boldsymbol{\Pi} : \boldsymbol{D} \\ &- \frac{1}{T}\sum_{K=1}^{N} \vec{\Phi}_K \cdot \left[(\mathrm{grad}\, \mu_K)_T - \vec{f}'_K\right] + \frac{1}{T}\vec{E}' \cdot \vec{J} + \frac{\rho}{T}\sum_{\alpha=1}^{\kappa} A_\alpha \dot{X}'_\alpha \end{aligned} \tag{4.142}$$

This equation can be interpreted as the barycentric entropy balance equation

$$\boxed{\rho \frac{\mathrm{D}s}{\mathrm{D}t} = -\mathrm{div}\,\vec{\Phi}_{(\mathrm{s})} + \pi_{(\mathrm{s})}} \tag{4.143}$$

in which the barycentric entropy flux is defined by

$$\boxed{\vec{\Phi}_{(\mathrm{s})} = \left(\vec{\Phi}_{(\mathrm{q})}/T\right) + \sum_{K=1}^{N} \vec{\Phi}_K s_K} \tag{4.144}$$

containing the reduced heat flux and the flow of the partial entropies with respect to the barycentric velocity. In (4.143) the entropy production per unit of time and unit of volume is defined by

$$\pi_{(\mathrm{s})} = \vec{\Phi}_{(\mathrm{q})} \cdot \mathrm{grad}\,T^{-1} - T^{-1} \sum_{K=1}^{N} \vec{\Phi}_K \cdot \left[(\mathrm{grad}\,\mu_K)_T - \vec{f}'_K\right]$$
$$- T^{-1}\boldsymbol{\Pi} : \boldsymbol{D} + T^{-1}\vec{E}' \cdot \vec{J} + \rho T^{-1} \sum_{\alpha=1}^{\kappa} A_\alpha \dot{X}'_\alpha \tag{4.145}$$

The spatial description of the barycentric entropy balance equation (4.143) becomes

$$\frac{\partial}{\partial t}(\rho s) = -\mathrm{div}\left(\vec{\Phi}_{(\mathrm{s})} + \rho \vec{v} s\right) + \pi_{(\mathrm{s})} \tag{4.146}$$

with

$$\vec{\Phi}_{(\mathrm{s})} + \rho \vec{v} s = \left(\vec{\Phi}_{(\mathrm{q})}/T\right) + \sum_{K=1}^{N} \rho_K \vec{v}_K s_K \tag{4.147}$$

in which $\rho s = \sum_{K=1}^{N} \rho_K s_K$ has been used.

Therefore the entropy flux consists of (reduced) energy transport and of convective transport of entropy. This seems to be a logical result, in which the production term is such that

$$\boxed{\pi_{(\mathrm{s})} \geq 0} \tag{4.148}$$

The uncompensated heat produced by irreversible processes is given by the dissipation function. The dissipation function $\mathfrak{D}$ can now be defined by

$$\boxed{\mathfrak{D} = T\pi_{(\mathrm{s})}} = -T^{-1}\vec{\Phi}_{(\mathrm{q})} \cdot \mathrm{grad}\,T - \sum_{K=1}^{N} \vec{\Phi}_K \cdot \left[(\mathrm{grad}\,\mu_K)_T - \vec{f}'_K\right]$$
$$- \boldsymbol{\Pi} : \boldsymbol{D} + \vec{E}' \cdot \vec{J} + \rho \sum_{\alpha=1}^{\kappa} A_\alpha \dot{X}'_\alpha \geq 0 \tag{4.149}$$

The heat produced by the electric current and by diffusion can be combined. Now

$$\vec{J} = \sum_{K=1}^{N} \rho_K \vec{V}_K q_K = \sum_{K=1}^{N} \vec{\Phi}_K q_K$$

so that (4.149) can be written as

$$\boxed{\begin{aligned}\mathfrak{D} = -T^{-1}\vec{\Phi}_{(q)} \cdot \operatorname{grad} T - \sum_{K=1}^{N} \vec{\Phi}_K \cdot \left[(\operatorname{grad} \mu_K)_T - \vec{f}'_K - q_K \vec{E}'\right] \\ - \boldsymbol{\Pi} : \boldsymbol{D} + \rho \sum_{\alpha=1}^{\kappa} A_\alpha \dot{X}'_\alpha \geq 0\end{aligned}} \tag{4.150}$$

in which the equality sign holds for equilibrium and the inequality sign holds if there is no equilibrium. The further details of the thermodynamics of the irreversible processes lead to inconvenient formulas, so that in the subsequent discussions the phenomena will be handled in groups.

From (4.150) and the balance equations it follows that the process fluxes can be defined as

Scalar fluxes: the κ reaction velocities $\{\dot{X}'_\alpha\}$.
Vector fluxes: the reduced heat flux $\vec{\Phi}_{(q)}$ and the N mass fluxes $\vec{\Phi}_K$.
Tensor fluxes: the viscous stress tensor $\boldsymbol{\Pi}$.

The conjugate process forces associated with the above fluxes are then

Scalar forces: the κ affinities $\{A_\alpha\}$.
Vector forces: $-T^{-1}\operatorname{grad} T$ and $-\{(\operatorname{grad} \mu_K)_T - \vec{f}'_K - q_K \vec{E}'\}$.
Tensor forces: the rate of deformation tensor $-\boldsymbol{D}$.

The processes of even and odd order are carefully distinguished in this arrangement. The even-order processes are the scalar and the tensor processes, while the odd-order processes are the vector processes. It is remarked that for the vector processes, gravity does not cause diffusion; only the non-conservative forces may cause diffusion. The acceleration of the gravitation is the same for the lighter and the heavier molecules. The variation of composition with height is due to the atmospheric pressure gradient.

The phenomenological equations have to be formulated after a choice has been made for the forces and the fluxes. This will be discussed extensively in the applications to diffusion and relaxing phenomena in Chapters 6 and 7 of this book. However, to illustrate the procedures used in the thermodynamics of irreversible processes, the processes of even order in complete systems are now briefly discussed.

4.10. PROCESSES OF EVEN ORDER

As a linear constitutive theory, the linearizations in the thermodynamics of irreversible processes do not arise so much in the formulation of the dissipation function (4.150), but in the formulation of the phenomenological equations. As summarized in TIP axiom V, the process fluxes are therefore homogeneous linear functions of the process forces,

Normal fluids are isotropic if the fluid is not subjected to electromagnetic fields. Hence, the Curie symmetry principle can be applied in the formulation of the linear phenomenological equations. This means that only fluxes of even (odd) order can depend on the forces of even (odd) order. The dissipation function of the processes of even order is, according to (4.150),

$$\mathfrak{D} = -\boldsymbol{\Pi} : \boldsymbol{D} + \rho \sum_{\alpha=1}^{\kappa} A_\alpha \dot{X}'_\alpha \geq 0 \tag{4.151}$$

The phenomenological equations for processes of even order are given, analogously to $(3.43)_1$ and $(3.43)_3$, by

$$\begin{aligned} \rho \frac{\mathrm{D}X'_\alpha}{\mathrm{D}t} &= \sum_{\beta=1}^{\kappa} L_{\alpha\beta} A_\beta - L_{\alpha\mathrm{v}}(\mathbf{I} : \boldsymbol{D}) \\ \boldsymbol{\Pi} &= \left(\sum_{\beta=1}^{\kappa} L_{\mathrm{v}\beta} A_\beta \right) \mathbf{I} - L_{(1)}(\mathbf{I} : \boldsymbol{D})\mathbf{I} - L_{(2)}\boldsymbol{D} - L_{(3)}\boldsymbol{D}^{\mathrm{T}} \end{aligned} \tag{4.152}$$

By using the symmetry property of the rate of deformation tensor $\boldsymbol{D} = \boldsymbol{D}^{\mathrm{T}}$, with the substitution of $\mathbf{I} : \boldsymbol{D} = \operatorname{div} \vec{v}$, and the replacement of $\boldsymbol{D}$ by its deviator $\boldsymbol{D} = \mathring{\boldsymbol{D}} + \frac{1}{3}(\operatorname{div} \vec{v})\,\mathbf{I}$, the equations in (4.152) become

$$\begin{aligned} \rho \frac{\mathrm{D}X'_\alpha}{\mathrm{D}t} &= \sum_{\beta=1}^{\kappa} L_{\alpha\beta} A_\beta - L_{\alpha\mathrm{v}} \operatorname{div} \vec{v} \\ \boldsymbol{\Pi} &= \left(\sum_{\beta=1}^{\kappa} L_{\mathrm{v}\beta} A_\beta \right) \mathbf{I} - \kappa\,(\operatorname{div} \vec{v})\mathbf{I} - 2\eta \mathring{\boldsymbol{D}} \end{aligned} \tag{4.153}$$

in which the two new constitutive coefficients κ and η have been introduced.

Apply finally TIP axiom VI. In the absence of an external magnetic field, the Onsager Casimir reciprocal relations hold with respect to an inertial system. Assume that with respect to that inertial system The A_β are even variables for all β, while $\operatorname{div} \vec{v}$ is an odd variable. Then the Onsager Casimir reciprocal relations are

$$L_{\alpha\beta} = L_{\beta\alpha} \qquad \text{and} \qquad L_{\alpha\mathrm{v}} = -L_{\mathrm{v}\alpha} \equiv \lambda_\alpha \tag{4.154}$$

The phenomenological equations therefore become

$$
\begin{aligned}
\rho\frac{\mathrm{D}X'_\alpha}{\mathrm{D}t} &= \sum_{\beta=1}^{\kappa} L_{\alpha\beta}A_\beta - \lambda_\alpha \operatorname{div}\vec{v} \\
\boldsymbol{\Pi} &= -\left(\sum_{\beta=1}^{\kappa}\lambda_\beta A_\beta\right)\mathbf{I} - \kappa\,(\operatorname{div}\vec{v}\,)\,\mathbf{I} - 2\eta\mathring{\boldsymbol{D}}
\end{aligned}
\tag{4.155}
$$

in which $L_{\alpha\beta}$ denote the chemical coefficients, κ the dilatational viscosity coefficient, η the viscosity coefficient of Newton and λ_α the chemical viscosity coefficient.

The dissipation function becomes with (4.154)

$$
\mathfrak{D} = \sum_{\alpha=1}^{\kappa}\sum_{\beta=1}^{\kappa} L_{\alpha\beta}A_\alpha A_\beta + \kappa\,(\operatorname{div}\vec{v}\,)^2 + 2\eta\mathring{\boldsymbol{D}} : \mathring{\boldsymbol{D}} \geq 0 \tag{4.156}
$$

With the assumption that the expansion flow ($\operatorname{div}\vec{v} \neq 0, \mathring{\boldsymbol{D}} = \mathbf{0}$), the deviatoric flow ($\operatorname{div}\vec{v} = 0, \mathring{\boldsymbol{D}} \neq \mathbf{0}$) and the scalar physico-chemical processes, can be controlled independently from each other, then the dissipation function shows that

$$
\|L_{\alpha\beta}\| \text{ is positive definite} \qquad \kappa \geq 0 \qquad \eta \geq 0 \tag{4.157}
$$

Since the introduction by Newton*, the existence of the viscosity coefficient η has been generally accepted for fluids with a relatively simple molecular structure (not macromolecules), in which η depends on first examination only on the temperature T

$$
\begin{aligned}
\text{gases: } & d\eta/dT > 0 \\
\text{liquids: } & d\eta/dT < 0
\end{aligned}
$$

The existence of the volume viscosity has to be coupled with scalar internal processes, according to Mandelstam and Leontovitsch†. The time scales of these scalar processes are much shorter than those associated with the exchange of momentum. An example in gases is thermal relaxation, because the vibrations of the polyatomic molecules are not in equilibrium with the translation of the molecules. Only for monatomic molecules does the relation of Stokes‡ $\kappa = 0$ hold with certainty.

---

* Newton I. 1687. *Philosophia Naturalis Principia Mathematica.* London.

† Mandelstam L.J. and Leontovitsch M.A. 1937. To the theory of sound absorption in fluids (in Russian). *Journal of Experimental and Theoretical Physics (USSR)*, 7: 438–449.

‡ Stokes G.G. 1845. On the theories of the internal friction of fluids in motion, and of the equilibrium and motion of elastic solids. *Transactions of the Cambridge Philosophical Society*, 8: 287–319.

In fluids it can be assumed that $\kappa = \mathcal{O}(\eta)^*$. The bulk viscosity κ can be measured with ultrasonic vibrations. The effect of the bulk viscosity in deviatoric flows is mostly of minor importance[†].

Meixner[‡] explains, parallel to Mandelstam and Leontovitsch, that the chemical viscosities are caused by fast progressing chemical reactions (fast compared with the exchange of momentum). Usually, the chemical viscosities are difficult to measure and are often neglected. No direct coupling between the chemical reactions and the flux of momentum is then found.

4.11. EXERCISES

Exercise 4.1. Derive from $(4.17)_2$ and (4.20)

$$dv/v = \alpha T(dT/T) - \kappa_T(\Delta P)(dp/\Delta P) + \sum_{K=1}^{N-1} \frac{v_K - v_N}{v}\, dw_K \tag{4.158}$$

For $|dv|/v \ll 1$ a fluid flow can be considered as incompressible. Show that the incompressibility conditions are:

(1) $|\alpha T| \ll 1$ or $|\Delta T|/T \ll 1$. Discuss whether the first condition is satisfied for liquids, and the second for gases where ΔT denotes a characteristic temperature difference in the flow field.
(2) $\kappa_T(\Delta P) \ll 1$, where ΔP denotes a characteristic pressure difference in the flow field. Use $\kappa_T = \gamma\kappa_s$, where $\gamma = C_P/C_V$ is the ratio of heat capacities (isentropic index), and assume that for the isentropic compressibility $\kappa_s = 1/\rho a^2$. The pressure difference ΔP is caused by the characteristic velocity V. Show that ΔP is of the order ρV^2, and that $\kappa_T(\Delta P) \approx \gamma\, Ma^2 \ll 1$ is in practice satisfied for $Ma < 0.1$. Estimate the upper bound for V for a liquid and for a gas at normal pressure and temperature.
(3) The condition $|v_K - v_N|/v \ll 1$ for $K = 1, 2, \cdots, N-1$ means that the partial volumes of the components must not differ greatly for using the

---

* Rosenhead L. 1954. The second coefficient of viscosity: a brief review of fundamentals. *Proceedings of the Royal Society*, A226: 1–6. For normal fluids it found that $1.7\eta \le \kappa \le 4.7\eta$; Emanuel G. 1992. Effect of bulk viscosity on a hypersonic boundary layer. *Physics of Fluids*, A4(3): 491–495; Tisza L. 1942. Supersonic absorption and Stokes' viscosity relation. *Physical Review*, 61: 531–536, gives for $\kappa/\eta \approx 2 \times 10^3$ for CO_2 and N_2O based on the experimental results obtained by Kneser H.O. 1933. Schallabsorption in mehratomigen Gasen. *Annalen der Physik*, 16: 337–349.

† Kuiken G.D.C. 1984. Wave propagation in fluid lines. *Applied Scientific Research*, 41: 69–91.

‡ Meixner J. 1951/1952. Strömungen von fluïden Medien mit inneren Umwandlungen und Druckviskosität. *Zeitschrift für Physik*, 131: 456–469.

incompressible approximation of the flow of a mixture. Show that for ideal gas mixtures this condition becomes

$$|v_K - v_N|/v = m|m_K^{-1} - m_N^{-1}| \ll 1 \qquad \text{for} \qquad K = 1, 2, \cdots, N-1$$

so that the molar masses of the components must be of the same order of magnitude.

(4) With (4.158) only the 'static' incompressibility conditions can be derived. Sudden changes in a fluid may cause the propagation of compressible waves. If the time scale $t^{(\mathrm{m})}$ of the macroscopic changes is small enough, compressible flow phenomena can show up. However, if the time scale of the macroscopic phenomena is large with respect to the time scale of the propagation of the compressible waves, the incompressible approximation can be used. Show that this holds for

$$Sr\, Ma \ll 1$$

where $Sr = L/t^{(\mathrm{m})}V$ is the Strouhal number, and estimate the order of magnitude of the Strouhal number for applying the incompressibility approximation for time-dependent flows.

Exercise 4.2. Show that the momentum balance equation (4.98) becomes with the substitution of (4.122) and (4.127)

$$\rho\frac{\mathrm{D}\vec{v}}{\mathrm{D}t} = \rho\vec{f} - \operatorname{div}\widetilde{\mathbf{P}} = \rho\vec{f} - \operatorname{grad} p - \operatorname{div}\boldsymbol{\Pi} \tag{4.159}$$

and that the substitution of the phenomenological equation $(4.155)_2$ yields the generalized Navier–Stokes equations

$$\rho\frac{\mathrm{D}\vec{v}}{\mathrm{D}t} = \rho\vec{f} - \operatorname{grad} p + 2\operatorname{div}\eta\mathring{\boldsymbol{D}} + \operatorname{grad}(\kappa\operatorname{div}\vec{v}\,) + \sum_{\beta=1}^{\kappa}\operatorname{grad}\,(\lambda_\beta A_\beta) \tag{4.160}$$

Assume that there is only one chemical process σ, and show that the diffusion equation (4.56) becomes with the substitution of (4.67)

$$\rho\frac{\mathrm{D}w_K}{\mathrm{D}t} = -\operatorname{div}\vec{\Phi}_K + \nu_{\sigma K} m_K \frac{\mathrm{D}X'_\sigma}{\mathrm{D}t} \tag{4.161}$$

and that with the substitution of the phenomenological equation from $(4.155)_1$ this diffusion equation is

$$\rho\frac{\mathrm{D}w_K}{\mathrm{D}t} = -\operatorname{div}\vec{\Phi}_K + \nu_{\sigma K} m_K\,(L_{\sigma\sigma}A_\sigma - \lambda_\sigma\operatorname{div}\vec{v}\,) \tag{4.162}$$

Discuss the fact that $\mathrm{D}w_K/\mathrm{D}t$ and $\vec{\Phi}_K$ are approximately zero by comparing the time scales of the internal scalar process with the macroscopic time scales. Show that the Navier–Stokes equation (4.160) then becomes

$$\rho\frac{\mathrm{D}\vec{v}}{\mathrm{D}t} = \rho\vec{f} - \operatorname{grad} p + 2\operatorname{div}\eta\overset{\circ}{\boldsymbol{D}} + \operatorname{grad}\{(\kappa_\sigma + \kappa)\operatorname{div}\vec{v}\}$$

with

$$\kappa_\sigma = \lambda_\sigma^2/L_{\sigma\sigma} > 0$$

Explain why $\kappa_\sigma > 0$, and that the dilatational viscosity coefficient can be understood in terms of scalar internal processes. Discuss whether the existence of the dilatational viscosity implies the existence of the chemical viscosity λ_σ.

Exercise 4.3. An arbitrary one-component fluid in equilibrium can sustain only spherically symmetric pressure stresses. In equilibrium, for the entropy

$$s^{(0)} = s(u,\rho) \tag{4.163}$$

If the fluid is not in equilibrium, then local equilibrium is defined by the entropy $s^{(0)}$, which is calculated as the equilibrium entropy for the true values of the density ρ and the internal energy u, so that again (4.163) applies locally. The differential formulation of (4.163) is

$$ds^{(0)} = \frac{1}{T}\,du - \frac{p}{\rho^2 T}\,d\rho \tag{4.164}$$

which defines the temperature and the thermodynamic pressure as equilibrium quantities. The thermal energy equation (4.123) is

$$\rho\frac{\mathrm{D}u}{\mathrm{D}t} = -\operatorname{div}\vec{\Phi}_{(\mathrm{u})} - \mathbf{P}:\boldsymbol{D} \tag{4.165}$$

Assume

$$\mathbf{P} = -P\mathbf{I} + \overset{\circ}{\boldsymbol{P}} \qquad \text{with} \qquad P = -\tfrac{1}{3}\mathbf{I}:\mathbf{P} \qquad \text{and} \qquad \mathbf{I}:\overset{\circ}{\boldsymbol{P}} = 0$$

and show that the balance equation for the entropy $s^{(0)}$ is given by

$$\begin{aligned}\rho\frac{\mathrm{D}s^{(0)}}{\mathrm{D}t} &= -\operatorname{div}(\vec{\Phi}_{(\mathrm{u})}/T) + \vec{\Phi}_{(\mathrm{u})}\cdot\operatorname{grad}(1/T)\\ &\qquad - \frac{1}{T}(P-p)I_D + \frac{1}{T}\overset{\circ}{\boldsymbol{P}}:\overset{\circ}{\boldsymbol{D}}\end{aligned} \tag{4.166}$$

where $I_D = \operatorname{div}\vec{v} = \mathbf{I}:\mathbf{D}$, and show that the changes of state are reversible if

$$\operatorname{grad} T = \vec{0} \qquad I_D = 0 \qquad \text{and} \qquad \overset{\circ}{\boldsymbol{D}} = \mathbf{0} \tag{4.167}$$

so that locally there is equilibrium or quasi-equilibrium, for which

$$\vec{\Phi}_{(\mathrm{u})} = \vec{0} \qquad \text{and} \qquad \mathbf{P} = -p\mathbf{I} \tag{4.168}$$

In a nonequilibrium system the true entropy is

$$s = s^{(0)} + s'(p, T, \operatorname{grad} p, \operatorname{grad} T, \mathring{\boldsymbol{D}}, I_D) \tag{4.169}$$

where s' may also depend on the time and spatial derivatives of the given arguments. Show that objectivity requires that s' depends only on objective scalars formed from the arguments of s'.

The local equilibrium is assumed to be stable, so that the entropy is a maximum in equilibrium. Show that the first approximation $s^{(1)}$ of s' is definite negative and is given by

$$s^{(1)} = -s_{TT}(\operatorname{grad} T)^{2\cdot} - s_{TP}(\operatorname{grad} T)\cdot(\operatorname{grad} p) - s_{PP}(\operatorname{grad} p)^{2\cdot} - s_{VV} I_D^2 - s_{DD}\mathring{\boldsymbol{D}} : \mathring{\boldsymbol{D}} \tag{4.170}$$

from which it follows that the coefficients are given by

$$\left.\begin{array}{lll} s_{TT} \geq 0 & s_{PP} \geq 0 & s_{TT}s_{PP} - s_{TP}^2 \geq 0 \\ & s_{VV} \geq 0 & s_{DD} \geq 0 \end{array}\right\} \tag{4.171}$$

Use the conditions for local equilibrium and show that $s_{PP} = 0$ and $s_{TP} = 0$, so that (4.170) reduces to

$$s^{(1)} = -s_{TT}(\operatorname{grad} T)^{2\cdot} - s_{VV} I_D^2 - s_{DD}\mathring{\boldsymbol{D}} : \mathring{\boldsymbol{D}} \tag{4.172}$$

so that $\operatorname{grad} p$ does not enter into the first approximation. Discuss whether from (4.172) it follows that local equilibrium applies to the first order deviations inclusive and that then the entropy

$$s = s^{(0)} \tag{4.173}$$

shows that the fundamental equation of Gibbs applies, and that (4.166) shows that the dissipation is given as a sum of the products of forces and fluxes.

Show that the entropy production outside local equilibrium follows from the entropy balance

$$\rho \frac{\mathrm{D}s}{\mathrm{D}t} = -\operatorname{div}(\vec{\Phi}_{(\mathrm{u})}/T) + \pi_{(\mathrm{s})} \tag{4.174}$$

with the entropy production

$$\mathfrak{D} = T\pi_{(\mathrm{s})} = \frac{1}{T}\left(\vec{\Phi}_{(\mathrm{u})} + 2s_{TT}\frac{\mathrm{D}}{\mathrm{D}t}\operatorname{grad} T\right)\cdot \operatorname{grad} T - \left(P - p + 2s_{VV}\frac{\mathrm{D}}{\mathrm{D}t} I_D\right) I_D - \left(\mathring{\boldsymbol{P}} + 2s_{DD}\frac{\mathrm{D}}{\mathrm{D}t}\mathring{\boldsymbol{D}}\right) : \mathring{\boldsymbol{D}} \tag{4.175}$$

Derive the corresponding constitutive equations and discuss that (4.176) applies only through terms of first order

$$\mathfrak{D} = T\pi_{(\mathrm{s})} = \sum \text{force} \times \text{flux} \tag{4.176}$$

Exercise 4.4. Derive the diffusion equations for the mixed diffusion flux and the molar diffusion flux given in table 4.1.

Exercise 4.5. Derive the Maxwell electromagnetic field equations. Compare a standard textbook derivation with the derivation given in appendix A. Decide which explanation of the Maxwell electromagnetic equations you prefer.

Exercise 4.6. The substitution of the condition for an incompressible fluid that the density is constant suggests that the specific heat at constant volume should be used in the temperature equation. For many liquids the difference between the specific heats at constant volume and constant pressure is negligible. For many polymeric liquids the difference can be large however, and then it becomes important which specific heat is used in the incompressible approximation. Discuss whether for non-uniformly heated liquids in the incompressible approximation the specific heat at constant pressure should be used instead of the specific heat at constant volume.

5 Statistical Foundation of the Onsager Casimir Reciprocal Relations for Homogeneous Systems

Summary: The reciprocal relations are derived, using the solutions of the Fokker–Planck equation that satisfies the microscopic reversibility and the regression axiom. The solutions are Gaussian distributions, and the Boltzmann distribution is one of possible solutions. Not all solutions result in the reciprocal relations. The Boltzmann distribution is one of the solutions that does yields the reciprocal relations.

5.1. INTRODUCTION

The theoretical basis for the reciprocal relations stated in postulate VI is discussed in this chapter. The symmetry postulate is an essential element of the thermodynamics of irreversible processes. Sometimes it is called the fourth law of thermodynamics. A blind acceptance of this postulate is, in fact, a denial of the great physical meaning of the TIP.

Occasionally, the Onsager Casimir reciprocal relations are found experimentally or the conjecture of the existence of these relations has been confirmed by tests. Also the kinetic theory of dilute gases gave expressions for the thermal diffusion and Dufour effect before the rise of the TIP. For very special cases the Onsager Casimir reciprocal relations are derived based on molecular structural models.

In 1931 Onsager* published pioneering articles that have stimulated the development of the TIP. He applied the invariance of the equations of motion for the atoms and molecules with respect to time reversal (the transformation $t \to -t$) at the level of fluctuation theory, and by doing so he made a principal decision, by which the transition from molecular reversibility to microscopic reversibility can be made. It is important to remark that Onsager did not use a particular molecular model. As a consequence, the results and limitations of the theory are valid for all materials, so that the theory can be related to the continuum theory.

Subsequent authors have built on the work of Onsager. For example, Casimir† emphazised the essential difference between the 'even' and 'odd'

* Onsager L. 1931. Reciprocal relations in irreversible processes, I. *Physical Review*, 37: 405–426; Reciprocal relations in irreversible processes, II. *Physical Review*, 38: 2265–2279.

† Casimir H.B.G. 1945. On the Onsager principle of microscopic reversibility. *Review of Modern Physics*, 17: 343–350.

quantities, while in a further stage electromagnetic fields were considered. The contributions of Onsager and Casimir can be split into three main subjects, namely

(1) The general statistical theory of fluctuations.
(2) The transition from molecular reversibility to microscopic reversibility.
(3) The laws controlling the regression of the fluctuations.

In the literature on the symmetry relations the first item is fairly generally accepted. However, the second and the third items have been subject to criticism and discussion. The statistics of the fluctuations given in the literature are quite generally based on the Boltzmann entropy postulate: *the entropy of a system is proportional to the logarithm of the probability of the corresponding state of the system*, although the foundation of this postulate is doubtful. The connection between the probability of a state and its entropy is suggested from the consideration that the entropy of two independent systems is additive, while the probability of the occurrence of the mutual state is determined by the product of the probabilities of each state separately. In mixing two gases, the mixed state is the most probable one. The entropy of the mixed state is larger than the sum of the entropies of both states separately. It is further assumed that there exists a relation between the entropy and the probability of the state of the system. For the combined system one deduces from the additivity of the entropy and from the statement that the probability is equal to the product of the separate probabilities, that the relation has to be logarithmic. Specification of the *Boltzmann constant* in this relation follows from the application of the supposed logarithmic law to ideal gases. Although the Boltzmann postulate (engraved on his tombstone) is hardly doubted by most physicists due to its proven usefulness, it has no rational foundation. To quote Khinchin*: 'All existing attempts to give a general proof of this postulate must be considered as an aggregate of logical and mathematical errors superimposed on a general confusion in the definition of the basic quantities. In most serious treatises on that subject (for example: Fowler R.H. 1936. *Statistical Mechanics.* Cambridge) the authors refuse to accept this postulate, indicating that it cannot be proved, and cannot be given a sensible formulation even on the basis of the exact notions of thermodynamics.'

By solving the Fokker–Planck equation for the probability distribution† in a homogeneous system, it turns out that it is not necessary to accept in advance the Boltzmann postulate for the derivation of the symmetry relations. In the formulation of this Fokker–Planck equation no assumptions regarding

* Khinchin A.I. 1949. *Mathematical Foundations of Statistical Mechanics.* Dover Publ. Inc., New York, p. 142.

† Kuiken G.D.C. 1977. The derivation of the Onsager Casimir reciprocity relations without using Boltzmann's postulate. *Journal of Non-Equilibrium Thermodynamics*, 2: 153–168.

microscopic reversibility are used. If microscopic reversibility is considered, then it follows from this derivation that more than one type of Gaussian distribution is acceptable, still guaranteeing the symmetry relations. The conditions for accepting the Boltzmann postulate also result.

The assumptions, proposed by Onsager and Casimir for the items (2) and (3), are as a matter of fact generalities of regularities that have already proven their usefulness in special physical problems. Furthermore, these assumptions appear not to be in contradiction with modern experience. Besides, as far as realizable, the consequences of these assumptions have been tested experimentally. As always in each physical theory, one has to look carefully at the circumstances under which the assumptions involved are declared to be true.

To apply the fluctuation theory to advantage, Onsager considered a system that consists of homogeneous phases with a homogeneous temperature distribution and that is almost in equilibrium. The temperature differences are thought to be very minor. Onsager called such a system an *aged system.* Such a system is kept in a compartment that is completely isolated from other systems. A system in such a confinement becomes aged, if it has been isolated sufficiently long. *Globally*, this system is in equilibrium; however, *locally* it may deviate from the equilibrium state due to stochastic motions of the atoms and molecules. This reasoning is based on considering ΔV's slightly smaller than the ΔV's for which the thermodynamic quantities can be defined (a discussion over the conditions and magnitudes of the ΔV's needed to define locally the thermodynamic quantities can be found in Chapter 1). The macroscopic quantities are defined using physical volume-elements ΔV's, while here small variations of the local equilibrium quantities are considered. As such, the equilibrium state is thought to be the most probable state, consistent with the imposed boundary conditions. Suppose that there is a small but finite chance that variations of the most probable state take place. These variations are caused by the stochastic motions of the atoms and molecules and as a consequence are also stochastic. These stochastic variations of the equilibrium state are called *fluctuations.* If a fluctuation is initiated then there is a tendency to return to the equilibrium state. *The return to the equilibrium state can be interpreted as an irreversible process.* Intuitively, it is conjectured that there is a connection between the progress of the macroscopic irreversible processes. This conjecture is the crux of the general statistical foundation of the thermodynamics of irreversible processes, for which the fluctuation theory used in this sense is still considered as belonging to the continuum theory. The thermodynamic functions are as such regarded as stochastic functions according to the fluctuations occurring.

From the above discussion it follows that the TIP can be seen as an extrapolation of classical thermodynamics (TIP = classical thermodynamics

+ fluctuation theory). This extrapolation is probably correct only in the first approximation and correct only for average values. The contribution of Onsager is the generalization of the fluctuation theory, the introduction of microscopic reversibility and above all a statement about *how* the subsystem returns to the equilibrium state after a fluctuation has occurred, the so-called regression axiom. For the proper appreciation of these contributions it is necessary to study the fluctuation theory. From this theory it will be shown that the regression axiom and the assumption of the microscopic reversibility are not sufficient to decide the symmetry relations of Onsager, as proposed in literature*, where the Boltzmann postulate is used in advance.

5.2. DESCRIPTION OF STOCHASTIC PROCESSES

A position vector $\vec{\zeta}(t) = [\zeta_1(t), \zeta_2(t), \cdots, \zeta_l(t)]$ in the l-dimensional ζ space, represents a stochastic process. This means that $\vec{\zeta}$ depends on t, but incompletely defined. The value at a particular point of time is an 'event' $(\vec{\zeta}, t)$. The behavior of the stochastic process $\vec{\zeta}(t)$ has to be described by statistical methods. For this purpose the concept of the 'ensemble' can be formally applied.

Ensemble. An *ensemble* exists of a very large number of systems that all satisfy the same macroscopic conditions. Fluctuations are of a microscopic nature. The microscopic state may differ from system to system, provided that each state of fluctuation is in concordance with the macroscopic constraints.

A system of the ensemble can be represented by a point mapped in the ζ space. Since usually the fluctuating state of each system changes with time, this point describes a path in this space. An ensemble is therefore represented in the ζ space by a cloud of points, and each point passes through trajectories in this space. At a time t it can be expected that a certain state of the system can be realized with a certain probability. A small volume $d\vec{\zeta}$ in the ζ space is given by

$$d\vec{\zeta} = \prod_{k=1}^{l} d\zeta_k$$

A small volume $d\vec{\zeta}$ around a point $\vec{\zeta}$ in ζ space is denoted by $\left[\vec{\zeta}, \vec{\zeta} + d\vec{\zeta}\right]$. Let the probability that $\vec{\zeta} \in \left[\vec{\zeta}, \vec{\zeta} + d\vec{\zeta}\right]$ at time t be given by

$$W_1(\vec{\zeta}, t)\, d\vec{\zeta}$$

* De Groot S.R. 1974. The Onsager relations; theoretical basis. In *Foundations of Continuum Thermodynamics*, Edited by Delgado Domingos J.J, Nina M.N.R. and Whitelaw J.H. MacMillan Press Ltd, London, 1974. See pp. 159–183.

If the statistical discussions are based on the concept of an ensemble, then $W_1 d\vec{\zeta}$ is the number of points of the ensemble that are found in $\left[\vec{\zeta}, \vec{\zeta} + d\vec{\zeta}\right]$ at time t.

The probability W_1 as a function of $\vec{\zeta}$ and t is in general *not* sufficient to characterize a stochastic process, since this probability says nothing about possible correlations between the events at different times. To establish the correlations between the various events, the various times $t^{(1)}, t^{(2)}, \cdots, t^{(n)}$ are fixed on the time scale. Now the following definitions can be given

$$
\begin{aligned}
W_1(\vec{\zeta}^{(i)}, t^{(i)})\, d\vec{\zeta}^{(i)} &= \text{the probability that} \\
&\quad \vec{\zeta} \in \left[\vec{\zeta}^{(i)}, \vec{\zeta}^{(i)} + d\vec{\zeta}^{(i)}\right] \qquad \text{for} \qquad t = t^{(i)}
\end{aligned}
$$

$$
\begin{aligned}
W_2(\vec{\zeta}^{(i)}, t^{(i)}; \vec{\zeta}^{(j)}, t^{(j)})\, d\vec{\zeta}^{(i)} d\vec{\zeta}^{(j)} &= \text{the probability that} \\
&\quad \vec{\zeta} \in \left[\vec{\zeta}^{(i)}, \vec{\zeta}^{(i)} + d\vec{\zeta}^{(i)}\right] \qquad \text{for} \qquad t = t^{(i)} \\
\text{and} &\quad \vec{\zeta} \in \left[\vec{\zeta}^{(j)}, \vec{\zeta}^{(j)} + d\vec{\zeta}^{(j)}\right] \qquad \text{for} \qquad t = t^{(j)}
\end{aligned}
$$

and so on, until finally

$$
\begin{aligned}
W_n(\vec{\zeta}^{(1)}, t^{(1)}; \vec{\zeta}^{(2)}, t^{(2)}; \cdots; \vec{\zeta}^{(n)}, t^{(n)}) \prod_{i=1}^{n} d\vec{\zeta}^{(i)} &= \text{the probability that} \\
&\quad \vec{\zeta} \in \left[\vec{\zeta}^{(1)}, \vec{\zeta}^{(1)} + d\vec{\zeta}^{(1)}\right] \qquad \text{for} \qquad t = t^{(1)} \\
\text{and} &\quad \vec{\zeta} \in \left[\vec{\zeta}^{(2)}, \vec{\zeta}^{(2)} + d\vec{\zeta}^{(2)}\right] \qquad \text{for} \qquad t = t^{(2)} \\
&\quad \vdots \\
\text{and} &\quad \vec{\zeta} \in \left[\vec{\zeta}^{(n)}, \vec{\zeta}^{(n)} + d\vec{\zeta}^{(n)}\right] \qquad \text{for} \qquad t = t^{(n)}
\end{aligned}
$$

The probabilities W_i for $i \geq 2$ are called the *joint probabilities.* It is assumed that a stochastic process can be characterized completely by a sequence of probabilities $W_i (i = 1, 2, \cdots, n)$. It is to be expected that n must be large in a given time interval. A stochastic process becomes more complicated as n increases. The probability function W_i has to satisfy the following three conditions

(1) W_i may not be negative: $W_i \geq 0 \quad \text{for} \quad i = 1, 2, \cdots, n$
(2) W_i is normalized to unity

$$
\int\!\!\int \cdots \int W_i(\vec{\zeta}^{(1)}, t^{(1)}; \vec{\zeta}^{(2)}, t^{(2)}; \cdots; \vec{\zeta}^{(i)}, t^{(i)}) \prod_{k=1}^{i} d\vec{\zeta}^{(k)} = 1
$$

(3) At each instant of time W_i has some value

$$W_i(\vec{\zeta}^{(1)}, t^{(1)}; \vec{\zeta}^{(2)}, t^{(2)}; \cdots; \vec{\zeta}^{(i)}, t^{(i)}) = \int W_{i+1}(\vec{\zeta}^{(1)}, t^{(1)}; \cdots; \vec{\zeta}^{(i)}, t^{(i)}; \vec{\zeta}^{(i+1)}, t^{(i+1)})\, d\vec{\zeta}^{(i+1)}$$

The integrations in the items (2) and (3) are done over the whole ζ space. A stochastic process is—due to item three—characterized by W_1 and W_n, if n is the largest index in the sequence of probability functions that determines the stochastic process with sufficient (that is, desired) accuracy.

Besides the joint probabilities, the *conditional* or *transition probabilities* are very important. A transition probability can be defined by

$$\begin{aligned} P_{1,1}(\vec{\zeta}^{(i)}, t^{(i)} \mid \vec{\zeta}^{(j)}, t^{(j)})\, d\vec{\zeta}^{(i)} = &\text{ the probability that} \\ &\vec{\zeta} \in \left[\vec{\zeta}^{(i)}, \vec{\zeta}^{(i)} + d\vec{\zeta}^{(i)}\right] \quad \text{for} \quad t = t^{(i)} \\ \text{if} \quad &\vec{\zeta} = \vec{\zeta}^{(j)} \qquad\qquad\qquad \text{for} \quad t = t^{(j)} \end{aligned}$$

This definition can be generalized

$$P_{m,n}(\vec{\zeta}^{(1)}, t^{(1)}; \cdots; \vec{\zeta}^{(m)}, t^{(m)} \mid \vec{\zeta}'^{(1)}, t^{(1)}; \cdots; \vec{\zeta}'^{(n)}, t^{(n)}) \prod_{k=1}^{m} d\vec{\zeta}^{(k)} =$$

the probability that simultaneously

$$\begin{aligned} &\vec{\zeta} \in \left[\vec{\zeta}^{(i)}, \vec{\zeta}^{(i)} + d\vec{\zeta}^{(i)}\right] \quad \text{for} \quad t = t^{(i)} \quad (i = 1, 2, \cdots, m) \\ \text{if} \quad &\vec{\zeta} = \vec{\zeta}'^{(j)} \qquad\qquad\qquad \text{for} \quad t = t'^{(j)} \quad (j = 1, 2, \cdots, n) \end{aligned}$$

These probabilities are not negative and are normalized to unity. The conditions imposed to the conditional probabilities are

(1) The $P_{n,m}$ may not be negative: $P_{m,n} \geq 0$.
(2) $P_{n,m}$ is normalized to unity

$$\iint \cdots \int P_{m,n}(\vec{\zeta}^{(1)}, t^{(1)}; \cdots; \vec{\zeta}^{(m)}, t^{(m)} \mid \vec{\zeta}'^{(1)}, t^{(1)}; \cdots; \vec{\zeta}'^{(n)}, t^{(n)}) \prod_{k=1}^{m} d\vec{\zeta}^{(k)} = 1$$

(3) At each instant of time, W_n has some value

$$\begin{aligned} W_n(\vec{\zeta}^{(1)}, t^{(1)}; \cdots; \vec{\zeta}^{(n)}, t^{(n)}) = &\, W_{n-1}(\vec{\zeta}^{(1)}, t^{(1)}; \cdots; \vec{\zeta}^{(n-1)}, t^{(n-1)}) \\ &\times P_{1,n-1}(\vec{\zeta}^{(n)}, t^{(n)} \mid \vec{\zeta}^{(1)}, t^{(1)}; \cdots; \vec{\zeta}^{(n-1)}, t^{(n-1)}) \end{aligned}$$

The equality in item (3) can of course be further generalized. From the condition in item (3) for the transition probabilities it follows that

$$W_1(\vec{\zeta}, t) = \int W_2(\vec{\zeta}', t' \mid \vec{\zeta}, t)\, d\vec{\zeta}'$$

This can be written using the definition of the transition probability $P_{1,1}$ as

$$W_1(\vec{\zeta}, t) = \int W_1(\vec{\zeta}', t')\, P_{1,1}(\vec{\zeta}, t \mid \vec{\zeta}', t')\, d\vec{\zeta}' \tag{5.1}$$

which is a linear integral equation for the probability W_1 with the transition probability $P_{1,1}$ as kernel function. Several integral equations of this kind can be derived.

5.2.1. Fundamental types of stochastic processes. *A stochastic process is stationary* if the mechanism of this process does not systematically depend on time. This implies that the probabilities W_n are invariant with respect to translations in time, so that

$$W_1(\vec{\zeta}, t) = W_1(\vec{\zeta})$$

and

$$W_i(\vec{\zeta}^{(1)}, t^{(1)}; \vec{\zeta}^{(2)}, t^{(2)}; \cdots; \vec{\zeta}^{(i)}, t^{(i)}) = \\ W_i(\vec{\zeta}^{(1)}; \vec{\zeta}^{(2)}, t^{(2)} - t^{(1)}; \cdots; \vec{\zeta}^{(i)}, t^{(i)} - t^{(1)}) \quad \text{for} \quad i \geq 2$$

Due to the relation between the W and P probabilities, the transition probabilities $P_{n,m}$ can equally depend only on time intervals and cannot depend on time itself in a stationary process; for example, one has for the transition probability $P_{1,1}$

$$P_{1,1}(\vec{\zeta}^{(1)}, t^{(1)} \mid \vec{\zeta}^{(2)}, t^{(2)}) = \\ P_{1,1}(\vec{\zeta}^{(1)} \mid \vec{\zeta}^{(2)}, t^{(2)} - t^{(1)}) = P_{1,1}(\vec{\zeta}^{(1)}, t^{(1)} - t^{(2)} \mid \vec{\zeta}^{(2)})$$

A process is purely random if the events at the various instants of time are independent of each other, so that

$$W_n(\vec{\zeta}^{(1)}, t^{(1)}; \vec{\zeta}^{(2)}, t^{(2)}; \cdots; \vec{\zeta}^{(n)}, t^{(n)}) = \prod_{k=1}^{n} W_1(\vec{\zeta}^{(k)}, t^{(k)})$$

A purely random stochastic process is therefore completely determined by W_1, and W_1 can be calculated by the methods developed in statistical thermodynamics.

A stochastic process of Markoff is defined by the condition that each transition process to some event is completely determined by this event and the immediate preceding event, so that

$$P_{1,n-1}(\vec{\zeta}^{(n)}, t^{(n)} \mid \vec{\zeta}^{(1)}, t^{(1)}; \vec{\zeta}^{(2)}, t^{(2)}; \cdots ; \vec{\zeta}^{(n-1)}, t^{(n-1)}) = P_{1,1}(\vec{\zeta}^{(n)}, t^{(n)} | \vec{\zeta}^{(n-1)}, t^{(n-1)}) \quad \text{if} \quad t^{(1)} \leq t^{(2)} \leq \cdots \leq t^{(n-1)} \leq t_{(n)}$$

This definition stems from M.S. Green*.

In a Markoff process the transition to $(\vec{\zeta}^{(n)}, t^{(n)})$ is correlated only with $(\vec{\zeta}^{(n-1)}, t^{(n-1)})$, but not with $(\vec{\zeta}^{(i)}, t^{(i)})$ for $i = 1, 2, \cdots, n-2$. This process therefore has a short memory, so that it is sometimes called *random flight* or *random walk*. For a Markoff process

$$W_n(\vec{\zeta}^{(1)}, t^{(1)}; \vec{\zeta}^{(2)}, t^{(2)}; \cdots ; \vec{\zeta}^{(n)}, t^{(n)}) = W_{n-1}(\vec{\zeta}^{(1)}, t^{(1)}; \cdots ; \vec{\zeta}^{(n-1)}, t^{(n-1)})\, P_{1,1}(\vec{\zeta}^{(n)}, t^{(n)} \mid \vec{\zeta}^{(n-1)}, t^{(n-1)})$$

Applying this repeatedly results in

$$W_n(\vec{\zeta}^{(1)}, t^{(1)}; \vec{\zeta}^{(2)}, t^{(2)}; \cdots ; \vec{\zeta}^{(n)}, t^{(n)}) = W_1(\vec{\zeta}^{(1)}, t^{(1)}) P_{1,1}(\vec{\zeta}^{(1)}, t^{(1)} \mid \vec{\zeta}^{(2)}, t^{(2)}) \cdots P_{1,1}(\vec{\zeta}^{(n)}, t^{(n)} \mid \vec{\zeta}^{(n-1)}, t^{(n-1)})$$

This shows that a Markoff process is completely determined by $P_{1,1} \equiv P_2$ and W_1, or alternatively by W_1 and W_2.

A Markoff process obeys the *master equation* of Smoluchowski. Start from

$$W_2(\vec{\zeta}^{(1)}, t^{(1)}; \vec{\zeta}^{(3)}, t^{(3)}) = \int W_3(\vec{\zeta}^{(1)}, t^{(1)}; \vec{\zeta}^{(2)}, t^{(2)}; \vec{\zeta}^{(3)}, t^{(3)})\, d\vec{\zeta}^{(2)}$$

Use of the transition probabilities gives

$$W_1(\vec{\zeta}^{(1)}, t^{(1)}) P_2(\vec{\zeta}^{(3)}, t^{(3)} \mid \vec{\zeta}^{(1)}, t^{(1)}) = \int W_1(\vec{\zeta}^{(1)}, t^{(1)})\, P_2(\vec{\zeta}^{(2)}, t^{(2)} \mid \vec{\zeta}^{(1)}, t^{(1)})\, P_2(\vec{\zeta}^{(3)}, t^{(3)} \mid \vec{\zeta}^{(2)}, t^{(2)})\, d\vec{\zeta}^{(2)}$$

and dividing by the common factor W_1 results in

$$P_2(\vec{\zeta}^{(3)}, t^{(3)} \mid \vec{\zeta}^{(1)}, t^{(1)}) = \int P_2(\vec{\zeta}^{(2)}, t^{(2)} \mid \vec{\zeta}^{(1)}, t^{(1)})\, P_2(\vec{\zeta}^{(3)}, t^{(3)} \mid \vec{\zeta}^{(2)}, t^{(2)})\, d\vec{\zeta}^{(2)}$$

---

* Green M.S. 1952. Markoff random processes and the statistical mechanics of time-dependent phenomena. *Journal of Chemical Physics*, 20: 1281–1295.

which is called the *master equation of Smoluchowski.* The master equation is a linear integral equation for $P_2(\vec{\zeta}^{(3)}, t^{(3)} \mid \vec{\zeta}, t)$ considered as a function of $\vec{\zeta}^{(3)}$ with the kernel $P_2(\vec{\zeta}^{(2)}, t^{(2)} \mid \vec{\zeta}, t)$. This kernel has an eigenfunction W_1 with the eigenvalue unity, as is readily seen from

$$W_1(\vec{\zeta}^{(2)}, t^{(2)}) = \int W_2(\vec{\zeta}^{(1)}, t^{(1)}; \vec{\zeta}^{(2)}, t^{(2)})\, d\vec{\zeta}^{(1)}$$

or

$$W_1(\vec{\zeta}^{(2)}, t^{(2)}) = \int W_1(\vec{\zeta}^{(1)}, t^{(1)})\, P_2(\vec{\zeta}^{(2)}, t^{(2)} \mid \vec{\zeta}^{(1)}, t^{(1)})\, d\vec{\zeta}^{(1)}$$

This integral equation has been derived already, and the two integral equations are sufficient to satisfy the requirements for W_n.

Stationary Markoff processes occur in aged systems, so that

$$P_2(\vec{\zeta}^{(3)}, t^{(3)} - t^{(1)} \mid \vec{\zeta}^{(1)}) = \int P_2(\vec{\zeta}^{(2)}, t^{(2)} - t^{(1)} \mid \vec{\zeta}^{(1)})\, P_2(\vec{\zeta}^{(3)}, t^{(3)} - t^{(2)} \mid \vec{\zeta}^{(2)})\, d\vec{\zeta}^{(2)}$$

To simplify the notation, put

$$t^{(2)} - t^{(1)} = t \qquad t^{(3)} - t^{(2)} = \tau \qquad \Rightarrow \qquad t^{(3)} - t^{(1)} = t + \tau$$

Call

$$\vec{\zeta}^{(3)} = \vec{\zeta} \qquad \vec{\zeta}^{(2)} = \vec{\zeta}' \qquad \vec{\zeta}^{(1)} = \vec{\zeta}_0$$

and the master equation can be written as

$$\boxed{P_2(\vec{\zeta}, t + \tau \mid \vec{\zeta}_0) = \int P_2(\vec{\zeta}', t \mid \vec{\zeta}_0)\, P_2(\vec{\zeta}, \tau \mid \vec{\zeta}')\, d\vec{\zeta}'} \tag{5.2}$$

From this equation the Fokker–Planck (FP-equation) is derived according to the method of De Groot and Mazur*. The aim is to transform the master equation into

$$\frac{\partial}{\partial t} P_2(\vec{\zeta}, t \mid \vec{\zeta}_0) = \mathcal{L}[P_2(\vec{\zeta}, t \mid \vec{\zeta}_0)]$$

with the boundary condition

$$\lim_{t \to 0} P_2(\vec{\zeta}, t \mid \vec{\zeta}_o) = \delta(\vec{\zeta} - \vec{\zeta}_0) = \prod_{\kappa=1}^{N} \delta(\zeta_\kappa - \zeta_{\kappa 0})$$

where $\mathcal{L}$ denotes a linear operator.

---

* De Groot S.R. and Mazur P. 1962. *Non-Equilibrium Thermodynamics.* North-Holland Publ. Co., Amsterdam, pp. 114–115. See also Van Kampen N.G. 1990. *Stochastic Processes in Physics and Chemistry.* North-Holland Publ. Co., Amsterdam.

5.2.2. Fokker–Planck equation. The transition of the linear integral equation (5.2) into a differential equation proves to be possible if P_2 varies slowly as a function of time relative to the time scales of the fluctuations. It is supposed therefore that the time can be divided into intervals of the order τ and such that a macroscopic quantity ϕ (considered as an average quantity that can be calculated using W_1 and P_2) is practically a linear function over the interval τ, while the stochastic variable $\vec{\zeta}$ varies rapidly during this interval.

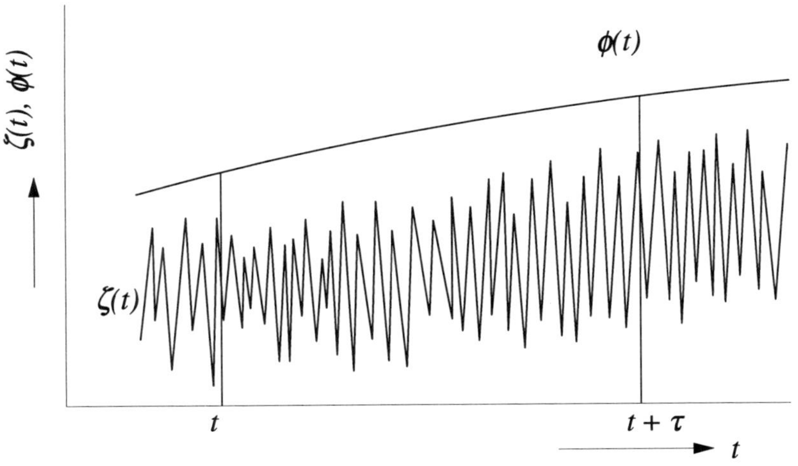

Figure 5.1. Illustration of the conditions for the transition of the linear integral equation into the Fokker–Planck equation. $\zeta(t)$ is a stochastic quantity and $\phi(t)$ the corresponding macroscopic quantity.

Figure 5.1 illustrates these assumptions: there exist time intervals τ that are infinitesimal from a macroscopic point of view, but are large from a microscopic point of view. Macroscopically we therefore have

$$\frac{d\phi}{dt} = \lim_{\tau \to 0} \frac{1}{\tau} \{\phi(t+\tau) - \phi(t)\}$$

The assumed conditions can be formulated more precisely. By introducing $\Delta\vec{\zeta} = \vec{\zeta}(t') - \vec{\zeta}(t) = \vec{\zeta}' - \vec{\zeta}$ and $\tau = t' - t$ the conditions read mathematically

$$\lim_{\tau \to 0} \frac{1}{\tau} \int \Delta\vec{\zeta}\, P_2(\vec{\zeta}', \tau \mid \vec{\zeta})\, d\vec{\zeta}' = \vec{\xi}(\vec{\zeta}, t) \tag{5.3}$$

$$\lim_{\tau \to 0} \frac{1}{\tau} \int \Delta\vec{\zeta}\, \Delta\vec{\zeta}\, P_2(\vec{\zeta}', \tau \mid \vec{\zeta})\, d\vec{\zeta}' = 2\, \mathbf{Q}(\vec{\zeta}, t) \tag{5.4}$$

$$\lim_{\tau \to 0} \frac{1}{\tau} \int \Delta\vec{\zeta}\,\Delta\vec{\zeta} \cdots \Delta\vec{\zeta}\, P_2(\vec{\zeta}', \tau \mid \vec{\zeta})\, d\vec{\zeta}' \approx 0 \tag{5.5}$$

for three or more factors $\Delta\vec{\zeta}$, which means that the correlations between $\vec{\zeta}$ and $\vec{\zeta}'$ are negligible. The limit $\tau \to 0$ has to be understood in a macroscopic sense, where τ is large in respect to the time scale of the fluctuations $\vec{\zeta}$ but small macroscopically, as sketched in figure 5.1. The factor 2 in (5.4) is added to avoid factors $\frac{1}{2}$ in the subsequent formulas. From the definition of the second moment the symmetry property

$$\mathbf{Q} = \mathbf{Q}^{\mathrm{T}} \tag{5.6}$$

follows. The assumed conditions (5.3)–(5.5) can also be understood if P_2—considered as a function of $\Delta\vec{\zeta}$—is a sharply peaked function, so that only for small values of $\Delta\vec{\zeta}$ does the transition probability P_2 have a significant value. A peaked distribution function is in accordance with the supposition that the physical volume element $(\Delta V)^\star$ defines uniquely a macroscopic quantity. Conditions (5.5) are plausible for sharply peaked distributions, and it is conjectured that the requirements (5.3)–(5.5) are closely related to a linearization procedure. Only the first and the second moment are then proportional to τ in the limit $\tau \to 0$, while the higher moments are at least of the order τ^2. It follows from the Pawula theorem* as matter of fact that

* From the inequality $\int\int [f(\zeta)g(\zeta') - f(\zeta')g(\zeta)]^2 P(\zeta)P(\zeta')\, d\zeta\, d\zeta' \geq 0$, in a one-dimensional space, where f and g are arbitrary functions, there follows for nonnegative P the generalized Schwartz inequality

$$\left[\int f(\zeta')g(\zeta')P(\zeta')\, d\zeta'\right]^2 \leq \int f(\zeta')^2 P(\zeta')\, d\zeta' \int g(\zeta')^2 P(\zeta')\, d\zeta'$$

Substitution of $f(\zeta) = (\Delta\zeta)^n$, $g(\zeta') = (\delta\zeta)^{n+m}$ and $P(\zeta') = P_2(\zeta', \tau \mid \zeta)$ with $(n, m \geq 0)$ yields for the moments $M_n = \int (\delta\zeta)^n P_2(\zeta', \tau \mid \zeta)\, d\zeta$ the inequality

$$M_{2n+m}^2 \leq M_{2n}\, M_{2n+2m}$$

For $m = 0$, the inequality reduces to $M_{2n}^2 = M_{2n}^2$ that is always satisfied for all n. For $n = 0$, the inequality reduces to $M_m^2 \leq M_{2m}$. If it is assumed that in the limit $\tau \to 0$ all moments are proportional to τ; then $M_i \approx \tau D^{(i)}$ applies. From this assumption no restrictions are found for the $D^{(i)}$ for $n = 0$. In the limit $\tau \to 0$ for the approximations $D^{(i)}$ with $(n \geq 1, m \geq 1)$ the inequality

$$\left[D^{(2n+m)}\right]^2 \leq D^{(2n)} D^{(2n+2m)}$$

applies. If $D^{(2n)} = 0$, then also $D^{(2n+m)}$, so that then $D^{(2n)}$ and all higher moments are zero. If $D^{(2n+2m)} = 0$, then $D^{(2n+m)}$ has to be zero. In repeated applications for all values either the suppositions (5.3)–(5.5) hold or that *all* higher moments are also proportional to τ in the limit $\tau \to 0$.

The vector $\vec{\xi}$ in (5.3) can be interpreted as the *regression of the fluctuation* at the time t, defined by

$$\vec{\xi}(\vec{\zeta}, t) = \lim_{\tau \to 0} \frac{1}{\tau} \left\{ \langle \vec{\zeta}', \tau \mid \vec{\zeta} \rangle - \vec{\zeta} \right\} \tag{5.7}$$

with the introduction of the *conditional expectation value* of the fluctuation

$$\langle \vec{\zeta}', \tau \mid \vec{\zeta} \rangle = \frac{1}{\tau} \int \vec{\zeta}' \, P_2(\vec{\zeta}', \tau \mid \vec{\zeta}) \, d\vec{\zeta}' \tag{5.8}$$

The regression of the fluctuations $\vec{\xi}$ is usually a stochastic quantity that can be interpreted as a drift velocity in ζ space with respect to the expectation value (5.8).

Assume for the derivation of the Fokker–Planck equation that $R(\vec{\zeta})$ is an *arbitrary* function of $\vec{\zeta}$ that for $|\vec{\zeta}| \to \infty$ goes sufficiently fast to zero. Consider for the discretization of the time derivative the following limit of a difference quotient

$$\begin{aligned}
&\int \frac{\partial}{\partial t} P_2(\vec{\zeta}, t \mid \vec{\zeta}_0) R(\vec{\zeta}) \, d\vec{\zeta} \\
&\quad = \lim_{\tau \to 0} \frac{1}{\tau} \int \left[P_2(\vec{\zeta}, t + \tau \mid \vec{\zeta}_0) - P_2(\vec{\zeta}, t \mid \vec{\zeta}_0) \right] R(\vec{\zeta}) \, d\vec{\zeta} \\
&\quad = \lim_{\tau \to 0} \frac{1}{\tau} \left[\int P_2(\vec{\zeta}', t + \tau \mid \vec{\zeta}_0) R(\vec{\zeta}') \, d\vec{\zeta}' - \int P_2(\vec{\zeta}, t \mid \vec{\zeta}_0) R(\vec{\zeta}) \, d\vec{\zeta} \right]
\end{aligned}$$

Substitution of the Smoluchowski equation (5.2) then yields

$$\begin{aligned}
&\int \frac{\partial}{\partial t} P_2(\vec{\zeta}, t \mid \vec{\zeta}_0) R(\vec{\zeta}) \, d\vec{\zeta} = \\
&\qquad \lim_{\tau \to 0} \frac{1}{\tau} \left[\iint P_2(\vec{\zeta}, t \mid \vec{\zeta}_0) \, P_2(\vec{\zeta}', \tau \mid \vec{\zeta}) R(\vec{\zeta}') \, d\vec{\zeta} \, d\vec{\zeta}' \right. \\
&\qquad\qquad \left. - \int P_2(\vec{\zeta}, t \mid \vec{\zeta}_0) R(\vec{\zeta}) \, d\vec{\zeta} \right]
\end{aligned} \tag{5.9}$$

---

of n and m it follows finally that if $D^{(2r)} = 0$ for $r \geq 1$ that all $D^{(i)}$ with $i \geq 3$ then have to be zero. Conversely, not accepting (5.5) implies that all moments have to be proportional to τ for small values of τ. For further details see Pawula R.F. 1967. Approximation of the linear Boltzmann equation by the Fokker–Planck equation. *Physical Review*, 162: 186–188; and Risken H. 1989. *The Fokker–Planck Equation*. Springer-Verlag, Berlin, pp. 70–71.

Next, expand the arbitrary auxiliary function $R(\vec{\zeta}')$ in a Taylor series around the point $\vec{\zeta}$

$$R(\vec{\zeta}') = R(\vec{\zeta} + \Delta\vec{\zeta}) = \\ R(\vec{\zeta}) + \frac{\partial R}{\partial \vec{\zeta}} \cdot \Delta\vec{\zeta} + \tfrac{1}{2}\left(\frac{\partial}{\partial \vec{\zeta}}\frac{\partial R}{\partial \vec{\zeta}}\right) : \left(\Delta\vec{\zeta}\,\Delta\vec{\zeta}\right) + \mathcal{O}(\Delta\vec{\zeta})^3 \quad (5.10)$$

With the substitution of (5.10) into (5.9) and after integration over $\vec{\zeta}'$ the contribution of $R(\vec{\zeta})$ cancels against the last term in (5.9), so that

$$\int \frac{\partial}{\partial t} P_2(\vec{\zeta}, t \mid \vec{\zeta}_0) R(\vec{\zeta})\, d\vec{\zeta} = \lim_{\tau \to 0} \frac{1}{\tau} \left[\iint P_2(\vec{\zeta}, t \mid \vec{\zeta}_0)\, P_2(\vec{\zeta}', \tau \mid \vec{\zeta}) \right. \\ \left. \times \left\{ \frac{\partial R}{\partial \vec{\zeta}} \cdot \Delta\vec{\zeta} + \tfrac{1}{2}\left(\frac{\partial}{\partial \vec{\zeta}}\frac{\partial R}{\partial \vec{\zeta}}\right) : \left(\Delta\vec{\zeta}\,\Delta\vec{\zeta}\right) + \mathcal{O}(\Delta\vec{\zeta})^3 \right\} d\vec{\zeta}\, d\vec{\zeta}' \right] \quad (5.11)$$

Integrate now over $\vec{\zeta}'$ and use the assumptions (5.3)–(5.5) so that terms of the order of $(\Delta\vec{\zeta})^3$ can be neglected. This results in

$$\int \frac{\partial}{\partial t} P_2(\vec{\zeta}, t \mid \vec{\zeta}_0) R(\vec{\zeta})\, d\vec{\zeta} = \\ \int P_2(\vec{\zeta}, t \mid \vec{\zeta}_0) \left[\vec{\xi}(\vec{\zeta}) \cdot \frac{\partial R}{\partial \vec{\zeta}} + \mathbf{Q}(\vec{\zeta}) : \left(\frac{\partial}{\partial \vec{\zeta}}\frac{\partial R}{\partial \vec{\zeta}}\right) \right] \quad (5.12)$$

Integration by parts with the condition that the arbitrary auxiliary function $R(\vec{\zeta})$ tends to zero for $\vec{\zeta} \to \infty$ sufficiently fast to guarantee that the contributions at the limits are zero, yields

$$\int \frac{\partial}{\partial t} P_2(\vec{\zeta}, t \mid \vec{\zeta}_0) R(\vec{\zeta})\, d\vec{\zeta} = \\ \int \left[-\frac{\partial}{\partial \vec{\zeta}} \cdot \left(P_2\, \vec{\xi}\right) + \left(\frac{\partial}{\partial \vec{\zeta}}\frac{\partial}{\partial \vec{\zeta}}\right) : \left(P_2\, \mathbf{Q}\right) \right] R(\vec{\zeta})\, d\vec{\zeta}_0 \quad (5.13)$$

Since $R(\vec{\zeta})$ is an arbitrary auxiliary function, the required Fokker–Planck*

* The one-dimensional Fokker–Planck equation was derived for the first time by Fokker A.D. 1914. Die mittlere Energie rotierender elektrischer Dipole im Strahlungsfeld. *Annalen der Physik*, Vierte Folge, 43: 810–820, and extensively discussed by Planck M. 1917. Über einen Satz der statistischen Dynamik und seine Erweiterung in der Quantentheorie. *Sitzberichte der Preussische Akademie der Wissenschaften*, Jahrgang 1917, Erster Halbband: 324–341. For the three-dimensional case, this equation was derived for the first time by Chandrasekhar S. 1943. Stochastic problems in physics and astronomy. *Review of Modern Physics*, 15: 1–89; for higher dimensions by Prigogine I. and Mazur P. 1953. Sur l'extension de la thermodynamique aux phénomènes irréversibles liés aux degrés de liberté internes. *Physica*, 19: 241-254 and by Meixner J. 1957. Zur statistischen Thermodynamik irreversibler Prozesse. *Zeitschrift für Physik*, 149: 624–646.

equation is inferred from (5.13)

$$\boxed{\begin{aligned}&\frac{\partial P}{\partial t} = \left[-\frac{\partial}{\partial \vec{\zeta}} \cdot \vec{\xi} + \left(\frac{\partial}{\partial \vec{\zeta}}\frac{\partial}{\partial \vec{\zeta}}\right) : \mathbf{Q}\right] P \\ &\text{with} \quad P = P_2(\vec{\zeta}, t \mid \vec{\zeta}_0)\end{aligned}} \tag{5.14}$$

The initial condition for P_2 follows from the definition of P_2 and the normalization of P_2. It is given by

$$\lim_{t \to 0} P_2(\vec{\zeta}, t \mid \vec{\zeta}_0) = \delta(\vec{\zeta} - \vec{\zeta}_0) = \prod_{K=1}^{N} \delta(\zeta_K - \zeta_{K0}) \tag{5.15}$$

The Fokker–Planck equation describes the evolution of P_2 in ζ space. For constant values of the drift vector $\vec{\xi}$ and the diffusion coefficient $\mathbf{Q}$, (5.14) is similar to a diffusion equation for the transition probability P_2, in which in the definition of the convective flux the convective velocity is now given by the drift vector $\vec{\xi}$. As time proceeds, the initial delta function (5.15) is broadened by 'diffusion'. The widening is determined by the probability W_1. For stationary Markoff processes it is to be expected that the probability of $(\vec{\zeta}, t)$ is not influenced by $(\vec{\zeta}, 0)$ any more as $t \to \infty$, so that in this limit the transition probability is given by the probability W_1.

Analogous to the transition probability P_2, a Fokker–Planck equation for the probability W_1 can be derived. The conditions (5.3)–(5.5) apply equally well here, but it is now not necessary that the stochastic process be a Markoff process. To appreciate this, consider the expression

$$\begin{aligned}&\int \frac{\partial}{\partial t} W_1(\vec{\zeta}, t) R(\vec{\zeta})\, d\vec{\zeta} \\ &= \lim_{\tau \to 0} \frac{1}{\tau} \left[\int W_1(\vec{\zeta}', t+\tau) R(\vec{\zeta}')\, d\vec{\zeta}' - \int W_1(\vec{\zeta}, t) R(\vec{\zeta})\, d\vec{\zeta} \right] \\ &= \lim_{\tau \to 0} \frac{1}{\tau} \left[\int W_1(\vec{\zeta}, t+\tau) P_2(\vec{\zeta}', t' \mid \vec{\zeta}, t) R(\vec{\zeta}')\, d\vec{\zeta}\, d\vec{\zeta}' - \int W_1(\vec{\zeta}, t) R(\vec{\zeta})\, d\vec{\zeta} \right]\end{aligned}$$

where (5.1) has been used. After the substitution of the Taylor expansion for R in (5.10), integration with application of the conditions (5.3)–(5.5), and integration by parts, the Fokker–Planck equation for the probability W_1 follows

$$\boxed{\frac{\partial W_1}{\partial t} = \left[-\frac{\partial}{\partial \vec{\zeta}} \cdot \vec{\xi} + \left(\frac{\partial}{\partial \vec{\zeta}}\frac{\partial}{\partial \vec{\zeta}}\right) : \mathbf{Q}\right] W_1} \tag{5.16}$$

It can be remarked that for stationary stochastic processes the relation $\partial W_1/\partial t = 0$ holds for the probability W_1, while for the transition probability P_2 it should be noted that $\partial P_2/\partial t \neq 0$.

5.3. STABLE NEUTRAL EQUILIBRIUM OF HOMOGENEOUS SYSTEMS

The theory of thermodynamic fluctuations is concerned with the fluctuations around the equilibrium state. For the derivation of the reciprocal relations, it is of primary importance to consider aged systems. These systems are not disturbed and are completely isolated. From the macroscopic point of view these systems are in equilibrium. So also the physical volume element is macroscopically fixed and is in thermal equilibrium. The fundamental equation of Gibbs (4.5) contains all the thermodynamic knowledge about the system in equilibrium. Suppose that the work is controlled by K independent extensive variables Y_κ (to be specified by quasi-static and adiabatic experiments). Then the fundamental equation of Gibbs reads

$$dU = T\,dS + \sum_{k=1}^{K} P_\kappa\,dY_\kappa + \sum_{\kappa=1}^{N} \mu_\kappa\,dM_\kappa \tag{5.17}$$

where P_κ are the intensive thermodynamic variables conjugate to the extensive thermodynamic variables Y_κ.

In the entropy representation the Gibbs fundamental equation is given by

$$dS = \frac{1}{T}dU - \sum_{k=1}^{K} \frac{P_\kappa}{T}dY_\kappa - \sum_{\kappa=1}^{N} \frac{\mu_\kappa}{T}dM_\kappa \tag{5.18}$$

For brevity in notation, the generalized extensive variables Z_κ and the generalized intensive variables F_κ are introduced as

$$Z_\kappa = \begin{cases} U & \text{for} \quad \kappa = 0, \\ Y_\kappa & \text{for} \quad \kappa = 1, 2, \cdots, K, \\ M_\kappa & \text{for} \quad \kappa = K+1, \cdots, K+N \end{cases} \tag{5.19}$$

$$F_\kappa = \begin{cases} 1/T & \text{for} \quad \kappa = 0, \\ -P_\kappa/T & \text{for} \quad \kappa = 1, 2, \cdots, K, \\ -\mu_\kappa/T & \text{for} \quad \kappa = K+1, \cdots, K+N \end{cases} \tag{5.20}$$

so that (5.18) can be written as

$$dS = \sum_{\kappa=0}^{r} F_\kappa\,dZ_\kappa \tag{5.21}$$

From (5.21) it follows that

$$S = S(Z_\kappa) \qquad F_\kappa = \frac{\partial S}{\partial Z_\kappa} \qquad \text{for } \kappa = 0, \cdots, K+N = r \tag{5.22}$$

It is supposed that for the homogeneous system the number of degrees of freedom $r + 1$ is finite. The macroscopic variables defined by the physical volume elements $(\Delta V)^{\star}$, defined in Chapter 1, will show fluctuations if they are defined by volume elements $(\Delta V)^{-} < (\Delta V)^{\star}$, where the superscript $-$ denotes that the volume element smaller than the physical volume element. Now suppose that the functions of state, which are equilibrium quantities, are also useful quantities for the elementary volumes $(\Delta V)^{-}$. With this assumption, these quantities are not thermodynamic equilibrium quantities, but are assumed to be stochastic quantities that fluctuate around their definite equilibrium value defined by the matter in physical volume elements $(\Delta V)^{\star}$ and by $\Delta t^{\star}$. The fluctuations are not observable in the global system.

The reservoir concept* can be used for the study of the equilibrium of homogeneous systems and its fluctuations. The aged system is seen in this method as a reservoir that acts as the environment that executes external actions on the system $(\Delta V)^{-}$. To each extensive variable Z_K is assigned for this purpose a reservoir R_K, which can perform on the system an intensive force F_K through a junction. The $r+1$ reservoirs operate independently from each other—this can be done if the introduced generalized extensive variables Z_K are independent—and together make up the reservoir system R. The reservoirs have to satisfy the following properties

(1) To each phase of $(\Delta V)^{-}$ corresponds a phase of R that is very large with respect to the corresponding phase of $(\Delta V)^{-}$.
(2) The generalized force F_K is exerted by the reservoir $R_K \in R$ on $(\Delta V)^{-}$, where the capacity of R_K is so large that in the reservoir $F_K = F_K^{+}$ remains constant during these actions on $(\Delta V)^{-}$.
(3) The total system $V_{(\mathrm{tot})} = (\Delta V)^{-} \cup R$ is completely isolated in the sense that this system does not exchange energy and matter with its environment. Then the generalized extensive variables Z_K are constant.
(4) The total system is in equilibrium. This equilibrium is neutral stable.
(5) Variations in Z_K occur only through interactions between $(\Delta V)^{-}$ and R. The extensive variables Z_K can vary in both directions.

Examples of reservoirs are the heat reservoir, for which in an exchange of heat between $(\Delta V)^{-}$ and the reservoir, the temperature of the reservoir remains constant; the volume reservoir, for which in a change of the volume of R_V the pressure in R_V remains constant, and the mass reservoir, for which in an exchange of mass of component K the chemical potential μ_K remains constant in the reservoir.

The extensive variables of the total system are given by

$$S_{(\mathrm{tot})} = S + S_R \qquad \text{and} \qquad Z_{(\mathrm{tot})_K} = Z_K + Z_R \tag{5.23}$$

* See for example Schottky W., Ulich H. and Wagner C. 1929. *Thermodynamik*. Springer Verlag, Berlin. Reprint 1973, pp. 450–453.

in which the unsubscripted quantities relate to the system $(\Delta V)^-$. If the system $(\Delta V)^-$ is in equilibrium with the reservoir system R, then

$$\delta S_{(\text{tot})} = \delta S + \delta S_R = 0 \tag{5.24}$$

Since for the total system the extensive variables are constant, the variation (5.24) holds for

$$\delta Z_{(\text{tot})_K} = \delta Z_K + (\delta Z_R)_K = 0 \tag{5.25}$$

The entropy change of the system is

$$\delta S = \sum_{K=0}^{r} F_K \, \delta Z_K \tag{5.26}$$

and in the reservoir system

$$\delta S_R = \sum_{K=0}^{r} F_K^+ \, (\delta Z_R)_K = - \sum_{K=0}^{r} F_K^+ \, \delta Z_K \tag{5.27}$$

where F_K^+ is the constant intensive thermostatic quantity of the reservoir R_K, and the second equality follows from (5.25). Substitution of (5.26) and (5.27) in (5.24) results in

$$\delta S_{(\text{tot})} = \sum_{K=0}^{r} (F_K - F_K^+) \, \delta Z_K = 0 \tag{5.28}$$

Since the Z_K's have to be chosen such that these variables are mutually independent, the usual equilibrium condition follows from (5.28)

$$F_K = F_K^+ \tag{5.29}$$

F_K^+ is constant in the reservoir R_K, but for the subsystem $(\Delta V)^-$, the quantities F_K and Z_K may undergo small fluctuations around their definite values. For the total system classical thermodynamics applies, where for a natural process from state I to state II

$$\Delta S = S_{II} - S_I \geq 0 \tag{5.30}$$

As the total system is an aged system and as a consequence the system is in a stable neutral equilibrium, a sufficient condition for stability with respect to these fluctuations is that

$$\Delta S_{(\text{tot})} = \Delta S + \Delta S_R \leq 0 \tag{5.31}$$

The inequality sign means that no spontaneous natural processes occur due to the fluctuations in the subsystem $(\Delta V)^-$that transform the total system from state I into state II. The total system stays in state I and is in stable equilibrium. If the equality sign applies then the total system is in neutral or indifferent equilibrium and similarly no spontaneous processes cause the system to change its state.

For the reservoir system F_K^+ is constant, so that from (5.27) it can be seen that

$$\Delta S_R = -\sum_{K=0}^{r} F_K^+ \,\delta Z_K \tag{5.32}$$

F_K and Z_K fluctuate around the equilibrium state in the system $(\Delta V)^-$. A Taylor expansion of the entropy S around this state yields

$$\begin{aligned} S - S^+ &= \Delta S \\ &= \sum_{K=0}^{r} \left(\frac{\partial S}{\partial Z_K}\right)^+ \delta Z_K + \frac{1}{2}\sum_{K=0}^{r}\sum_{M=0}^{r}\left(\frac{\partial^2 S}{\partial Z_K \partial Z_M}\right)^+ \delta Z_K\,\delta Z_M + \cdots \\ &= \sum_{K=0}^{r} F_K^+\,\delta Z_K - \frac{1}{2}\sum_{K=0}^{r}\sum_{M=0}^{r} g_{KM}\,\delta Z_K\,\delta Z_M + \cdots \end{aligned} \tag{5.33}$$

where from (5.22) we have that

$$F_K^+ = \left(\frac{\partial S}{\partial Z_K}\right)^+ \qquad \text{for} \quad K = 0, \cdots, r \tag{5.34}$$

and

$$g_{KM} = -\left(\frac{\partial^2 S}{\partial Z_K \partial Z_M}\right)^+ = g_{MK} \qquad \text{for} \quad K, M = 0, \cdots, r \tag{5.35}$$

Substitution of (5.32) and (5.33) into (5.31) gives

$$\Delta S_{(\mathrm{tot})} = -\frac{1}{2}\sum_{K=0}^{r}\sum_{M=0}^{r} g_{KM}\,\delta Z_K\,\delta Z_M \le 0 \tag{5.36}$$

if the fluctuations around the equilibrium state are not too large.

The total system is isolated and no transport of entropy to and from the total system can occur. For deviations from the equilibrium state, the entropy change (5.36) has to be accomplished *within* the total system. From the stability of the equilibrium state it is found that a symmetric and positive definite matrix g_{KM} exists that may be used to define the metric in the space of the thermodynamic variables Z_K

$$(dZ)^2 = g_{KM}\,dZ_K\,dZ_M \tag{5.37}$$

Here it is remarked that the space of the thermodynamic variables Z_K^+ does not usually have a metric, but that with (5.37) a metric can be defined in the neighborhood of each stable equilibrium state. There is no need to add higher order terms to the expansion in (5.37), since in the thermodynamics of irreversible processes the theory is restricted to small fluctuations around the equilibrium state.

To describe these fluctuations it is advantageous to introduce for the first order deviations from equilibrium the *Onsager variables*, in short the *O-variables*, by

$$\zeta_K(t) = \delta Z_K = Z_K(t) - Z_K^+ \qquad \text{for} \quad K = 0, \cdots, r \tag{5.38}$$

so that the condition (5.36) for neutral stable equilibrium of the total system can be written as

$$\Delta S_{(\text{tot})} = -\frac{1}{2} \sum_{K=0}^{r} \sum_{M=0}^{r} g_{KM} \zeta_K \zeta_M \leq 0 \tag{5.39}$$

This shows that the O-variables make up a metric space: the *Onsager space* or shortly the *O-space*, in which the metric is given by g_{KM}. Since g_{KM} depends only on the equilibrium state, g_{KM} is considered in the O-space to be a constant matrix. The O-space is therefore a Euclidean space, which has the point $\vec{Z}^+$ as center. An O-space can be constructed in each stable equilibrium state. Cartesian coordinates can be used in the Euclidean O-space. In the O-space, (5.39) might be written as

$$\Delta S_{(\text{tot})} = -\tfrac{1}{2} \vec{\zeta} \cdot \mathbf{g} \cdot \vec{\zeta} \leq 0 \tag{5.40}$$

with

$$\mathbf{g} = \mathbf{g}^{\mathrm{T}} \tag{5.41}$$

The O-variables represent a stochastic process. Analogous to the O-variable $\vec{\zeta}(t)$, the *Onsager force*, or *O-force*, can be introduced for the system $(\Delta V)^-$ as

$$\vec{K}(t) = \vec{F}(t) - \vec{F}^+ \tag{5.42}$$

For the O-force an expression analogous to that for F_K in (5.22) can be found. Consider therefore (5.31) with (5.36)

$$\Delta S_{(\text{tot})} = \Delta S + \Delta S_R = S - S^+ + \Delta S_R = -\tfrac{1}{2} \vec{\zeta} \cdot \mathbf{g} \cdot \vec{\zeta} \leq 0 \tag{5.43}$$

Substitution of (5.38) into (5.32) gives

$$\Delta S_R = -\vec{F}^+ \cdot (\vec{Z} - \vec{Z}^+) = \vec{F}^+ \cdot \vec{\zeta} \tag{5.44}$$

and (5.43) can be written as

$$S = S^{+} + \vec{F}^{+} \cdot \vec{\zeta} - \tfrac{1}{2}\vec{\zeta} \cdot \mathbf{g} \cdot \vec{\zeta} \leq 0 \tag{5.45}$$

From (5.22) and (5.45) it follows that

$$\vec{F} = \frac{\partial S}{\partial \vec{Z}} = \frac{\partial S}{\partial \vec{\zeta}} = \vec{F}^{+} - \mathbf{g} \cdot \vec{\zeta} \tag{5.46}$$

so that for the O-force

$$\vec{K} = -\mathbf{g} \cdot \vec{\zeta} = -\vec{\zeta} \cdot \mathbf{g} = \frac{\partial \Delta S_{(\text{tot})}}{\partial \vec{\zeta}} \tag{5.47}$$

The entropy difference (5.36) in the insulated total system arises from perturbations of the stable neutral equilibrium. Since the system stays in neutral stable equilibrium, the entropy production—the rate of entropy per unit of time—in the total system becomes with (5.43)

$$\Pi_{(\mathrm{S})} = \frac{\partial \Delta S_{(\text{tot})}}{\partial t} = -\vec{\zeta} \cdot \mathbf{g} \cdot \dot{\vec{\zeta}} \geq 0 \tag{5.48}$$

The entropy production is according to the thermodynamics of irreversible processes directly related to irreversible processes occurring in the system. Irreversible processes proceed according to Planck with finite velocity. This implies that in the occurrence of irreversible processes the quantities of state change with a finite velocity. In the homogeneous reservoirs, the forces $(F_R)_K = F_K^{+}$ are constants, so that no irreversible processes take place *within* the reservoir system. Therefore, irreversible processes can be found only in the subsystem $(\Delta V)^-$ and/or at the junctions between $(\Delta V)^-$ and the reservoir system R.

Irreversible processes within $(\Delta V)^-$ are relaxation processes, which are described by *internal* or *hidden* variables of state. Examples of those processes are the flow of viscoelastic fluids, homogeneous chemical reactions, thermal excitation of internal molecular degrees of freedom (rotation and/or vibration of polyatomic molecules), propagation of dislocations in crystals, and so on. Irreversible processes at the junctions between $(\Delta V)^-$ and R are the transport processes, which occur only at the junctions due to the homogeneity of $(\Delta V)^-$ and R. The system $(\Delta V)^-$ remains in good approximation homogeneous in the presence of such transport processes if those transport processes in the junctions are slow in relation to the corresponding transport processes *in* $(\Delta V)^-$.

With this consideration one can say that the entropy production (5.48) relates to the system $(\Delta V)^-$, if the junctions between $(\Delta V)^-$ and R are considered as belonging to the 'boundary of $(\Delta V)^-$ being a part of the system $(\Delta V)^-$'. The significance of (5.48) can be illustrated as follows. The system $(\Delta V)^-$ is in equilibrium if the total system is in equilibrium. The system $(\Delta V)^-$ is thrown out of equilibrium by actions of the environment, symbolized by the reservoir system. Due to these actions $\vec{Z}^+ \to \vec{Z}$ and the entropy of the total system decreases with (5.36). Spontaneous processes occur now in the system $(\Delta V)^-$and strive to recover the equilibrium state. For these spontaneous processes $S_{(\mathrm{tot})}$ increases again. This accounts for the inequality sign in (5.48), which also dictates the direction of the spontaneous processes, so that the spontaneous processes are irreversible. It is clear that the spontaneous irreversible processes oppose the actions that R exerts on $(\Delta V)^-$. Entropy production means therefore that the system $(\Delta V)^-$ resists these actions that are 'external' as seen from the system $(\Delta V)^-$. Fluctuations are accordingly closely related with dissipation.

This above discussion explains the usefulness of the idea of the reservoir system. Interest is in the first place concentrated on the behaviour of $(\Delta V)^-$, if $(\Delta V)^-$ is thrown out of equilibrium by the environment. Interest is not concentrated on the behavior of the environment itself. The environment is now represented by the reservoir system, such that no irreversible processes can occur *in* the reservoir. The entropy production (5.48) is thus concerned only with the system $(\Delta V)^-$. The entropy production $\Pi_{(\mathrm{S})}$ is zero in equilibrium. This condition defines in addition the equilibrium processes or reversible processes. Nonetheless, $\dot{\vec{\zeta}}$ can be set equal to $\vec{0}$ in both cases. Two cases can still be discriminated in equilibrium; namely, *unconstrained equilibrium* or *true equilibrium*, in which $\vec{\zeta} = \vec{0}$, and *constrained equilibrium* or *false equilibrium*, in which $\vec{\zeta} \neq \vec{0}$. The 'generalized force' or 'affinity'—see (5.47)—of the process is unequal to zero in the last case, but the process can not progress with a finite velocity by some constraint. Examples of such a process are found for chemical reactions for which the affinity of the chemical reaction is not zero, but for which the reaction rate does not differ noticeably from zero; by adding a catalyst the reaction speeds up remarkably, so that a catalyst cancels the constraint.

The equilibrium is assumed to be a true equilibrium in the thermodynamics of irreversible processes. With the introduction of the *Onsager flux*—in short the O-flux—as

$$\vec{J} = \dot{\vec{\zeta}} \tag{5.49}$$

and with the substitution of (5.47), the entropy production (5.48) can be written as

$$\boxed{\Pi_{(\mathrm{S})} = \frac{\partial \Delta S_{(\mathrm{tot})}}{\partial t} = -\vec{\zeta} \cdot \mathbf{g} \cdot \dot{\vec{\zeta}} = \vec{K} \cdot \vec{J} \geq 0} \tag{5.50}$$

There are no assumptions made about the microscopic structure of a substance in the continuum theory. The relations between the forces and fluxes are described by constitutive equations in the continuum theory. In the thermodynamics of irreversible processes, the relation between the flux and the force is given by a linear approximation

$$\vec{J} = \mathbf{L} \cdot \vec{K} \tag{5.51}$$

This phenomenological equation is not a general linear relation between the flux and the force, since such a relation is given by a convolution integral. The linear relation (5.51) holds only for slow variations of the variables of state, and on that account, (5.51) is to be considered as a low frequency approximation for the linear approximation of the functional relation between the flux and the force. This approximation has been shown of great value in practise, since all linear relaxation processes and transport processes can be described by this approximation. Possible memory-effects are introduced by 'hidden' thermodynamic variables. However, for a complete set of fluxes and forces the relation is instantaneous.

The phenomenological matrix $\mathbf{L}$ depends neither on $\vec{\zeta}$ nor t in the low frequency approximation, but only on the equilibrium values of $\vec{Z}^{+}$. Substitution of (5.51) into (5.50) yields

$$\boxed{\Pi_{(\mathrm{S})} = \vec{K} \cdot \mathbf{L} \cdot \vec{K} \geq 0} \tag{5.52}$$

so that the symmetric part of $\mathbf{L}$ is positive definite. Next, the symmetry of $\mathbf{L}$ has to be shown.

5.4. MICROSCOPIC REVERSIBILITY

The term *microscopic reversibility* is rather confusing, since no reversibility in the sense of macroscopic thermodynamics is implied by this nomenclature. In the previous section we have just discussed the fact that from the macroscopic viewpoint the fluctuations are irreversible. Microscopic reversibility relates the probability density of trajectories of the O-variables with the probability density of the corresponding time-reversed trajectories of these O-variables.

The concept of *molecular reversibility* is easier. The reversibility of the molecular motions is a direct consequence of Hamilton equations of motion. Suppose that the system consists of a large number of molecules subject to conservative force fields. The Hamiltonian for these molecules is

$$H = E_{(\mathrm{kin})} + E_{(\mathrm{pot})} \tag{5.53}$$

where $E_{(\mathrm{kin})}$ is the kinetic energy and $E_{(\mathrm{pot})}$ is the potential energy of the molecules. H is a function of the momenta p_k and the coordinates q_k, in which the Hamiltonian H is an even function of the momenta

$$H(q_k, p_k) = H(q_k, -p_k) \tag{5.54}$$

The Hamiltonian equations of motion read

$$\dot{p}_k = -\frac{\partial H}{\partial q_k} \qquad \text{and} \qquad \dot{q}_k = \frac{\partial H}{\partial p_k} \tag{5.55}$$

These equations of motion are invariant under the transformation $t \to -t$, since for time reversal the transformation $q_k \to q_k$ and $p_k \to -p_k$ are valid. If a solution of the Hamiltonian form of the equations of motion is known, a second solution follows therefore by time reversal. In this regard it is remarked that these considerations apply only if the molecular motions are described with respect to an inertial system. Casimir* pointed out that it has been assumed that the motions are not subjected to an external magnetic field $\vec{B}$. Figure 5.2 shows the corresponding trajectories.

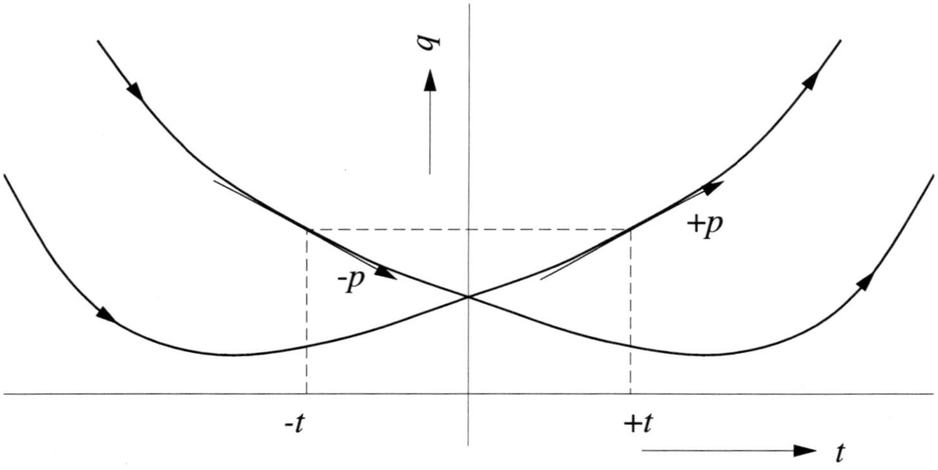

Figure 5.2. Coordinate of a trajectory and its corresponding inverse trajectory for a one-dimensional motion, sketched as a function of time, in which $p \sim dq/dt$ is a parameter that can be given along the trajectory.

Particles with an electrical charge will have the tendency to execute circular motions around the lines of force in the presence of an external magnetic field. The direction of these motions is coupled with the direction of the magnetic field. The direction of the rotations of these circular motions has to be reversed for a reversal of the path, so that $\vec{B}$ has to be replaced by $-\vec{B}$. Therefore, the trajectories of the molecular motions in an inertial system can be reversed for the transformation

$$t \to -t \qquad \text{and} \qquad \vec{B} \to -\vec{B} \tag{5.56}$$

* Casimir H.B.G. 1945. On the Onsager principle of microscopic reversibility. *Review of Modern Physics*, 17: 343–350.

Clearly, $\vec{B}$ in (5.56) is the externally applied magnetic induction. The internal electromagnetic fields produced by the charged particles themselves within the system are reversed at the very instant that the motions of the particles reverse. Also the externally applied electromagnetic fields have to be reversed for a reversal of the molecular motions, so that obviously $\vec{B}$ represents the externally applied electromagnetic field. It is remarked that an external magnetic field is not shielded and is caused by bodies not part of the system being considered. Aging of the system can be described only if it is assumed that $\vec{B}$ is constant in time.

A macroscopic quantity Z can be considered as a statistical average of a molecular property. This assumption applies equally for an Onsager variable $\zeta = Z - Z^+$. The O-variable ζ can be seen as a 'function' of the momentum coordinates of the molecules, and ζ can be an even or an odd function of the momentum coordinates. The Onsager variable ζ is in the first case an even variable, and in the second case an odd variable.

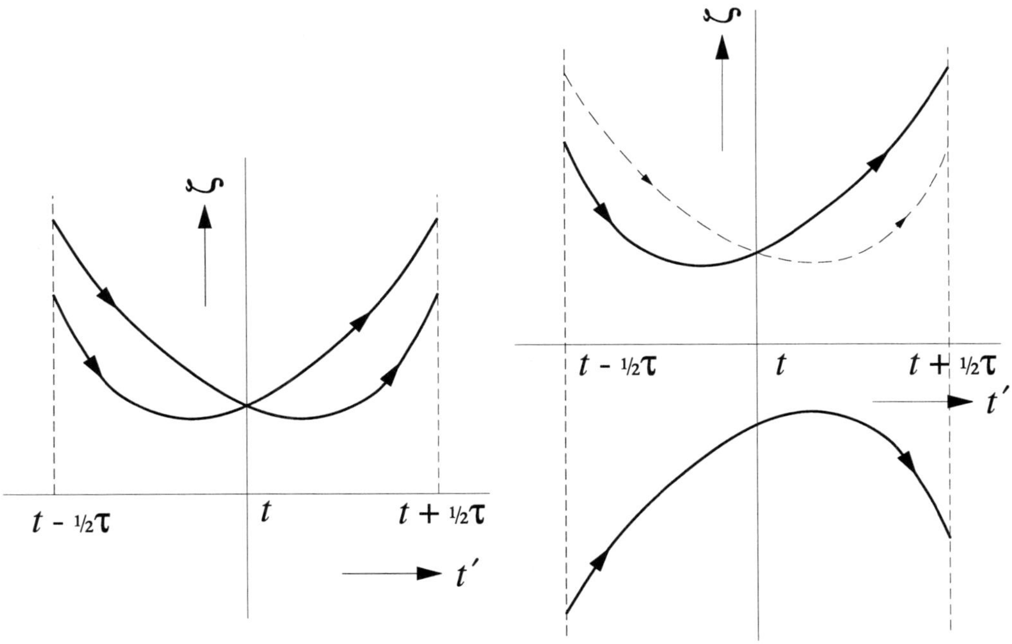

Figure 5.3. On the left-hand side is sketched a trajectory and its reversed trajectory for an even Onsager variable ζ as a function of t' at a fixed time t. The full lines on the right-hand side give a sketch of the trajectories of an odd Onsager variable.

If ζ is an even variable, then with each trajectory in the (ζ, t) plane there is a corresponding inverse trajectory that is obtained by reflection with respect to the ζ axis as depicted in the left-hand side of figure 5.3. It is clear that ζ is now an even function of time. Hence, $\zeta(-t) = \zeta(t)$. If ζ is an odd variable, the inverse trajectory will be obtained by reflection with respect to the ζ axis (dashed line in the right-hand side of figure 5.3 *and* a reflection with respect to the t' axis.

Obviously, reversal of the molecular trajectories is mechanically possible. The existence of the inverse paths of the Onsager variables is concluded from the existence of the inverse molecular trajectories. However, no assumption is introduced with this conclusion so far. An assumption is introduced if the same probability is assigned to the corresponding inverse trajectory as is assigned to a given trajectory. For molecular motions this statement is known as the principle of detailed balance*. For the statistical description, an ensemble of systems is introduced, in which the molecular state of each system of the ensemble is compatible with the imposed macroscopic conditions. The molecular states of the various systems in the ensemble differ usually. *Molecular reversibility* now denotes that a certain phase trajectory has the same probability as its corresponding inverse trajectory that is derived from the given trajectory by the transformation (5.56). This is a generally accepted principle. This principle of molecular reversibility is closely connected with the principle of detailed balance.

As previously remarked, the Onsager variable is a molecular property, which is averaged over a part of the phase space and is represented by a point in the Onsager space. This point will traverse a path in the Onsager space with the passage of time. *Microscopic reversibility* denotes that for *fluctuations in aged systems* the trajectories of the Onsager variables have the same probability as the corresponding inverse trajectories. The extrapolation of the molecular reversibility to the microscopic reversibility is a hypothesis that is, however, in accordance with the continuum view of matter. There are no specifications made from the continuum point of view of matter about the microscopic structure of matter. As a consequence, microscopic reversibility may be considered as a 'chaos' principle. It indicates that $(\Delta V)^-$ is structureless.

First consider an even Onsager variable to examine the consequences of the principle of microscopic reversibility. Figure 5.4 gives an illustration of the

---

* Watanabe S. 1955. Symmetry of physical laws, Part 1. Symmetry in space-time and balance theorems. *Review of Modern Physics*, 27: 26–39. In §5: "The mirage of coordinates involves not only mirages of positions of particles, but also the mirage of shapes of particles and mirage of boundary. It is now clear that, if the molecules are spherical (or points) and spinless and the boundary is symmetrical with regard to mirage, then the classical theorem of detailed balance can be deduced, without discussion of collision processes, simply by the assumption that the distribution function is independent of position. This shows that the classical theorem of detailed balance is based on the 'chaos' hypothesis in regard of the positions of molecules (chaos in angular momentum is a result of chaos in positions). ... Only if the boundary has a symmetry with regard to mirage, can we have the classical theorem of detailed balance on assumption of two kinds of chaos, one regarding position, the other regarding the orientation of molecules."

principle of microscopic reversibility. The probability density of the number of trajectories of the system of the ensemble that passes at the instant of time $t+\frac{1}{2}\tau$ the interval $[\zeta, \zeta+d\zeta]$, and at the instant of time $t-\frac{1}{2}\tau$ the interval $[\zeta', \zeta'+d\zeta']$, respectively, is equal to the probability density of the number of trajectories that passes at the instant of time $t-\frac{1}{2}\tau$ the interval $[\zeta', \zeta'+d\zeta']$, and at the instant of time $t+\frac{1}{2}\tau$ the interval $[\zeta, \zeta+d\zeta]$, respectively.

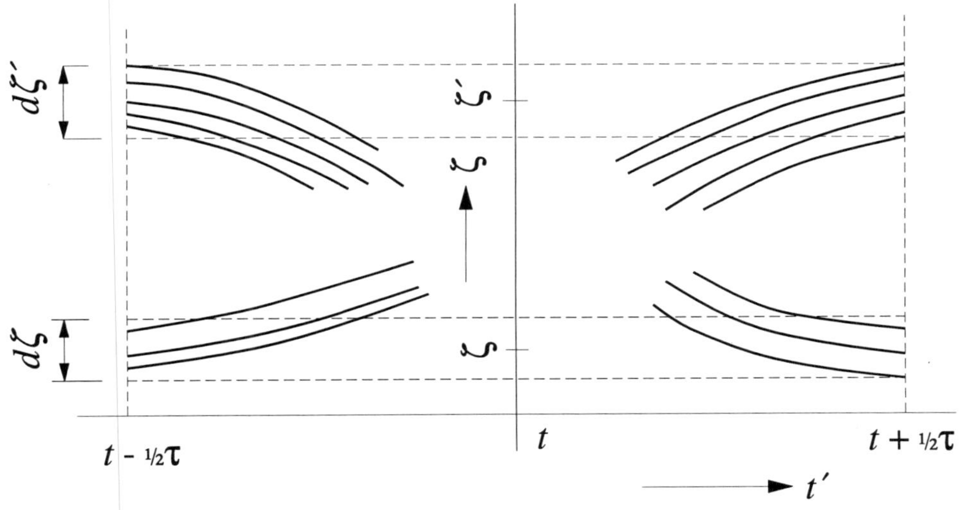

Figure 5.4. Determination of the symmetry property of W_2 for an even Onsager variable ζ.

In terms of a formula the microscopic reversibility is expressed as

$$W_2(\zeta, t-\tfrac{1}{2}\tau; \zeta', t+\tfrac{1}{2}\tau) = W_2(\zeta', t-\tfrac{1}{2}\tau; \zeta, t+\tfrac{1}{2}\tau) \tag{5.57}$$

For stationary stochastic processes, (5.57) becomes

$$W_2(\zeta; \zeta', \tau) = W_2(\zeta'; \zeta, \tau) \tag{5.58}$$

Analogously, microscopic reversibility for odd Onsager variables is formulated as

$$W_2(\zeta, t-\tfrac{1}{2}\tau; \zeta', t+\tfrac{1}{2}\tau) = W_2(-\zeta', t-\tfrac{1}{2}\tau; -\zeta, t+\tfrac{1}{2}\tau) \tag{5.59}$$

For stationary stochastic processes, (5.59) becomes

$$W_2(\zeta; \zeta', \tau) = W_2(-\zeta'; -\zeta, \tau) \tag{5.60}$$

Both formulations (5.58) and (5.60) of the microscopic reversibility for the even and odd Onsager variables can be taken together and generalized by the introduction of the diagonal matrix $\boldsymbol{\epsilon}$ in the Onsager space. For an element ϵ_{KK} on the diagonal $\epsilon_{KK} = \epsilon_{(K)} = +1(-1)$, if ζ_K is an even (odd) Onsager variable. Then the mathematical formulation of the principle of the microscopic reversibility for stationary stochastic processes becomes

$$W_2(\vec{\zeta};\vec{\zeta}',\tau)_{+\vec{B}} = W_2(\boldsymbol{\epsilon}\cdot\vec{\zeta}';\boldsymbol{\epsilon}\cdot\vec{\zeta},\tau)_{-\vec{B}} \tag{5.61}$$

in which a change in the direction of the magnetic field is denoted by subscripts. The microscopic reversibility specifies with (5.61) a property of W_2, but it does not determine W_2 completely. The condition for W_1 is found with the integration of (5.61) over $\vec{\zeta}'$

$$\begin{aligned} W_1(\vec{\zeta}) &= \int W_2(\vec{\zeta};\vec{\zeta}',\tau)_{+\vec{B}}\,d\vec{\zeta}' \\ &= \int W_2(\boldsymbol{\epsilon}\cdot\vec{\zeta}';\boldsymbol{\epsilon}\cdot\vec{\zeta},\tau)_{-\vec{B}}\,d\vec{\zeta}' = W_1(\boldsymbol{\epsilon}\cdot\vec{\zeta})_{-\vec{B}} \end{aligned} \tag{5.62}$$

This condition expresses mathematically the requirement that for a time reversal the stationary probability distribution W_1 has to be invariant.

Finally, the microscopic reversibility may also be expressed in a *correlation matrix* $\mathbf{R}$ that is defined by

$$\mathbf{R}(\tau)_{+\vec{B}} = \langle\vec{\zeta}(t)\,\vec{\zeta}(t+\tau)\rangle_{+\vec{B}} \tag{5.63}$$

Definition (5.63) is specified with the substitution of $\vec{\zeta}(t+\tau) = \vec{\zeta}'$ by

$$\mathbf{R}(\tau)_{+\vec{B}} = \iint \vec{\zeta}\vec{\zeta}'\,W_2(\vec{\zeta};\vec{\zeta}',\tau)_{+\vec{B}}\,d\vec{\zeta}\,d\vec{\zeta}' \tag{5.64}$$

Substitution of (5.61) and applying the change of variables $\vec{\zeta} \to \boldsymbol{\epsilon}\cdot\vec{\zeta}$ and $\vec{\zeta}' \to \boldsymbol{\epsilon}\cdot\vec{\zeta}'$ in the integral expression, gives for the correlation matrix

$$\begin{aligned} \mathbf{R}(\tau)_{+\vec{B}} &= \iint \boldsymbol{\epsilon}\cdot(\vec{\zeta}\vec{\zeta}')\cdot\boldsymbol{\epsilon}\,W_2(\vec{\zeta}';\vec{\zeta},\tau)_{-\vec{B}}\,d\vec{\zeta}\,d\vec{\zeta}' \\ &= \iint \boldsymbol{\epsilon}\cdot(\vec{\zeta}\vec{\zeta}')\cdot\boldsymbol{\epsilon}\,W_2(\vec{\zeta}';\vec{\zeta},\tau)_{-\vec{B}}\,d\vec{\zeta}'\,d\vec{\zeta} \end{aligned}$$

so that microscopic reversibility implies the property

$$\boxed{\mathbf{R}(\tau)_{+\vec{B}} = \boldsymbol{\epsilon}\cdot\mathbf{R}^{\mathrm{T}}(\tau)_{-\vec{B}}\cdot\boldsymbol{\epsilon}} \tag{5.65}$$

for the correlation matrix. With the substitution of

$$W_2(\vec{\zeta};\vec{\zeta}',\tau) = W_1(\vec{\zeta})P_2(\vec{\zeta}',\tau \mid \vec{\zeta}) \tag{5.66}$$

into (5.64), the correlation matrix might also be given by

$$\mathbf{R}(\tau) = \iint \vec{\zeta}\vec{\zeta}' W_1(\vec{\zeta}) P_2(\vec{\zeta}', \tau \mid \vec{\zeta}) \, d\vec{\zeta} \, d\vec{\zeta}'$$
$$= \int \vec{\zeta} \langle \vec{\zeta}', \tau \mid \vec{\zeta} \rangle W_1(\vec{\zeta}) \, d\vec{\zeta} \tag{5.67}$$

in which, in the second line, the definition of the conditional expectation value (5.8) is substituted. For $\tau = 0$ for the correlation matrix applies

$$\mathbf{R}(0) = \langle \vec{\zeta}\vec{\zeta} \rangle = \int \vec{\zeta}\vec{\zeta} W_1(\vec{\zeta}) \, d\vec{\zeta} = \mathbf{R}^{\mathrm{T}}(0) \tag{5.68}$$

which is equal to the variance of the probability distribution W_1.

5.5. REGRESSION AXIOM OF THE FLUCTUATION THEORY

The lack of knowledge of W_2 has to be supplemented with a postulate. This postulate concerns the progress of the fluctuations as a function of time. The central idea is that the progress of the irreversible processes for times $t' > t$ does not depend on how the state at time t has arisen. It is possible that this state has arisen by accidental fluctuations, although it is also possible that this state has originated by a systematic change of variables. The processes, that drive the system into its most probable equilibrium state, have to be formulated in terms of the average regression of the fluctuations in the first situation. In the second situation, the processes should be described by the phenomenological equations. In both cases the mechanisms are different, and the common element has apparently to be found in the irreversibility of the processes. Onsager and Casimir assumed that both descriptions are equivalent. This suggests that during the irreversible processes that drive the system back into the equilibrium state, it is impossible to decide whether the deviation of the equilibrium state is caused by spontaneous fluctuations or artificially by external actions. The processes have no memory of how the deviation of the state of equilibrium is generated. This signifies that a Markoff character can be assigned to these processes.

The phenomenological course of the process is described by (5.51). With the application of (5.47) and (5.49) this phenomenological equation can be formulated in the Onsager variables $\vec{\zeta}(t)$

$$\dot{\vec{\zeta}}(t) = \vec{J} = \mathbf{L} \cdot \vec{K} = -\mathbf{L} \cdot \mathbf{g} \cdot \vec{\zeta} = -\mathbf{M} \cdot \vec{\zeta}(t) \tag{5.69}$$

where the matrix $\mathbf{M}$ is defined by

$$\mathbf{M} = \mathbf{L} \cdot \mathbf{g} \tag{5.70}$$

Written in differences, (5.69) becomes

$$\lim_{\tau\to 0}\frac{1}{\tau}\left[\vec{\zeta}(t+\tau)-\vec{\zeta}(t)\right] = -\mathbf{M}\cdot\vec{\zeta}(t) \tag{5.71}$$

It is clear that τ is again a time interval that is long from the microscopic point of view but short from the macroscopic point of view. For time scales $\tau \ll \tau_{(\mathrm{macro})}$, (5.71) is the equivalent of the regression axiom, if $\tau_{(\mathrm{macro})}$ is a measure for the time, in which the deviations of equilibrium state are macroscopically observable. The progress of the fluctuations for time scales τ as a function of time is determined by laws, which are completely analogous with the macroscopic laws, according to the regression axiom. From (5.71), it turns out in addition that $\vec{\zeta}(t+\tau)$ is completely determined by $\vec{\zeta}(t)$ and τ, which is typical for a stochastic Markoff process.

The regression of the fluctuations (5.7) is found in the Fokker–Planck equation (5.14) or (5.16). By multiplication of (5.71) with $P_2(\vec{\zeta}',\tau \mid \vec{\zeta})\,d\vec{\zeta}'$, denoting $\vec{\zeta}' = \vec{\zeta}(t+\tau)$, and integration over $d\vec{\zeta}'$, the regression (5.7) is controlled with the regression axiom (5.71) by

$$\vec{\xi}(\vec{\zeta},t) = -\mathbf{M}\cdot\vec{\zeta}(t) \tag{5.72}$$

This formulation of the regression axiom follows from (5.71), but the reverse is not true, so that (5.72) is not equivalent to (5.71).

Similarly, by multiplication of $P_2(\vec{\zeta},t \mid \vec{\zeta}_\circ)\,d\vec{\zeta}$ by $\vec{\zeta} = \vec{\zeta}_\circ$ for $t=0$ and by integration over $d\vec{\zeta}$, the regression axiom for the average regression of the fluctuations can be found from (5.72). Using definition (5.8) this yields

$$\langle\vec{\xi},t \mid \vec{\zeta}_\circ\rangle = -\mathbf{M}\cdot\langle\vec{\zeta},t \mid \vec{\zeta}_\circ\rangle \tag{5.73}$$

The first member of (5.73) can be interpreted as follows. Consider

$$\frac{d}{dt}\langle\vec{\zeta},t \mid \vec{\zeta}_\circ\rangle = \lim_{\tau\to 0}\frac{1}{\tau}\left[\int \vec{\zeta}' P_2(\vec{\zeta}',t+\tau \mid \vec{\zeta}_\circ)\,d\vec{\zeta}' - \int \vec{\zeta} P_2(\vec{\zeta},t \mid \vec{\zeta}_\circ)\,d\vec{\zeta}\right]$$

Substitution of (5.2) into the first integral gives

$$\frac{d}{dt}\langle\vec{\zeta},t \mid \vec{\zeta}_\circ\rangle = \lim_{\tau\to 0}\frac{1}{\tau}\left[\iint (\vec{\zeta}+\Delta\vec{\zeta})P_2(\vec{\zeta}',\tau \mid \vec{\zeta})P_2(\vec{\zeta},t \mid \vec{\zeta}_\circ)\,d\vec{\zeta}\,d\vec{\zeta}' - \int \vec{\zeta} P_2(\vec{\zeta},t \mid \vec{\zeta}_\circ)\,d\vec{\zeta}\right]$$

The last term cancels the contribution of $\vec{\zeta}$ in the double integral in the integration over $d\vec{\zeta}'$. The contribution of $\Delta\vec{\zeta}$ with the application of the condition (5.3) results in

$$\frac{d}{dt}\langle\vec{\zeta},t \mid \vec{\zeta}_\circ\rangle = \int \vec{\xi} P_2(\vec{\zeta},t \mid \vec{\zeta}_\circ)\,d\vec{\zeta} = \langle\vec{\xi},t \mid \vec{\zeta}_\circ\rangle \tag{5.74}$$

With this result (5.73) becomes

$$\boxed{\frac{d}{dt}\langle\vec{\zeta},t \mid \vec{\zeta}_\circ\rangle = -\mathbf{M}\cdot\langle\vec{\zeta},t \mid \vec{\zeta}_\circ\rangle} \tag{5.75}$$

The axiom (5.75) has an 'averaged' character. In this expression, $\langle\vec{\zeta},t \mid \vec{\zeta}_\circ\rangle$ is therefore called the *averaged regression of the fluctuations.*

Casimir remarked that the phenomenological equations (5.51) are valid only within the framework of the continuum theory. Due to the linear character of the equations, these equations are appropriate only if the deviations from the equilibrium are small from the macroscopic point of view. However, the continuum theory is applied for deviations from the equilibrium that are large with regard to the effective (root mean square) values of the fluctuations. Given the last consideration, it is clear that the regression axiom signifies that the macroscopic laws are extrapolated into the region of the microscopic fluctuations. At first sight, with reference to the linear character of the macroscopic relations (5.51), this seems permissible, but there may be a question of pseudo-linearity. Actually, this appears to happen, since Callen* showed, using statistical studies, that $\vec{\xi}$ yields the null vector in the mathematical limit $\tau \to 0$ if $\mathbf{M}$ is constant. This means that $\tau \gg \tau_{(\mathrm{micro})}$, if $\tau_{(\mathrm{micro})}$ denotes the time scale of the fluctuations. It is invariably presupposed in the macroscopic theory that the scales of time and space reduce to zero macroscopically, but remain from the physical point of view small and finite.

Regression axioms were introduced ad hoc for particular problems of the fluctuation theory in the period before Onsager. A well-known example is the theory of Einstein† for the Brownian motion of small particles in a fluid. The force exerted by the fluid on the fluctuating particle is modeled in this theory by the macroscopic law of Stokes. Studies of this type were very successful. Onsager generalized such ad hoc hypotheses, and accordingly contributed a general foundation to the linear theory of irreversible processes.

In conclusion, it is found that the regression axiom can be used successfully, provided the constitutive equation of the system is linear. This implies that the linear relaxation processes and the linear transport processes can be explained on the basis of the Markoff processes. Of course, this need not to be true for systems with an essential nonlinear behavior.

---

* Callen H.B. 1947. Ph.D. Thesis. M.I.T.; Callen H.B. 1948. The application of the Onsager's reciprocal relations to thermoelectric, thermomagnetic, and galvanomagnetic effects. *Physical Review*, 73: 1349–1358; Callen H.B. 1952. A note on the adiabatic thermomagnetic effects. *Physical Review*, 85: 16–19.

† Einstein A. 1906. Zur Theorie der Brownschen Bewegung. *Annalen der Physik*, 19: 371–381.

5.6. DETERMINATION OF W_1 FROM THE FOKKER–PLANCK EQUATION

Knowledge of W_1 is required to obtain further consequences of the statistics of the fluctuations and the determination of the conditions for which the reciprocal relations apply. It is not necessary in this respect to start* with the Boltzmann entropy relation $S = k \ln W_1$. The probability W_1 describes a stationary stochastic process for aged systems, in which $\vec{\xi}$ is specified by the regression axiom (5.72). $\mathbf{Q}$ is a constant matrix for aged systems. Then the probability distribution W_1 is easily solved from the Fokker–Planck equation (5.16) for W_1. The FP-equation (5.16) simplifies with the substitution of the regression axiom (5.72) for constant matrices $\mathbf{M}$ and $\mathbf{Q}$ to

$$\frac{\partial}{\partial \vec{\zeta}} \cdot (W_1 \mathbf{M} \cdot \vec{\zeta}) + \left(\mathbf{Q} : \frac{\partial}{\partial \vec{\zeta}} \frac{\partial}{\partial \vec{\zeta}}\right) W_1 = 0 \tag{5.76}$$

With the computation of the first term in the left-hand side, this equation becomes

$$\mathbf{M} : \left(\mathbf{I}\, W_1 + \vec{\zeta}\, \frac{\partial W_1}{\partial \vec{\zeta}}\right) + \left(\mathbf{Q} : \frac{\partial}{\partial \vec{\zeta}} \frac{\partial}{\partial \vec{\zeta}}\right) W_1 = 0 \tag{5.77}$$

The solution of (5.76) that satisfies the boundary conditions $W_1 \to 0$ for $\vec{\zeta} \to \pm\infty$, is the Gaussian probability distribution

$$W_1 = \Omega \exp(-\tfrac{1}{2} \mathbf{V}^{-1} : \vec{\zeta}\vec{\zeta}) \tag{5.78}$$

where Ω is a constant normalization factor determined by $\int W_1\, d\vec{\zeta} = 1$, and $\mathbf{V}$ is a symmetric matrix that is calculated from (5.78) and (5.77). Differentiating (5.78) yields

$$\begin{aligned} \frac{\partial W_1}{\partial \vec{\zeta}} &= -\mathbf{V}^{-1} \cdot \vec{\zeta}\, W_1 \\ \left(\frac{\partial}{\partial \vec{\zeta}} \frac{\partial}{\partial \vec{\zeta}}\right) W_1 &= -\left(\mathbf{V}^{-1} - (\mathbf{V}^{-1} \cdot \vec{\zeta})(\mathbf{V}^{-1} \cdot \vec{\zeta})\right) W_1 \end{aligned} \tag{5.79}$$

and the substitution into (5.77) gives

$$(\mathbf{M} - \mathbf{Q}.\mathbf{V}^{-1}) : \left(\mathbf{I} - \vec{\zeta}(\mathbf{V}^{-1} \cdot \vec{\zeta})\right) = 0 \tag{5.80}$$

* It is usually supposed that W_1 is defined by the Boltzmann entropy relation. See for instance De Groot S.R. and Mazur P. 1962. *Non-Equilibrium Thermodynamics*. North-Holland Publ. Co., Amsterdam, p. 91. W_1 is then fixed for aged systems by $W_1 = \Omega \exp(\Delta S/k) = \Omega \exp\big(-(1/2k)\mathbf{g} : \vec{\zeta}\vec{\zeta}\big)$.

This equation has to be satisfied for all $\vec{\zeta}$, from which it is concluded that (5.78) is indeed a solution of (5.76) if $\mathbf{V}$ satisfies

$$\mathbf{V} = \mathbf{V}^{\mathrm{T}} = \mathbf{M}^{-1} \cdot \mathbf{Q} = \mathbf{g}^{-1} \cdot \mathbf{L}^{-1} \cdot \mathbf{Q} \tag{5.81}$$

Substitution of (5.70) for $\mathbf{M}$ yields the right-hand equality. The symmetry of $\mathbf{L}$ cannot be concluded from (5.81).

Even if $\mathbf{Q}$ is specified and even if is known that $\mathbf{L}$ is symmetric, the distribution is not yet completely known. The Gaussian probability distribution W_1 is then fixed if the variance and the normalization factor are also known. The variance $\mathbf{R}(0)$ defined by (5.68) can be calculated first. Using $(5.79)_1$ it follows that

$$\mathbf{V}^{-1} \cdot \langle \vec{\zeta}\vec{\zeta} \rangle = \int (\mathbf{V}^{-1} \cdot \vec{\zeta})\, \vec{\zeta}\, W_1 \, d\vec{\zeta} = -\int \frac{\partial W_1}{\partial \vec{\zeta}}\, \vec{\zeta}\, d\vec{\zeta}$$

Integration by parts, and using the boundary conditions that $W_1 \to 0$ for $\vec{\zeta} \to \pm\infty$, yields

$$\mathbf{V}^{-1} \cdot \langle \vec{\zeta}\vec{\zeta} \rangle = \int \left(\frac{\partial}{\partial \vec{\zeta}} \vec{\zeta} \right) W_1 \, d\vec{\zeta} = \int \mathbf{I}\, W_1 \, d\vec{\zeta} = \mathbf{I}$$

so that the variance (or $\mathbf{R}(0)$) for the Gaussian probability distribution W_1 in (5.78) is given by

$$\mathbf{R}(0) = \langle \vec{\zeta}\vec{\zeta} \rangle = \mathbf{V} = \mathbf{V}^{\mathrm{T}} \tag{5.82}$$

The normalization factor Ω is found by diagonalizing the symmetric matrix $\mathbf{V}$ with $r+1$ principal values and performing the integration in the principal directions. These operations yield

$$\Omega = \sqrt{\frac{|\mathbf{V}^{-1}|}{(2\pi)^{r+1}}} \tag{5.83}$$

Microscopic reversibility has led to the condition (5.62) for W_1. Application on (5.78) gives

$$\vec{\zeta} \cdot \mathbf{V}^{-1} \cdot \vec{\zeta} = \vec{\zeta} \cdot \boldsymbol{\epsilon} \cdot \mathbf{V}^{-1} \cdot \boldsymbol{\epsilon} \cdot \vec{\zeta}$$

or

$$\vec{\zeta} \cdot \left(\mathbf{V}^{-1} - \boldsymbol{\epsilon} \cdot \mathbf{V}^{-1} \cdot \boldsymbol{\epsilon} \right) \cdot \vec{\zeta} = 0 \qquad \text{for} \quad \forall \vec{\zeta} \tag{5.84}$$

from which follows that no cross terms between the even and odd variables are found. If the direction of the magnetic field is denoted by subscripts, then it is concluded from (5.84) that

$$(\mathbf{V})_{+\vec{B}} = \boldsymbol{\epsilon} \cdot (\mathbf{V}^{\mathrm{T}})_{-\vec{B}} \cdot \boldsymbol{\epsilon} \tag{5.85}$$

The events in the Onsager spaces for the odd and the even Onsager variables are obviously statistically independent of each other.

For the subsequent specification of W_1, microscopic reversibility has to be used again. . In the formulation of (5.65), the correlation matrix can be calculated through (5.67) by the average regression of the fluctuations. This average regression of the fluctuations is subject to the regression axiom, and is determined by the differential equation (5.75). The solution of this differential equation with the initial condition $\langle\vec{\zeta}, 0 \mid \vec{\zeta}_\circ\rangle = \vec{\zeta}_\circ$ is given by

$$\langle\vec{\zeta}, t \mid \vec{\zeta}_\circ\rangle = \exp(-\mathbf{M}t) \cdot \vec{\zeta}_\circ \tag{5.86}$$

where the operator $\exp(-\mathbf{M}t)$ is defined by

$$\exp(-\mathbf{M}t) = \sum_{n=0}^{\infty} \frac{(-\mathbf{M}t)^n}{n!} \tag{5.87}$$

with

$$\begin{aligned} &\mathbf{M}^n = \mathbf{M}^{n-1} \cdot \mathbf{M} = \mathbf{M} \cdot \mathbf{M}^{n-1} \qquad \text{for} \qquad n = 2, 3, \cdots \\ &\mathbf{M}^1 = \mathbf{M} \qquad \text{and} \qquad \mathbf{M}^0 = \mathbf{I} \end{aligned} \tag{5.88}$$

Since the symmetry properties of $\mathbf{M}$ are unknown, the powers of $\mathbf{M}$ are defined by (5.88). By mathematical induction it follows that

$$(\mathbf{M}^n)^{\mathrm{T}} = (\mathbf{M}^{\mathrm{T}})^n \tag{5.89}$$

Solution (5.86) shows that the average regression of the fluctuations for $t \to \infty$ relaxes to zero. Substitution of (5.86) into (5.67) results in

$$\begin{aligned} \mathbf{R}(\tau) &= \int \vec{\zeta}_\circ \langle\vec{\zeta}', \tau \mid \vec{\zeta}_\circ\rangle W_1(\vec{\zeta}_\circ)\, d\vec{\zeta}_\circ \\ &= \int \vec{\zeta}_\circ \exp(-\mathbf{M}\tau) \cdot \vec{\zeta}_\circ W_1(\vec{\zeta}_\circ)\, d\vec{\zeta}_\circ \\ &= \int \vec{\zeta}_\circ\, \vec{\zeta}_\circ W_1(\vec{\zeta}_\circ)\, d\vec{\zeta}_\circ \cdot \exp(-\mathbf{M}^{\mathrm{T}}\tau) \\ &= \mathbf{R}(0) \cdot \exp(-\mathbf{M}^{\mathrm{T}}\tau) \end{aligned}$$

for the left-hand side of (5.65). For the right-hand side of (5.65) this substitution yields

$$\begin{aligned} \boldsymbol{\epsilon} \cdot \mathbf{R}^{\mathrm{T}}(\tau) \cdot \boldsymbol{\epsilon} &= \boldsymbol{\epsilon} \cdot \int \langle\vec{\zeta}', \tau \mid \vec{\zeta}_\circ\rangle \vec{\zeta}_\circ W_1(\vec{\zeta}_\circ)\, d\vec{\zeta}_\circ \cdot \boldsymbol{\epsilon} \\ &= \boldsymbol{\epsilon} \cdot \exp(-\mathbf{M}\tau) \cdot \int \vec{\zeta}_\circ \vec{\zeta}_\circ W_1(\vec{\zeta}_\circ)\, d\vec{\zeta}_\circ \cdot \boldsymbol{\epsilon} \\ &= \boldsymbol{\epsilon} \cdot \exp(-\mathbf{M}\tau) \cdot \mathbf{R}(0) \cdot \boldsymbol{\epsilon} \end{aligned}$$

so that the correlation expression (5.65) of the microscopic reversibility yields the symmetry condition

$$\mathbf{R}(0) \cdot \exp(-\mathbf{M}^{\mathrm{T}}\tau) = \boldsymbol{\epsilon} \cdot \exp(-\mathbf{M}\tau) \cdot \mathbf{R}(0) \cdot \boldsymbol{\epsilon} \tag{5.90}$$

in which $\mathbf{R}(0)$ is given by (5.82) and (5.81)

$$\mathbf{R}(0) = \mathbf{V} = \mathbf{M}^{-1} \cdot \mathbf{Q} = \mathbf{V}^{\mathrm{T}} = (\mathbf{M}^{-1} \cdot \mathbf{Q})^{\mathrm{T}} = \mathbf{Q}^{\mathrm{T}} \cdot (\mathbf{M}^{-1})^{\mathrm{T}} \tag{5.91}$$

The symmetry condition (5.90), with the substitution of (5.91) becomes

$$\mathbf{Q}^{\mathrm{T}} \cdot (\mathbf{M}^{-1})^{\mathrm{T}} \cdot \exp(-\mathbf{M}^{\mathrm{T}}\tau) = \boldsymbol{\epsilon} \cdot \exp(-\mathbf{M}\tau) \cdot \mathbf{M}^{-1} \cdot \mathbf{Q} \cdot \boldsymbol{\epsilon} \tag{5.92}$$

Since (5.92) holds for arbitrary times τ, it is concluded that the coefficients of τ^n in the definition (5.87) of the operator $\exp(-\mathbf{M}\tau)$ have to satisfy the symmetry conditions

$$\mathbf{Q}^{\mathrm{T}} \cdot (\mathbf{M}^{\mathrm{T}})^{n-1} = \boldsymbol{\epsilon} \cdot \mathbf{M}^{n-1} \cdot \mathbf{Q} \cdot \boldsymbol{\epsilon} \qquad \text{for} \quad n = 0, 1, 2, \cdots \tag{5.93}$$

By elimination of the diagonal matrix $\boldsymbol{\epsilon}$ (+1 for even O-variables and −1 for odd O-variables, $\boldsymbol{\epsilon} \cdot \boldsymbol{\epsilon} = 1$) on the right-hand side of (5.93), the symmetry conditions can be written as

$$\boxed{\mathbf{M}^{n-1} \cdot \mathbf{Q} = \boldsymbol{\epsilon} \cdot (\mathbf{M}^{n-1} \cdot \mathbf{Q})^{\mathrm{T}} \cdot \boldsymbol{\epsilon}} \qquad \text{for} \quad n = 0, 1, 2, \cdots \tag{5.94}$$

which is an equation for the as yet unknown $\mathbf{Q}$.

With the determination of $\mathbf{Q}$ the probability distribution W_1 is now known. If (5.94) is satisfied, then W_1 satisfies the conditions implied by the assumptions of microscopic reversibility and the regression axiom. These two axioms are, however, not sufficient to confirm to the Onsager Casimir symmetry relations

$$\boxed{\mathbf{L}(+\vec{B}) = \boldsymbol{\epsilon} \cdot \mathbf{L}^{\mathrm{T}}(-\vec{B}) \cdot \boldsymbol{\epsilon}} \tag{5.95}$$

Not every solution of (5.94) for $\mathbf{Q}$ results in the symmetry relations. For example, $\mathbf{Q} = \mathbf{I}$ and $\mathbf{Q} = \mathbf{M}$ are solutions of (5.94). For either solution it follows that $\mathbf{M}$ is symmetric, from which it cannot however be concluded that $\mathbf{L}$ is also symmetric. This conclusion cannot be drawn, because the product of two symmetric matrices ($\mathbf{M}$ and $\mathbf{g}$) need not inevitably be symmetric, so that if $\mathbf{M}$ is symmetric, the symmetry of $\mathbf{L} = \mathbf{M} \cdot \mathbf{g}^{-1}$ does not necessarily follow.

5.7. ONSAGER CASIMIR RECIPROCAL RELATIONS

First, it is remarked that from (5.94) for $n = 1$ it follows that

$$\mathbf{Q} = \boldsymbol{\epsilon} \cdot \mathbf{Q}^{\mathrm{T}} \cdot \boldsymbol{\epsilon} \tag{5.96}$$

$\mathbf{Q}$ is therefore symmetric, and no terms with products of even and odd variables occur in the second moment of the transition probability. This conforms with (5.85), since it was already concluded that the events in the Onsager spaces of even and odd Onsager variables are statistically independent of each other.

For $n = 0$ the symmetry condition (5.94) becomes, with the substitution of $\mathbf{M} = \mathbf{L} \cdot \mathbf{g}$

$$(\mathbf{L} \cdot \mathbf{g})^{-1} \cdot \mathbf{Q} = \boldsymbol{\epsilon} \cdot \left((\mathbf{L} \cdot \mathbf{g})^{-1} \cdot \mathbf{Q}\right)^{\mathrm{T}} \cdot \boldsymbol{\epsilon} \tag{5.97}$$

The Onsager Casimir reciprocal relations (5.95) are found for three basic solutions* of (5.97)

$$\boxed{\mathbf{Q} = k_{\mathrm{S}}\, \mathbf{g}^{-1} \qquad \text{or} \qquad \mathbf{Q} = k_{\mathrm{B}}\, \mathbf{L} \qquad \text{or} \qquad \mathbf{Q} = k_{\mathrm{M}}\, \mathbf{M} \cdot \mathbf{L}} \tag{5.98}$$

where k_{S}, k_{B}, and k_{M} are to be specified positive constants, and $\mathbf{Q}$ is symmetric by (5.96). The constants k_{S}, k_{B}, and k_{M} are positive, because all matrices are positive definite. The solutions (5.98) also satisfy (5.94) for all other values of n. According to the discussion given after the formulas (5.71) and (5.75) it is noted that values of $n \geq 2$ in (5.94) are not of use within the framework of this theory, because τ has to satisfy

$$\tau_{(\mathrm{micro})} \ll \tau \ll \tau_{(\mathrm{macro})} \tag{5.99}$$

and it must also be true that contributions of $\mathcal{O}(\tau^2)$ can be neglected.

Furthermore, it is pointed out that for all linear combinations of the three basic solutions (5.98), the reciprocal relations (5.95) result. Thus, a large class of Gaussian probability distributions yields the Onsager Casimir symmetry relations, provided that the assumptions of microscopic reversibility and the regression axiom are accepted. The variances (the square of the standard deviation) of the Gaussian distribution (5.78) for the three basic solutions follow by substitution of (5.98) into (5.91)

$$\mathbf{V} = k_{\mathrm{S}}\, \mathbf{g}^{-1} \cdot \mathbf{L}^{-1} \cdot \mathbf{g}^{-1} \quad \text{or} \quad \mathbf{V} = k_{\mathrm{B}}\, \mathbf{g}^{-1} \quad \text{or} \quad \mathbf{V} = k_{\mathrm{M}}\, \mathbf{L} \tag{5.100}$$

To discuss the significance of the solutions (5.98), consider the Fokker–Planck equation (5.14) for the transition probability P_2. The matrix $\mathbf{Q}$ can be

* Kuiken G.D.C. 1977. The derivation of the Onsager Casimir reciprocity relations without using Boltzmann's postulate. *Journal of Non-Equilibrium Thermodynamics*, 2: 153–168. In this article only the first solution is given.

considered in this FP-equation* as a 'diffusion coefficient' that determines the speed of transformation for the transition probability P_2 into the probability distribution W_1. For the solution $(5.98)_2$, this speed is proportional with the phenomenological coefficient in (5.51). The application of this macroscopic linear relation between the forces and the fluxes is by that extended, and is now used also for microscopic variables and the fluctuating Onsager variables. Substitution of $(5.98)_2$ into (5.81) results in the Boltzmann probability distribution

$$\boxed{W_1 = \Omega_{\mathrm{B}} \exp\left(-\frac{1}{2k_{\mathrm{B}}}\vec{\zeta}\cdot\mathbf{g}\cdot\vec{\zeta}\right) = \Omega_{\mathrm{B}} \exp(\Delta S_{(\mathrm{tot})}/k_{\mathrm{B}})} \tag{5.101}$$

Use was made of (5.40) to get the right-hand side equality, and the constant k_{B} can be identified with the Boltzmann constant ($k_{\mathrm{B}} = 1.380\,658 \times 10^{-23}$ J K^{-1}). The normalization factor Ω_{B} follows from the substitution of $(5.100)_2$ into (5.83)

$$\Omega_{\mathrm{B}} = \sqrt{\frac{|\mathbf{g}|}{(2\pi\, k_{\mathrm{B}})^{r+1}}} \tag{5.102}$$

The most probable state is found for $\vec{\zeta} = \vec{0}$ and that is the equilibrium state for which the probability $W_1 = \Omega_{\mathrm{B}}$, with Ω_{B} given by (5.102).

The Boltzmann postulate is therefore a useful probability distribution to derive the Onsager reciprocal relations, but that is not the only one. The Boltzmann postulate implies the extra assumption that the macroscopic phenomenological equations can be extrapolated to the stochastic processes in $(\Delta V)^-$. The variance $(5.100)_2$ for the Boltzmann distribution (5.101) indicates

* For aged systems this equation becomes

$$\frac{\partial P_2}{\partial t} = \mathbf{M} : \left(\mathbf{I}\,P_2 + \vec{\zeta}\,\frac{\partial P_2}{\partial \vec{\zeta}}\right) + \left(\mathbf{Q} : \frac{\partial}{\partial \vec{\zeta}}\frac{\partial}{\partial \vec{\zeta}}\right) P_2$$

for which the solution is given by

$$P_2 = P_2(\vec{\zeta}, t \mid \vec{\zeta}_\circ) = \Omega \exp\left[-\tfrac{1}{2}\mathbf{W}^{-1}(\vec{\zeta} - \langle\vec{\zeta}, t \mid \vec{\zeta}_\circ\rangle)(\vec{\zeta} - \langle\vec{\zeta}, t \mid \vec{\zeta}_\circ\rangle)\right]$$

The normalization constant Ω is given by (5.83) if in this expression the variance $\mathbf{V}$ is replaced by the variance $\mathbf{W}$. The variance $\mathbf{W}$ is defined by

$$\mathbf{W}(t) = \mathbf{V} - \exp(-\mathbf{M}t)\cdot\mathbf{V}\cdot\exp(-\mathbf{M}^{\mathrm{T}}t)$$

$\mathbf{W}$ relaxes to $\mathbf{V}$ for $t \to \infty$, while it follows from (5.86) that $\langle\vec{\zeta}, t \mid \vec{\zeta}_\circ\rangle$ also relaxes to the null vector, so that

$$\lim_{t\to\infty} P_2(\vec{\zeta}, t \mid \vec{\zeta}_\circ) = W_1(\vec{\zeta})$$

that in the stationary state the spread of the Onsager variables is obviously determined only by the metric of the Onsager space and not by the processes occurring in the Onsager space due to the appearing fluctuations.

Reversely, the fluctuations appearing give rise to irreversible processes in $(\Delta V)^-$. The solution $(5.98)_1$ indicates that the above 'diffusion process' for P_2 is also determined by the matrix $\mathbf{g}$ that specifies the local metric of the Onsager space. The driving force for the irreversible processes is then proportional to the fluctuations of the thermodynamic variables. Substitution of $(5.98)_1$ into the variance (5.81) yields for the probability distribution

$$W_1 = \Omega_S \exp\left[-\frac{1}{2k_S}\vec{\zeta}\cdot\mathbf{g}\cdot\mathbf{L}\cdot\mathbf{g}\cdot\vec{\zeta}\right]$$

$$= \Omega_S \exp\left[-\frac{1}{2k_S}(-\vec{\zeta}\cdot\mathbf{g})\cdot(-\mathbf{M}\cdot\vec{\zeta})\right] \tag{5.103}$$

$$= \Omega_S \exp\left[-\frac{1}{2k_S}(-\vec{\zeta}\cdot\mathbf{g}\cdot\dot{\vec{\zeta}})\right] \tag{5.104}$$

$$= \Omega_S \exp\left[-\frac{1}{2k_S}\Pi_{(S)}\right] \tag{5.105}$$

where (5.104) follows from the substitution of (5.69) into (5.103), and (5.105) from the substitution of (5.50) into (5.104). The normalization factor (5.83) with the substitution of $(5.100)_1$ then becomes

$$\Omega_S = \sqrt{\frac{|\mathbf{g}\cdot\mathbf{L}\cdot\mathbf{g}|}{(2\pi\, k_S)^{r+1}}} \tag{5.106}$$

If no fluctuations of the equilibrium state take place, then $\vec{\zeta} = \vec{0}$, and also $\Pi_{(S)} = 0$ then the equilibrium state is found from the maximum probability Ω_S. The equilibrium state is the most probable state. It is the mean value of the Gaussian distribution and is given by the normalization factor (5.106). Macroscopically the type of fluctuation around the equilibrium state cannot be observed, and although different variances are found, the most probable state, the equilibrium state, has to be the same. This result in the condition $\Omega_B = \Omega_S$, and the constant k_S is specified by

$$k_S = k_B\left(\frac{|\mathbf{g}\cdot\mathbf{L}\cdot\mathbf{g}|}{|\mathbf{g}|}\right)^{\frac{1}{r+1}} \tag{5.107}$$

The probability of finding a fluctuation of the equilibrium state can also be described from the microscopic point of view by using the dissipation. The dissipation Π_S is zero for reversible processes. That is, $\Pi_S = 0$ defines the most probable state, which is the equilibrium state. All equilibrium states at the state surface are equally probable, in the sense that the probability is determined only by the metric of the Onsager space and the phenomenological coefficient $\mathbf{L}$ that is considered constant. Since $k_S > 0$, the probability distribution (5.105) shows that for the dissipation associated with the spontaneous fluctuations

$$\Pi_{(S)} \geq 0 \tag{5.108}$$

This microscopic result is in agreement with the condition for the entropy production (5.52), which is derived using phenomenological considerations. If the material is locally stable and dissipative processes occur by fluctuations, then (5.108) with (5.105) guarantees that the material remains locally stable. The fluctuations are local, and the variance of the equilibrium distribution of the fluctuations $(5.100)_1$ is now also determined by phenomenological coefficient $\mathbf{L}$. The probability distribution shows that local stability demands also that the local entropy production is positive definite. Comparison of the distributions (5.102) and (5.105) shows that there must exist a relation between the dissipation and the correlation function $\mathbf{R}$ ($\mathbf{R}(0) = \mathbf{V}$) of the spontaneous fluctuations. The general relation is known as the fluctuation dissipation theorem*.

The third solution $(5.98)_3$ for a Gaussian probability distribution, which results in the reciprocal relations, shows that the time evolution of the transition probability P_2 now also depends on $\mathbf{g}$ as well as on $\mathbf{L}$. Substitution of $(5.98)_3$ into (5.81) yields the probability distribution

$$\begin{aligned} W_1 &= \Omega_M \exp\left[-\frac{1}{2k_M}\vec{\zeta}\cdot\mathbf{L}^{-1}\cdot\vec{\zeta}\right] \\ &= \Omega_M \exp\left[-\frac{1}{2k_M}(-\vec{\zeta}\cdot\mathbf{g})\cdot(-\mathbf{M}^{-1}\cdot\vec{\zeta})\right], \end{aligned} \tag{5.109}$$

where $\mathbf{L} = \mathbf{M}\cdot\mathbf{g}$ has been used. The normalization constant now becomes

$$\Omega_M = \sqrt{\frac{1}{(2\pi\, k_M)^{r+1}\mathbf{L}}} \tag{5.110}$$

and the constant k_M is chosen to be

$$k_M = k_B\left(\frac{1}{|\mathbf{L}||\mathbf{g}|}\right)^{\frac{1}{r+1}} \tag{5.111}$$

to guarantee that the equilibrium state is the most probable state with the same probability as the two previously discussed basic solutions of the Fokker–Planck equation, from which the reciprocal relations can also be deduced.

---

* De Groot S.R. and Mazur P. 1962. *Non-Equilibrium Thermodynamics.* North-Holland Publ. Co., Amsterdam. See chapter VIII.

If (5.103) is compared with (5.109), then it is noticed that the difference is that $\mathbf{M}^{-1}$ is now found in the probability distribution instead of $\mathbf{M}$. It can now be understood that a distribution for which in the regression axiom (5.71) the 'flux' and the 'force' are reversed, will also result in the reciprocal relations.

Linear combinations of Gaussian distributions are also Gaussian distributions. Linear combinations of the basic solutions (5.98) are probability distributions for which the reciprocal relations can be deduced. The constant should be determined by the condition that the most probable state is the equilibrium state with the probability Ω_B and consistent with the macroscopic conditions imposed. Clearly, the Boltzmann distribution is a solution of the Fokker–Planck equation that satisfies both microscopic reversibility and the regression axiom. It is found that other distributions of the fluctuations also result in the reciprocal relations, but not all Gaussian distributions that satisfy microscopic reversibility yield the reciprocal relations. For example, for $\mathbf{Q} = \mathbf{M}$ the Gaussian distribution $W_1 = \Omega \exp(-\frac{1}{2}\vec{\zeta} \cdot \mathbf{I} \cdot \vec{\zeta})$ follows, but from this distribution the reciprocal relations cannot be deduced*. The experimental verification of the reciprocal relations for a variety of physical processes has been accomplished by Miller and others†.

Sometimes it is necessary or advantageous to transform the process forces and the process fluxes linearly. An example is the derivation of the molar description of diffusion and of heat conduction from the barycentric description. It is therefore important to examine to what extent the Onsager Casimir reciprocal relations remain valid under linear transformations. This means that allowed transformations have to be done separately in the Onsager

---

* De Groot S.R. 1974. The Onsager relations; theoretical basis. In *Foundations of Continuum Thermodynamics*, eds, Delgado Domingos J.J., Nina M.N.R. and Whitelaw J.H. MacMillan Press Ltd, London, pp. 159–183. On page 161: 'The probability $f(\mathbf{a})$ $(= W_1(\vec{\zeta}))$ is supposed to fulfil a *central limit theorem* that means it has the Gaussian form (5.101).' Assuming the Boltzmann distribution, the author stated on page 163: '*Theorem:* The property of microscopic reversibility has as a consequence the symmetry of the matrix $\mathbf{L} = \mathbf{L}^\mathrm{T}$ or, in other words, the validity of the *Onsager reciprocal relations* between the transport coefficients.' From our discussion it follows that contradictions to this theorem can be found.

† Miller D.G. 1960. Thermodynamics of irreversible processes, the experimental verification of the Onsager reciprocal relations. *Chemical Reviews*, 60: 15–17; Woolf L.A., Miller D.G. and Gosting L.J. 1962. Isothermal diffusion measurements on the system H_2O-Glycine-KCl at 25 °C; Tests of the Onsager reciprocal relations. *Journal of the American Chemical Society*, 84: 317–331; Miller D.G. 1974. The Onsager relations; experimental evidence. In *Foundations of Continuum Thermodynamics*, eds, Delgado Domingos J.J., Nina M.N.R. and Whitelaw J.H. MacMillan Press Ltd, London, pp. 185–214; See also Mason E.A. 1974. The Onsager reciprocal relations—experimental evidence; Discussion Paper. In *Foundations of Continuum Thermodynamics*, eds, Delgado Domingos J.J., Nina M.N.R. and Whitelaw J.H. MacMillan Press Ltd, London, pp. 215–227.

spaces of even and odd variables. Once formulated the reciprocal relations have to be preserved, then the forces and the fluxes cannot be interchanged since for even $\vec{\zeta}$ the time derivative $d\vec{\zeta}/dt$ becomes odd. The Onsager Casimir reciprocal relations hold, but have to be formulated anew if fluxes and forces are interchanged. An example is discussed in the derivation of the Maxwell-Stefan diffusion in the next chapter.

From (5.105) is observed that the probability distribution is invariant for linear transformations, if the entropy also remains invariant under the same linear transformation. A coordinate transformation in the Onsager space of even and odd variables is a transformation of the variables ζ_K that describe the state of the system. The entropy has to remain invariant in the entropy representation, since changes in the entropy imply that either the state of the system or the system itself is changed. The dissipation has to remain unchanged by the transformation of the variables, so that invariance of the entropy is maintained. This requirement can be obtained from (5.105), from which it also follows that the bilinear form of the entropy production has to be invariant.

There has been a lot of confusion in literature about the admissible transformation rules. A summary has been given by Meixner*. He remarked that it is not surprising that the Onsager Casimir reciprocal relations remain invariant for a special class of linear transformations†, just as the Hamilton equations are invariant for special transformations (the canonical transformations of the phase space coordinates and momenta).

In diffusion and other applications of the thermodynamics of irreversible processes, it is frequently advantageous to use linearly dependent fluxes and forces. It is of course always possible to obtain an independent set of fluxes and forces by eliminating the dependences using the linear relations that state the dependences between the fluxes and forces. By doing so, one may wonder if the validity of the Onsager Casimir reciprocal relations is still preserved.

The symmetry relations derived here apply for homogeneous systems, so that the results are not directly applicable to transport processes in inhomogeneous systems. For *inhomogeneous systems*, in which transport processes occur, the Onsager Casimir reciprocal relations can be derived from the results obtained for the homogeneous systems. To this purpose, the inhomogeneous systems are divided into a large number of physical volume elements $(\Delta V)^\star$that are considered homogeneous subsystems. The theory

---

* Meixner J. 1973. Consistency of the Onsager-Casimir reciprocal relations. *Advances in Molecular Relaxation Processes*, 5: 319–331.

† Meixner J. *loc. cit.*: 'The admitted transformations therefore have to make all the new fluxes odd or even with respect to time reversal and to keep the bilinear form of the entropy production invariant, and not only to keep the quadratic form of the entropy production form-invariant.'

developed for homogeneous systems is then applied to each subsystem, and the interaction between the subsystems is accounted for. Then the Onsager Casimir reciprocal relations are deduced again after some mathematical operations, provided that the phenomenological tensors do not depend on the size and shape of the body, and that in vacuum no transport processes occur.

5.8. EXERCISES

Exercise 5.1. In a relative reference system that rotates with angular velocity $\vec{\Omega}$ with respect to an inertial system the reciprocal relations (5.95) have to be replaced by

$$\mathbf{L}(+\vec{B},+\vec{\Omega}) = \boldsymbol{\epsilon}\cdot\mathbf{L}^{\mathrm{T}}(-\vec{B},-\vec{\Omega})\cdot\boldsymbol{\epsilon}$$

Discuss the conditions for which the dependence of reciprocal relations on the angular velocity for a dilute gas in a relative reference system can be neglected.

Exercise 5.2. Assume the Boltzmann probability distribution (5.101) from the start and derive the Onsager symmetry relations.

Exercise 5.3. Show that the solution of the Fokker–Planck equation

$$\frac{\partial P_2}{\partial t} = \mathbf{M} : \left(\mathbf{I}\,P_2 + \vec{\zeta}\,\frac{\partial P_2}{\partial \vec{\zeta}}\right) + \left(\mathbf{Q} : \frac{\partial}{\partial \vec{\zeta}}\frac{\partial}{\partial \vec{\zeta}}\right) P_2$$

is given by

$$P_2 = P_2(\vec{\zeta},t \mid \vec{\zeta}_\circ) = \Omega \exp\left[-\tfrac{1}{2}\mathbf{W}^{-1}(\vec{\zeta} - \langle\vec{\zeta},t \mid \vec{\zeta}_\circ\rangle)(\vec{\zeta} - \langle\vec{\zeta},t \mid \vec{\zeta}_\circ\rangle)\right]$$

where the normalization constant Ω is given by

$$\Omega = \sqrt{\frac{|\mathbf{W}^{-1}|}{(2\pi)^{r+1}}}$$

with the variance $\mathbf{W}$ defined by

$$\mathbf{W}(t) = \mathbf{V} - \exp(-\mathbf{M}t)\cdot\mathbf{V}\cdot\exp(-\mathbf{M}^{\mathrm{T}}t)$$

Show that for $t \to \infty$ the solution relaxes to

$$\lim_{t\to\infty} P_2(\vec{\zeta},t \mid \vec{\zeta}_\circ) = \Omega \exp(-\tfrac{1}{2}\mathbf{V}^{-1} : \vec{\zeta}\vec{\zeta}) = W_1(\vec{\zeta})$$

Exercise 5.4. Meixner has shown that for linear transformations, which transform the fluxes and forces of definite parities to others of definite parities and leaving the bilinear form of the entropy production invariant, the Onsager reciprocal relations are preserved. Consider two fluxes X_i and two forces Y_i, for which the entropy production is given by

$$\Pi_{(\mathrm{S})} = X_1 Y_1 + X_2 Y_2$$

This entropy production is invariant for the fluxes X_1, X_2, and the forces $Y_1 + \alpha X_2$, $Y_2 - \alpha X_1$, where α is an arbitrary constant. Discuss for which even and odd parities of X_i, Y_i the forces $Y_1 + \alpha X_2$, $Y_2 - \alpha X_1$ have definite parities for which the Onsager reciprocal relations are preserved.

6 Multicomponent Diffusion

Summary: In many real mass transfer processes, the slowness of the multicomponent diffusion determines the process rate. A thorough knowledge of the basics of multicomponent diffusion is needed for the calculation of mass transfer processes and these basics are given in this chapter. The Maxwell–Stefan description is emphasized for reasons outlined in the chapter. The Fick description for binary systems is derived from the Maxwell–Stefan description. Estimates are given for binary and ternary Maxwell–Stefan diffusivities. Fick type multicomponent diffusivities for liquids are discussed. These diffusivities depend on the composition and concentration of the components of the entire mixture in contrast to the concentration-independent Maxwell–Stefan diffusivities. Molar diffusion flux with respect to the mass average velocity is also considered. Finally, the conditions are investigated for which the most simple diffusion equations are found.

6.1. PROCESSES OF ODD ORDER

Material isomorphism has shown that the vector processes do not interact with the scalar and the tensor processes for materials with certain symmetry properties. The vector processes are consequently less complicated to discuss, and are examined in this chapter. Dissipation is always the starting point for the analysis of natural phenomena in the theory of the thermodynamics of irreversible processes. Suppose that no electromagnetic field acts on the system. Electromagnetic fields obey the Maxwell–Lorentz transformations and are not Galilean invariant. Without electromagnetic fields it is easy to show that all the descriptions of diffusion are equivalent. For electrolyte solutions $\vec{E}'$ is of major importance and should then not be neglected. Substitution of $\vec{E}' = \vec{0}$ into the dissipation function $\mathfrak{D}$ given in (4.150) yields for the dissipation for vector processes without external electromagnetic fields acting on the system

$$\boxed{\mathfrak{D}^{(\mathrm{v})} = -T^{-1}\vec{\Phi}_{(\mathrm{q})} \cdot \operatorname{grad} T - \sum_{K=1}^{N} \rho_K \vec{V}_K \cdot \left[(\operatorname{grad}\mu_K)_T - \vec{f}'_K\right] \geq 0} \tag{6.1}$$

where $\vec{\Phi}_K = \rho_K \vec{V}_K$ is substituted. If no external electromagnetic fields act on the system, then the choice of the reference velocity is arbitrary. According to $(4.17)_3$

$$(\,d\mu_K)_T = v_K\, dp + (\,d\mu_K)_{T,p}$$

Applying the principle of local equilibrium yields the relation

$$(\operatorname{grad}\mu_K)_T = v_K \operatorname{grad} p + (\operatorname{grad}\mu_K)_{T,p}$$

Introduce the (mechanical) *diffusion vector** now as

$$\boxed{\vec{d}_K = \frac{\rho_K}{p}\left[(\operatorname{grad}\mu_K)_{T,p} + (v_K - v)\operatorname{grad}p + \left(\vec{f}' - \vec{f}'_K\right)\right]} \tag{6.2}$$

with $v = \sum_{K=1}^{N} w_K\, v_K$. The Gibbs–Duhem relation (4.24) is

$$\sum_{K=1}^{N} \rho_K\,(\operatorname{grad}\mu_K)_{T,p} = \vec{0} \tag{6.3}$$

so that with $\rho\, v = \sum_{K=1}^{N} \rho_K v_K$ and $\rho\vec{f}' = \sum_{K=1}^{N} \rho_K \vec{f}'_K$ it follows that

$$\sum_{K=1}^{N} \vec{d}_K = \vec{0} \tag{6.4}$$

which shows that the set of mechanical diffusion vectors is linearly dependent.

With the substitution of the mechanical diffusion vector, the dissipation function for vector processes (6.1) may be written in several ways

$$\begin{aligned} \mathfrak{D}^{(v)} &= -T^{-1}\vec{\Phi}_{(q)}\cdot\operatorname{grad}T + \sum_{K=1}^{N} \vec{d}_K\cdot\left(-p\,\vec{V}_K\right) \geq 0 \\ \mathfrak{D}^{(v)} &= -T^{-1}\vec{\Phi}_{(q)}\cdot\operatorname{grad}T + \sum_{K=1}^{N} \vec{\Phi}_K\cdot\left(\frac{-p}{\rho_K}\,\vec{d}_K\right) \geq 0 \end{aligned} \tag{6.5}$$

Products of fluxes and forces occur in the dissipation function. The above equivalent expressions of the dissipation function emphasize again that there is a choice in the definition of the flux and force. The various descriptions of the multicomponent diffusion depend on this choice.

It is now easy to show that the reference velocity in the description of diffusion is arbitrary, since due to $\sum_{K=1}^{N} \vec{d}_K = \vec{0}$ it follows that

$$p\sum_{K=1}^{N} \vec{V}_K\cdot\vec{d}_K = p\sum_{K=1}^{N} (\vec{V}_K \pm \vec{v}_\star)\cdot\vec{d}_K$$

* The dimensions of μ_K are $m^2\,s^{-2}$, so that the dimensions of d_K are m^{-1}. For ideal systems the diffusion vector $\vec{d}_K$ becomes

$$\vec{d}_K = \operatorname{grad}x_K + (x_K - w_K)\,\frac{\operatorname{grad}p}{p} + \frac{\rho_K}{p}\left(\vec{f}' - \vec{f}'_K\right)$$

The introduction of $\vec{d}_K$ is misleading because this quantity cannot easily be calculated for liquids for instance. The diffusion vector $\vec{d}_K$ is inspired by ideal systems. For nonideal systems, it is often advantageous to use activities.

in which $\vec{v}_\star$ is an arbitrary velocity. The conclusion from this is that the diffusion can be formulated with respect to any more or less arbitrary reference velocity, so that each rational description of diffusion for isotropic bodies (and not only for fluids) is acceptable. This is not applicable therefore only for mechanical* equilibrium, as Prigogine† has suggested. This arbitrariness in the choice of the diffusion velocity is a general aspect of diffusion. This freedom might cause much confusion. However, for solving the diffusion equations, the reference velocity has to be chosen such that the reference velocity can be calculated easily by means of the continuity equation and the equations of motion. This condition suggests the choice $\vec{v}_\star = \vec{0}$. Then the reference velocity of the diffusion is the barycentric velocity.

6.2. TYPES OF DIFFUSION DESCRIPTIONS

Three fundamentally different ways to descibed multicomponent diffusion are

(1) Description of Maxwell–Stefan‡. Description of diffusion where fluxes and forces are mixed§. Generalized by Merk for thermal diffusion.
(2) Description of Chapman–Cowling¶ and Hirschfelder–Curtiss–Bird‖. All chemical components are treated on an equal footing.
(3) Description with a preference for some particular component, as for example the barycentric description by De Groot and Mazur#.

---

* Mechanical equilibrium means that, on a molecular scale, the exchange of momentum proceeds much faster than the exchange of matter and heat. This might be the case for liquids with reasonable accuracy, but the molecular exchange of momentum, matter and heat as determined for gases is of the same order (by the Schmidt number and the Prandtl number). The existence of mechanical equilibrium in gases for heat and mass transfer is not physically realizable, and the application of this mechanical equilibrium has to be avoided.

† Prigogine I. 1947. *Étude Thermodynamique des Phénomèmes Irréversibles.* Desoer, Liège; De Groot S.R. and Mazur P. 1962. *Non-Equilibrium Thermodynamics.* North-Holland Publ. Co., Amsterdam. See pp. 43–45 and 239.

‡ Maxwell J.C. 1860. Illustrations of the dynamical theory of gases. Part 1. On the motions and collisions of perfectly elastic spheres. *Philosophical Magazine*, 19(4): 19–32; Part 2. On the process of diffusion of two or more kinds of moving particles among one another. *Philosophical Magazine*, 20: 21–37; Stefan J. 1871. Über das Gleichgewicht und die Bewegung, insbesondere die Diffusion von Gasmengen. *Sitzungsberichte Akademie Wissenschaften Wien*, 63(2): 63–124.

§ Merk H.J. 1957. Stofoverdracht in laminaire grenslagen door gedwongen convectie. Dissertation, Delft University of Technology (in Dutch); Lightfoot E.N. 1974. *Transport Phenomena and Living Systems.* John Wiley & Sons, Inc., New York. See Ch. III.

¶ Chapman S. and Cowling T.G. 1939. *The Mathematical Theory of Non-Uniform Gases.* Cambridge University Press, Cambridge (third revised printing 1970).

‖ Hirschfelder J.O., Curtiss C.F. and Bird R.B. 1954. *Molecular Theory of Gases and Liquids.* John Wiley & Sons, Inc., New York (fourth printing 1967). See Ch. 11; Curtiss C.F. 1968. Symmetric gaseous diffusion coefficients. *Journal of Chemical Physics*, 49: 2917–2919.

De Groot S.R. and Mazur P. 1962. *Non-Equilibrium Thermodynamics.* North-Holland Publ. Co., Amsterdam.

The first method is universally applicable, and this method is also the starting point for deriving the other descriptions of diffusion in this book. The second method is particularly useful for gas mixtures, if there is no need for a preference of a particular component. The diffusion coefficients are developed in this method from the kinetic theory of gases. It is useful to strive for results obtained with irreversible thermodynamics that are consistent with the results obtained from the kinetic theory. The third method is in particular used for dilute solutions, in which the solvent is eliminated.

6.3. VECTOR TRANSPORT PROCESSES ACCORDING TO MAXWELL AND STEFAN

6.3.1. Phenomenological equations. A method to describe diffusion in multicomponent, isotropic systems, in which the Onsager reciprocal relations can be expressed directly in terms of the diffusion coefficients, is given by Maxwell* and Stefan. Maxwell suggested for dilute gases a particular form for the diffusion equations. His formulation was followed by Stefan for liquids, so that we speak of the Maxwell–Stefan equations for the diffusion. Maxwell stressed the fact that diffusion arises by *equal and opposite forces* that are proportional to the velocity differences of the components. This formulation is independent of the chosen reference velocity.

The process fluxes are defined in the Maxwell–Stefan formulation of diffusion as

$$\vec{\Phi}_{(\mathrm{q})} \qquad \text{and} \qquad \left\{\vec{d}_K\right\}$$

This description is called mixed, since the diffusion vectors are defined as fluxes, while the mass balance equations suggest $\vec{\Phi}_K$ as fluxes. It follows from the dissipation $(6.5)_1$ that the conjugate process forces are then, respectively

$$-T^{-1}\mathrm{grad}\, T \qquad \text{and} \qquad \left\{-p\,\vec{V}_K\right\}$$

The phenomenological equations for an isotropic system are therefore

$$\vec{d}_K = -\sum_{L=1}^{N} L_{KL}\, p\,\vec{V}_L - L_{KT}\, T^{-1}\mathrm{grad}\, T \tag{6.6}$$

$$\vec{\Phi}_{(\mathrm{q})} = -L_{TT}T^{-1}\mathrm{grad}\, T - \sum_{L=1}^{N} L_{TL}\, p\,\vec{V}_L \tag{6.7}$$

Since grad T is an even variable, and $\vec{V}_K$ an odd one (see Chapter 5), the Onsager-Casimir reciprocal relations read

$$L_{KL} = L_{LK} \qquad \text{and} \qquad L_{KT} = -L_{TK} \tag{6.8}$$

---

* Maxwell J.C. 1890. *Scientific Papers*. Cambridge University Press, Cambridge, Vol. 2: 625.

After substitution of the phenomenological equations (6.6) and (6.7) into the dissipation function (6.5), the dissipation function becomes, by also using the symmetry relations (6.8),

$$\mathfrak{D}^{(\mathrm{v})} = L_{TT}T^{-2}(\operatorname{grad} T)^{2} + p^2 \sum_{K=1}^{N}\sum_{L=1}^{N} L_{KL}\vec{V}_K \cdot \vec{V}_L \geq 0 \tag{6.9}$$

It is assumed that the process forces can be applied independently of each other, so that it follows that

$$L_{TT} \geq 0 \tag{6.10}$$

From the condition that $\sum_{K=1}^{N} \vec{d}_K = \vec{0}$, the linear dependences

$$\sum_{K=1}^{N} L_{KL} = 0 \qquad \text{and} \qquad \sum_{K=1}^{N} L_{KT} = 0 \tag{6.11}$$

result. According to (6.11), the form $\vec{V}_K \cdot \vec{V}_L$ is positive *semi*definite, so that in any case the condition det $|L_{KL}| = 0$ holds. Further conclusions are for the time being postponed. The linear dependences (6.11) and the reciprocal relations (6.8) imply that there are $\frac{1}{2}N(N-1)$ possible independent L_{KL}.

L_{KK} can be solved from $(6.11)_1$. By using the symmetry relation it follows that

$$L_{KK} = -\sum_{L=1}^{N}{}' L_{LK} = -\sum_{L=1}^{N}{}' L_{KL} \tag{6.12}$$

where the accent $'$ in the $\sum$ sign denotes that the summation has to be performed over all values $L \neq K$. With the substitution of (6.12) into (6.6), it is found that the diffusion vector can be written independently of the choice of the reference velocity for the diffusion

$$\boxed{\vec{d}_K = p\sum_{L=1}^{N}{}' L_{KL}(\vec{V}_K - \vec{V}_L) - L_{KT}T^{-1}\operatorname{grad} T} \tag{6.13}$$

The dissipation function (6.9) becomes with the substitution of (6.12)

$$\mathfrak{D}^{(\mathrm{v})} = L_{TT}T^{-2}(\operatorname{grad} T)^{2} + p^2 \sum_{K=1}^{N}\sum_{L=1}^{N}{}' L_{KL}\vec{V}_K \cdot (\vec{V}_L - \vec{V}_K) \geq 0 \tag{6.14}$$

where the accent in the $\sum$ sign can be dropped, since the term with $L = K$ vanishes automatically in the summation.

By exchanging the order of the summation and applying the reciprocal relation $(6.8)_1$, it is found for the double summation term in (6.14) that

$$\sum_{K=1}^{N}\sum_{L=1}^{N} L_{KL}\vec{V}_K\cdot(\vec{V}_L-\vec{V}_K)=\sum_{K=1}^{N}\sum_{L=1}^{N} L_{LK}\vec{V}_L\cdot(\vec{V}_K-\vec{V}_L)$$
$$=\sum_{K=1}^{N}\sum_{L=1}^{N} L_{KL}\vec{V}_L\cdot(\vec{V}_K-\vec{V}_L)=-\sum_{K=1}^{N}\sum_{L=1}^{N} L_{KL}\vec{V}_L\cdot(\vec{V}_L-\vec{V}_K)$$
$$=-\tfrac{1}{2}\sum_{K=1}^{N}\sum_{L=1}^{N} L_{KL}(\vec{V}_L-\vec{V}_K)^{2\cdot} \tag{6.15}$$

With the help of this result the dissipation function (6.9) can be written as

$$\mathfrak{D}^{(\mathrm{v})}=L_{TT}T^{-2}(\operatorname{grad}T)^{2\cdot}-\tfrac{1}{2}p^2\sum_{K=1}^{N}\sum_{L=1}^{N} L_{KL}(\vec{V}_L-\vec{V}_K)^{2\cdot}\geq 0 \tag{6.16}$$

or

$$\boxed{\mathfrak{D}^{(\mathrm{v})}=L_{TT}T^{-2}(\operatorname{grad}T)^{2\cdot}-p^2\sum_{K=1}^{L-1}\sum_{L=2}^{N} L_{KL}(\vec{V}_L-\vec{V}_K)^{2\cdot}\geq 0} \tag{6.17}$$

In isothermal situations, a sufficient condition for satisfying the inequality is that

$$-L_{KL}\geq 0 \qquad \text{for} \qquad K\neq L \tag{6.18}$$

This is not a necessary condition, because only $N-1$ velocity *differences* are independent. By eliminating one of the velocity differences $\vec{V}_L-\vec{V}_K$ a less stringent constraint for the phenomenological coefficients is obtained. For example for an isothermal three-component system the dissipation (6.17) becomes

$$\mathfrak{D}^{(\mathrm{v})}=-p^2\left(L_{12}V_{21}^2+L_{13}V_{31}^2+L_{23}V_{32}^2\right)\geq 0$$

where for brevity $\vec{V}_{KL}\equiv\vec{V}_K-\vec{V}_L$ and $V_{KL}^2\equiv(\vec{V}_K-\vec{V}_L)^{2\cdot}$. Elimination of $\vec{V}_{32}$ with $\vec{V}_{32}=\vec{V}_{31}-\vec{V}_{21}$, and dividing the result by $\vec{V}_{31}^2$ yields the constraint

$$-(L_{12}+L_{23})\left(\frac{V_{21}}{V_{31}}\right)^2+2L_{23}\left(\frac{V_{21}}{V_{31}}\right)-(L_{13}+L_{23})\geq 0 \tag{6.19}$$

which is a quadratic expression. Necessary conditions for satisfying the inequality are that

$$-L_{21}-L_{23}=L_{22}\geq 0 \qquad\text{and}\qquad -L_{31}-L_{32}=L_{33}\geq 0 \tag{6.20}$$

where (6.12) is used (elimination of $\vec{V}_{31} = \vec{V}_{32} - \vec{V}_{12}$ would give the third necessary condition $-L_{12} - L_{13} = L_{11} \geq 0$). If the necessary conditions (6.20) are satisfied and if also the discriminant is negative, then a sufficient condition is obtained, and the inequality is satisfied for all values of V_{12} and V_{13}. This yields for the three-component system the constraint

$$L_{22}L_{33} - L_{23}^2 \geq 0 \tag{6.21}$$

or written differently

$$L_{21}L_{13} + L_{23}(L_{12} + L_{13}) \geq 0 \tag{6.22}$$

which shows that there is only one sufficient constraint (6.22) for a three-component system irrespective of the choice of independent velocity difference. The condition (6.22) is satisfied for (6.18), but it also shows that it is possible that one or two phenomenological coefficients $-L_{KL}$ in a three-component system might be negative, without violating the condition that the dissipation has to be positive. Inequality (6.19) shows that even if the sufficient condition (6.21) or (6.22) is not satisfied, the dissipation might still be positive if the necessary conditions (6.20) are satisfied in processes for which either $V_{12} \ll V_{13}$ or $V_{13} \ll V_{12}$, but such checks are rarely needed.

The sufficient condition (6.22) shows that $-L_{23}$ need not be positive and can be negative subject to the constraint

$$-L_{23} \geq \frac{L_{12}L_{13}}{L_{12} + L_{13}} < 0 \tag{6.23}$$

if for example it is assumed for the two other diffusion coefficients $-L_{12} > 0$, and $-L_{13} > 0$.

The constraints for an N-component system are found along similar lines. The generalization of the inequalities (6.20) becomes

$$-L_{KK} \geq 0 \qquad \text{for} \qquad N \geq 3 \tag{6.24}$$

which are N necessary constraints. The generalization of the constraint (6.21) becomes

$$L_{KK}L_{LL} - L_{KL}^2 \geq 0 \qquad \text{for} \qquad N \geq 3 \tag{6.25}$$

or written in the form (6.22)

$$(L_{KK} + L_{KL})(L_{LL} + L_{KL}) - L_{KL}\left[(L_{KK} + L_{KL}) + (L_{LL} + L_{KL})\right] \geq 0 \tag{6.26}$$

which yields $\frac{1}{2}N(N-1)$ independent constraints for $N \geq 4$. Clearly, it is not forbidden by the dissipation function for multicomponent systems that some phenomenological coefficients $-L_{LK}$ can be negative. Finally, note that the sign of $L_{KT} = -L_{TK}$ is not prescribed in any way.

Irreversible thermodynamics provides us with statements about the nature of the coefficients, but it does not give information about the magnitude of the coefficients. These values have to come from measurements on particular multicomponent fluids. For the further development of the theory the constitutive equations (6.6) and (6.7) have to be specified.

6.3.2. Diffusion coefficients. Maxwell* proposed that diffusion is described by the velocity differences between the molecules of different species, and that those differences give rise to forces from the 'friction' between the molecules. Maxwell considered a chemical potential gradient caused by friction, and this friction is proportional to the number of molecules. This force and others are found in the mechanical diffusion vector. One might add an electrical force to the diffusion vector by not assuming that $\vec{E}' = \vec{0}$, but this is here omitted, because electromagnetic fields might cause the fluid to behave anisotropically, and this has consequences for the symmetry relations. The *diffusion coefficients of Maxwell–Stefan* can therefore be defined as follows

$$Đ_{KL} = \frac{-x_K x_L}{p L_{KL}} \tag{6.27}$$

where the $x_K = c_K/c$ are the already defined mole fractions.

From the symmetry relations $(6.8)_1$ and the inequalities (6.18), the relations

$$Đ_{KL} \geq 0 \qquad \text{with} \qquad Đ_{KL} = Đ_{LK} \tag{6.28}$$

are found for the diffusion coefficients. From the necessary conditions (6.24) it is found that the diagonal elements of the matrix of diffusion coefficients have to be positive

$$Đ_{KK} \geq 0 \qquad \text{for} \qquad N \geq 3 \tag{6.29}$$

Thus, the Maxwell–Stefan diffusion coefficients satisfy simple symmetry relations. Since the diffusion coefficients have to be determined experimentally in a phenomenological theory, simple symmetry relations are then of great advantage.

Satisfying all the inequalities in (6.28) yields a sufficient condition for the dissipation function to be positive definite, but it might be too strong a constraint, particularly for electrolyte solutions. For binary systems the Maxwell–Stefan diffusivity has to be positive, but for multicomponent systems

* Maxwell J.C. 1890. *Scientific Papers*, Cambridge University Press, Cambridge, Vol. 2: 625.

negative diffusivities are not forbidden. Negative diffusivities have been found so far only in electrolyte solutions*.

The necessary constraints (6.29) result in less strict conditions than found in (6.28). For the Maxwell–Stefan diffusivities in an N-component system the conditions become

$$\sum_{L=1}^{N}{}' \frac{x_L}{Đ_{LK}} \geq 0 \qquad \text{for} \qquad N \geq 3 \tag{6.30}$$

where the prime denotes that the summation is to be performed for $L \neq K$. In addition to these constraints the sufficient condition (6.25) is needed to guarantee the positive definiteness of the dissipation function. Possible negative Maxwell–Stefan diffusivities $Đ_{KL}$ are allowed if they satisfy

$$Đ_{KL}^2 \geq Đ_{KK} Đ_{LL} \qquad \text{for} \qquad N \geq 3 \tag{6.31}$$

As the diagonal elements $Đ_{KK}$ are commonly not used in diffusion theories, it is useful to express (6.31) in terms of the diffusivities $Đ_{KL}$, yielding the conditions

$$\sum_{K=1}^{N}{}' \frac{x_K}{Đ_{KL}} \sum_{L=1}^{N}{}' \frac{x_L}{Đ_{KL}} - \frac{x_K x_L}{Đ_{KL}^2} \geq 0 \qquad \text{for} \qquad N \geq 3 \tag{6.32}$$

If the sufficient condition is not satisfied, a positive dissipation function can still result with a negative diffusivity $Đ_{KL}$, but then the diffusion velocity differences $\vec{V}_{KL}$ in the process have to be known, which then have to be much smaller than the other velocity differences.

The conditions for the diffusivities have now been examined and we have to specify the constitutive equations. The diffusion vector (6.13) becomes after insertion of the Maxwell–Stefan diffusion coefficients (6.27)

$$\vec{d}_K = \sum_{L=1}^{N} \frac{x_K x_L}{Đ_{KL}} (\vec{V}_L - \vec{V}_K) - L_{KT} T^{-1} \operatorname{grad} T \tag{6.33}$$

The term with L_{KT} points out that for all $\vec{d}_K = \vec{0}$ diffusion velocities $\vec{V}_K$ can arise due to temperature gradients. This phenomenon is called *thermal diffusion* or the *Soret effect*†. The Soret effect is important in engineering

---

* Kraaijeveld G. and Wesselingh J.A. 1993. Negative Maxwell–Stefan diffusion coefficients. *Industrial Engineering and Chemistry Research*, 32: 738–742; Kraaijeveld G., Wesselingh J.A., and Kuiken G.D.C. 1994. Comments on 'negative diffusion coefficients'. *Industrial Engineering and Chemistry Research*, 33(3): 750–751.

† Soret Ch. 1893. Propagation de la chaleur dans les cristaux. *Comptes rendus des séances de la Société de Physique et d'Histoire Naturelle de Genève*, 10: 15–16, *Archives des Sciences Physiques et Naturelles (Genève)*, 29: 4.

practise whenever large temperature gradients coexist with large molecular weight differences*. Thermal diffusion in its purest form appears if $\vec{d}_K = \vec{0}$ and $\operatorname{grad} T \neq 0$. For all $\vec{d}_K = \vec{0}$, the mass flux vector $\vec{\Phi}_K$ is therefore determined by

$$\vec{\Phi}_K = \rho_K \vec{V}_K = -\rho D_K^{(T)} \frac{\operatorname{grad} T}{T} \qquad (\forall \vec{d}_K = \vec{0}) \tag{6.34}$$

in which the phenomenological coefficients $D_K^{(T)}$ are introduced. These coefficients are defined as the barycentric coefficients of thermal diffusion†. Due to the fact that $\sum_{K=1}^{N} \rho_K \vec{V}_K = \vec{0}$, these thermal diffusion coefficients satisfy

$$\sum_{K=1}^{N} D_K^{(T)} = 0 \tag{6.35}$$

For $\{\vec{d}_K = \vec{0}\}$—that is for a homogeneous distribution of the pressure and concentration, and in the absence of external forces—thermal diffusion can occur with $\vec{V}_K = -(D_K^{(T)}/w_K)T^{-1}\operatorname{grad} T$, where from (6.33) and with the substitution of (6.34), it follows that

$$-L_{KT} = +L_{TK} = \sum_{L=1}^{N} \frac{x_K x_L}{Đ_{KL}} \left(\frac{D_L^{(T)}}{w_L} - \frac{D_K^{(T)}}{w_K} \right) \tag{6.36}$$

With the specification of the coefficient L_{KT}, the Maxwell–Stefan description of diffusion (6.33) can be written as

$$\boxed{\vec{d}_K = \sum_{L=1}^{N} \frac{x_K x_L}{Đ_{KL}} (\vec{V}_L^{\diamond} - \vec{V}_K^{\diamond})} \tag{6.37}$$

where $\vec{V}_K^{\diamond}$ is defined by

$$\boxed{\vec{V}_K^{\diamond} = \vec{V}_K + \frac{D_K^{(T)}}{w_K} \frac{\operatorname{grad} T}{T}} \tag{6.38}$$

The equations (6.37) are considered as a generalization of the Maxwell diffusion equations, since Maxwell formulated equations of this type without

* Rosner D.E. 1980. Thermal (Soret) diffusion effects on interfacial mass transport rates. *PCH Physicochemical Hydrodynamics*, 1: 159–185.

† This definition differs slightly from the $D_K^T = \rho D_K^{(T)}$ in Hirschfelder, Curtiss and Bird. Here care has been taken that the physical dimension of $D_K^{(T)}$ in (6.34) is equal to that of $Đ_{KL}$, namely, m^2/s. The definition is in accordance with the definition used in the third edition of Chapman S. and Cowling T.G. 1939. *The Mathematical Theory of Non-Uniform Gases*. Cambridge University Press, Cambridge (third revised printing 1970).

taking into account thermal diffusion, nonideal states, pressure diffusion and diffusion by external forces. Equations (6.37) have also been derived for dilute gases by Curtiss and Hirschfelder* by using the kinetic theory of gases. These authors remarked that there is some experimental evidence that the equations derived for dilute gases apply quite well for dense gases and liquids. With the phenomenological derivation using the thermodynamics of irreversible processes it has now been shown that equations (6.37) are valid for isotropic liquids as well and are not just restricted to dilute gases. Although irreversible thermodynamics does not enable us to calculate the values of the diffusion coefficients, it is on the other hand an advantage of the theory that the fluids need not be specified, and that therefore the results are also not limited to special substances.

6.3.3. Heat conductivity coefficients. With reference to the phenomenological equation (6.7) for $\vec{\Phi}_{(\mathrm{q})}$, the usual heat conductivity coefficient can be defined by

$$\lambda = L_{TT}/T \geq 0 \tag{6.39}$$

where the inequality follows from (6.10) and from the condition that $T > 0$ ('absolute temperature'). The phenomenological equation (6.7) becomes with (6.36) and (6.39)

$$\vec{\Phi}_{(\mathrm{q})} = -\lambda \operatorname{grad} T + p \sum_{K=1}^{N} \sum_{L=1}^{N}{}' \frac{x_K x_L}{\mathit{Đ}_{KL}} \left(\frac{D_K^{(T)}}{w_K} - \frac{D_L^{(T)}}{w_L} \right) \vec{V}_K \tag{6.40}$$

By interchanging the order of the summation and by using the symmetry of $\mathit{Đ}_{KL}$, it can be shown that

$$\sum_{K=1}^{N} \sum_{L=1}^{N}{}' \frac{x_K x_L}{\mathit{Đ}_{KL}} \frac{D_K^{(T)}}{w_K} \vec{V}_K = \sum_{K=1}^{N} \sum_{L=1}^{N}{}' \frac{x_K x_L}{\mathit{Đ}_{KL}} \frac{D_L^{(T)}}{w_L} \vec{V}_L$$

Substitution of this result into (6.40) yields

$$\vec{\Phi}_{(\mathrm{q})} = -\lambda \operatorname{grad} T + p \sum_{K=1}^{N} \sum_{L=1}^{N}{}' \frac{x_K x_L}{\mathit{Đ}_{KL}} \frac{D_L^{(T)}}{w_L} (\vec{V}_L - \vec{V}_K) \tag{6.41}$$

The second term in the right-hand side represents a flow of heat due to diffusion. This effect is called the *Dufour effect*†. As can be seen from (6.41),

* Hirschfelder J.O., Curtiss C.F. and Bird R.B. 1954. *Molecular Theory of Gases and Liquids*. John Wiley & Sons, Inc., New York (fourth printing 1967). See p. 718, eq. (11.2–54).

† Dufour L. 1872. *Archives des Sciences Physiques et Naturelles (Genève)*, 45: 9; See Dufour L. 1873. Ueber die Diffusion der Gase durch poröse Wände und die sie begleitenden Temperaturveränderungen. *Annalen der Physik*, 28(5): 490–492. The cross-effects of Soret and Dufour are complementary *cross-effects*. The Dufour effect can usually be neglected in liquids, whereas in gases the Soret-effect and the Dufour effect are of the same order of magnitude, and both effects have to be accounted for.

this effect is found only if the diffusion velocities $\vec{V}_K$ differ from each other. If the diffusion velocities are equal, then there is obviously no Dufour effect, since then $\vec{V}_K = \vec{V} = \vec{0}$ for all K.

If all $\vec{d}_K$ are zero, and only thermal diffusion is considered for which the mass flux $\vec{\Phi}_K$ is given by (6.34), then the substitution into (6.41) yields

$$\vec{\Phi}_{(q)} = -\lambda \operatorname{grad} T - \frac{p}{T} \sum_{K=1}^{N} \sum_{L=1}^{N} \frac{x_K x_L}{\text{Đ}_{KL}} \frac{D_L^{(T)}}{w_L} \left(\frac{D_L^{(T)}}{w_L} - \frac{D_K^{(T)}}{w_K} \right) \operatorname{grad} T$$

or

$$\vec{\Phi}_{(q)} = -\lambda \operatorname{grad} T - \frac{1}{2} \frac{p}{T} \sum_{K=1}^{N} \sum_{L=1}^{N} \frac{x_K x_L}{\text{Đ}_{KL}} \left(\frac{D_L^{(T)}}{w_L} - \frac{D_K^{(T)}}{w_K} \right)^2 \operatorname{grad} T$$

or

$$\boxed{\vec{\Phi}_{(q)} = -\lambda' \operatorname{grad} T \qquad \text{for all} \quad \vec{d}_K = \vec{0}} \tag{6.42}$$

with

$$\boxed{\lambda' = \lambda + \frac{1}{2} \frac{p}{T} \sum_{K=1}^{N} \sum_{L=1}^{N} \frac{x_K x_L}{\text{Đ}_{KL}} \left(\frac{D_L^{(T)}}{w_L} - \frac{D_K^{(T)}}{w_K} \right)^2} \tag{6.43}$$

which reduces to formula (7.4–65) in Hirschfelder, Curtiss and Bird* for a dilute, monatomic gas in a first order approximation.

Because of $\text{Đ}_{KL} \geq 0$ in (6.28) for non-electrolyte solutions, it is easily seen that

$$\lambda' \geq \lambda \geq 0 \tag{6.44}$$

It follows from the difference between λ' and λ that in a multicomponent system with diffusion, the heat conductivity coefficient cannot be defined unambiguously, but that the definition depends on the circumstances under which the heat conduction takes place. The difference between λ' and λ can be illustrated from a physical point of view as follows. Suppose that for $t < 0$ the system is in equilibrium, so that the distributions of the pressure, the temperature and the concentrations are homogeneous and stationary. A temperature gradient is applied to this system for $t > 0$ that causes at time $t = 0^+$ not only heat conduction, but also separation by thermal diffusion. The corresponding heat conductivity coefficient for this effect is λ'. The diffusion vectors $\vec{d}_K \neq \vec{0}$ for all K due to separation, and ordinary diffusion is developed for $t > 0$ because of the fact that $\vec{d}_K \neq \vec{0}$ until ordinary diffusion compensates the thermal diffusion, so that $\vec{V}_K = \vec{0}$ for all K. This stage is reached after

---

* Hirschfelder J.O., Curtiss C.F. and Bird R.B. 1954. *Molecular Theory of Gases and Liquids.* John Wiley & Sons, Inc., New York (fourth printing 1967). See §7.4 and for non-dilute gases Ch. 11.

some time (formally for $t \to \infty$). A stationary state is approached as $t \to \infty$, if the applied temperature gradient is also stationary, for which case the heat conductivity coefficient has become equal to λ. In this final situtation the heat conduction is no longer enhanced by thermal diffusion, so that $\lambda' > \lambda > 0$. It is now clear that in the general relation for $\vec{\Phi}_{(q)}$ the heat conductivity coefficient λ has to be introduced, as has been done in (6.39).

6.3.4. Maxwell–Stefan equations. The Maxwell–Stefan equations do not depend on a special choice of the reference velocity, and therefore they are a proper starting point for other descriptions of multicomponent diffusion. All components are treated equivalently in the Maxwell–Stefan description of the diffusion. This is certainly correct for gases. Furthermore, it is shown* that in the first approximation of the kinetic theory of gases for dilute monatomic gases

$$Đ_{AB} \approx [\mathbb{D}_{AB}]_1 \tag{6.45}$$

in which $\mathbb{D}_{AB}$ is the diffusion coefficient of a binary mixture consisting of the components A and B. If the molecules consist of smooth, hard spheres, then for instance

$$[\mathbb{D}_{AB}]_1 = \frac{3}{16}\frac{1}{\sigma_{AB}^2}\frac{1}{p}\sqrt{\frac{2}{\pi}\frac{m_A + m_B}{m_A m_B}(kT)^3} \tag{6.46}$$

where $\sigma_{AB} = \frac{1}{2}(\sigma_A + \sigma_B)$ with σ_K being the diameter of the molecule K. The binary diffusion coefficient is in the first approximation independent of the composition of the gas. Only in the second and higher approximations does the diffusion coefficient become concentration dependent. The differences between $\mathbb{D}_{AB}$ and $[\mathbb{D}_{AB}]_1$ are small for thermally stable gases and are usually of the order of a few percent, so that for practical purposes and for dilute gases, the common diffusion coefficients can be assumed to be independent of the concentration of the components. Moreover, (6.45) applies only in the first approximation of the kinetic theory of gases, so $Đ_{AB} \neq \mathbb{D}_{AB}$ for the exact values of these quantities.

For N-component ideal gases, the diffusivities $Đ_{KL}$ are independent of the composition of the gas, and are equal to the diffusivity $Đ_{KL}$ of the binary gas pair KL. The multicomponent diffusion coefficients are often unknown, but for gases can be estimated by means of (6.45), where the binary diffusivities can be calculated with the kinetic theory.

The binary diffusivities for liquids can be estimated from interpolation formulas†, which will be discussed in the forthcoming sections. In the Maxwell–Stefan description the multicomponent diffusivities $Đ_{KL}$ can then

---

* Curtiss C.F. and Hirschfelder J.O. 1949. Transport properties of multicomponent gas mixtures. *Journal of Chemical Physics*, 17: 550–555.

† Rutten Ph.W.M. 1992. ***Diffusion in Liquids.*** Delft University Press, Delft. Thesis, has compiled the known data.

be estimated from the measured binary diffusivities $Đ_{AB}$. From the practical point of view, it is also important to realize that in an N-component system only $N(N-1)/2$ different Maxwell–Stefan diffusivities are required as a result of the simple symmetry relations.

In summary, it is remarked that all descriptions of diffusion are equivalent. However, there is a preference for the Maxwell–Stefan description of diffusion, because

- The Maxwell–Stefan description is independent of the choice of the reference velocity.
- All components in the Maxwell–Stefan description are treated on an equal footing.
- The Maxwell–Stefan description is in agreement with the results found with the kinetic theory of dilute monatomic gases.
- The Maxwell–Stefan diffusivities $Đ_{KL}$ are binary diffusivities, which are, for ideal mixtures and many nonideal mixtures, independent of the concentration of the components in the multicomponent system.

The mass and molar diffusion fluxes are important in practice, and can be derived from the Maxwell–Stefan description. For the obtaining the Maxwell–Stefan multicomponent diffusivities the binary diffusivities are important, and besides, those diffusivities are the simplest to measure. For a good understanding of multicomponent diffusion, binary diffusion has to be understood too.

6.4. APPLICATIONS TO BINARY SYSTEMS

The Maxwell–Stefan equations applies to all continuous, *isotropic* systems. The results for binary systems are most easily illustrated and interpreted, for which the relations between the barycentric mass, and the two molar descriptions of the diffusion can be derived with ease from one another. Consider therefore first a binary of components A and B. To solve the mass balance equations the diffusion flux has to be known, and for solving the temperature equation the heat flux has to be known. The mass flux is discussed in §6.4.1., and later the heat flux in §6.4.2.

6.4.1. Mass transport. The diffusion vector (6.2) becomes for the binary system considered

$$\vec{d}_A = -\vec{d}_B = \frac{\rho_A}{p}\left[(\operatorname{grad}\mu_A)_{T,p} + (v_A - v)\operatorname{grad}p + \left(\vec{f}' - \vec{f}'_A\right)\right] \tag{6.47}$$

In this diffusion vector

$$v_A - v = (w_A + w_B)v_A - w_A v_A - w_B v_B = w_B\,(v_A - v_B)$$
$$\vec{f}' - \vec{f}'_A = w_A\vec{f}'_A + w_B\vec{f}'_B - (w_A + w_B)\vec{f}'_A = -w_B\left(\vec{f}'_A - \vec{f}'_B\right)$$

Define the thermodynamic correction factor Γ as

$$p\Gamma = \rho_A \left(\frac{\partial \mu_A}{\partial x_A}\right)_{T,p} = \rho_B \left(\frac{\partial \mu_B}{\partial x_B}\right)_{T,p} \tag{6.48}$$

where the last equality follows from the Gibbs–Duhem relation (6.3). For an ideal system $\Gamma = 1$, so that Γ is a measure the deviation from the ideal behavior of the binary system.

With the introduction of the thermodynamic correction factor Γ it follows that

$$\frac{\rho_A}{p}\left(\operatorname{grad}\mu_A\right)_{T,p} = \frac{\rho_A}{p}\left(\frac{\partial \mu_A}{\partial x_A}\right)_{T,p} \operatorname{grad} x_A = \Gamma \operatorname{grad} x_A \tag{6.49}$$

The mechanical diffusion vector (6.47) can be written as

$$\vec{d}_A = -\vec{d}_B = \Gamma \operatorname{grad} x_A + \rho w_A w_B \left(v_A - v_B\right) \frac{\operatorname{grad} p}{p} - \rho w_A w_B \frac{\vec{f}'_A - \vec{f}'_B}{p} \tag{6.50}$$

With the neglect of electromagnetic fields, the mechanical diffusion vector for a binary system is determined by concentration gradients, the pressure gradient and non-conservative external forces.

The diffusion equation (4.56) is a differential expression in w_K. For a binary system, it is easy to express $\operatorname{grad} x_A$ in $\operatorname{grad} w_A$. By using $m = m_A x_A + m_B x_B$ and $x_A + x_B = 1$ it follows that

$$dw_A = d\frac{m_A}{m} x_A = \frac{m_A}{m}\left[dx_A - x_A \frac{dm}{m}\right]$$

and

$$\frac{dm}{m} = \frac{m_A - m_B}{m}\, dx_A$$

and from the last two relations, it is found that

$$dw_A = \frac{m_A m_B}{m^2}\, dx_A \tag{6.51}$$

so that (6.50) can also be written as

$$\vec{d}_A = \frac{m^2}{m_A m_B} \Gamma \operatorname{grad} w_A + \rho\, w_A w_B \left(v_A - v_B\right) \left\{ \frac{\operatorname{grad} p}{p} - \frac{\vec{f}'_A - \vec{f}'_B}{p\left(v_A - v_B\right)} \right\} \tag{6.52}$$

From (6.37) it follows for a binary system that

$$\vec{V}^{\diamond}_B - \vec{V}^{\diamond}_A = \frac{Ð_{AB}}{x_A x_B} \vec{d}_A \tag{6.53}$$

where according to (6.38):

$$\vec{V}_B^{\diamond} - \vec{V}_A^{\diamond} = \vec{V}_B - \vec{V}_A + \left(\frac{D_B^{(T)}}{w_B} - \frac{D_A^{(T)}}{w_A} \right) \frac{\operatorname{grad} T}{T}$$

With the relation (6.35), which is $\sum_{K=1}^{N} D_K^{(T)} = 0$, this becomes

$$\vec{V}_B^{\diamond} - \vec{V}_A^{\diamond} = \vec{V}_B - \vec{V}_A - \frac{D_A^{(T)}}{w_A w_B} \frac{\operatorname{grad} T}{T}$$

For binary systems, according to (4.48), it follows that $\rho_A \vec{V}_A + \rho_B \vec{V}_B = \vec{0}$, so that

$$\vec{V}_B^{\diamond} - \vec{V}_A^{\diamond} = -\frac{\vec{V}_A}{w_B} - \frac{D_A^{(T)}}{w_A w_B} \frac{\operatorname{grad} T}{T} = \frac{Ð_{AB}}{x_A x_B} \vec{d}_A \tag{6.54}$$

in combination with (6.53). From the last equality we have

$$\vec{\Phi}_A = \rho_A \vec{V}_A = -\rho \frac{w_A w_B}{x_A x_B} Ð_{AB} \vec{d}_A - \rho D_A^{(T)} \frac{\operatorname{grad} T}{T} \tag{6.55}$$

Finally, the diffusion flux (6.55) becomes, with the relation (6.52) for the diffusion vector

$$\boxed{\vec{\Phi}_A = -\rho \mathbb{D}_{AB} \operatorname{grad} w_A - \rho \mathbb{D}_A^{(p)} \left\{ \frac{\operatorname{grad} p}{p} - \frac{\vec{f}'_A - \vec{f}'_B}{p\,(v_A - v_B)} \right\} - \rho \mathbb{D}_A^{(T)} \frac{\operatorname{grad} T}{T}} \tag{6.56}$$

in which for the binary thermal diffusivity $\mathbb{D}_A^{(T)} = D_A^{(T)}$ is valid. The binary diffusivity $\mathbb{D}_{AB}$ in (6.56) is defined by

$$\mathbb{D}_{AB} = \Gamma Ð_{AB} \tag{6.57}$$

and the binary pressure diffusivity $\mathbb{D}_A^{(p)}$ is defined by

$$\mathbb{D}_A^{(p)} = \rho Ð_{AB} \frac{m_A m_B}{m^2} w_a w_B \,(v_A - v_B) \tag{6.58}$$

According to (6.56), which can be interpreted as the *barycentric Fick diffusion*, diffusion occurs in a binary system due to concentration differences (ordinary diffusion), pressure differences (pressure diffusion), external forces, and temperature differences (thermal diffusion). For binary systems also the thermal diffusion ratio $k_A^{(T)}$ and the thermal diffusion factor $\alpha_A^{(T)}$ are introduced by

$$\mathbb{D}_A^{(T)} = k_A^{(T)} \mathbb{D}_{AB} = \alpha_A^{(T)} w_A w_B \mathbb{D}_{AB} \tag{6.59}$$

The relative importance of thermal diffusion with respect to ordinary diffusion can be expressed in these quantities. For binary systems there exists the relation

$$\vec{v} = w_A \vec{v}_A + w_B \vec{v}_B = w_A(\vec{v}_A - \vec{v}_B) + \vec{v}_B = w_B(\vec{v}_B - \vec{v}_A) + \vec{v}_A \tag{6.60}$$

and for the barycentric diffusion flux

$$\vec{\Phi}_A = \rho_A \vec{V}_A = \rho_A(\vec{v}_A - \vec{v}) = \rho_A w_B(\vec{v}_A - \vec{v}_B) = \rho w_A w_B(\vec{v}_A - \vec{v}_B) \tag{6.61}$$

With the substitution of these results into (6.56), it follows that the thermal diffusion ratio is defined such that component A for $k_A^{(T)} > 0$ moves from the hot region to the cold region, and for $k_A^{(T)} < 0$ in the reverse direction. A change of sign occurs, for instance, in a neon-ammonia gas-mixtures. The heavier neon molecules move to the cold side if their concentration is more than 75%, whereas for lower concentrations the ammonia molecules move to the cold side*.

Analogously, the pressure diffusivity $\mathbb{D}_A^{(p)}$ and the pressure diffusion factor $\alpha_A^{(p)}$ are introduced as

$$\mathbb{D}_A^{(p)} = k_A^{(p)} \mathbb{D}_{AB} = \alpha_A^{(p)} w_A w_B \mathbb{D}_{AB} \tag{6.62}$$

with

$$k_A^{(p)} = \frac{\rho}{\Gamma} \frac{m_A m_B}{m^2} w_A w_B \, (v_A - v_B) \tag{6.63}$$

$$\alpha_A^{(p)} = \frac{\rho}{\Gamma} \frac{m_A m_B}{m^2} \, (v_A - v_B) \tag{6.64}$$

The definitions above of the binary diffusion quantities are in accord with those given by Chapman and Cowling and by Hirschfelder, Curtiss and Bird†. It can be remarked furthermore that for ideal systems $v_A = 1/(cm_A)$, so that (6.64) becomes

$$\alpha_A^{(p)} = \frac{m_B - m_A}{m} \tag{6.65}$$

The pressure diffusion is thus proportional to the difference of the molar masses of the diffusing components, in which the heavier components have the tendency to diffuse to the regions of increasing pressure. It turns out that, according to the kinetic theory of gases, the thermal diffusion is also

* Chapman S. and Cowling T.G. 1939. *The Mathematical Theory of Non-Uniform Gases.* Cambridge University Press, Cambridge (third revised printing 1970). See pp. 274–275.

† Hirschfelder J.O., Curtiss C.F. and Bird R.B. 1954. *Molecular Theory of Gases and Liquids.* John Wiley & Sons, Inc., New York (fourth printing 1967). See pp. 516–520.

proportional to the difference of the molar masses of the diffusing components. Pressure diffusion and thermal diffusion are both appropriate for separating isotopes. Molecular diffusion processes are slow processes. The pressure gradient in ultracentrifugation is built up by centrifugal forces, whereas the diffusion due to external Coriolis forces can be neglected in practice.

The barycentric diffusion flux (6.56) is expressed, with the above definitions, in the weight fractions

$$\boxed{\vec{\Phi}_A = -\rho I\!\!D_{AB}\left(\vec{d}^{\circ}_A + \alpha_A^{(T)} w_A w_B \text{grad } \ln T\right)} \tag{6.66}$$

in which the barycentric diffusion vector $\vec{d}^{\circ}_A$, expressed in the weight fractions, is given by

$$\vec{d}^{\circ}_A = \text{grad } w_A + \alpha_A^{(p)} w_A w_B \left\{\frac{\text{grad } p}{p} - \frac{\vec{f}'_A - \vec{f}'_B}{p\,(v_A - v_B)}\right\} \tag{6.67}$$

Substitution of the barycentric mass flux $\vec{\Phi}_A$ into the diffusion equation (4.56) yields a second order differential equation for the weight fraction w_A.

The mass diffusion fluxes with respect to the average barycentric velocity can be transformed into the molar descriptions of the diffusion. Starting from the mass barycentric description of the diffusion, it is for binary systems easy to derive the mass diffusion flux with respect to the molar average velocity and the molar diffusion flux with respect to the molar average velocity. For this purpose, (6.66) can be expressed in the molar concentration

$$\vec{\Phi}_A = -\rho I\!\!D_{AB}\frac{m_A m_B}{m^2}\left(\vec{d}^{\star}_A + \alpha_A^{(T)} x_A x_B \text{grad } \ln T\right) \tag{6.68}$$

where the molar diffusion vector $\vec{d}^{\star}_A$ is defined by

$$\vec{d}^{\star}_A = \text{grad } x_A + \alpha_A^{(p)} x_A x_B \left\{\frac{\text{grad } p}{p} - \frac{\vec{f}'_A - \vec{f}'_B}{p\,(v_A - v_B)}\right\} \tag{6.69}$$

In fluid dynamics, one uses also the molar diffusion flux relative to the average barycentric velocity. For this molar diffusion flux

$$\vec{\Phi}'_A = c_A \vec{V}_A = \frac{c_A}{\rho_A}\vec{\Phi}_A \tag{6.70}$$

so that with (6.68) and $\rho = c\,m$, the phenomenological equation for molar diffusion flux relative to the mass average velocity can be written as

$$\boxed{\vec{\Phi}'_A = c_A \vec{V}_A = -c\,D'_{AB}\left(\vec{d}^{\star}_A + \alpha_A^{(T)} x_A x_B \text{grad } \ln T\right)} \tag{6.71}$$

with the introduction of the diffusivity D'_{AB} used in the molar diffusion flux relative to the mass average velocity

$$D'_{AB} = \frac{m_B}{m} I\!D_{AB} \tag{6.72}$$

The diffusivity D'_{AB} is not symmetric, but satisfies the symmetry relation

$$m_A D'_{AB} = m_B D'_{BA} \tag{6.73}$$

It is clear—starting from (6.68)—that the laws of diffusion can be transformed into other forms. The molar velocity is needed for a transformation into a consistent* molar description of the diffusion in binary systems. For binary systems, the molar velocity becomes

$$\vec{v}^\star = \sum_{K=1}^{N} x_K \vec{v}_K = x_A \vec{v}_A + x_B \vec{v}_B \tag{6.74}$$

and the molar diffusion flux relative to the average molar velocity

$$\begin{aligned}\vec{\Phi}^\star_A &= c_A \left(\vec{v}_A - \vec{v}^\star\right) = c_A \left(\vec{v}_A - x_A \vec{v}_A - x_B \vec{v}_B\right) = c_A x_B \left(\vec{v}_A - \vec{v}_B\right) \\ &= c\, x_A x_B \left(\vec{v}_A - \vec{v}_B\right) = \frac{c}{\rho} \frac{m^2}{m_A m_B} \vec{\Phi}_A = \frac{m}{m_B} \vec{\Phi}'_A\end{aligned} \tag{6.75}$$

where the relations for the barycentric diffusion flux follow by using (6.61) and the definitions of the weight fractions and the molar fractions. The relation for the molar diffusion flux relative to the mass average velocity follows by using (6.70). The generalization of the Fick diffusion law is found by the substitution of (6.68) into (6.75), yielding

$$\boxed{\vec{\Phi}^\star_A = c_A \vec{V}^\star_A = -c\, I\!D_{AB} \left(\vec{d}^\star_A + \alpha^{(T)}_A x_A x_B \operatorname{grad} \ln T\right)} \tag{6.76}$$

With this result, it is shown that $I\!D_{AB}$ can be identified with the *diffusion coefficient of Fick.* The Fick diffusivity is found in the barycentric description (6.66) as well as in the molar description (6.76). Therefore, the Fick diffusivity is of great importance and most measurements give values for the Fick diffusion coefficient.

An example of a Fick diffusion coefficient is depicted in figure 6.1. The example shows also that in nonideal binary systems the Fick diffusivity varies

---

* Consistent means that the equations can be derived from each other, and that the mass balance equations are formulated with the proper time derivative. In the past, much confusion has arisen because the barycentric time derivative and the molar time derivative have been used in the wrong places.

considerably with the concentration, and exhibits for nonideal systems a minimum at intermediate concentrations. With increasing nonideality the minimum may become zero at the critical demixing point or may even become negative, causing the mixture to be split into two liquid phases (an example is the benzene-water system).

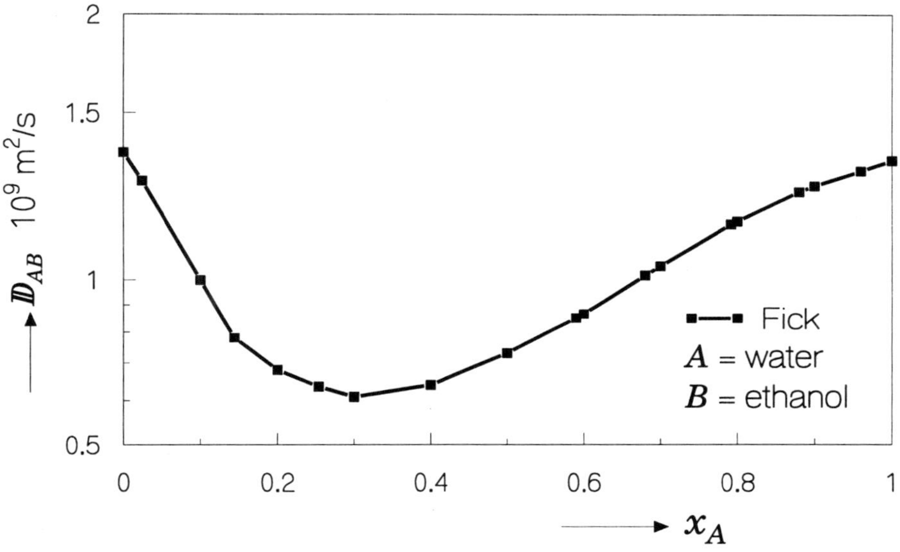

Figure 6.1. The concentration dependence of the Fick diffusivity for a binary mixture of water-ethanol at 40 °C as a function of the molar fraction of water. The Fick diffusivities are those measured by Tyn and Calus*.

The behavior of the Fick diffusion coefficient in nonideal systems is quite complicated. It will be pointed out that the Maxwell–Stefan coefficient behaves quite well, and we have already found that for binary systems it is always a positive number. The relation between the Fick diffusivity $\mathbb{D}_{AB}$ and the Maxwell–Stefan diffusivity $Đ_{AB}$ is given by (6.57).

From (6.76) and (6.69) it follows that in the molar description of the diffusion the mole fraction is the preferable diffusion potential, and not the molar concentration that is used in the older literature. The partial pressure p_A of component A in the molar description can be defined by $p_A/p = x_A$, so that for diffusion under constant pressure also the partial pressure can be considered as a diffusion potential in the molar description. One has to be meticulous with regard to the consistency of the description of the diffusion with regard to the diffusion potentials introduced.

* Tyn M.T. and Calus W.F. 1975. Temperature and concentration dependence of mutual diffusion coefficients of some binary liquid systems. *Journal of Chemical Engineering Data*, 20: 310–316.

A disadvantage of the molar description is that the mass velocity $\vec{v}$ has to be calculated from the momentum balance equation, and that the molar velocity $\vec{v}^\star$ cannot be easily calculated from $\vec{v}$, even for binary systems. It can also be remarked that the mass remains constant for chemical reactions, but that the number of moles need *not* be constant, so that in the barycentric description use can be made of $\sum_{K=1}^{N} \pi_K = 0$, whereas in the consistent molar description the relation $\sum_{L=1}^{N} \pi_L/m_L \neq 0$ does not hold in general.

6.4.2. Heat transport. With the momentum equations and the diffusion equations, the isothermal mass transport can be determined for a body in a flow field. However, mass transport is seldom isothermal in practice. Mass transport processes are often accompanied by heat effects that can be caused by latent heat (evaporation, condensation), dissolution heat (absorption, extraction), chemical reaction heat (conversion, combustion) and finally, inhomogeneous concentration distributions can cause heat flows (Dufour effect).

The thermal energy equation has to be solved for the calculation of the thermal effects, for which a phenomenological equation for the reduced heat flux has to be derived. The heat flux, given by (6.41), becomes for binary systems

$$\vec{\Phi}_{(q)} = -\lambda \operatorname{grad} T + p \frac{x_A x_B}{Đ_{AB}} \left[\frac{D_A^{(T)}}{w_A} \left(\vec{V}_A - \vec{V}_B \right) + \frac{D_B^{(T)}}{w_B} \left(\vec{V}_B - \vec{V}_A \right) \right]$$

or with $D_A^{(T)} + D_B^{(T)} = 0$, $I\!D_{AB} = \Gamma Đ_{AB}$, and (6.59)

$$\vec{\Phi}_{(q)} = -\lambda \operatorname{grad} T + p \Gamma x_A x_B \alpha_A^{(T)} \left(\vec{V}_A - \vec{V}_B \right) \tag{6.77}$$

The various formulations of the reduced heat flux for the various diffusion fluxes follow by the repeated substitution of (6.75) into (6.77)

$$\begin{aligned} \vec{\Phi}_{(q)} &= -\lambda \operatorname{grad} T + \frac{p}{\rho} \Gamma \frac{m^2}{m_A m_B} \alpha_A^{(T)} \vec{\Phi}_A \\ \vec{\Phi}_{(q)} &= -\lambda \operatorname{grad} T + \frac{p}{c} \Gamma \frac{m}{m_B} \alpha_A^{(T)} \vec{\Phi}'_A \\ \vec{\Phi}_{(q)} &= -\lambda \operatorname{grad} T + \frac{p}{c} \Gamma \alpha_A^{(T)} \vec{\Phi}^\star_A \end{aligned} \tag{6.78}$$

In each of these three equations the second term on the right-hand sides represents the Dufour effect, which generally can be neglected only in liquids*.

* Rosner D.E. 1980. Thermal (Soret) diffusion effects on interfacial mass transport rates. *PCH Physicochemical Hydrodynamics*, 1: 159–185.

6.5. IDEAL SYSTEMS

6.5.1. Gases. For binary systems it has been shown in the previous section that the various descriptions follow from the general Maxwell–Stefan equations for multicomponent diffusion. The relations between the various diffusivities also follow. The weak point in the previous discussion is the introduction of the diffusion vector

$$\vec{d}_K = \frac{\rho_K}{p}\left[(\operatorname{grad}\mu_K)_{T,p} + (v_K - v)\operatorname{grad} p + \left(\vec{f}' - \vec{f}'_K\right)\right] \tag{6.79}$$

For *binary systems*, and for *ideal systems*, can this diffusion vector be expressed in the concentrations. Ideal systems are in the first place ideal gas mixtures that have to be sufficiently dilute that the collisions between the gas molecules are predominantly binary collisions. An ideal system is defined by the fact that in equilibrium the components do not influence each other, so that each component behaves as if the component is the only component at the same volume at the same temperature. This is possible only if the components are sufficiently diluted. Then each component satisfies the experimental thermal laws of Boyle and Gay Lussac. The temperature and the volume act as independent variables in this description, so that the free-energy representation seems appropriate for an ideal fluid.

If for a sufficient dilution only the component K is present in pure form, then, according to Boyle and Gay Lussac,

$$p = R_{(\mathrm{A})}\, c_K\, T \tag{6.80}$$

where $R_{(\mathrm{A})} = 8.314\,471$ J mol^{-1} K^{-1} is the Avogradro molar gas constant*.

The free-energy $df = -s\,dT - p\,dv + \sum_{K=1}^{N} \mu_K\, dw_K$ reduces for the single pure component K with $N = 1$ to

$$df_K = -s_K\, dT - p\, dv_K \tag{6.81}$$

For the pure component K the relation

$$v_K = 1/\rho_K = 1/\left(c_K m_K\right)$$

applies. Substitution yields for (6.81)

$$df_K = -s_K\, dT + p\frac{dc_K}{m_K\, c_K^2}$$

* This value is 4.7 ppm smaller than the 1986 value and measured with an uncertainty of 1.7 ppm, which is five times smaller than the 1986 uncertainty. Cohen E.R. and Taylor B.N. 1992. The fundamental physical constants. *Physics Today Buyers' Guide*, BG1992: 9–13.

so that with (6.80)

$$\left(\frac{\partial f_K}{\partial c_K}\right)_T = \frac{p}{m_K\, c_K^2} = \frac{R_{(\mathrm{A})}\, T}{m_K\, c_K}$$

Integration, using $R_K = R_{(\mathrm{A})}/m_K$ also yields

$$f_K = \phi_K(T) + R_K\, T \ln c_K \tag{6.82}$$

Yet, as always for ideal gases, $\phi_K(T)$ is a function of T not further specified. Hence, an ideal gas is only thermally determined but calorically undetermined. The indeterminacy is removed if the specific heats are known as functions of the temperature.

The result (6.82) applies for the pure component K, which component is considered to represent an ideal system. The relation holds by definition in an ideal fluid if f_K is considered as the partial, specific free-energy of the component K, so that the free-energy of the system becomes

$$F = \sum_{K=1}^{N} M_K f_K = \sum_{K=1}^{N} M_K \left[\phi_K(T) + R_K\, T \ln c_K\right] \tag{6.83}$$

in which $c_K m_K = \rho_K = M_K/V$ or

$$c_K = M_K/(m_K\, V) \tag{6.84}$$

Relation (6.83) expresses the additivity of the free-energy of an ideal system. The partial free-energy f_K depends only on the molar concentration c_K, so that in this respect the ideal multicomponent system is also called a 'mixture'. From the free-energy representation and (6.83) there follows

$$\mu_K = \left(\frac{\partial F}{\partial M_K}\right)_{T,V,M'} = \mu_K^{\circ}(T) + R_K\, T \ln c_K \tag{6.85}$$

with

$$\mu_K^{\circ}(T) = \phi_K(T) + R_K\, T \tag{6.86}$$

According to the free-energy representation and (6.83)

$$p = -\left(\frac{\partial F}{\partial V}\right)_{T,M} = \sum_{K=1}^{N} \rho_K\, R_K\, T$$

so that the partial pressures p_K satisfy

$$\begin{aligned} p &= \sum_{K=1}^{N} p_K && \text{with} \quad p_K = R_{(\mathrm{A})}\, c_K\, T = \rho_K\, R_K\, T \\ p &= R_{(\mathrm{A})}\, c\, T = R\, \rho\, T && \text{with} \quad R = R_{(\mathrm{A})}/m \end{aligned} \tag{6.87}$$

From (6.85) and $(6.87)_2$ it is found that

$$\mu_K^{(\mathrm{id})}(T,p,x_K) = \mu_K^{+}(T,p) + R_K\, T \ln x_K \tag{6.88}$$

with

$$\mu_K^{+}(T,p) = \mu_K^{\circ}(T) + R_K\, T \ln \frac{p}{R_{(\mathrm{A})}\, T} \tag{6.89}$$

These relations for ideal mixtures suggest the use of mole fractions. Since convincing arguments can be made for the use of the mass fractions as well as for the mole fractions, both fractions are used.

From the Gibbs representation and (6.88) it is concluded that for an ideal mixture that

$$d\mu_K = -s_K\, dT + v_K\, dp + R_K\, T\, dx_K / x_K \tag{6.90}$$

For ideal gases it follows for instance that $p = \rho_K (\partial \mu_K / \partial x_K)_{T,p}$ holds from this result. The partial volume found in (6.90) follows from $(6.87)_2$ and (6.84)

$$p = R_{(\mathrm{A})}\, c T = \frac{R_{(\mathrm{A})}\, T}{V} \sum_{K=1}^{N} \frac{M_K}{m_K} \quad \rightarrow \quad V = \frac{R_{(\mathrm{A})}\, T}{p} \sum_{K=1}^{N} \frac{M_K}{m_K}$$

From the last equation it is concluded that

$$v_K = \left(\frac{\partial V}{\partial M_K} \right)_{T,p,M'} = \frac{R_{(\mathrm{A})}\, T}{p\, m_K} = \frac{1}{c\, m_K} \tag{6.91}$$

from which for the *partial* quantities it does *not* follow that $v_K = 1/\rho_K$, since $\rho_K = c_K m_K$. The diffusion vector (6.79) becomes, with the last equation of the equations (6.91)

$$\vec{d}_K^{(\mathrm{id})} = \operatorname{grad} x_K + (x_K - w_K) \frac{\operatorname{grad} p}{p} + \rho_K \frac{\vec{f}' - \vec{f}_K'}{p} \tag{6.92}$$

The partial pressure is additive in an ideal mixture, for which from (6.87) it follows

$$p_K / p = x_K \tag{6.93}$$

The partial pressure can be *defined* for nonideal systems by (6.93), so that $(6.87)_{1,1}$ still applies. The diffusion vector can also be expressed by (6.93) in the partial pressures. Now it follows from (6.87)–(6.89) that

$$\mu_K^{(\mathrm{id})} = \mu_K^{\circ}(T) + R_K\, T \ln c_K \tag{6.94}$$

or with $(6.87)_2$

$$\mu_K^{(\mathrm{id})} = \tilde{\mu}_K(T) + R_K\, T \ln p_K \tag{6.95}$$

with

$$\tilde{\mu}_K(T) = \mu_K^\circ(T) - R_K\, T \ln\left(R_{(A)}\, T\right) \tag{6.96}$$

It is dangerous to derive the diffusion vector $\vec{d}_K^{(\mathrm{id})}$ by the substitution of (6.95) into (6.79), since $\mu_K^{(\mathrm{id})}$ has to be defined at constant temperature T and at constant pressure $p = \sum_{K=1}^{N} p_K$. It is better to start from (6.92) and to apply (6.93)

$$\vec{d}_K^{(\mathrm{id})} = \frac{\operatorname{grad} p_K}{p} - w_K \frac{\operatorname{grad} p}{p} + \rho_K \frac{\vec{f}' - \vec{f}'_K}{p} \tag{6.97}$$

It is clear that the term with the pressures is needed, because $\sum_{K=1}^{N} \vec{d}_K = \vec{0}$. The simple expression (6.97) has in the past induced many authors to introduce the partial pressure as a diffusion potential. For real gases, which are not ideal gases, one can try to maintain the structure of (6.95) and (6.97) by introducing at the suggestion of Lewis* the fugacities φ_K of the components as

$$\mu_K = \tilde{\mu}_K(T) + R_K\, T \ln \varphi_K(T, p) \tag{6.98}$$

where $\tilde{\mu}_K$ is given by (6.96), so that the fugacity φ_K is defined by (6.98). From (6.95) and (6.98) follows

$$\mu_K = \mu_K^{(\mathrm{id})} + R_K\, T \ln\left(\varphi_K / p_K\right) \tag{6.99}$$

All gases become ideal gases at sufficiently low pressures. In the Gibbs representation the relation

$$\lim_{p \to 0} \mu_K = \mu_K^{(\mathrm{id})}$$

holds, and the fugacity becomes equal to the pressure, so that

$$\lim_{p \to 0} \varphi_K / p_K = 1 \tag{6.100}$$

The fugacity can be considered as a new intensive thermodynamic quantity of state that depends on T, p and the composition of the gas. The effects of the intermolecular forces on the thermodynamic properties of a gas are incorporated in the fugacity.

From $(4.17)_3$ it follows in the representation of Gibbs

$$\left(\frac{\partial \mu_K}{\partial p}\right)_{T,\ \text{composition gas}} = v_K \tag{6.101}$$

* Lewis G.N. 1901. Laws of physico-chemical change. *Proceedings of the American Academy of Arts and Sciences*, 37: 49–69; Lewis G.N. 1901. Das Gesetz physiko-chemischer Vorgänge. *Zeitschrift für physikalische Chemie*, 38: 205–226.

From this it is found that

$$\mu_K = \mu_K^{(\mathrm{id})} + \lim_{p_0 \to 0} \int_{p_0}^{p} \left(v_K - v_K^{(\mathrm{id})} \right) dp \tag{6.102}$$

since in this representation the ideal state is formally attained at low pressures. The integrand $v_K - v_K^{(\mathrm{id})}$ can further be expressed in terms of the virial coefficients (see for details for instance Prigogine and Defay*). The fugacity can hereby also be calculated directly. From (6.99) and (6.102) there results

$$R_K T \ln (\varphi_K / p_K) = \lim_{p_0 \to 0} \int_{p_0}^{p} \left(v_K - v_K^{(\mathrm{id})} \right) dp \tag{6.103}$$

The quotient φ_K / p_K is called the fugacity coefficient†.

The application of the fugacity (6.98) for the calculation of the diffusion vector is not straightforward, since the independent variables $\{T, p, x_K\}$ have to be transformed‡ into the Gibbs representation into $\{T, p, \varphi_K\}$. In the comparison of a real gas with an ideal gas, one can also start from (6.88), and write for a real gas

$$\mu_K = \mu_K^{+}(T, p) + R_K T \ln \alpha_K \tag{6.104}$$

in which μ_K^{+} is the same as in (6.88), so that

$$\mu_K = \mu_K^{(\mathrm{id})} + R_K T \ln (\alpha_K / x_K) \tag{6.105}$$

The chemical *activities* α_K are defined by (6.104). The activity can be considered again as new intensive functions of state for a real gas depending on the temperature T, the pressure p, and all the mole fractions $\{x_K\}$ (Gibbs representation), in which the intermolecular forces are taken into account. From (6.99) and (6.105) it follows that

$$\gamma_K \equiv \frac{\alpha_K}{x_K} = \frac{\varphi_K}{p_K} \tag{6.106}$$

---

* Prigogine I. and Defay R. 1954. *Chemical Thermodynamics*. Longmans Green & Co., London, p. 139.

† Reid R.C., Prausnitz J.M. and Poling B.E. 1988. *The Properties of Gases and Liquids*. McGraw-Hill Book Co., New York. See p. 248.

‡ For example:

$$\mu = \mu_K(T, p, x_L) = \mu_K \left[T, p, \varphi_L(T, p, x_K) \right]$$

and

$$\left(\frac{\partial \mu_K}{\partial T} \right)_{p,x} = \left(\frac{\partial \mu_K}{\partial T} \right)_{p,\varphi} + \sum_{L=1}^{N} \left(\frac{\partial \mu_K}{\partial \varphi_L} \right)_{T,p} \left(\frac{\partial \varphi_L}{\partial T} \right)_{p,x}$$

where γ_K is called the *activity coefficient*. With (6.93) it is seen that the activity is related to the fugacity by

$$\alpha_K = \frac{\varphi_K}{p} \tag{6.107}$$

In the representation of Gibbs one can write analogously to (6.100)

$$\lim_{p \to 0} \frac{\alpha_K}{x_K} = 1 \tag{6.108}$$

Expression (6.105) implies

$$\mu_K = \mu_K^{(\mathrm{id})} + R_K T \ln(\varphi_K / p_K) \tag{6.109}$$

6.5.2. Liquids. A solution is a condensed phase (liquid or solid) that consists of several components. In principle there is a major difference between dissolved molecules—even in dilute solutions—and gas molecules, although corresponding formulas exist between solutions and gases. Dissolved molecules are subjected to strong intermolecular forces exerted by the molecules of the solvent. This is not the case in a gas, since then the intermolecular forces are effective only during the intermolecular collisions. This difference is experimentally illustrated by the fact that the heat of solution of a solid usually does not differ too much from the heat of fusion, but differs considerably from the heat of evaporation, which can be twice as much as the heat of solution. The molar concentration in a solution is at the same temperature much larger than that in a gas.

Despite the great fundamental differences between solutions and gases, *some* laws for solutions are analogous to those for gases. This has been expressed, perhaps for the first time, explicitly by Van 't Hoff (1887), who investigated the osmotic pressure and concluded that the osmotic pressure of a solute has the same value as the pressure that would be exerted by the undissolved substance if this substance in the gaseous phase was confined in the same volume at the same temperature. If the solution is sufficiently dilute, the osmotic pressure is described by an equation reminiscent of that for an ideal gas.

This notion has led to the definition of ideal solutions as a special case of ideal systems. The chemical potential of an ideal solution is therefore according to (6.88)

$$\mu_K^{(\mathrm{id})} = \mu_K^{+}(T, p) + R_K T \ln x_K \tag{6.110}$$

This is however different from the relation for ideal gases since now μ_K^{+} is not given by (6.89).

If the concentrations of all chemical components present are of the same order, then there is no difference between the solvent and the solute. If there is one component with a much higher concentration than those of the other components, then this component is called the *solvent* (subscript s). Experimentally it is found that dilute solutions, in which the molar concentrations are sufficiently close to zero, behave like ideal solutions. How small these concentrations have to be in practice, to guarantee that the solute is ideal with required accuracy, depends to a large extent on the nature of the solvent and the dissolved substances.

In electrolytes, deviations from the ideal behavior can occur even in very dilute solutions. This can be linked to the large spatial range of the electromagnetic forces.

Solutions of molecules with about the same chemical structure can behave ideally over a vast region of concentrations, if these molecules are of 'normal' size (not macromolecules). The principal difference between ideal gases and ideal solutions is caused by the short time of influence between the molecules in gases and the large intermolecular times of influence in liquids. Therefore, $\mu_K^+(T,p)$ is for ideal gases quite different from the corresponding quantities for ideal solutions.

The difference between ideal gases and ideal solutions can be illustrated especially for the pressure dependence of $\mu_K^+(T,p)$. Define for each component a (partial) compressibility coefficient

$$\kappa_K = -\frac{1}{v_K}\left(\frac{\partial v_K}{\partial p}\right)_T \tag{6.111}$$

The compressibility κ_K for a liquid is to a first approximation independent of the pressure. In that case it is found from (6.111) that

$$v_K(T,p) = v_K(T,0)\,(1-\kappa_K\,p) \tag{6.112}$$

The specific volume v_K does not depend on the concentration in an ideal system. After all, from the differential expression $(4.11)_3$ for the Gibbs function it follows that

$$\left(\frac{\partial \mu_K}{\partial p}\right)_{T,M} = \left(\frac{\partial V}{\partial M_K}\right)_{T,p,M'} = v_K \tag{6.113}$$

With this result it is found from (6.110) that for *all* ideal systems (both gases and liquids)

$$\left(\frac{\partial \mu_K}{\partial p}\right)_{T,M} = \left(\frac{\partial \mu_K^+(T,p)}{\partial p}\right)_T = v_K^{(\mathrm{id})} \tag{6.114}$$

so that $v_K^{(\mathrm{id})}$ depends only on the temperature and the pressure. From (6.112), which is already based on this statement, and from (6.114) it follows that

$$\mu_K^+(T,p) = \tilde{\mu}_K(T) = p\left(1-\tfrac{1}{2}\kappa_K^{(\mathrm{id})}\,p\right)v_K^{(\mathrm{id})}(T,0) \tag{6.115}$$

The chemical potential $\mu_K^+(T,p)$ depends quite differently on the pressure in ideal gas mixtures. This can be understood briefly as follows: the isothermal compressibility for an ideal gas mixture is given by $1/p$, whereas for liquids the compressibility is so insignificant, that the compressibility can be neglected in first the approximation for not too high pressures. Summing up: for ideal gases (6.113) applies, but the form of dependency of $\mu^+(T,p)$ is entirely different for gases than it is for liquids. The introduction of fugacities for liquids is pointless for the description of the deviations from the ideal behavior.

Activities can be introduced by means of (6.113) analogous to (6.104) to describe the deviations of the ideal behaviour of solutions. For the solvent $x_s \approx 1$ holds, so that it is not very useful to express the mass transport in $\operatorname{grad} x_s$.

6.6. NONIDEAL SYSTEMS

For nonideal binary systems the relations for the mass fluxes are rather complicated, but they may be simplified formal if activities are introduced for the diffusion potentials. Also for multicomponent systems the introduction of activities results in simplifications*, since in general one cannot derive macroscopic balance equations in which only the concentration of one component is found explicitly, as is possible for binary systems. The deviations from ideal systems can be expressed in the diffusion coefficients of binary systems, whereas the relation (6.57) between the diffusivities of Fick and of Maxwell–Stefan can also be used. The thermodynamic factor in this relation can be expressed in terms of the activities, so that the Maxwell–Stefan diffusivities can be calculated from those of Fick.

For the calculation of the mass fluxes and the diffusion vector, the chemical potential μ_K can be considered to depend on the temperature, the pressure, and the activities α_K, so that

$$d\mu_K = \left(\frac{\partial \mu_K}{\partial \alpha_K}\right)_{T,p} d\alpha_K + \left(\frac{\partial \mu_K}{\partial p}\right)_{\alpha_K,T} dp + \left(\frac{\partial \mu_K}{\partial T}\right)_{p,\alpha_K} dT$$

while from (6.104) it follows that for the partial derivative

$$\left(\frac{\partial \mu_K}{\partial \alpha_K}\right)_{T,p} = \frac{R_K T}{\alpha_K} \tag{6.116}$$

For nonideal systems we have

$$\begin{aligned} \frac{\rho_K}{p} (\operatorname{grad} \mu_K)_{T,p} &= \frac{\rho_K R_K T}{p\,\alpha_K} (\operatorname{grad} \alpha_K)_{T,p} \\ &= \frac{p_K}{p\,\alpha_K} (\operatorname{grad} \alpha_K)_{T,p} = \frac{x_K}{\alpha_K} (\operatorname{grad} \alpha_K)_{T,p} \end{aligned} \tag{6.117}$$

* Merk H. 1957. Stofoverdracht in laminaire grenslagen door gedwongen convectie. Ph.D. dissertation. Delft University of Technology (in Dutch).

When we make use of the relation $p_K = \rho_K R_K T$ $(6.87)_1$ and $x_K = p_K/p$ (6.93), the diffusion vector (6.79) for nonideal systems, with the gradient of the activity as the chemical driving force, can be written as

$$\vec{d}_K = \frac{x_K}{\alpha_K}(\operatorname{grad}\alpha_K)_{T,p} + (\rho_K v_K - w_K)\frac{\operatorname{grad} p}{p} + \rho_K \frac{\vec{f}' - \vec{f}'_K}{p} \tag{6.118}$$

The description of diffusion is very complicated in systems with more than two components. The diffusion coefficients are generally unknown or inaccurately known. We do have moderate good information on binary systems, however. The Maxwell–Stefan diffusivities $Đ_{AB}$ can be estimated with (6.45) where $\mathbb{D}_{AB}$ has been extensively* investigated experimentally and theoretically.

The diffusivity $Đ_{AB}$ is independent of the concentration for ideal gases. Also for ideal liquids it turns out that the Maxwell–Stefan diffusivities are almost independent of the concentration. The Maxwell–Stefan diffusivities can be considered as constants for ideal liquids within reasonable bounds, whereas this is certainly not the case for the Fick diffusivities, as figure 6.1 illustrates. The Fick diffusivities $\mathbb{D}_{AB}$ are the commonly measured diffusivities, while for the solution of the balance equations the barycentric description is easier to use; hence it is important to be able to derive the Maxwell–Stefan diffusion coefficients from the Fick diffusivities. The relation between the two diffusivities is given by (6.57)

$$\frac{\mathbb{D}_{AB}}{Đ_{AB}} = \Gamma \tag{6.119}$$

This relation has been derived by Merk†. Sometimes it is thought that the Fick and the Maxwell–Stefan diffusivities can be transformed into one other only for systems that can be described by the concentration gradients‡ as the chemical driving force exclusively. The factor Γ is called the *thermodynamic correction factor.*

The mechanical diffusion vector in (6.50) is expressed in the mole fractions x_A with the introduction of the thermodynamic correction factor Γ, in which the deviations from the ideal system are specified. The factor Γ can be related

---

* Chapman S. and Cowling T.G. 1939. *The Mathematical Theory of Non-Uniform Gases.* Cambridge University Press, Cambridge (third revised printing 1970), p. 263; and Hirschfelder J.O., Curtiss C.F. and Bird R.B. 1954. *Molecular Theory of Gases and Liquids.* John Wiley & Sons, Inc., New York (fourth printing 1967.), p. 579; Taylor R. and Krishna R. 1993. *Multicomponent Mass Transfer.* John Wiley & Sons, Inc., New York.

† Merk H. 1957. Stofoverdracht in laminaire grenslagen door gedwongen convectie. Ph.D. dissertation, Delft University of Technology (in Dutch).

‡ Rutten Ph.W.M. 1992. *Diffusion in Liquids.* Delft University Press, Delft. Thesis. See p. 24.

to the activities by using (6.116), $(6.87)_1$ and the definition of the partial pressure (6.93) for nonideal systems

$$\Gamma = \frac{\rho_A}{p}\left(\frac{\partial \mu_A}{\partial x_A}\right)_{T,p} = \frac{\rho_A}{p}\left(\frac{\partial \mu_A}{\partial \alpha_A}\right)_{T,p}\left(\frac{\partial \alpha_A}{\partial x_A}\right)_{T,p}$$
$$= \frac{p_A}{p\,\alpha_A}\left(\frac{\partial \alpha_A}{\partial x_A}\right)_{T,p} = x_A\frac{\partial \ln \alpha_A}{\partial x_A} = \frac{\partial \ln \alpha_A}{\partial \ln x_A} \tag{6.120}$$

Chemical engineers have a preference for the use the activity coefficient γ_A, which is defined by $\alpha_A = \gamma_A x_A$. With the substitution of this definition into (6.120), the quantity Γ can be calculated with

$$\Gamma = 1 + \frac{\partial \ln \gamma_A}{\partial \ln x_A} \tag{6.121}$$

By means of the relation (6.119), the Maxwell–Stefan diffusivities can now be calculated straightforwardly from the Fick diffusivities if the activity coefficients are known. Simple models for the calculation of the activity coefficients are the Van Laar* model and the Margules model†, for which the empirical parameters can be found in the literature‡.

Also by differentiation of the UNIQUAC§ (UNIversal QUAsi Chemical) or

---

* The activity coefficient in the Van Laar model is given by

$$\ln \gamma_i = A_{ij}\left(\frac{A_{ji}x_j}{A_{ij}x_i + A_{ji}x_j}\right)^2$$

for which $i \neq j$ can have the values 1 or 2 (A or B), and the coefficients A_{ij} are empirical parameters. By differentiation with the condition $x_1 + x_2 = 1$ one obtains for the thermodynamic factor $\Gamma = 1 - 2x_1x_2A_{12}^2A_{21}^2/(A_{12}x_1 + A_{21}x_2)^3$.

† The activity coefficient in the Margules model is given by

$$\ln \gamma_i = [A_{ij} + 2\,(A_{ji} - A_{ij})\,x_i]\,x_j^2$$

and the thermodynamic factor becomes $\Gamma = 1 + 2x_1x_2\,[x_1(A_{12} - 2A_{21}) + x_2(A_{21} - 2A_{12})]$.

‡ Gmehlin J. and Onken U. 1977. *Vapour-Liquid Equilibrium Data Collection.* Volume 1, part 1, Dechema, Frankfurt a/d Main. In these Dechema handbooks the empirical parameters for numerous binary systems and models can be found. The parameters for the Van Laar, the Margules, the UNIQUAC and the NRTL models are given. These parameters are obtained from experimentally determined values of the thermodynamic quantities. Errors of several percent may be realistically expected. See also Sprensen J.M. and Arlt W. 1977. *Liquid-Liquid Equilibrium Data Collection.* Dechema, Frankfurt a/d Main; Novak J.P., Matous J. and Pick J. 1987. *Liquid-Liquid equilibria.* Elsevier. Amsterdam; Reid R.C., Prausnitz J.M. and Poling B.E. 1987. The Properties of Gases & Liquids. McGraw-Hill Book Co., New York. See Ch. 11.

§ Anderson T.F. and Prausnitz J.M. 1978. Application of the UNIQUAC equation to calculation of multicomponent phase equilibria. *Industrial Chemical Engineering, Process Design & Development*, 17: 552–561, 561–567.

the NRTL* (Non Random Two Liquid) relations for the excess Gibbs energy the activity coefficients can be calculated†. Rutten‡ has optimized the values of the parameters A_{ij} for more than one hundred binary systems for use with the UNIQUAC or NRTL model to get a good description of the activity coefficients.

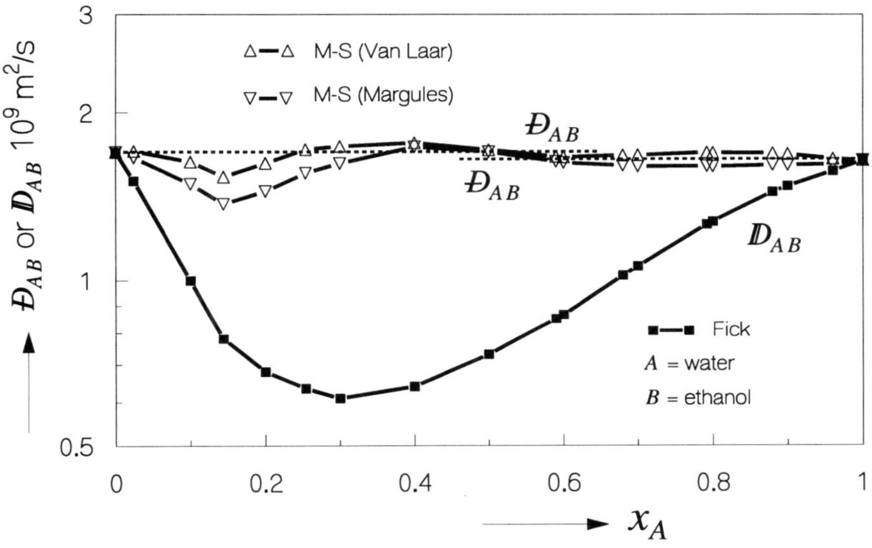

Figure 6.2. The concentration (in)dependence of the Maxwell–Stefan diffusivity for a binary mixture of water-ethanol at 40 °C as a function of the mole fraction of water. The Fick diffusivities are those measured by Tyn and Calus§. The Maxwell–Stefan diffusivities are calculated with the Van Laar model, where A_{12}=1.5922 and A_{21}=0.9836; and with the Margules model, where A_{12}=1.4286 and A_{21}=0.8608. The dashed lines extrapolates the diffusivity at infinite dilution that is already a very good approximation to the Maxwell–Stefan diffusivity for all values of the concentration.

The uppermost two curves in figure 6.2 depict the Maxwell–Stefan diffusion coefficients $Đ_{AB}$ as calculated¶ by using the Van Laar model and the Margules model. The lowest curve depicts the measured diffusivities $\mathbb{D}_{AB}$ as reported

* Renon H. and Prausnitz J.M. 1968. Local compositions in thermodynamic excess functions for liquid mixtures. *A.I.Ch.E. Journal*, 14: 135–144.

† Reid R.C., Prausnitz J.M. and Poling B.E. 1988. *The Properties of Gases and Liquids*. McGraw-Hill Book Co., New York. See chapter 8-5: 251–259.

‡ Rutten Ph.W.M. 1992. *Diffusion in Liquids*. Delft University Press, Delft. Thesis.

§ Tyn M.T. and Calus W.F. 1975. Temperature and concentration dependence of mutual diffusion coefficients of some binary liquid systems. *Journal of Chemical Engineering Data*, 20: 310–316.

¶ Mees P.A.J. and Muller F.L. 1989. Het verband tussen de diffusiecoëfficiënten van Fick en Maxwell–Stefan. Unpublished notes in Dutch. T.U. Delft.

by Tyn and Calus*. The results in this figure show that for the water/ethanol binary system at 40°C the diffusion coefficient at infinite dilution is already a very good approximation for the Maxwell–Stefan diffusivities of the solution. For this water/ethanol system a linear approximation or a logarithmic approximation (a straight line) is a reasonably good approximation for the Maxwell–Stefan diffusivities for all concentrations, while this is not so for the Fick diffusivities.

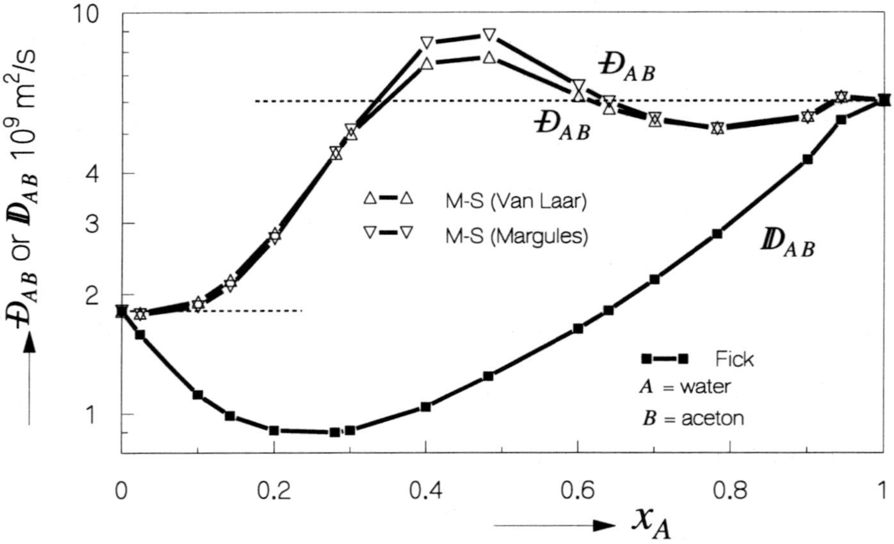

Figure 6.3. The concentration (in)dependence of the Maxwell–Stefan diffusivity for a binary mixture of water-acetone at 45 °C as a function of the mole fraction of water. The Fick diffusivities are those measured by Tyn and Calus. The Maxwell–Stefan diffusivities are calculated with the Van Laar model, where A_{12}=1.9404 and A_{21}=1.4891; and with the Margules model, where A_{12}=1.9125 and A_{21}=1.4757. The two dashed lines extrapolate the diffusivity at infinite dilution that is already a reasonable first approximation of the Maxwell–Stefan diffusivity for not too large values of the concentration.

A logarithmic interpolation† has been suggested by Vignes, who has shown that in binary mixtures a logarithmic interpolation usually suffices in practice,

* Tyn M.T. and Calus W.F. 1975. Temperature and concentration dependence of mutual diffusion coefficients of some binary liquid systems. *Journal of Chemical Engineering Data*, 20: 310–316.

† Vignes A. 1966. Diffusion in binary solutions. *Industrial and Engineering Chemistry, Fundamentals*, 5: 189–199; Siddiqi M.A., Krahn W. and Lucas K. 1987. Mutual diffusion coefficients in some binary liquid mixtures. *Journal of Chemical Engineering Data*, 32: 48–50. See also chapter 11 in Reid R.C., Prausnitz J.M. and Poling B.E. 1988. *The Properties of Gases and Liquids*. McGraw-Hill Book Co., New York, for a review of interpolation formulas.

certainly for ideal or nearly ideal mixtures*. Due to nonlinear intermolecular forces and concentration dependences the liquid diffusivities $Đ_{AB} \neq Đ_{BA}$. Figure 6.2 suggests that for the water/ethanol system the diffusivities can be approximated by an ideal system diffusivities, provided that the Maxwell–Stefan diffusivity is used and the activity is used instead of the chemical potential. The Maxwell–Stefan diffusivity for the binary system water/ethanol can be considered to be independent of the concentration. However, for temperatures above 60°C the deviations from the ideal behavior increase rapidly, and the Maxwell–Stefan diffusivities can no longer be approximated as a concentration independent constant. The diffusion coefficients for infinite dilution can still provide a good estimate for the Maxwell–Stefan diffusion coefficient for nonideal systems for not too high concentrations, and in many cases a logarithmic interpolation between both infinite dilute diffusivities is a practical estimate.

From the results in figure 6.3 for the binary water/acetone at 45°C, it is found that the infinitely dilute diffusivities are good approximations up to moderate concentrations of the Maxwell–Stefan diffusivity, while here a logarithmic interpolation deviates too much from the measured values. For these strong nonideal systems, one has to be aware of deviations from linear behavior, and usually empirical relations for the estimation of the concentration dependence of the diffusivities have to be used. Those empirical relations in general lack a sound theoretical basis.

6.7. ESTIMATE OF THE MAXWELL–STEFAN DIFFUSIVITY

6.7.1. Binary diffusion coefficients. The concentration dependence of the diffusivity can be considerable, and for solving the balance equations these variations have to be known. If the diffusivities have not been measured over the whole range of the concentrations, one has to rely on interpolation schemes. In the interpolation schemes the diffusivities at infinite dilution are often used, which will be denoted by

$$Đ_{IJ}^{J\to 1} \qquad \text{if} \qquad x_I \to 0 \qquad x_J \to 1 \tag{6.122}$$

This notation applies also for multicomponent diffusivities at infinite dilution. If both components I and J are dilute in an almost pure K component the notation becomes $Đ_{IJ}^{K\to 1}$.

Let us now discuss several ways of estimating $Đ_{IJ}^{J\to 1}$. We begin with the Stokes–Einstein relation. Small isolated spheres submerged in a liquid are submitted to Brownian motion. The friction of the particle in the liquid

* Dullien F.A.L. 1971. Statistical test of Vignes' correlation of liquid-phase diffusion coefficients. *Industrial and Engineering Chemistry, Fundamentals*, 10: 41–49.

is given by the Stokes law*. Einstein† has used this law to calculate the mean-square displacement of the particle. The mean-square displacement turns out to increase linearly with time with the proportionality constant being the Stokes–Einstein diffusivity $k_B T/6\pi\eta R$, where R is the radius of the particle. For dilute suspensions of hard spheres the diffusion coefficient increases slightly with the volume fraction, with a factor $1 + 1.45\phi$, where ϕ is the volume fraction‡. The Stokes–Einstein diffusivity is applicable to particles falling in a solvent. For liquid diffusion the sizes of the solute and solvent molecules are of the same order, and the calculations from hydrodynamics may not be suitable. However, a good estimate can still be obtained for liquid diffusivities, if the number 6 in the original Stokes–Einstein diffusivity is optimized to a number n_{SE} derived from data on binary systems. Rutten§ shows that the variation in that number can be minimized if the generalized Stokes–Einstein diffusivity (6.122) is modified as

$$Đ_{IJ}^{J\to 1} = \frac{k_B T}{n_{SE}\pi\eta_J R_I}\frac{R_J}{R_I} \tag{6.123}$$

to account for an expected dependence of the ratio of the solvent molecule radius to the solute molecule radius. The number n_{SE} is smaller than 6 and about 3.5. For nonassociating binary organic mixtures the number is 3.54 with a relative standard deviation of 11.3% in a data set of 106 binary systems. The same number, 3.53, is found for associating organic mixtures with a relative standard deviation of 8.81% for a data set of 33 diffusivities. The number is a little higher for the self-diffusivities of organic species, 3.65, with only 4.01% standard deviation. The number becomes 3.47 with a relative standard deviation of 7.91% for polar organic species in water, where hydrogen bonding occurs. The bonding is accounted for by taking the volume from which the radius R_j is calculated twice as large. The radius is calculated from the Van der Waals volume as listed by Edward¶ and Bondi‖. The same satisfactory estimates of binary infinite dilution diffusivities can be obtained using the

---

* Stokes G.G. 1856. *Transactions of the Cambridge Philosophical Society*, 9: 5.

† Einstein A. 1905. Über die von der molekularkinetischen Theorie der Wärme geforderte Bewegung von in ruhenden Flüssigkeiten suspendierten Teilchen. *Annalen der Physik*, 17: 549–560.

‡ Batchelor G.K. 1976. Brownian diffusion of particles with hydrodynamic interactions. *Journal of Fluid Mechanics*, 74: 1–29. Russell W.B., Saville D.A. and Schowalter W.R. 1989. *Colloidal Dispersions*. Cambridge University Press, Cambridge. See chapter 13.

§ Rutten Ph.W.M. 1992. *Diffusion in Liquids*. Delft University Press, Delft. Thesis. See chapter 5.

¶ Edward J.T. 1970. Molecular volumes and the Stokes–Einstein equation. *Journal of Chemical Education*, 47: 261–270.

‖ Bondi A. 1964. Van der Waals volumes and radii. *Journal of Physical Chemistry*, 68: 441–451.

critical volume as listed in Reid, Prausnitz and Poling*. The relative standard deviation of the diffusivities is then almost the same as those calculated with the Van der Waals volumes, provided that for the Stokes–Einstein numbers n_{SE}, is taken the number 2.40 for nonassociating binary organic mixtures, 2.42 for associating organic mixtures, 2.08 for self-diffusivities of organic species and 2.35 for organic species in water. The radii needed in (6.123) are calculated from the critical volumes by

$$R_I = \left(\frac{3V_I}{4\pi N_{\mathrm{A}}}\right)^{\frac{1}{3}}$$

where $N_{\mathrm{A}} = 6.022\,1367 \times 10^{23}\ \mathrm{mol}^{-1}$ is the Avogadro constant. The viscosities needed in (6.123) can be found for instance in the Landolt–Börnstein† series, or for low pressures in Reid, Prausnitz and Poling‡, and for high pressures in Stephan and Lucas§.

The estimates of the diffusivities with the modified Stokes–Einstein relation (6.123) are much more accurate than estimated with the famous Wilke–Chang relation¶. Applied to the same data set of diffusivities, the relative standard deviations for the Wilke–Chang relation are 15.3%, 31.6%, 13.6% and 13.5% respectively.

The Vignes interpolation formula for concentrated binary solutions is adequate for nearly ideal solutions. Leffler and Cullinan‖ have given a modified version of the Vignes interpolation formula as

$$Đ_{IJ} = \frac{1}{\eta}\left(Đ_{IJ}^{I\to 1}\eta_I\right)^{x_I}\left(Đ_{IJ}^{J\to 1}\eta_J\right)^{x_J} \tag{6.124}$$

which takes into account the mixture viscosity η and the pure component viscosities η_I. The diffusivity for moderately ideal solutions is predicted with an accuracy of the order of 10% by this scheme. The scheme reduces to the Vignes interpolation formula if the viscosities are equal. The Leffler–Cullinan

* Reid R.C., Prausnitz J.M. and Poling B.E. 1988. *The Properties of Gases and Liquids.* McGraw-Hill Book Co., New York. See pp. 656–672.

† Borchers H., Hausen H., Hellwege K-H., Schäfer K. and Schmidt E. 1969. *Landolt–Börnstein Zahlenwerte und Funktionen.* Vol. IIb (Transportphänomene). Springer Verlag, Berlin.

‡ Reid R.C., Prausnitz J.M. and Poling B.E. 1988. *The Properties of Gases and Liquids.* McGraw-Hill Book Co., New York.

§ Stephan K. and Lucas K. 1979. *Viscosity of Dense Fluids.* Plenum Press, New York.

¶ Wilke C.R. and Chang P. 1955. Correlation of diffusion coefficients in dilute solutions. *A.I.Ch.E. Journal*, 1: 264–270.

‖ Leffler J. and Cullinan H.T 1970. Variation of liquid diffusion coefficients with composition. *Industrial and Chemical Engineering, Fundamentals*, 9: 84–93.

scheme overcorrects for the viscosity. It is found by Rutten that only the linear interpolation with the viscosity correction

$$Ð_{IJ} = \frac{1}{\eta}\left(Ð_{IJ}^{I\to 1}\eta_I x_I + Ð_{IJ}^{J\to 1}\eta_J x_J\right) \tag{6.125}$$

suggested by Hartley and Crank* correlates values of the diffusivity within the experimental error.

Both the Leffler–Cullinan and the Hartley–Crank schemes give estimates with an accuracy of about 10% for nonideal nonassociating systems for which the thermodynamic correction factor Γ deviates more than 20% from unity. For associating mixtures, like the one given in figure 6.3, these schemes both give large errors.

6.7.2. Ternary diffusion coefficients. Ternary diffusion is much more complicated than binary diffusion. There is also a lack of data on ternary diffusion. From the discussion above it has been seen already that the Maxwell–Stefan diffusivities are a much less complicated functions of the concentration than are the Fick diffusivities. To obtain estimates for ternary diffusion coefficients, the interpolation formulas of Leffler–Cullinan (6.124) can be extended for ternary diffusion

$$Ð_{IJ} = \frac{1}{\eta}\left(Ð_{IJ}^{I\to 1}\eta_I\right)^{x_I}\left(Ð_{IJ}^{J\to 1}\eta_J\right)^{x_J}\left(Ð_{IJ}^{K\to 1}\eta_K\right)^{x_K} \tag{6.126}$$

or the Hartley–Crank relation (6.125)

$$Ð_{IJ} = \frac{1}{\eta}\left(Ð_{IJ}^{I\to 1}\eta_I x_I + Ð_{IJ}^{J\to 1}\eta_J x_J + Ð_{IJ}^{K\to 1}\eta_K x_K\right) \tag{6.127}$$

Both formulas reduce to the binary formulas for $x_K = 0$. For the complete estimate of the ternary diffusion for nonideal† liquid mixtures, six diffusivities at infinite dilution are needed and three diffusivities of the type $Ð_{IJ}^{K\to 1}$. Only the three diffusivities $Ð_{IJ}^{K\to 1}$ cannot be obtained from the binary diffusivities. The diffusivity of dilute I and dilute J in pure K has to be estimated. For almost ideal mixtures, Rutten‡ found that the estimate is quite good if the empirical relation

$$Ð_{IJ}^{K\to 1} = \left(Ð_{IJ}^{I\to 1}Ð_{IJ}^{J\to 1}\frac{\sqrt{\eta_I\eta_J}}{\eta_K}\right)^{\frac{1}{2}} \tag{6.128}$$

* Hartley G.S. and Crank J. 1949. Some fundamental definitions and concepts in diffusion processes. *Transactions of the Faraday Society*, 45: 801–818.

† For ideal mixtures the Maxwell–Stefan diffusivities are independent of concentration, and the three diffusivities $Ð_{12}$, $Ð_{13}$, and $Ð_{23}$ describe the system completely. In nonideal liquid mixtures the diffusivities $Ð_{IJ}^{K\to 1}$ and $Ð_{IJ}^{K\to 1}$ are needed for the linear or logarithmic schemes.

‡ Rutten Ph.W.M. 1992. *Diffusion in Liquids.* Delft University Press, Delft. Thesis. See chapter 5.

is used.

The non-electrolyte multicomponent diffusion coefficients are reduced to binary diffusivities which are positive definite, so that the Maxwell–Stefan multicomponent diffusivities are also positive definite. The concentration-independent Maxwell–Stefan diffusion coefficients can always be reduced to a number of binary systems, and so these diffusivities are positive definite. In electrolyte systems a cation is always accompanied by an anion, and a binary system with water and an anion is impossible. Multicomponent electrolyte systems cannot therefore be reduced to a set of binary sytems, but can be reduced to a set of ternary systems. Negative diffusion coefficients can therefore exist in ternary systems and are consistent with the thermodynamics of irreversible processes, if they satisfy the conditions (6.30) and (6.32)*.

6.8. DIFFUSION DESCRIPTION BY AN INDEPENDENT SET OF FLUXES AND FORCES

If the velocity field has to be calculated by means of the momentum balance, then one is quickly convinced that the barycentric velocity is the right choice for the reference velocity for diffusion. Molecular considerations and/or the use of ideal systems lead to mole fractions, so that a molar description with respect to the average mass velocity of the diffusion results. This description of diffusion will be discussed in the next section.

In the kinetic theory of gases† all components are treated on equal footing, and the starting point is the usual choice $\vec{\Phi}_K$ for the fluxes, which are defined by the divergence term in the mass balance equations. The conjugate forces follow then from the dissipation function $(6.5)_2$. This results in the definition of the usual diffusion coefficients and relations between the diffusivities. This approach is appropriate for gases, but for solutions the solvent is considered as special and, a common practice is to transform the dependent set of diffusion fluxes into an independent set of diffusion fluxes.

The molar and mass diffusion coefficients for an independent set of diffusion fluxes are now discussed in this section, where mainly the arguments of De Groot and Mazur‡ will be followed. In this derivation the solvent is considered as a special component, and the solvent diffusion flux will be used to obtain an independent set of diffusion fluxes. In irreversible thermodynamics the dissipation and the diffusion vector, given in boxed equations (6.2) and (6.1), are the starting point if

---

* Kraaijeveld G., Wesselingh J.A. and Kuiken G.D.C. 1994. Comments on 'negative diffusion coefficients'. *Industrial and Chemical Engineering Reseach*, 33(3): 750–751.

† Hirschfelder J.O., Curtiss C.F. and Bird R.B. 1954. *Molecular Theory of Gases and Liquids.* John Wiley & Sons, Inc., New York (fourth printing 1967). See chapter 11.2, p. 704.

‡ De Groot S.R. and Mazur P. 1962. *Non-Equilibrium Thermodynamics.* North-Holland Publ. Co., Amsterdam. See Ch. XI.

(1) There is no external magnetic field present.
(2) The system is not polarized.
(3) No vector internal processes, causing relaxation phenomena, occur.
(4) The system is isotropic.

The last condition implies that the symmetry principle of Curie holds, and that the vector processes can be discussed independently from the scalar and tensor processes. Hence, one can start by combining (6.1) and (6.2) to get

$$\mathfrak{D}^{(\mathrm{v})} = -\frac{1}{T}\vec{\Phi}_{(\mathrm{q})} \cdot \operatorname{grad} T - \sum_{K=1}^{N} \vec{\Phi}_K \cdot \vec{d}'_K \geq 0 \tag{6.129}$$

in which the diffusion vector $\vec{d}'_K = (p/\rho_K)\,\vec{d}_K$ is given by

$$\vec{d}'_K = (\operatorname{grad}\mu_K)_{T,p} + (v_K - v)\operatorname{grad} p + \left(\vec{f}' - \vec{f}'_K\right) \tag{6.130}$$

and the barycentric diffusion flux is given as

$$\vec{\Phi}_K = \rho_K\,(\vec{v}_K - \vec{v}) \tag{6.131}$$

This flux and force combination would also be the choice in a straightforward procedure with the method of the thermodynamics of irreversible processes. The resulting diffusion coefficients are the diffusion coefficients that one usually encounters in the literature. In the derivation of De Groot and Mazur* the component N is considered as the solvent and the solvent diffusion flux is used in setting up linearly independent fluxes and forces, although this is not strictly necessary according to Meixner.

From the definition of $\vec{v}$ it follows that

$$\sum_{K=1}^{N} \vec{\Phi}_K = \vec{0} \tag{6.132}$$

and from the Gibbs–Duhem relation that

$$\sum_{K=1}^{N} \rho_K \vec{d}'_K = \vec{0} \tag{6.133}$$

With (6.132) and (6.133) it is found that

$$\begin{aligned}\sum_{K=1}^{N} \vec{\Phi}_K \cdot \vec{d}'_K &= \sum_{K=1}^{N-1} \vec{\Phi}_K \cdot \vec{d}'_K + \vec{\Phi}_N \cdot \vec{d}'_N \\ &= \sum_{K=1}^{N-1} \vec{\Phi}_K \cdot \vec{d}'_K + \sum_{K=1}^{N-1}\sum_{L=1}^{N-1} \vec{\Phi}_K \cdot \frac{\rho_L}{\rho_N}\vec{d}'_L\end{aligned}$$

---

* De Groot S.R. and Mazur P. 1962. *Non-Equilibrium Thermodynamics.* North-Holland Publ. Co., Amsterdam. See Ch. XI.

The substitution of

$$\vec{X}_K = \sum_{L=1}^{N-1} \left(\delta_{KL} + \frac{\rho_L}{\rho_N} \right) \vec{d}'_L \tag{6.134}$$

which is an independent process force, yields

$$\sum_{K=1}^{N} \vec{\Phi}_K \cdot \vec{d}'_K = \sum_{K=1}^{N-1} \vec{\Phi}_K \cdot \vec{X}_K$$

and for the dissipation (6.129)

$$\mathfrak{D}^{(\mathrm{v})} = -\frac{1}{T}\vec{\Phi}_{(\mathrm{q})} \cdot \operatorname{grad} T - \sum_{K=1}^{N-1} \vec{\Phi}_K \cdot \vec{X}_K \geq 0 \tag{6.135}$$

The coefficients in the definition of the linear independent process forces

$$A_{KL} = \delta_{KL} + \frac{\rho_L}{\rho_N} \tag{6.136}$$

can be considered as a $(N-1) \times (N-1)$ matrix $[A]$ that consists of a diagonal matrix and one row, so that (6.134) can be written as

$$\vec{X}_K = \sum_{L=1}^{N-1} A_{KL} \vec{d}'_L \tag{6.137}$$

or as

$$[\vec{X}] = [A] \cdot [\vec{d}'_L] \tag{6.138}$$

in which $[\vec{X}]$ is a column consisting of $N-1$ vectors $\vec{X}_1, \vec{X}_2, \cdots, \vec{X}_{N-1}$, and $[\vec{d}'_L]$ is a column matrix of the vectors $[\vec{d}'_L]$.

The following phenomenological equations correspond to (6.135)

$$\begin{aligned} \vec{\Phi}_K &= -\sum_{L=1}^{N-1} L''_{KL} \vec{X}_L - L''_{KT} \frac{\operatorname{grad} T}{T} \\ \vec{\Phi}_{(\mathrm{q})} &= -\sum_{L=1}^{N-1} L''_{TL} \vec{X}_L - L''_{TT} \frac{\operatorname{grad} T}{T} \end{aligned} \tag{6.139}$$

Since $\vec{X}_K$ and $\operatorname{grad} T$ are even quantities, the reciprocity relations read

$$L''_{KL} = L''_{LK} \qquad \text{and} \qquad L''_{KT} = L''_{TK} \tag{6.140}$$

When these symmetry properties and (6.139) are used, the dissipation (6.135) becomes

$$\mathfrak{D}^{(\mathrm{v})} = L''_{TT}\left(\frac{\operatorname{grad} T}{T}\right)^2 + 2\sum_{K=1}^{N-1} L''_{KT}\vec{X}_K \cdot \frac{\operatorname{grad} T}{T} + \sum_{K=1}^{N-1}\sum_{L=1}^{N-1} L''_{KL}\vec{X}_K \cdot \vec{X}_L \geq 0 \tag{6.141}$$

The matrix L''_{KL} is therefore symmetrical and positive semidefinite, and the coefficient $L''_{TT} \geq 0$.

If the concentration distributions are homogeneous, all $\vec{d}_K = \vec{0}$, then only thermal diffusion and heat conduction occur. Suppose therefore that

$$\lambda' = L''_{TT}/T \geq 0 \tag{6.142}$$

and

$$L''_{TK} = L''_{KT} = \rho\, D_K^{(T)} \tag{6.143}$$

Elimination of the coefficient of thermal diffusion of the solvent by

$$\sum_{K=1}^{N-1} D_K^{(T)} = -D_N^{(T)} \tag{6.144}$$

and substitution in the phenomenological equations yield

$$\begin{aligned} \vec{\Phi}_K &= -\sum_{L=1}^{N-1} L''_{KL}\vec{X}_L - \rho\, D_K^{(T)}\frac{\operatorname{grad} T}{T} \\ \vec{\Phi}_{(\mathrm{q})} &= -\rho\sum_{L=1}^{N-1} D_L^{(T)}\vec{X}_L - \lambda' \operatorname{grad} T \end{aligned} \tag{6.145}$$

Isothermal diffusion will be considered for the identification of L''_{KL}. In the matrix notation and with use of (6.138) one finds for isothermal diffusion

$$[\vec{\Phi}] = -[L''] \cdot [A] \cdot [\vec{d}\,] \tag{6.146}$$

in which, according to (6.130), and by neglecting the contributions of the non-conservative fields, the diffusion vector is

$$[\vec{d}\,] = (\operatorname{grad}[\mu])_{T,p} + ([v] - v[1]) \operatorname{grad} p \tag{6.147}$$

According to the Gibbs fundamental equation is

$$dg = -s\, dT + v\, dp + \sum_{K=1}^{N-1} (\mu_K - \mu_N)\, dw_K$$

From this equation the following Maxwell relations, among others, are found

$$\mu_{KL} - \mu_{NL} = \mu_{LK} - \mu_{NK} \tag{6.148}$$

with

$$\mu_{KL} = \left(\frac{\partial \mu_K}{\partial w_L}\right)_{T,p} \tag{6.149}$$

The Gibbs–Duhem equation reads

$$\sum_{M=1}^{N} \rho_M \, (d\mu_M)_{T,p} = 0 \qquad \text{or} \qquad \sum_{M=1}^{N-1} \rho_M \mu_{MK} = -\rho_N \, \mu_{NK}$$

Eliminate μ_{NL} and μ_{NK} with this last equation from (6.148)

$$\sum_{M=1}^{N-1} A_{KM}\mu_{ML} = \sum_{M=1}^{N-1} A_{LM}\mu_{MK} \tag{6.150}$$

or

$$[G] = [G]^{\mathrm{T}} \tag{6.151}$$

with

$$[G] = [A] \cdot [\mu] \tag{6.152}$$

For isothermal diffusion, (6.146) and (6.147) give

$$[\vec{\Phi}] = -[L''] \cdot [A] \cdot [\mu] \cdot \operatorname{grad}[w] = -[L''] \cdot [G] \cdot \operatorname{grad}[w]$$

Suppose

$$\rho\,[D] = [L''] \cdot [G] \tag{6.153}$$

so that

$$[\vec{\Phi}] = -\rho\,[D] \cdot \operatorname{grad}[w] \tag{6.154}$$

or

$$\vec{\Phi}_K = -\rho \sum_{L=1}^{N-1} D_{KL} \operatorname{grad} w_L \tag{6.155}$$

These are the $N-1$ barycentric diffusion equations. The barycentric diffusion coefficients (6.153) depend on the composition of the entire system and the concentration of the components, contrary to the concentration-independent Maxwell–Stefan diffusivities. The symmetry relations are also complicated. According to $(6.140)_1$ and (6.151), the barycentric diffusion coefficients have to satisfy

$$[D] \cdot [G]^{-1} = [G]^{-1} \cdot [D]^{\mathrm{T}} \tag{6.156}$$

These are $\frac{1}{2}[(N-1)^2-(N-1)] = \frac{1}{2}(N-1)(N-2)$ relations for the $(N-1)^2$ diffusivities D_{KL}, so that there are $\frac{1}{2}N(N-1)$ independent diffusion coefficients. However, the relations are not as simple so in the case of Maxwell–Stefan diffusivities, and in practice $(N-1)^2$ diffusivities have to be measured.

Sometimes a description with the molar diffusion flux relative to the average mass velocity of the components may be desirable. In this molar description the reference velocity is the barycentric velocity and the composition of the system is expressed in the mole fractions $[x] = \{x_1, x_2, \cdots, x_{N-1}\}$, and (6.155) has to be transformed into a form with the gradients of the mole fractions. Now

$$dw_L = \sum_{M=1}^{N-1} \frac{\partial w_L}{\partial x_M} dx_M \tag{6.157}$$

in which with $x_L = c_L/c$ and $\rho_L = c_L m_L$

$$\frac{\partial w_L}{\partial x_M} = m_L \frac{\partial x_L/m}{\partial x_M} = \frac{m_L}{m}\left[\delta_{LM} - \frac{x_L}{m}\frac{\partial m}{\partial x_M}\right]$$

Since $m = \sum_{M=1}^{N} x_M m_M = \sum_{M=1}^{N-1} x_M (m_M - m_N) + m_N$, it follows that

$$\frac{\partial m}{\partial x_M} = m_M - m_N$$

Substitution gives

$$\frac{\partial w_L}{\partial x_M} = \frac{m_L}{m}\left[\delta_{LM} + x_L \frac{m_N - m_M}{m}\right] \tag{6.158}$$

and (6.157) becomes

$$dw_L = \sum_{M}^{N-1} \frac{m_L}{m}\left[\delta_{LM} + x_L \frac{m_N - m_M}{m}\right] dx_M$$

or

$$d[w] = [B] \cdot d[x] \tag{6.159}$$

with

$$B_{KL} = \frac{m_K}{m}\left(\delta_{KL} + x_K \frac{m_N - m_L}{m}\right) = \frac{\partial w_K}{\partial x_L} \tag{6.160}$$

From (6.154) and (6.159) it follows that

$$[\vec{\Phi}] = -\rho\,[D] \cdot [B] \cdot \operatorname{grad}[x] = -[L''] \cdot [A] \cdot [\mu] \cdot [B] \cdot \operatorname{grad}[x] \tag{6.161}$$

with

$$[\mu] \cdot [B] = \left(\frac{\partial[\mu]}{\partial[w]}\right) \cdot \left(\frac{\partial[w]}{\partial[x]}\right) = \frac{\partial[\mu]}{\partial[x]}$$

Suppose we define $[\mu]'$ by

$$[\mu]' = [\mu] \cdot [B] = \frac{\partial[\mu]}{\partial[x]} \tag{6.162}$$

Then the diffusion fluxes (6.161) become

$$[\vec{\Phi}] = -\rho\,[D''] \cdot \mathrm{grad}\,[x] \tag{6.163}$$

with

$$\rho[D''] = [L''] \cdot [A] \cdot [\mu]' \tag{6.164}$$

It is clear that these diffusivities also depend on the composition of the entire system. The molar diffusion flux with respect to the average barycentric velocity follows in a straightforward manner from (6.163)

$$\vec{\Phi}'_K = \frac{c_K}{\rho_K}\vec{\Phi}_K = -c\frac{m}{m_K}\sum_{L=1}^{N-1} D''_{KL}\mathrm{grad}\,x_L \tag{6.165}$$

De Groot and Mazur* generalize this type of diffusion coefficients. For instance, if in (6.163) the molar fluxes relative to the molar average velocity $\vec{\Phi}^\star_K$ are used and if ρ is replaced by the concentration c, the generalized molar diffusivities with respect to the average molar velocity are defined. These are also called after Fick, and one usually refers to these Fick multicomponent diffusion coefficients in the literature. Obtaining the relation between the Maxwell–Stefan diffusion coefficients and the Fick diffusion coefficients is not an easy task. Further details and applications of the above description of diffusion can be found in De Groot and Mazur.

6.9. MOLAR DIFFUSION RELATIVE TO THE MASS AVERAGE VELOCITY WITH HEAT TRANSPORT

Besides the Maxwell–Stefan descriptions, a number of other descriptions of diffusion can be found in the literature. The mass diffusion flux with respect to the mass average velocity is one of the possible descriptions, and if for this description the mass balance equations are formulated with the substantial derivative, a consistent barycentric description of mass diffusion result. This

* De Groot S.R. and Mazur P. 1962. *Non-Equilibrium Thermodynamics.* North-Holland Publ. Co., Amsterdam. See p. 244.

description and analogous the consistent molar description* are from the theoretical point of view the most elegant.

For the calculations and the measurements in flows the molar diffusion flux relative to the mass average velocity is often implied. For measuring the diffusivities with the Taylor dispersion method†, the barycentric velocity is used in the mass balance equation, and the molar concentration is measured. A square pulse of solution is injected at the entrance of a long capillary. During the flow through the capillary the initial square pulse is flattened, and the diffusivity is measured from the distribution of the solution at the end of the long capillary. In principle, molar diffusion with respect to the mass average velocity is measured in the Taylor dispersion method, but for the very long capillary it turns out that at the end of the capillary $\vec{v} \approx \vec{v}^\star$ and the differences between the molar diffusion with respect to the mass average velocity and molar diffusion with respect to the molar average velocity can be neglected.

From the dissipation function (6.5) it is seen that the following vector process fluxes can be defined

$$\vec{\Phi}_{(q)} \qquad \text{and} \qquad \left\{\vec{\Phi}'_K = c_K \vec{V}_K\right\} \tag{6.166}$$

where the molar diffusion flux with respect to the mass average velocty is given by

$$\vec{\Phi}'_K = c_K \vec{V}_K \tag{6.167}$$

The molar description relative to the mass average velocity is in this section considered, since it is found in practice and it serves as an example for the method used in the thermodynamics of irreversible processes. In the molar diffusion description relative to the mass average velocity, the volume concentration is expressed in moles/m^3, whereas for solving the momentum equations in multicomponent systems, the choice of the barycentric velocity as the reference velocity is convenient.

---

* Merk H.J. 1959. The macroscopic equations for simultaneous heat and mass transfer in isotropic, continuous and closed systems. *Applied Scientific Research*, A8: 73–99.

† Taylor G.I. 1953. Dispersion of soluble matter in solvent flowing slowly through a tube. *Proceedings of the Royal Society*, A219: 186–203; Taylor G.I. 1954. Conditions under which dispersion of a solute in a stream of solvent can be used to measure molecular diffusion. *Proceedings of the Royal Society*, A225: 473–477; Aris R. 1956. On the dispersion of a solute in a fluid flowing through a tube. *Proceedings of the Royal Society*, A235: 67–77; Alizadeh A., Nieto de Castro C.A. and Wakeham W.A. 1980. The theory of the Taylor dispersion technique for liquid diffusivity measurements. *International Journal of Thermophysics*, 1(3): 243–284; Baldauf W. and Knapp H. 1983. Measurements of diffusivities in liquids by the dispersion method. *Chemical Engineering Science*, 38(7): 1031–1037; Price W.E. 1988. Theory of the Taylor dispersion technique for three-component-system diffusion measurements. *Journal of the Chemical Society, Faraday Transactions 1*, 84(7): 2431–2439.

The use of c_K is often practical, particularly if chemical reactions occur and/or the system does not deviate too much from an ideal system. Due to $\sum_{K=1}^{N} \rho_K \vec{V}_K = \vec{0}$, the molar diffusion fluxes with respect to the mass average velocity are linearly dependent

$$\sum_{K=1}^{N} m_K \vec{\Phi}'_K = \vec{0} \tag{6.168}$$

The process forces conjugate to the process fluxes in (6.166) are

$$-T^{-1}\mathrm{grad}\, T \qquad \text{and} \qquad \left\{-p\, \vec{d}_K / c_K\right\}$$

and the phenomenological equations therefore become

$$\begin{aligned} \vec{\Phi}'_K &= -\sum_{L=1}^{N} \frac{p}{c_L} L'_{KL} \vec{d}_L - L'_{KT} T^{-1}\mathrm{grad}\, T \\ \vec{\Phi}_{(\mathrm{q})} &= -L'_{TT} T^{-1}\mathrm{grad}\, T - \sum_{L=1}^{N} \frac{p}{c_L} L'_{TL} \vec{d}_L \end{aligned} \tag{6.169}$$

Because of (6.168) the phenomenological coefficients have to satisfy

$$\sum_{K=1}^{N} m_K\, L'_{KL} = 0 \qquad \text{and} \qquad \sum_{K=1}^{N} m_K\, L'_{KT} = 0 \tag{6.170}$$

Since grad T and $\vec{d}_K$ are even quantities (see Chapter 5), the reciprocal relations read

$$L'_{KL} = L'_{LK} \qquad \text{and} \qquad L'_{KT} = L'_{TK} \tag{6.171}$$

Using that the diffusion vectors ($\sum_{K=1}^{N} \vec{d}_K = \vec{0}$) are not independent of each other, the phenomenological equation $(6.169)_1$ can be written as

$$\vec{\Phi}'_K = {\sum_{L=1}^{N}}' p \left(\frac{L'_{KK}}{c_K} - \frac{L'_{KL}}{c_L} \right) \vec{d}_L - L'_{KT} T^{-1}\mathrm{grad}\, T \tag{6.172}$$

This result leads to the following definitions of the molar multicomponent diffusion coefficients relative to the mass average velocity

$$D'_{KL} = p \left(\frac{L'_{KK}}{x_K} - \frac{L'_{KL}}{x_L} \right) \tag{6.173}$$

so that $D'_{KK} = 0$.

The mechanical diffusion vectors $\vec{d}_K$ are zero for a spatial homogeneous concentration and pressure distribution, but thermal diffusion can occur, as has been discussed in subsection 6.3.2. Then, according to (6.34)

$$\vec{\Phi}'_K = c_K \vec{V}_K = \frac{c_K}{\rho_K}\vec{\Phi}_K = -c\,\frac{\rho}{c}\frac{c_K}{\rho_K} D_K^{(T)} \frac{\operatorname{grad} T}{T}$$

or

$$\vec{\Phi}'_K = -c\,\frac{m}{m_K} D_K^{(T)} \frac{\operatorname{grad} T}{T}$$

so that

$$L'_{KT} = L'_{TK} = c\,\frac{m}{m_K} D_K^{(T)} \tag{6.174}$$

with the same meaning of the multicomponent thermal diffusivities as in subsection 2, where $(6.170)_2$ is in accord with (6.35).

The diffusion flux (6.172) becomes, with (6.173) and (6.174)

$$\vec{\Phi}'_K = c\left[\sum_{L=1}^{N}{}' D'_{KL}\vec{d}_L - \frac{m}{m_K} D_K^{(T)} \frac{\operatorname{grad} T}{T}\right] \tag{6.175}$$

It should be remembered that the accent on the summation sign means that the summation has to be done over all values of L except $L = K$. For a binary system the familiar minus sign in front of the diffusivity is obtained.

Since λ' is the heat conduction coefficient for $\{\vec{d}_K = \vec{0}\}$, the coefficient

$$L'_{TT}/T = \lambda' \tag{6.176}$$

in the equation $(6.169)_2$. When this result is used and with the substitution of (6.174), the heat flux vector $(6.169)_2$ becomes

$$\vec{\Phi}_{(\mathrm{q})} = -\lambda' \operatorname{grad} T - p \sum_{K=1}^{N} \frac{D_K^{(T)}}{w_K}\vec{d}_K \tag{6.177}$$

By substitution of (6.175) and (6.177), the dissipation is found to be

$$\mathfrak{D}^{(\mathrm{v})} = \lambda' T^{-1} (\operatorname{grad} T)^{2\cdot} + 2\frac{p}{T}\sum_{K=1}^{N} \frac{D_K^{(T)}}{w_K}\vec{d}_K \cdot (\operatorname{grad} T) - p \sum_{K=1}^{N}\sum_{L=1}^{N}{}' \frac{D'_{KL}}{x_K}\vec{d}_K \cdot \vec{d}_L \geq 0 \tag{6.178}$$

from which it follows that $\lambda' \geq 0$ (see also (6.44)).

6.9.1. Onsager Casimir reciprocal relations. The reciprocal relations are more complicated for the molar diffusivities with respect to the mass average velocity than for the Maxwell–Stefan diffusivities.

From $L'_{KL} = L'_{LK}$ (6.171) and from $\sum_{K=1}^{N} m_K L'_{KL} = 0$ (6.170) it follows that

$$\sum_{L=1}^{N} m_L L'_{KL} = 0 \tag{6.179}$$

Express D'_{KL} by solving (6.173) for L'_{KL}

$$L'_{KL} = \frac{x_L}{x_K} L'_{KK} - \frac{x_L}{p} D'_{KL} \tag{6.180}$$

and substitution this into (6.179) yields

$$\sum_{L=1}^{N} m_L L'_{KL} = 0 = \frac{m}{x_K} L'_{KK} - \frac{1}{p} \sum_{L=1}^{N} \frac{\rho_L}{c} D'_{KL}$$

or

$$L'_{KK} = \frac{x_K}{p} \sum_{L=1}^{N} w_L D'_{KL}$$

With this (6.180) becomes

$$L'_{KL} = \frac{x_L}{p} \sum_{M=1}^{N} w_M D'_{KM} - \frac{x_L}{p} D'_{KL} \tag{6.181}$$

from which, by using the symmetry relation $(6.171)_1$ it is finally found that

$$x_L D'_{KL} - x_K D'_{LK} = x_L \sum_{M=1}^{N} w_M D'_{KM} - x_K \sum_{M=1}^{N} w_M D'_{LM} \tag{6.182}$$

The Onsager Casimir reciprocal relations are expressed with this result for the molar multicomponent diffusion coefficients with respect to the mass average velocity D'_{KL}.

Equation (6.182) implies $\frac{1}{2}[(N-1)^2 - (N-1)] = \frac{1}{2}(N-1)(N-2)$ relations between the $(N-1)^2$ multicomponent diffusivities D'_{KL}, so that there are at most $\frac{1}{2}N(N-1)$ independent multicomponent diffusivities. This corresponds to the $\frac{1}{2}N(N-1)$ independent Maxwell–Stefan diffusivities $\mathit{Đ}_{KL}$.

Other relations can be derived from (6.181) and $(6.170)_1$, for example

$$\sum_{K=1}^{N} m_K L'_{KL} = \frac{x_L}{p} \sum_{K=1}^{N} \sum_{P=1}^{N} m_K w_P D'_{KP} - \frac{x_L}{p} \sum_{K=1}^{N} m_K D'_{KL} = 0$$

and similarly for L'_{KM}, so that the following expressions

$$\sum_{K=1}^{N}{}' m_K D'_{KL} = \sum_{K=1}^{N}{}' m_K D'_{KM} \tag{6.183}$$

are obtained for each L and M, where the accent is added to the summation sign because $D'_{LL} = 0$.

6.9.2. Transformation of the molar diffusivities with respect to the mass average velocity to the Maxwell–Stefan diffusivities. Since the Maxwell–Stefan diffusivities are generally better known, measured, and/or estimated than the other types of diffusivities in multicomponent systems, it is useful to relate the molar multicomponent diffusivities with respect to the mass average velocity D'_{KL} to those of Maxwell–Stefan. From (6.175) it follows that

$$\vec{V}_K = \frac{1}{x_K} \sum_{L=1}^{N}{}' D'_{KL} \vec{d}_L - \frac{D_K^{(T)}}{w_K} \frac{\operatorname{grad} T}{T}$$

or with (6.38)

$$\vec{V}_K^{\diamond} = \frac{1}{x_K} \sum_{P=1}^{N}{}' D'_{KP} \vec{d}_P \tag{6.184}$$

The diffusion vector (6.37) becomes by substitution and taking account of the term $K \neq L$ in the summation

$$\vec{d}_K = \sum_{L=1}^{N}{}' \frac{1}{Ð_{KL}} \left[\sum_{P=1}^{N}{}' \left\{ (x_K D'_{LP} - x_L D'_{KP}) \vec{d}_P \right\} - x_L D'_{KL} \vec{d}_L \right]$$

or with $\vec{d}_L = -\sum_{P=1}^{\prime N} \vec{d}_P$

$$\vec{d}_K = \sum_{L=1}^{N}{}' \sum_{P=1}^{N}{}' \frac{1}{Ð_{KL}} (x_K D'_{LP} - x_L D'_{KP} + x_L D'_{KL}) \vec{d}_P \tag{6.185}$$

Interchange the order of the summation, and then equation (6.185) is satisfied if

$$\sum_{L=1}^{N}{}' \frac{1}{Ð_{KL}} (x_K D'_{LP} - x_L D'_{KP} + x_L D'_{KL}) = \delta_{KP} \tag{6.186}$$

Replace P by Q in the above equation

$$\sum_{L=1}^{N}{}' \frac{1}{Ð_{KL}} (x_K D'_{LQ} - x_L D'_{KQ} + x_L D'_{KL}) = \delta_{KQ}$$

and subtract (6.186)

$$\sum_{L=1}^{N}{}' \frac{1}{Đ_{KL}} \left[x_K \left(D'_{LQ} - D'_{LP} \right) - x_L \left(D'_{KQ} - D'_{KP} \right) \right] = \delta_{KQ} - \delta_{KP} \qquad (6.187)$$

Now introduce the quantity F_{KL} that is completely expressed in the Maxwell–Stefan diffusivities

$$F_{KL} = \frac{x_K}{Đ_{KL}} \left(1 - \delta_{KL} \right) - \left(\sum_{R=1}^{N}{}' \frac{x_R}{Đ_{KR}} \right) \delta_{KL}, \qquad N > 2 \qquad (6.188)$$

which with the help of (6.187) can be written as

$$\sum_{L=1}^{N} F_{KL} \left(D'_{LQ} - D'_{LP} \right) = \delta_{KQ} - \delta_{KP} \qquad (6.189)$$

The inverse of the matrix $[F_{KL}]$ is defined by

$$\sum_{M=1}^{N} F^{-1}_{KM} F_{ML} = \delta_{KL} \qquad (6.190)$$

With this it follows from (6.189) that $D'_{LQ} - D'_{LP} = F^{-1}_{LQ} - F^{-1}_{LP}$, so that for $L = P$ it is found that

$$D'_{PQ} - D'_{PP} = F^{-1}_{PQ} - F^{-1}_{PP}$$

and with the substitution of $D'_{PP} = 0$ (substitute $K = L$ in (6.173)) one obtains for the diffusion coefficients D'_{KL}

$$D'_{KL} = F^{-1}_{KL} - F^{-1}_{KK} \qquad (6.191)$$

This is the desired result. If the molar diffusivities with respect to the molar average velocity are known, then the Maxwell–Stefan diffusivities can be calculated with, for instance, (6.186).

If we start from the Maxwell–Stefan equations, this method can be applied to other definitions of the multicomponent diffusion coefficients. It can be remarked in this respect that the numerous relations, derived by Hirschfelder et al. for the so called first approximation of the kinetic theory of gases for dilute gases, are easily generalized for general isotropic systems. This has been already noted by Merk*.

* Merk H.J. 1957. Stofoverdracht in laminaire grenslagen door gedwongen convectie. Dissertation, Delft University of Technology (in Dutch).

6.10. SIMPLIFIED DIFFUSION EQUATIONS

The description of the diffusion in systems with more than two components is very complicated. The multicomponent diffusivities for many systems are not well-known or else inaccurately known. For solutes with about the same Van der Waals radii in solvents for almost ideal mixtures the Maxwell–Stefan diffusivities are close to each other. The simplest assumption then is that for these systems the Maxwell–Stefan diffusivities are all equal, or

$$\text{Đ}_{KL} = \text{Đ} \qquad \text{for} \qquad K, L = 1, 2, \cdots, N \tag{6.192}$$

The Maxwell–Stefan binary diffusion coefficients are not equal for mixtures with components of different size and type and for strongly nonideal systems. In those cases it is recognized that $\text{Đ}_{KL} \neq \text{Đ}_{KM}$, and the approximation is not sound. In dilute ideal systems the Maxwell–Stefan diffusivities are equal to the Fick diffusivities $\text{Đ}_{KL} = I\!D_{KL}$. In the following discussion the simplification is not assumed, since we have noticed that the Fick diffusivities are strongly concentration dependent, contrary to the Maxwell–Stefan diffusivities.

With (6.192) it follows from (6.37) that

$$\sum_{L=1}^{N} x_L \left(\vec{V}_L - \vec{V}_K \right) = \frac{\text{Đ}}{x_K} \vec{d}_K + \sum_{L=1}^{N} x_L \left(\frac{D_K^{(T)}}{w_K} - \frac{D_L^{(T)}}{w_L} \right) \frac{\operatorname{grad} T}{T} \tag{6.193}$$

in which the mechanical diffusion vector (6.118) for nonideal systems can be written as

$$\vec{d}_K = \sum_{L=1}^{N} \Gamma_{KL} (\operatorname{grad} x_L)_{T,p} + (\rho_K v_K - w_K) \frac{\operatorname{grad} p}{p} + \rho_K \frac{\vec{f}' - \vec{f}'_K}{p} \tag{6.194}$$

with the introduction of the general thermodynamic correction factor

$$\Gamma_{KL} = x_K \frac{\partial \ln \alpha_K}{\partial x_L} = \delta_{KL} + x_K \frac{\partial \ln \gamma_K}{\partial x_L} \tag{6.195}$$

The thermal diffusion factors $\alpha_{KL}^{(T)}$ can be defined with this assumption by

$$\alpha_{KL}^{(T)} = \left(\frac{D_K^{(T)}}{w_K} - \frac{D_L^{(T)}}{w_L} \right) I\!D_K^{-1} \tag{6.196}$$

where the diffusion coefficients $I\!D_K$ are defined by

$$I\!D_K = \Gamma_{KK} \text{Đ} \tag{6.197}$$

The definition (6.196) reduces for binary systems to the definition of the thermal diffusion factor given by (6.59) with $\Gamma_1 = \Gamma$ and $I\!D_1 = I\!D$.

It is further assumed that the diagonal terms in (6.195) of the matrix of thermodynamic correction factors are dominant, and if (6.196) is substituted, then (6.193) becomes

$$\sum_{L=1}^{N} x_L \left(\vec{V}_L - \vec{V}_K\right) = \frac{D\!\!\!D_K}{x_K}\left(\vec{d}_K^{\star} + \sum_{L=1}^{N} x_K x_L \alpha_{KL}^{(T)} \frac{\operatorname{grad} T}{T}\right) \tag{6.198}$$

in which the molar diffusion vector is given by

$$\vec{d}_K^{\star} = (\operatorname{grad} x_K)_{T,p} + (\rho_K v_K - w_K)\frac{\operatorname{grad} p}{p} + \rho_K \frac{\vec{f}' - \vec{f}_K'}{p} \tag{6.199}$$

A direct expression for the mass flux is needed, expressed in the mole fractions or in the mass fractions, in order to solve diffusion equations. From the definition of the mole fraction it follows that

$$\sum_{L=1}^{N}{}' x_L = 1 - x_K$$

so that the left-hand side of (6.198) can be written as

$$\sum_{L=1}^{N} x_L \left(\vec{V}_L - \vec{V}_K\right) = \sum_{L=1}^{N} x_L \vec{V}_L - \vec{V}_K$$

From the definitions (4.48) of the barycentric velocity, the barycentric diffusion velocity (4.52) and the molar fluxes and velocities

$$\vec{\Phi}_K^{\star} = c_K\left(\vec{v}_K - \vec{v}^{\star}\right) \qquad \text{with} \qquad \vec{v}^{\star} = \sum_{K=1}^{N} x_K \vec{v}_K \tag{6.200}$$

it follows that

$$\vec{v}^{\star} = \vec{v} + \sum_{L=1}^{N} x_L \vec{V}_L$$

By substitution it is seen that the following relations hold

$$\sum_{L=1}^{N} x_L \left(\vec{V}_L - \vec{V}_K\right) = \vec{v}^{\star} - \vec{v}_K = -\frac{1}{c_K}\vec{\Phi}_K^{\star}$$

by means of which (6.198) can be written as

$$\vec{\Phi}_K^{\star} = -c D\!\!\!D_K \left(\vec{d}_K^{\star} + \sum_{L=1}^{N} x_K x_L \alpha_{KL}^{(T)} \frac{\operatorname{grad} T}{T}\right) \tag{6.201}$$

This result is completely analogous to (6.76) for a binary nonideal system and is obtained by assuming that all Maxwell–Stefan diffusivities are equal, and that the diagonal terms in the matrix of the generalized thermodynamic factor are dominant. The molar mass flux $\vec{\Phi}^\star_K$ is then as a matter of fact determined only by its own molar diffusion vector $\vec{d}^\star_K$ and in isothermal processes does not depend on the mole fractions x_L.

To arrive at an expression analogous to (6.201) for the barycentric mass flux it is necessary to suppose that the non-conservative force field can be neglected. In addition, in (6.199) the contribution $\rho_K v_K$ in the term with grad $\ln p$ can be simplified only if the components of the system behave as ideal gases, for which $\alpha_K = x_K$. For ideal systems it is true that $\rho_K v_K = x_K$, so that

$$
\begin{aligned}
\rho_K v_K - w_K = x_K - w_K &= x_K\left(1 - \frac{m_K}{m}\right) \\
&= \sum_{L=1}^{N} x_K\left(\frac{c_L m_L}{cm} - x_L\frac{m_K}{m}\right) \\
&= \sum_{L=1}^{N} x_K x_L \alpha^{(p)}_{KL}
\end{aligned}
\tag{6.202}
$$

For ideal multicomponent systems and with the neglect of possible non-conservative force fields, the diffusion vector (6.199) also becomes similar to the molar diffusion vector (6.69) for binary systems

$$
\vec{d}^\star_K = (\operatorname{grad} x_K)_{T,p} + \sum_{L=1}^{N} x_K x_L \alpha^{(p)}_{KL}\,\frac{\operatorname{grad} p}{p}
\tag{6.203}
$$

where now the pressure diffusion factors are defined by

$$
\alpha^{(p)}_{KL} = -\frac{m_K - m_L}{m}
\tag{6.204}
$$

This definition of the pressure diffusion factors is again similar to the definition (6.65) for a binary system.

Formula (6.201) is appropriate for applications in the molar description of the diffusion. For ideal systems it reduces to (6.203). The diffusion flux (6.201) has to be transformed for the barycentric description of the diffusion. This transformation can be done for ideal systems with the inclusion of the pressure diffusion. For nonideal systems the pressure diffusion has to be neglected. Now

$$
\begin{aligned}
&\sum_{L=1}^{N} x_L m_L \vec{\Phi}^\star_K - \sum_{L=1}^{N} x_K m_L \vec{\Phi}^\star_L \\
&\qquad = \sum_{L=1}^{N} \frac{\rho_L}{c} c_K\left(\vec{v}_K - \vec{v}^\star\right) - \sum_{L=1}^{N} x_K m_L c_L\left(\vec{v}_L - \vec{v}^\star\right) \\
&\qquad = x_K\rho\left(\vec{v}_K - \vec{v}^\star\right) - x_K\rho\left(\vec{v} - \vec{v}^\star\right) = \frac{m}{m_K}\vec{\Phi}_K
\end{aligned}
\tag{6.205}
$$

so that

$$\vec{\Phi}_K = \frac{m_K}{m} \sum_{L=1}^{N} m_L \left(x_L \vec{\Phi}_K^\star - x_K \vec{\Phi}_L^\star \right) \tag{6.206}$$

With the relations $w_K = m_K x_K / m$ and $m = \rho/c = \sum_{L=1}^{N} m_L x_L$ it follows that

$$\frac{\rho}{c}\, dw_K = \frac{m_K}{m} \sum_{L=1}^{N} m_L \left(x_L\, dx_K - x_K\, dx_L \right) \tag{6.207}$$

by which result for the pressure diffusion vector

$$\vec{d}_K^\circ = (\operatorname{grad} w_K)_{T,p} = \frac{c}{\rho}\frac{m_K}{m} \sum_{L=1}^{N} m_L \left(x_L \vec{d}_K^\star - x_K \vec{d}_L^\star \right) \tag{6.208}$$

analogous to (6.67) with neglect of possible non-conservative force fields and the pressure diffusion. Substitution of (6.201) into (6.206) yields

$$\vec{\Phi}_K = -\rho I\!\!D_K \left(\vec{d}_K^\circ + \sum_{L=1}^{N} w_K w_L \alpha_{KL}^{(T)} \frac{\operatorname{grad} T}{T} \right) \tag{6.209}$$

in which $\vec{d}_K^\circ$ is now given by (6.208) and where use is made of the following relation from (6.196)

$$\alpha_{KM}^{(T)} - \alpha_{LM}^{(T)} = \alpha_{KL}^{(T)}$$

With this (6.209) is completely analogous to (6.66). This simplified equation can be used for barycentric descriptions of diffusion. By neglecting the thermal diffusion the barycentric diffusion flux (6.209) can be simplified further to

$$\vec{\Phi}_K = -\rho I\!\!D_K \operatorname{grad} w_K \tag{6.210}$$

The underlying assumptions made in the derivation of this equation can be summarized as follows

(1) The Maxwell–Stefan diffusivities are equal.
(2) The diagonal terms in the thermodynamic correction factor matrix are dominant.
(3) Only conservative forces are applied to the system.
(4) The pressure diffusion is negligible.
(5) The thermal diffusion is negligible.

If (6.210) applies for the mass diffusion flux relative to the mass average velocity, then the macroscopic equations are very simple. Substitution of (6.210) into (4.56) yields

$$\rho \frac{\mathrm{D} w_K}{\mathrm{D}t} = -\operatorname{div} \left(\rho I\!\!D_K \operatorname{grad} w_K \right) + \pi_K \tag{6.211}$$

The transformation to the molar description with respect to the mass average velocity of (6.209) is possible only if the thermal diffusion is also neglected. With (6.210) and (6.207) it follows then that in the molar description with respect to the mass average velocity all mole fractions play a role. The simplified multicomponent diffusion law in the molar description with respect to the mass average velocity becomes

$$\vec{\Phi}'_K = -c\mathbb{D}_K \left(\operatorname{grad} x_K - x_K \sum_{L=1}^{N} \frac{m_L}{m} \operatorname{grad} x_L \right) \tag{6.212}$$

Since now the other mole fractions are also found in this diffusion flux, the resulting diffusion equations are now not decoupled as in the set of diffusion equations (6.211).

In the application of (6.211) for mass transfer problems with chemical reactions, the source term π_K may be eliminated in a manner indicated by Spalding*. He assumed that all diffusion coefficients are equal, $\mathbb{D}_K = \mathbb{D}$, and eliminated the source term as follows. For a chemical reaction the mass of a chemical element α is conserved. If $\gamma_{\alpha,K}$ is the weight fraction of element α in component K, then

$$\sum_{K=1}^{N} \gamma_{\alpha,K} \pi_K = 0 \tag{6.213}$$

By the introduction of

$$P_\alpha = \sum_{K=1}^{N} \gamma_{\alpha,K} w_K \tag{6.214}$$

it follows by summing of (6.211) over K and using (6.213) that

$$\rho \frac{\mathrm{D}P}{\mathrm{D}t} = -\operatorname{div}\,(\rho D \operatorname{grad} P) \tag{6.215}$$

in which P is a linear combination of the P_α's, for instance

$$P = AP_\alpha + BP_\beta + \cdots \tag{6.216}$$

By this line of reasoning it is possible to calculate the mass transport without the use of chemical kinetics. Spalding called quantities like P conserved quantities.

Merk† has discussed in his dissertation the boundary layer approximation of the diffusion equations and the mass transfer coefficients and has shown

---

* Spalding D.B. 1954. The calculation of mass transfer rates in absorption, vaporization, condensation and combustion processes. *Proceedings of the Institute of Mechanical Engineers*, 168: 545–567.

† Merk H.J. 1957. Stofoverdracht in laminaire grenslagen door gedwongen convectie. Dissertation, Delft University of Technology (in Dutch).

that for this application the pressure diffusion is negligible. For a description of other applications the textbook of Bird, Stewart and Lightfoot* can be referred to. For calculations with the Maxwell–Stefan description of the diffusion the books written by Krishna, Taylor and Wesselingh† may be consulted. Many applications can be found in this literature, together with algorithms to solve problems.

6.11. EXERCISES

Exercise 6.1. Discuss whether it is possible in a ternary mixture for two of the three Maxwell–Stefan diffusion coefficients to be negative.

Exercise 6.2. The Maxwell–Stefan diffusion has been derived in this chapter by considering the diffusion vector as a process flux, whereas the diffusion equation defines the barycentric diffusion flux as the process flux and from the dissipation the diffusion vector is the process force. Derive the flux equations analogously to the method used in section 6.9 for the molar diffusion flux relative to the average mass velocity. Invert‡ the equations and obtain the Maxwell–Stefan description of diffusion as is given in the equations (6.37).

Exercise 6.3. For quasi-stationary low Reynolds number flow the Navier-Stokes equations reduce to

$$\vec{0} = \rho\vec{f} - \operatorname{grad} p + \eta\Delta\vec{v}$$

and the diffusion coefficient for the momentum is given by η/ρ. Suppose that D denotes a characteristic diffusion coefficient for the mass transport. Show that the velocity gradients approach zero much faster than the concentration gradients if the Schmidt number

$$Sc = \eta/(\rho D) \gg 1$$

For solutions in liquids $Sc > 1$ and for gases it is $\mathcal{O}(1)$. Discuss whether the fluid friction term in the Navier-Stokes equation might be neglected far away from fixed walls and whether only then is mechanical equilibrium of the form

$$\vec{0} = \rho\vec{f} - \operatorname{grad} p$$

obtained.

---

* Bird R.B., Stewart W.E. and Lightfoot E.N. 1960. *Transport Phenomena*, John Wiley & Sons, New York, 1960. See chapter 22.

† Krishna, R. and Taylor R. 1986. Multicomponent mass transfer: theory and applications. In the *Handbook for Heat and Mass Transfer Operations*, Cheremisinoff N.P. (ed.), Gulf Publ. Corp., Houston, TX, Vol. II, Ch. 7, pp. 259–432; Wesselingh J.A. and Krishna R. 1990. *Mass Transfer*. Ellis Horwood, Ltd., Chichester; Taylor R. and Krishna R. 1993. *Multicomponent Mass Transfer*. John Wiley & Sons, New York.

‡ Merk H.J. 1959. The macroscopic equations for simultaneous heat and mass transfer in isotropic, continuous and closed systems. *Applied Scientific Research*, A8: 73–99.

Exercise 6.4. Discuss whether mechanical equilibrium can be obtained for viscous flows. Show that the Navier-Stokes equation in an inertial system

$$\rho\frac{\mathrm{D}\vec{v}}{\mathrm{D}t} = \rho\,\vec{f} - \operatorname{grad} p + \operatorname{grad}\,(\kappa\operatorname{div}\vec{v}) + \operatorname{div}\,(\mu\boldsymbol{D})$$

becomes in a system that rotates with a constant angular velocity $\vec{\Omega}$ ralative to the inertial system

$$\rho\frac{\mathrm{D}\vec{v}}{\mathrm{D}t} = \rho\,\vec{f} - 2\rho\,\vec{\Omega}\times\vec{v} - \rho\,\vec{\Omega}\times\left(\vec{\Omega}\times\vec{x}\right) - \operatorname{grad} p + \operatorname{grad}\,(\kappa\operatorname{div}\vec{v}) + \operatorname{div}\,(\mu\boldsymbol{D})$$

Discuss whether mechanical equilibrium can be obtained for an instationary flow in a rotating system if

$$Ro = \text{Rossby number} = \frac{\rho\vec{v}\cdot\operatorname{grad}\vec{v}\,|}{|\rho\,\Omega\times\vec{v}\,|} \approx \frac{V}{L\Omega} \ll 1$$

with V being a characteristic velocity of the relative system.

Discuss that the influence of the viscosity in mechanical equilibrium can be neglected if

$$\frac{|\mu\Delta\vec{v}\,|}{|\rho\,\vec{\Omega}\times\vec{v}\,|} \approx \frac{Ro}{Re} \ll 1$$

and that it is then not necessary that the Reynolds numbers $Re \ll 1$.

Exercise 6.5. Use mechanical equilibrium to show that a gravitational field produces pressure diffusion.

7 Rheology

Summary: The phenomenological responses of materials with relaxation are formulated with scalar, vector, and tensor internal processes. Elastic, anelastic and viscous types of deformation of the material do not all contribute to the thermodynamic potentials, the entropy production or the dissipation. After the formulation of the basic constitutive equations, the three types of internal processes are explored. Relaxation and retardation of heat waves are found for internal vector processes. In the discussion of the scalar processes the thermodynamic coefficients for both equilibrium and constrained processes are defined, and the dynamical equations are formulated. The dissipation function and the energy function are derived for the tensor processes found in linear viscoelastic materials. The four standard classes of linear rheological propertiess with discrete spectra are supplemented by degenerate classes possible for linear bodies with continuous spectra.

7.1. PHENOMENOLOGICAL EXPERIMENTS

The thermodynamics of irreversible processes is based, as every physical theory, on the observation of certain limited aspects of nature. The following discussions are based on responses in mechanical recovery experiments.

Suppose that a body or a system has a stress free state, in which the body or the system is in equilibrium. Such a state is chosen as the reference state. The body is loaded from the reference state. After some time the load is released, and the deformation is observed as a function of time. After the removal of the load, it appears that the body tries to return to the reference state. The way this happens, and the extent to which the material recovers, not only differ from material to material, but they are often also determined by the magnitude of the applied load and the geometrical configuration in the reference state of the material tested. For certain typical loadings the materials can be classified according to their behavior during recovery experiments.

To clarify the point, simple recovery experiments with a general body are considered, which are described by one component of the deformation, for example: uniaxial stretch, simple shear, isotropic volume changes. The deformation is then described by a scalar ϵ, different for the various types of simple recovery experiment, in which the deformation is defined with respect to the reference state.

Suppose that the load is released at the time $t = 0$. From numerous simple recovery experiments it has been shown that for $t > 0$ the deformation can be split in four parts:

(1) **Elastic part** $\epsilon^{(e)}$: this part of the deformation recovers instantaneously and is therefore reversible and is not accompanied by dissipation.
(2) **Anelastic part** $\epsilon^{(a)}$: this part of the deformation recovers, provided that one waits long enough. This means that the anelastic part recovers with finite velocity. According to Planck every natural process that proceeds without external actions is irreversible. The anelastic deformation is accordingly irreversible and dissipative.
(3) **Viscous part** $\epsilon^{(v)}$: this part of the deformation is not recovered at all and is therefore irreversible and dissipative. Viscous deformations are not determined by the magnitude of the load or the geometrical configuration of the material.
(4) **Plastic part** $\epsilon^{(p)}$: this part of the deformation is not recovered, but is found only if the load exceeds a certain critical magnitude.

The definition of the anelastic deformation in this list is strictly followed in this book. The notion of 'anelasticity' has been introduced by Zener*. Viscous and/or plastic deformations are not here classified as anelastic, as is done by some authors.

The process of the deformation in a recovery experiment† can be denoted by

$$\epsilon(t) = \epsilon^{(e)}\,[1 - H(t)] + \epsilon^{(a)}(t) + \epsilon^{(v)} H(t) \tag{7.1}$$

with

$$\lim_{t\to\infty} \epsilon^{(a)}(t) = 0 \tag{7.2}$$

in which $H(t)$ is the unit step function of Heaviside. The deformation is sketched in figure 7.1. The viscous contribution $\epsilon^{(v)}$ in (7.1) can be replaced by the plastic contribution $\epsilon^{(p)}$ if needed.

From the results of recovery experiments it follows that four fundamental rheological properties can be found, namely *elasticity*, *anelasticity* (or after-effect, or memory-effect), *viscosity* and *plasticity*.

It is found in many recovery experiments that $\epsilon^{(a)}$ after removal of the load decreases monotonically with time. If this happens for each type of load, then, by generalizing, it can be assumed that the anelastic deformation is a complete monotonic function of time, that is

$$(-1)^n \frac{d^n}{dt^n}\epsilon^{(a)}(t) \geq 0 \qquad \text{for} \qquad n = 0, 1, 2, \cdots \tag{7.3}$$

---

* Zener C. 1952. *Elasticity and Anelasticity of Metals.* University of Chicago Press, Chicago, Illinois.

† Recovery experiments have been introduced by Jenckel E. and Ueberreiter K. 1938. Über Polystyrolgläser verschiedener Kettenlänge. *Zeitschrift für physikalische Chemie*, A182: 361–383.

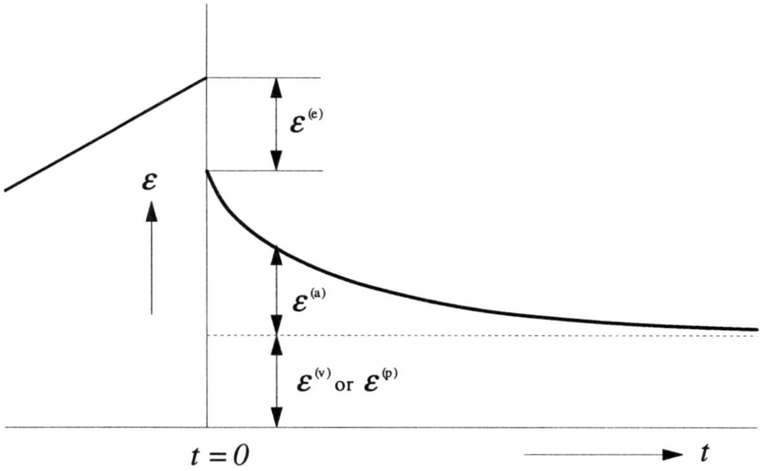

Figure 7.1. Sketch of the deformation ϵ in a recovery experiment. The elastic part of the deformation is denoted by $\epsilon^{(e)}$, the anelastic part by $\epsilon^{(a)}$, the viscous part by $\epsilon^{(v)}$, and the plastic part by $\epsilon^{(p)}$.

Processes with this monotonic behavior are called *creep processes*. Creep processes are caused by micro-processes or molecular processes, in which the mass inertia effects are neglected. It is clear that the neglect of mass inertia effects is not exactly correct for very short times after the load is taken off.

The fundamental behavior of materials can also be demonstrated by other experiments. Suppose that for $t < 0$ the body is in a stress free equilibrium state (the reference state), then two dual experiments can be distinguished

Static retardation experiments: At time $t = 0$ a known load is applied and kept constant for $t > 0$. The material strives for an equilibrium with the applied load, and the deformation ϵ is observed as a function of time. 'Static' means here that the load is kept constant for $t > 0$.

Static relaxation experiments: At time $t = 0$ a deformation is imposed, which is kept constant for $t > 0$. Now the stress is studied as a function of time.

Such experiments can also be executed "dynamically" by *small amplitude sinusoidal experiments*, in which the applied loads or deformations are harmonic functions of time. For the following discussion the above-mentioned four types of deformation observed with recovery experiments are sufficient.

7.2. RHEOLOGICAL MODELS OR IDEAL MODELS

7.2.1. Classification of non-plastic bodies. Experience teaches that the above-mentioned rheological properties (elasticity, anelasticity, viscosity, plasticity) occur in various degrees in different materials. This experience leads to the definition of models, that is of ideal bodies, in which only certain properties are found. The 'idealization' is a simplification of reality in which one concentrates attention on effects that are supposed to be significant for the relevant process and by neglecting secondary effects.

It is of course the intention that a 'model' or 'ideal body' can be realized in nature with sufficient accuracy (that is within the alleged accuracy of measurement). The success of this assumption is determined not only by the nature of the material, but also—above all—by the circumstances under which the material is studied. The realization of a model therefore concerns not only the specification of the materials, but also the external circumstances.

Consider now for the moment only materials that have no noticeable plastic properties. From the rheological point of view these materials have only the properties elasticity, anelasticity, and viscosity. Non-plastic bodies may now be classified according its decreasing resistance to a load, as follows

(1) **Rigid bodies**: $\epsilon^{(e)} = 0$, $\epsilon^{(a)} = 0$, $\epsilon^{(v)} = 0$.
These bodies are *rigid* under all kinds of loads, this means that no deformations occur in the body. It is clear that this is an approximation of the 'zeroth' order of the behavior of solid bodies, which is unrealistic if the load is sufficiently high and/or the measurements are performed with increased accuracy.

(2) **Elastic bodies**: $\epsilon^{(e)} \neq 0$, $\epsilon^{(a)} = 0$, $\epsilon^{(v)} = 0$.
The deformations of these bodies are reversible. Clearly this is a 'first' approximation of the behavior of solid bodies, and this approximation will fail if the deformations proceed sufficiently rapidly.

(3) **Anelastic bodies**: $\epsilon^{(e)} = 0$, $\epsilon^{(a)} \neq 0$, $\epsilon^{(v)} = 0$.
Kelvin* remarked that 'no change of volume or shape can be produced in any kind of matter without dissipation of energy'. If the deformations in solid bodies proceed sufficiently fast, then the deformations have to be dissipative. This thought corresponds to that of Planck. Kelvin probably introduced for the first time the notion of anelasticity, as it is here defined. He imagined that the material exists as an elastic matrix (skeleton), of which the pores are filled with a liquid.

(4) **Firmo-viscous bodies**: $\epsilon^{(e)} \neq 0$, $\epsilon^{(a)} \neq 0$, $\epsilon^{(v)} = 0$.
These bodies are inspired by the observation that numerous bodies show after the removal of the load an instantaneous recovery of the deformation in recovery experiments. The accuracy of the measurement plays a major part in these findings.

(5) **Viscoelastic bodies**: $\epsilon^{(e)} \neq 0$, $\epsilon^{(a)} \neq 0$, $\epsilon^{(v)} \neq 0$.
These bodies show viscous flow phenomena. Examples of this class

---

* Kelvin, Thompson Sir W. 1890. *On the Elasticity and Viscosity of Metals. Mathematical and Physical Papers*, III. Cambridge University Press, Cambridge. See p. 27.

are solid materials, consisting of linear polymers. With respect to fast varying loads these materials behave almost as elastic, while for slowly varying loading this material behaves almost as a viscous fluid. The time scales of the imposed stresses or strains are very important for the classification of a material.

(6) **Elastic fluids**: $\epsilon^{(e)} = 0$, $\epsilon^{(a)} \neq 0$, $\epsilon^{(v)} \neq 0$.

Bodies with $\epsilon^{(e)} = 0$, $\epsilon^{(v)} \neq 0$ can be called fluids. Examples of this class are molten polymers, polymer solutions, suspensions, and emulsions.

(7) **Elastico-viscous fluids**: $\epsilon^{(e)} \neq 0$, $\epsilon^{(a)} = 0$, $\epsilon^{(v)} \neq 0$.

These bodies were introduced by Maxwell* in his theory of the dynamics of gases. According to Maxwell, elastic effects have to occur in gases which for a given deviatoric deformation cause relaxation of the deviatoric stresses, such that for constant deviatoric deformation the deviatoric stresses finally disappear. This point of view is confirmed by the kinetic theory of gases.

(8) **Viscous fluids**: $\epsilon^{(e)} = 0$, $\epsilon^{(a)} = 0$, $\epsilon^{(v)} \neq 0$.

These are the 'normal' fluids with a relatively simple molecular structure. The ability to flow is already substantial for viscous fluids.

(9) **Ideal fluids**: these are the most mobile bodies and they are incapable of sustaining deviatoric stresses at finite deviatoric strains. Models that cannot sustain any stresses, have no physical meaning.

It is remarked that in the given nomenclature, Jeffreys† called the Maxwell bodies (i.e. '7') elastico-viscous in agreement with the list above, but he called the Kelvin bodies (i.e. '3') firmo-viscous, while we called these bodies anelastic.

7.2.2. Unification axioms of rheology. Reiner‡ postulated as the *first rheological axiom* that 'under isotropic pressure all materials behave in the same way; they are purely and simply elastic'. There are many exceptions to this axiom, as has been discussed by Reiner in his book. The axiom should have been formulated as follows: the volume deformations of all materials always contain an elastic part that under certain circumstances can be approximated by a rigid part. Even dilute gases satisfy this axiom. This means that $\epsilon^{(e)} = 0$ is related only to the deviatoric part of the deformation tensor.

As a *second rheological axiom* Reiner§, proposed that 'In reality ... every material possesses all rheological properties, although in varying degrees'. This is sometimes called the unification axiom. All materials possess in

---

* Maxwell J.C. 1868. On the dynamical theory of gases. *Philosophical Magazine*, 35: 129–145, 185–217. See also Maxwell J.C. 1867. On the dynamical theory of gases. *Philosophical Transactions of the Royal Society of London*, A157: 49–88.

† Jeffreys H. 1924. *The Earth.* Cambridge University Press, Cambridge (sixth edition 1976).

‡ Reiner M. 1960. *Deformation, Strain and Flow.* H.K. Lewis & Co., London. See p. 4.

§ Reiner M. 1960. *loc. cit.*, p. 11.

principle elastic, anelastic and viscous properties, so that the above given classification seems to be senseless. However, experience teaches that under certain conditions the rheological behavior of any material can be well described by one of the above-mentioned nine classes of behavior. Apparently, the 'conditions' are important, and they have to be specified in some way. This may be done by means of the following procedure.

When a body is deformed, the material points of a body move with respect to each other. This is possible only if in the material *internal processes* occur on a micro scale or on a molecular scale. An internal process α has in principle a *natural time* $t_{(\mathrm{n})}^{(\alpha)}$ that is a measure of the time needed for the internal process α to adjust to a new equilibrium after a change of the macroscopic conditions. The natural time $t_{(\mathrm{n})}^{(\alpha)}$ is therefore a measure of the time delay for the re-establishment of the equilibrium resulting from the internal process α.

Macroscopically, the internal processes with a delayed establishment of equilibrium can be interpreted as an after-effect (or memory-effect). The unification-axiom can now be formulated as follows: in all materials internal processes occur with a retarded return to equilibrium after a deformation; the materials differ in the nature of the internal processes, especially with respect to the number of internal processes and natural times.

Hence, the classification of the rheological bodies has to be related to the natural times. Consider first an elastic body. After freeing this body from all types of loadings it returns instantaneously to the reference state, so that it can be said that the *memory* of elastic bodies is infinitely long. For all internal processes involved in the deformation of an elastic body, it therefore holds that $t_{(\mathrm{n})}^{(\alpha)} = \infty$. As always, the statement '$t_{(\mathrm{n})}^{(\alpha)}$ is very large' is a relative judgment, related to the time scale of the observer. Suppose that the time scale of the macroscopic observer is given by $t^{(\mathrm{o})}$. This time scale can be considered as the interval between two macroscopic observations. From a continuous observation with chart recorders, the macroscopic time scale can be estimated by

$$t^{(\mathrm{o})} \sim f/(df/dt)$$

if f is the displacement of the chart.

A material is elastic if

$$t^{(\mathrm{o})} \ll \forall\, t_{(\mathrm{n})}^{(\alpha)} \qquad \text{(elastic)} \tag{7.4}$$

This means that the material behaves as an elastic body for an observer with the time scale $t^{(\mathrm{o})}$.

An anelastic body is perceived by an observer as a body with a delayed return to the equilibrium state, so that now

$$t^{(\mathrm{o})} \sim \forall\, t_{(\mathrm{n})}^{(\alpha)} \qquad \text{(anelastic)} \tag{7.5}$$

A viscous body never returns to its original reference configuration. Within the time scale of the observer such a body does not seem to have any memory of previous states. The body is considered viscous for

$$t^{(o)} \gg \forall\, t^{(\alpha)}_{(n)} \qquad \text{(viscous)} \tag{7.6}$$

It can also happen that the natural times of the various internal processes show spreading that can be distinguished for instance into three groups concentrated around $t^{(e)}$, $t^{(a)}$, $t^{(v)}$, respectively, with

$$t^{(v)} \ll t^{(a)} \ll t^{(e)}$$

For

$$t^{(v)} \ll t^{(o)} \approx t^{(a)} \ll t^{(e)} \tag{7.7}$$

the material behaves as a viscoelastic body for the observer with the time scale $t^{(o)}$. For $t^{(o)} \ll t^{(v)}$ the same material behaves as an elastic body, and by contrast for $t^{(o)} \gg t^{(e)}$ it behaves as a viscous body.

The classification of a material is therefore determined by the time scale of the macroscopic observation and the distribution of the natural times that are specific for a definite material. In general, the natural times depend on macroscopic quantities like temperature and pressure, so that the classification of a material can depend on the macroscopic circumstances insofar as these conditions influence the natural times of the internal processes.

From the discussion above, it follows that *every* material behaves as a viscous body at any load—however tiny—if $t^{(o)}$ is large enough. All materials subjected to gravitation will flow eventually. For some materials one has to wait a very long time, but this does not alter the principle. This thought was particularly advocated by Reiner under the motto *panta rhei*, attributed to Heraclitus*.

Reiner† postulated as the *third rheological axiom*: 'There is a hierarchy of ideal bodies, corresponding to the different rheological behavior of real materials, such that the rheological equation of the simple body (lower in the hierarchy) can be derived by putting one or the other of the constants of the rheological equation of the less simple body (higher in the hierarchy) equal to zero'. This means that bodies higher in the hierarchy of the ideal bodies include more internal processes.

The lowest bodies in the Reiner hierarchy are: (1) rigid bodies and ideal fluids, followed by (2) elastic bodies and viscous fluids, (3) anelastic bodies and elastic-viscous fluids, (4) firmo-viscous bodies and elastic fluids, and finally, (5) viscoelastic bodies, which are the highest bodies in the Reiner hierarchy.

---

* Quotation of Heraclitus (540–480 B.C.) by Plato as propagated by Cratylus.

† Reiner M. 1960. *Deformation, Strain and Flow.* H.K. Lewis & Co., London. See p. 124.

7.2.3. Plastic bodies. The existence of plasticity suggests the generalizing thought that there have to exist bodies, for which the response depends on the load. In first approximation one might think of materials of which the behavior on a macroscopic scale changes abruptly if the load exceeds a certain critical value. Critical value may be still too vague. It can be imagined that in six-dimensional space (the stress tensor is symmetrical) formed with the values of the stress tensor, five dimensional surfaces exist that separate regions in which the material has a different rheological behavior. Such a surface can be called a critical stress surface. Considering that internal mechanisms are responsible for the rise of deformations and internal stresses, a particular system of internal mechanisms is related to each behavior. If to such a system one or more internal mechanisms are added or removed, then another rheological behavior results. It is possible that in a body certain internal mechanisms are frozen, for example by steric hindrances resulting from the surroundings. If the load passes a critical stress surface, the hindering of one or more frozen internal mechanisms is removed, so that the release of those internal mechanisms induces processes that proceed with finite velocity. The internal structure of a material changes in passing a critical stress surface in the sense that in the regions on both sides of a critical stress surface different rheological constitutive equations apply.

It is here remarked that the constraints of the internal mechanisms are influenced not only by the mechanical load, but also by thermodynamic quantities like pressure and temperature. Plasticity (i.e., change of the rheological behavior by mechanical loads) is in fact the fourth fundamental element of rheology besides elasticity, anelasticity, and viscosity.

Plastic bodies are special cases of the above discussed bodies. Suppose that there is one critical stress surface, which divides the stress space into an inner region with 'small' stresses and an outer region with 'large' stresses (the outer region contains the infinity of the stress space). The body is plastic, if for the stresses in the inner region $\epsilon^{(v)} = 0$ applies, and for the stresses in the outer region $\epsilon^{(v)} \neq 0$ (now it is assumed that $\epsilon^{(v)} = \epsilon^{(p)}$). The inner region is the preplastic region and the outer region is the plastic region in which the material flows. The critical stress surface is now called the yield surface.

If the state of stress passes a yield surface, then the constraint of at least one internal process with very short natural time is relieved. In rheology this is sometimes spoken of 'breaking' the structure of a solid body, so that the body can flow plastically.

Evidently, it is also possible that the structure of a material can change gradually, so that there is not a sharp change but moreover a critical zone on both sides of the stress surface. In the literature concentration is particularly focused on plastic materials with one yield surface, the shape and the magnitude of which may depend on the prehistory of the material (deformation, work hardening, Bauschinger effect).

Table 7.1. Outline of the rheological material classes. The names in the preplastic and the plastic region refer to the non-plastic type of body.

Non-Plastic Bodies				
Type of body	Representative	$\epsilon^{(e)}$	$\epsilon^{(a)}$	$\epsilon^{(v)}$
1. **Rigid**	Euclid	0	0	0
2. **Elastic**	Hooke	$\neq 0$	0	0
3. **Anelastic**	Kelvin–Voigt	0	$\neq 0$	0
4. **Firmo-viscous**	Poynting–Thomson	$\neq 0$	$\neq 0$	0
5. **Visco-elastic**	Burgers	$\neq 0$	$\neq 0$	$\neq 0$
6. **Elastic fluids**	Jeffreys	0	$\neq 0$	$\neq 0$
7. **Elastico-viscous**	Maxwell	$\neq 0$	0	$\neq 0$
8. **Viscous fluids**	Newton	0	0	$\neq 0$
9. **Ideal fluids**	Pascal	Sustain only pressures		

Plastic Bodies			
Type of body	Representative	$\epsilon^{(p)} = 0$	$\epsilon^{(p)} \neq 0$
1. **Ideal**	St. Venant	Rigid	Ideal fluid
2. **Dynamic**	Bingham	Rigid	Viscous fluid
3. **Static**	Prandtl–Reuss	Elastic	Ideal fluid
4. **Complex**	Schofield–Scott-Blair	Firmo-viscous	Visco-elastic

7.2.4. Summary of the rheological material classes. A summary of idealized rheological response is given table 7.1. The proper names in the boxes refer to the representatives of a class. The names given in the preplastic and plastic region in the class of plastic bodies refer to the two constitutive equations that are separated from each other by the yield surface in the stress space. The first three types of body represent well-known simple models for plastic bodies, while the fouth type of plastic body has the most complex rheological behavior. Bird, Armstrong and Hassager* call the second and the fourth type of the plastic bodies *viscoplastic fluids.*

The theory of the thermodynamics of irreversible processes is a good starting point for the derivation of the constitutive equations of the linear bodies in the various classes, and also of the fundamental plastic bodies that according to Kluitenberg have to be treated quasi-linearly.

---

* Bird R.B., Armstrong R.C. and Hassager O. 1987. *Dynamics of Polymeric Liquids.* 2nd ed. John Wiley & Sons, New York. See p. 61.

7.3. TIP MODEL OF RHEOLOGICAL BODIES

7.3.1. Idealized assumptions. The treatment of the linear rheological bodies according to the thermodynamics of irreversible processes is based on the following assumptions

(1) The material is chemically simple in the sense that no detectable diffusion phenomena occurring.
(2) The material is rheological simple, so that the principle of local action applies.
(3) The material is not polarized (nonpolar, no couple stresses), so that the stress tensor is symmetric.
(4) The deformation can be split into an elastic, an anelastic, and a viscous (or plastic) part.
(5) The anelastic and the viscous deformations are caused by internal mechanisms, which can be described thermodynamically by hidden variables with a tensor character.
(6) The material is isotropic (this condition yields the most simple rheological equations of state).
(7) The material is from the rheological point of view linear.

Although these assumptions are restrictive, various models can be derived that are useful in practice and illustrative for rheological behavior.

7.3.2. Simple rheological bodies. Oldroyd* made the concept *simple rheological materials* clear in 1950, probably for the first time. In a slightly less general form the concept was introduced explicitly by Truesdell and Noll†. Practically all constitutive theories are based on this concept.

Simple rheological bodies satisfy per definition the *principle of local action*, according to which the response of a material in a material point is determined by the processes found in the infinitesimal material surroundings ('material ϵ-environment') of this point. Actions propagating in simple rheological bodies have to be passed from ϵ-environment to ϵ-environment. The passing of the actions is determined by the balance equations for mass, momentum, moment of momentum, energy, electrical charge, and the electromagnetic field equations of Maxwell. The principle of the local action implies that the macroscopic behavior as observed in the recovery experiments also hold *in-the-small*, that is for material ϵ-environments.

The concept introduced by Eckart‡ of *relaxability-in-the-small* can be

* Oldroyd J.G. 1950. On the formulation of rheological equations of state. *Proceedings of the Royal Society of London*, A200: 523–541.

† Truesdell C. and Noll W. 1965. The nonlinear field theories of mechanics. In the *Encyclopedia of Physics*, III/3. Springer-Verlag, Berlin.

‡ Eckart E. 1948. The thermodynamics of irreversible processes. IV. The theory of elasticity and anelasticity. *Physical Review*, 73: 373–382.

translated in fact as the *principle of the local relaxation.* According to Kluitenberg* the principle can be described as follows. Suppose that the body has a stress free reference state, which is an equilibrium state, so that in this state the temperature is homogeneous and stationary (say $T = T_\circ$ in the reference state). Consider now a material ϵ-environment or equivalently a physical volume-element. Imagine that such an element is cut from the body. If this is done while the body is in the reference state, no changes will occur in the physical element if this element is left isolated after it is cut out of the medium.

Suppose now that the element is cut out of the body when the body is not in a state of equilibrium. On the surface of the cutout element no mechanical load will be applied, but the temperature is brought to the value $T_\circ$. Now the deformation at $T = T_\circ$ of the cut-out element can be observed as a function of time, in which the stresses relax to zero. As for the macroscopic bodies, elastic, anelastic, viscous or plastic deformations, can also be observed for the cut out physical volume element. Along these lines of thought the macroscopic observations also apply locally (that is 'in-the-small').

Kluitenberg† pointed out that Eckart does not take into account the temperature. The above discussion which includes the temperature can be found in Kluitenberg‡.

From the arguments in this subsection it follows that the assumption under item (2) in the previous subsection implied the assumption under item (4).

7.3.3. Linear rheological bodies. The theory of the thermodynamics of irreversible processes can be applied without any reservation to bodies with a linear behavior.

A rheological body is linear if the relation between the stresses and the deformations satisfies the superposition principle of Boltzmann§, so that we also speak of Boltzmann bodies. For a body to be linear from the rheological point of view, it is necessary, but not sufficient, that successive deformations can be superimposed. The conditions for the additivity are discussed in this section. The deformation can be described locally, since the principle of local action applies. The deformation of a material ϵ-environment can be formulated by the deformation of all material line-elements of such an environment.

Suppose that a material point, as defined in Chapter 1, is located at time $t_\circ$ at the point $\vec{X}$ and that the configuration of all material points at time $t_\circ$

* Kluitenberg G.A. 1962. Thermodynamic theory of elasticity and plasticity. *Physica*, 28: 217–232.

† Kluitenberg G.A. 1962. Thermodynamic theory of elasticity and plasticity. *Physica*, 28: 217–232.

‡ Kluitenberg G.A. 1963. On the thermodynamics of viscosity and plasticity. *Physica*, 29: 633–652.

§ Boltzmann L. 1876. Zur Theorie der elastischen Nachwirkung. *Annalen der Physik und Chemie (Poggendorffer Annalen)*, Ergbd. 7: 624–654.

forms the undeformed reference configuration. The considered material point is located at time t at point $\vec{x}$ in the space (see figure 7.2). In the reference configuration, a material line element at $\vec{X}$ is given by $d\vec{X}$, and at time t at point $\vec{x}$ of the deformed configuration $\vec{\chi}(\vec{X}, t)$ by $d\vec{x}$. Now it can be written that

$$d\vec{x} = \boldsymbol{F} \cdot d\vec{X} \qquad \Longleftrightarrow \qquad d\vec{X} = \boldsymbol{F}^{-1} \cdot d\vec{x} \tag{7.8}$$

where $\boldsymbol{F}$ is the deformation gradient, which describes the deformation of a material ϵ-environment. Obviously,

$$\boldsymbol{F} = \frac{\partial \vec{x}}{\partial \vec{X}} \qquad \text{and} \qquad \boldsymbol{F}^{-1} = \frac{\partial \vec{X}}{\partial \vec{x}} \tag{7.9}$$

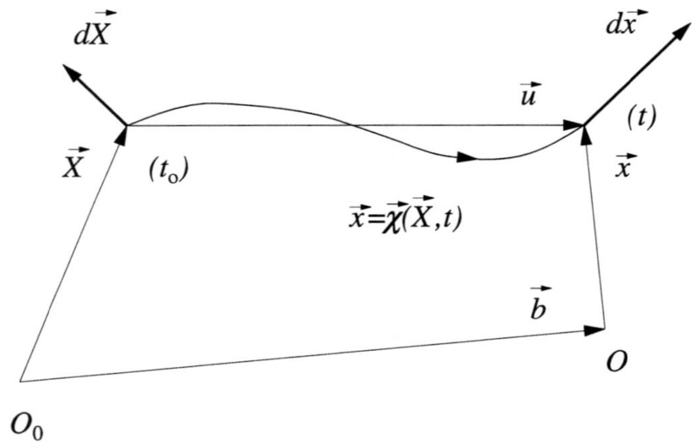

Figure 7.2. Material coordinates $\vec{X}$ with origin O_0 of the undeformed material and spatial coordinates $\vec{x}$ with origin O of the deformed material. The origin O is at the point $\vec{b}$ in the material coordinate system. The displacement of the point at $\vec{X}$ at time t_o to the point $\vec{x}$ at time t in the deformed configuration $\vec{\chi}$ is given by $\vec{u}$. The undeformed line element is $d\vec{X}$, the deformed line element is $d\vec{x}$.

From (7.8) it is found that

$$d\vec{X} \cdot d\vec{X} = d\vec{x} \cdot \boldsymbol{F}^{-1^{\mathrm{T}}} \cdot \boldsymbol{F}^{-1} \cdot d\vec{x} = d\vec{x} \cdot \mathbf{c} \cdot d\vec{x} \tag{7.10}$$

with

$$\mathbf{c} = \boldsymbol{F}^{-1^{\mathrm{T}}} \cdot \boldsymbol{F}^{-1} = \left(\frac{\partial \vec{X}}{\partial \vec{x}}\right)^{\mathrm{T}} \cdot \left(\frac{\partial \vec{X}}{\partial \vec{x}}\right) \tag{7.11}$$

the Cauchy deformation tensor, which is a measure of the pure deformation, defined as the displacement of the material points relative to each other.

From (7.8) it can also be seen that

$$d\vec{x} \cdot d\vec{x} = d\vec{X} \cdot \boldsymbol{F}^{\mathrm{T}} \cdot \boldsymbol{F} \cdot d\vec{X} = d\vec{X} \cdot \boldsymbol{C} \cdot d\vec{X} \tag{7.12}$$

with

$$\boldsymbol{C} = \boldsymbol{F}^{\mathrm{T}} \cdot \boldsymbol{F} = \left(\frac{\partial \vec{x}}{\partial \vec{X}}\right)^{\mathrm{T}} \cdot \left(\frac{\partial \vec{x}}{\partial \vec{X}}\right) \tag{7.13}$$

where $\boldsymbol{C}$ is the Green deformation tensor, which is likewise a proper measure of the pure deformation.

From (7.10) and (7.12) it follows that

$$(d\vec{x} \cdot d\vec{x}) - \left(d\vec{X} \cdot d\vec{X}\right) = 2\, d\vec{x} \cdot \boldsymbol{e} \cdot d\vec{x} = 2\, d\vec{X} \cdot \boldsymbol{E} \cdot d\vec{X} \tag{7.14}$$

with

$$\begin{aligned} \boldsymbol{e} &= \tfrac{1}{2}\left(\mathbf{I} - \boldsymbol{c}\right) \\ \boldsymbol{E} &= \tfrac{1}{2}\left(\boldsymbol{C} - \mathbf{I}\right) \end{aligned} \tag{7.15}$$

These are the *spatial*, and *material relative deformation tensors* (or *strain tensors*), respectively, where $\mathbf{I}$ denotes the unit tensor. From (7.14) the relation

$$\boldsymbol{E} = \left(\frac{\partial \vec{x}}{\partial \vec{X}}\right)^{\mathrm{T}} \cdot \boldsymbol{e} \cdot \left(\frac{\partial \vec{x}}{\partial \vec{X}}\right) = \boldsymbol{F}^{\mathrm{T}} \cdot \boldsymbol{e} \cdot \boldsymbol{F} \tag{7.16}$$

follows. The displacement vector $\vec{u}$ is sometimes introduced in the deformation theory

$$\vec{X} + \vec{u} = \vec{b} + \vec{x} \tag{7.17}$$

in which $\vec{b}$ is a constant vector that specifies the origin O of the spatial coordinate system relative to the origin O_0 of the material coordinate system. From (7.9) and (7.17) it follows that

$$\boldsymbol{F} = \mathbf{I} + \frac{\partial \vec{u}}{\partial \vec{X}} \qquad \text{and} \qquad \boldsymbol{F}^{-1} = \mathbf{I} - \frac{\partial \vec{u}}{\partial \vec{x}} \tag{7.18}$$

where $\partial\vec{u}/\partial\vec{X}$ is the material displacement gradient and $\partial\vec{u}/\partial\vec{x}$ the spatial displacement gradient. From (7.11) and $(7.18)_2$ it follows that

$$\boldsymbol{c} = \mathbf{I} - \left[\left(\frac{\partial \vec{u}}{\partial \vec{x}}\right)^{\mathrm{T}} + \left(\frac{\partial \vec{u}}{\partial \vec{x}}\right)\right] + \left(\frac{\partial \vec{u}}{\partial \vec{x}}\right)^{\mathrm{T}} \cdot \left(\frac{\partial \vec{u}}{\partial \vec{x}}\right)$$

or with $(7.15)_1$

$$\boldsymbol{e} = \boldsymbol{\epsilon} - \frac{1}{2}\left(\frac{\partial \vec{u}}{\partial \vec{x}}\right)^{\mathrm{T}} \cdot \left(\frac{\partial \vec{u}}{\partial \vec{x}}\right) \tag{7.19}$$

where

$$\epsilon = \frac{1}{2}\left[\left(\frac{\partial\vec{u}}{\partial\vec{x}}\right)^{\mathrm{T}} + \left(\frac{\partial\vec{u}}{\partial\vec{x}}\right)\right] \tag{7.20}$$

This is the spatial infinitesimal strain tensor. For $\boldsymbol{E}$ analogous results apply.

Consider now the two successive deformations depicted in figure 7.3 from $\vec{X}$ to $\vec{x}_1$ and next from $\vec{x}_1$ to $\vec{x}_2$

$$\begin{aligned} &\vec{X} \to \vec{x}_1 \quad \text{with} \quad \boldsymbol{c}_1 = \left(\frac{\partial\vec{X}}{\partial\vec{x}_1}\right)^{\mathrm{T}} \cdot \left(\frac{\partial\vec{X}}{\partial\vec{x}_1}\right) \\ &\vec{x}_1 \to \vec{x}_2 \quad \text{with} \quad \boldsymbol{c}_2 = \left(\frac{\partial\vec{x}_1}{\partial\vec{x}_2}\right)^{\mathrm{T}} \cdot \left(\frac{\partial\vec{x}_1}{\partial\vec{x}_2}\right) \end{aligned} \tag{7.21}$$

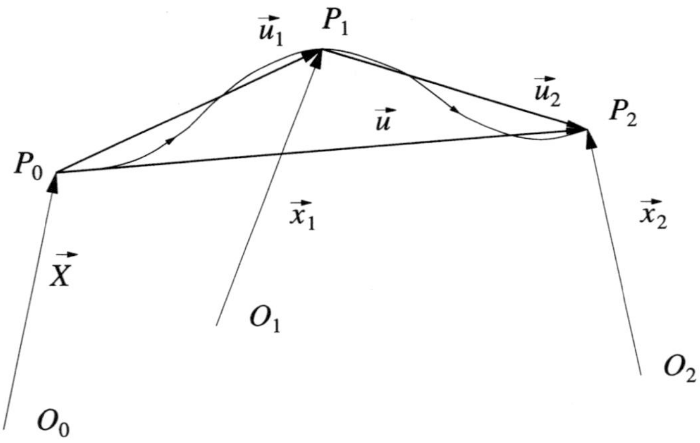

Figure 7.3. Two successive deformations. The displacement $\vec{u}$ of the point P_0 at point $\vec{X}$ to the point P_2 at $\vec{x}_2$ in the coordinate system with the origin O_2 is also the result of the displacements $\vec{u}_1 + \vec{u}_2$. The vectors $\vec{b}_1$ and $\vec{b}_2$ that specify the location of the origins O_1 and O_2 relative to O_0 are not drawn in the figure.

For the resulting deformation

$$\boldsymbol{c} = \left(\frac{\partial\vec{X}}{\partial\vec{x}_2}\right)^{\mathrm{T}} \cdot \left(\frac{\partial\vec{X}}{\partial\vec{x}_2}\right) \tag{7.22}$$

Now

$$\left(\frac{\partial\vec{X}}{\partial\vec{x}_2}\right) = \left(\frac{\partial\vec{X}}{\partial\vec{x}_1}\right) \cdot \left(\frac{\partial\vec{x}_1}{\partial\vec{x}_2}\right)$$

so that

$$\boldsymbol{c} = \left(\frac{\partial \vec{x}_1}{\partial \vec{x}_2}\right)^{\mathrm{T}} \cdot \left(\frac{\partial \vec{X}}{\partial \vec{x}_1}\right)^{\mathrm{T}} \cdot \left(\frac{\partial \vec{X}}{\partial \vec{x}_1}\right) \cdot \left(\frac{\partial \vec{x}_1}{\partial \vec{x}_2}\right)$$

or

$$\boldsymbol{c} = \left(\frac{\partial \vec{x}_1}{\partial \vec{x}_2}\right)^{\mathrm{T}} \cdot \boldsymbol{c}_1 \cdot \left(\frac{\partial \vec{x}_1}{\partial \vec{x}_2}\right) \tag{7.23}$$

The strain tensors are defined according to $(7.15)_1$ as

$$\boldsymbol{e} = \tfrac{1}{2}\left(\mathbf{I} - \boldsymbol{c}\right) \qquad \boldsymbol{e}_1 = \tfrac{1}{2}\left(\mathbf{I} - \boldsymbol{c}_1\right) \qquad \boldsymbol{e}_2 = \tfrac{1}{2}\left(\mathbf{I} - \boldsymbol{c}_2\right) \tag{7.24}$$

With the substitution of (7.23) the result is

$$\boldsymbol{e} = \tfrac{1}{2}\mathbf{I} - \tfrac{1}{2}\left(\frac{\partial \vec{x}_1}{\partial \vec{x}_2}\right)^{\mathrm{T}} \cdot \left(\mathbf{I} - 2\boldsymbol{e}_1\right) \cdot \left(\frac{\partial \vec{x}_1}{\partial \vec{x}_2}\right)$$

or

$$\boldsymbol{e} = \boldsymbol{e}_2 + \left(\frac{\partial \vec{x}_1}{\partial \vec{x}_2}\right)^{\mathrm{T}} \cdot \boldsymbol{e}_1 \cdot \left(\frac{\partial \vec{x}_1}{\partial \vec{x}_2}\right) \tag{7.25}$$

In general the relative deformation tensors $\boldsymbol{e}_1$ and $\boldsymbol{e}_2$ cannot be added linearly to each other. However, substitution of the the 'shifted' strain tensors defined by the relation (7.16)

$$\begin{aligned}
\boldsymbol{E} &= \left(\frac{\partial \vec{x}_2}{\partial \vec{X}}\right)^{\mathrm{T}} \cdot \boldsymbol{e} \cdot \left(\frac{\partial \vec{x}_2}{\partial \vec{X}}\right) \\
\boldsymbol{E}_1 &= \left(\frac{\partial \vec{x}_1}{\partial \vec{X}}\right)^{\mathrm{T}} \cdot \boldsymbol{e}_1 \cdot \left(\frac{\partial \vec{x}_1}{\partial \vec{X}}\right) \\
\boldsymbol{E}_2 &= \left(\frac{\partial \vec{x}_2}{\partial \vec{X}}\right)^{\mathrm{T}} \cdot \boldsymbol{e}_2 \cdot \left(\frac{\partial \vec{x}_2}{\partial \vec{X}}\right)
\end{aligned} \tag{7.26}$$

into (7.25) yields

$$\boldsymbol{E} = \boldsymbol{E}_1 + \boldsymbol{E}_2 \tag{7.27}$$

This means that after shifting the relative deformation tensors to the reference configuration in $P_\circ$, the deformation tensors can be added linearly. In the material description a superposition principle can always be formulated. From (7.25) it follows that this superposition principle does not necessarily apply in the spatial description. Bodies that can be described by linear constitutive equations in terms of the material coordinates, and for which then the superposition principle applies in the material description, but not in the spatial description, are called *quasi-linear*.

The superposition principle holds for linear bodies both in the material description and in the spatial description. The deformation tensor can be superposed in the spatial description if the displacement gradients are infinitesimal

$$|\partial\vec{u}/\partial\vec{x}| \ll 1 \tag{7.28}$$

From (7.18) it follows that

$$\frac{\partial\vec{u}}{\partial\vec{X}} = \frac{\partial\vec{u}}{\partial\vec{x}} \cdot \frac{\partial\vec{x}}{\partial\vec{X}} = \frac{\partial\vec{u}}{\partial\vec{x}} \cdot \left(\mathbf{I} + \frac{\partial\vec{u}}{\partial\vec{X}}\right) \tag{7.29}$$

From (7.28) and (7.29) it follows that

$$\left|\partial\vec{u}/\partial\vec{X}\right| \ll 1 \tag{7.30}$$

If quantities of the second order in the displacement gradients can be neglected, then (7.19) becomes

$$\boldsymbol{e} \approx \boldsymbol{\epsilon}$$

in which $\boldsymbol{\epsilon}$ is now a first order small quantity. The tensor $\boldsymbol{e}$ is then also called the infinitesimal strain tensor of Cauchy. From (7.16) and (7.18)

$$\boldsymbol{E} = \left[\left(\mathbf{I} + \frac{\partial\vec{u}}{\partial\vec{x}}\right)^{\mathrm{T}}\right] \cdot \boldsymbol{e} \cdot \left[\mathbf{I} + \frac{\partial\vec{u}}{\partial\vec{x}}\right]$$

Since $\boldsymbol{e}$ is small of the first order, the strain tensors are approximately equal

$$\boldsymbol{e} \approx \boldsymbol{E} \approx \boldsymbol{\epsilon} \tag{7.31}$$

If the displacement gradients are infinitesimal, then there is, up to the first order in the displacement gradients, no difference between the material description and the spatial description of the deformation. From (7.27) and (7.31) it follows that

$$\boldsymbol{\epsilon} = \boldsymbol{\epsilon}_1 + \boldsymbol{\epsilon}_2 \tag{7.32}$$

and the superposition principle for the deformations now applies in both descriptions. This conclusion is trivial if no essential difference exists between the material and the spatial description of the deformation.

The velocity is defined by

$$\vec{v} = \frac{\mathrm{D}\vec{x}}{\mathrm{D}t} = \frac{\mathrm{D}\vec{u}}{\mathrm{D}t} \tag{7.33}$$

The deformation rate tensor of Euler is

$$\boldsymbol{D} = \tfrac{1}{2}\left[\frac{\partial\vec{v}}{\partial\vec{x}} + \left(\frac{\partial\vec{v}}{\partial\vec{x}}\right)^{\mathrm{T}}\right] \tag{7.34}$$

Now is

$$\frac{\mathrm{D}}{\mathrm{D}t}\frac{\partial\vec{u}}{\partial\vec{x}} = \left(\frac{\partial}{\partial t} + \vec{v}\cdot\frac{\partial}{\partial\vec{x}}\right)\frac{\partial\vec{u}}{\partial\vec{x}}$$
$$= \frac{\partial}{\partial\vec{x}}\frac{\partial\vec{u}}{\partial t} + \frac{\partial}{\partial\vec{x}}\left(\vec{v}\cdot\frac{\partial}{\partial\vec{x}}\right)\vec{u} - \frac{\partial\vec{u}}{\partial\vec{x}}\cdot\frac{\partial\vec{v}}{\partial\vec{x}}$$
$$= \frac{\partial}{\partial\vec{x}}\frac{\mathrm{D}\vec{u}}{\mathrm{D}t} - \frac{\partial\vec{u}}{\partial\vec{x}}\cdot\frac{\partial\vec{v}}{\partial\vec{x}} = \frac{\partial\vec{v}}{\partial\vec{x}} - \frac{\partial\vec{u}}{\partial\vec{x}}\cdot\frac{\partial\vec{v}}{\partial\vec{x}}$$

Obviously it follows that

$$\boldsymbol{D} = \dot{\boldsymbol{\epsilon}} + \tfrac{1}{2}\left[\frac{\partial\vec{u}}{\partial\vec{x}}\cdot\frac{\partial\vec{v}}{\partial\vec{x}} + \left(\frac{\partial\vec{u}}{\partial\vec{x}}\cdot\frac{\partial\vec{v}}{\partial\vec{x}}\right)^{\mathrm{T}}\right] \tag{7.35}$$

If the displacement gradients are infinitesimal, then

$$\boldsymbol{D} \approx \dot{\boldsymbol{\epsilon}} \equiv \frac{\mathrm{D}\boldsymbol{\epsilon}}{\mathrm{D}t} \tag{7.36}$$

From (7.35) it can be concluded that the velocity gradients are small if $\dot{\boldsymbol{\epsilon}}$ and $\partial\vec{u}/\partial\vec{x}$ are small. In the theory of the linear bodies, (7.31) and (7.36) represent major simplifications.

7.4. THERMOSTATICS OF LINEAR RHEOLOGICAL BODIES

7.4.1. Work due to stresses. In the formulation of the Gibbs fundamental equation one has to know the work done during a quasi-static change on the matter being considered. In rheology it is assumed that this work is mechanical in nature.

Consider a volume element δV, bounded by the surface δA as depicted in figure 7.4. This volume element is subjected to a quasi-static deformation, in which the material points of δA undergo the virtual displacements $\vec{u}$. Suppose that $\vec{\sigma}$ is the traction exerted on the material in δA by surroundings. The traction can be assumed to be equal to its equilibrium value $\vec{\sigma}^{(\mathrm{eq})}$ for a quasi-static deformation. The work done therefore becomes

$$\delta W = \oiint_{\delta A}\left(\vec{\sigma}^{(\mathrm{eq})}\cdot d\vec{u}\right)dA$$

According to the fundamental stress theorem of Cauchy it follows that

$$\vec{\sigma}^{(\mathrm{eq})} = \vec{n}\cdot\boldsymbol{\sigma}^{(\mathrm{eq})}$$

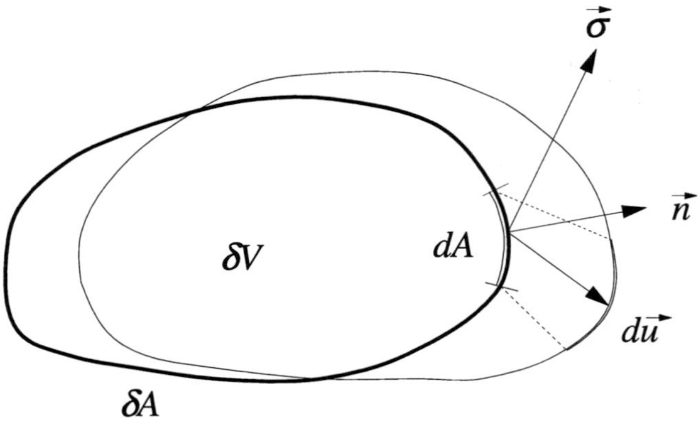

Figure 7.4. Forces exerted on a physical volume element δV. The thinner line gives the physical volume element after a virtual displacement $d\vec{u}$. The traction $\vec{\sigma}$ is positive for $\vec{n} \cdot \vec{\sigma} > 0$ in which $\vec{n}$ is the unit normal to the surface element dA of δV.

in which $\vec{n}$ is the unit normal to the surface and $\boldsymbol{\sigma}$ is the stress tensor. The work therefore becomes

$$\delta W = \oiint_{\delta A} \left(\vec{n} \cdot \boldsymbol{\sigma}^{(\mathrm{eq})} \cdot d\vec{u} \right) dA$$

For sufficiently small δV the equilibrium stress $\boldsymbol{\sigma}^{(\mathrm{eq})}$ can be assumed constant, so that

$$\delta W = \boldsymbol{\sigma}^{(\mathrm{eq})} : \oiint_{\delta A} (\vec{n}\, d\vec{u})\, dA$$

Application of the Gauss divergence theorem yields

$$\delta W = \boldsymbol{\sigma}^{(\mathrm{eq})} : \left(\frac{\partial}{\partial \vec{x}} d\vec{u} \right)^{\mathrm{T}} \delta V \tag{7.37}$$

The stress tensor is symmetrical if the medium is nonpolar, so that

$$\delta W = \boldsymbol{\sigma}^{(\mathrm{eq})} : \tfrac{1}{2} \left[\left(\frac{\partial}{\partial \vec{x}} d\vec{u} \right) + \left(\frac{\partial}{\partial \vec{x}} d\vec{u} \right)^{\mathrm{T}} \right] \delta V \tag{7.38}$$

The displacement of the material points of δA are along their paths, so that

$$d\vec{u} = \vec{v}\, dt \tag{7.39}$$

where t is the path parameter and $\vec{v}$ is tangent to the path line. If t is interpreted as the time, then $\vec{v}$ is the velocity. Substitution of (7.39) into (7.38) yields

$$\delta W = \boldsymbol{\sigma}^{(\mathrm{eq})} : \boldsymbol{D}\, dt\, \delta V$$

where for the infinitesimal displacement gradients, according to (7.36), it applies that

$$\boldsymbol{D}\, dt = d\boldsymbol{\epsilon}$$

The work is therefore

$$\delta W = \left(\boldsymbol{\sigma}^{(\mathrm{eq})} : d\boldsymbol{\epsilon}\right) \delta V \tag{7.40}$$

The changeover from (7.38) to (7.40) can also be done with

$$d\boldsymbol{\epsilon} = \tfrac{1}{2} d \left[\left(\frac{\partial}{\partial \vec{x}} \vec{u} \right) + \left(\frac{\partial}{\partial \vec{x}} \vec{u} \right)^{\mathrm{T}} \right] = \tfrac{1}{2} \left[\left(\frac{\partial}{\partial \vec{x}} d\vec{u} \right) + \left(\frac{\partial}{\partial \vec{x}} d\vec{u} \right)^{\mathrm{T}} \right]$$

The interchange of d and $\partial/\partial\vec{x}$ is allowed only for infinitesimal displacement gradients.

7.4.2. Fundamental thermostatic Gibbs equation. Because of (7.40), the fundamental equation of Gibbs (4.23) for a chemical simple material is noted as

$$du = T\, ds + \frac{1}{\rho} \boldsymbol{\sigma}^{(\mathrm{eq})} : d\boldsymbol{\epsilon} \tag{7.41}$$

where u is the internal energy per unit of mass and s the entropy per unit of mass. Now $\boldsymbol{\epsilon}$ has to be considered more closely. If plastic deformations are left out of consideration, then one has to distinguish between

(1) *Elastic deformations* $\boldsymbol{\epsilon}^{(\mathrm{e})}$, which are reversible. These deformations are therefore found in the thermostatic potentials, but *not* in the entropy production or the dissipation.
(2) *Anelastic deformations* $\boldsymbol{\epsilon}^{(\mathrm{a})}$, which are recovered with after-effects. These deformations are also found in the thermodynamic potentials *and* in the entropy production and the dissipation.
(3) *Viscous deformations* $\boldsymbol{\epsilon}^{(\mathrm{v})}$, which are not recovered. These deformations are therefore *not* found in the thermodynamic potentials, but are found in the entropy production and the dissipation.

The total deformation is the sum of the elastic, anelastic and viscous deformations

$$\boldsymbol{\epsilon} = \boldsymbol{\epsilon}^{(\mathrm{e})} + \boldsymbol{\epsilon}^{(\mathrm{a})} + \boldsymbol{\epsilon}^{(\mathrm{v})} \tag{7.42}$$

Since the viscous deformations are not allowed to appear in the thermostatic potentials, $\boldsymbol{\epsilon}$ in (7.41) has to be replaced by

$$\boldsymbol{\epsilon}' = \boldsymbol{\epsilon}^{(\mathrm{e})} + \boldsymbol{\epsilon}^{(\mathrm{a})} \tag{7.43}$$

so that then

$$\rho\, du = \rho T\, ds + \boldsymbol{\sigma}^{(\mathrm{eq})} : d\boldsymbol{\epsilon}' \tag{7.44}$$

The stress tensor $\boldsymbol{\sigma}$ can always be split into a deviatoric part and an isotropic part, in which the isotropic part is usually interpreted as the *pressure* p multiplied by the unit tensor $\mathbf{I}$

$$\boldsymbol{\sigma} = \mathring{\boldsymbol{\sigma}} - p\mathbf{I} \tag{7.45}$$

with

$$\text{tr } \mathring{\boldsymbol{\sigma}} = \mathring{\boldsymbol{\sigma}} : \mathbf{I} = 0 \qquad \text{and} \qquad p = -\tfrac{1}{3}\text{tr } \boldsymbol{\sigma} = -\tfrac{1}{3}\boldsymbol{\sigma} : \mathbf{I} \tag{7.46}$$

The deformation is similarly to be split into a deviatoric part and an isotropic part, in which the isotropic part is concerned with the volume changes

$$\boldsymbol{\epsilon} = \mathring{\boldsymbol{\epsilon}} + \tfrac{1}{3}I_\epsilon \mathbf{I} \tag{7.47}$$

with

$$\text{tr } \mathring{\boldsymbol{\epsilon}} = \mathring{\boldsymbol{\epsilon}} : \mathbf{I} = 0 \qquad \text{and} \qquad I_\epsilon = \frac{dV - dV_\circ}{dV_\circ} = \text{tr } \boldsymbol{\epsilon} = \boldsymbol{\epsilon} : \mathbf{I} \tag{7.48}$$

and using that in the continuum theory limit δV equals dV. Furthermore suppose

$$\begin{gathered} I_\epsilon^{(\mathrm{e})} = \text{tr } \boldsymbol{\epsilon}^{(\mathrm{e})} \qquad I_\epsilon^{(\mathrm{a})} = \text{tr } \boldsymbol{\epsilon}^{(\mathrm{a})} \qquad I_\epsilon^{(\mathrm{v})} = \text{tr } \boldsymbol{\epsilon}^{(\mathrm{v})} \\ I'_\epsilon = \text{tr } \boldsymbol{\epsilon}' = I_\epsilon^{(\mathrm{e})} + I_\epsilon^{(\mathrm{a})} \end{gathered} \tag{7.49}$$

From (7.45) and (7.47) it follows that

$$\begin{aligned} \boldsymbol{\sigma} : d\boldsymbol{\epsilon} &= \left(\mathring{\boldsymbol{\sigma}} - p\mathbf{I}\right) : \left(d\mathring{\boldsymbol{\epsilon}} + \tfrac{1}{3}\mathbf{I}\, dI_\epsilon\right) \\ &= \mathring{\boldsymbol{\sigma}} : d\mathring{\boldsymbol{\epsilon}} - p\, dI_\epsilon \end{aligned} \tag{7.50}$$

The Gibbs fundamental equation (7.44) becomes with this

$$\rho\, du = \rho T\, ds + \mathring{\boldsymbol{\sigma}}^{(\mathrm{eq})} : d\mathring{\boldsymbol{\epsilon}}' - p^{(\mathrm{eq})} dI'_\epsilon \tag{7.51}$$

From the perspectives of the thermodynamics of irreversible processes, the existence of anelastic deformations is related to the existence of internal processes with a delayed return to the equilibrium state. This indicates that the Gibbs fundamental equation (7.51) is still incomplete for the description of the thermodynamic potential, since the internal processes are not yet formulated. The internal processes can be taken into account by their extent of advancement or extent of change. Also, the tensor character of the internal processes has to be described with an extent of advancement.

An internal scalar process α is formulated by a scalar extent of advancement ξ_α, an internal vector process β is described by a vector extent of advancement

$\vec{\eta}_\beta$ and an internal tensor process γ by a tensor extent of advancement $\boldsymbol{\zeta}_\gamma$. Hence, the complete fundamental equation of Gibbs becomes

$$\rho\, du = \rho T\, ds - p^{(\mathrm{eq})} dI'_\epsilon + \mathring{\boldsymbol{\sigma}}^{(\mathrm{eq})} : d\mathring{\boldsymbol{\epsilon}}' - \sum_{\alpha=1}^{\kappa} A_\alpha\, d\xi_\alpha - \sum_{\beta=1}^{\mu} \vec{H}_\beta \cdot d\vec{\eta}_\beta - \sum_{\gamma=1}^{\nu} \mathring{\boldsymbol{P}}_\gamma : d\mathring{\boldsymbol{\zeta}}_\gamma \tag{7.52}$$

where A, $\vec{H}$, and $\mathring{\boldsymbol{P}}$ denote the *affinities* of the internal processes . Note that now the subscripts do not indicate tensor components, but the number of the various processes. The ν internal tensor quantities can, similar to (7.50), be split, for which the isotropic parts can possibly be subsumed into A and ξ.

The free energy of Helmholtz is

$$f = u - Ts \tag{7.53}$$

From (7.52) and (7.53)

$$\rho\, df = -\rho s\, dT - p^{(\mathrm{eq})} dI'_\epsilon + \mathring{\boldsymbol{\sigma}}^{(\mathrm{eq})} : d\mathring{\boldsymbol{\epsilon}}' - \sum_{\alpha=1}^{\kappa} A_\alpha\, d\xi_\alpha - \sum_{\beta=1}^{\mu} \vec{H}_\beta \cdot d\vec{\eta}_\beta - \sum_{\gamma=1}^{\nu} \mathring{\boldsymbol{P}}_\gamma : d\mathring{\boldsymbol{\zeta}}_\gamma \tag{7.54}$$

so that the free energy depends on the temperature, the elastic and anelastic deformations and the affinities of the internal processes.

$$f = f(T, I'_\epsilon, \mathring{\boldsymbol{\epsilon}}', \xi_\alpha, \vec{\eta}_\beta, \mathring{\boldsymbol{\zeta}}_\gamma) \tag{7.55}$$

For isotropic bodies one of the linearized thermostatic equations of state becomes

$$\mathring{\boldsymbol{\sigma}}^{(\mathrm{eq})} = \frac{\partial f}{\partial \mathring{\boldsymbol{\epsilon}}'} = c'_{\mathrm{ee}} \mathring{\boldsymbol{\epsilon}}' + \sum_{\gamma=1}^{\nu} c'_{\mathrm{e}\gamma} \mathring{\boldsymbol{\zeta}}_\gamma \tag{7.56}$$

It is found from observations that the elastic part of the deformation vanishes instantaneously by the cancellation of the load. This means that

$$\mathring{\boldsymbol{\sigma}}^{(\mathrm{eq})} = 0 \qquad \Longleftrightarrow \qquad \mathring{\boldsymbol{\epsilon}}^{(\mathrm{e})} = 0 \tag{7.57}$$

Equation (7.56) satisfies (7.57) if

$$c'_{\mathrm{ee}} \mathring{\boldsymbol{\epsilon}}^{(\mathrm{a})} + \sum_{\gamma=1}^{\nu} c'_{\mathrm{e}\gamma} \mathring{\boldsymbol{\zeta}}_\gamma = 0 \tag{7.58}$$

so that it can be noted that

$$\overset{\circ}{\boldsymbol{\epsilon}}^{(\mathrm{a})} = \sum_{\gamma=1}^{\nu} \overset{\circ}{\boldsymbol{\epsilon}}_{\gamma}^{(\mathrm{a})} \qquad \text{with} \qquad \overset{\circ}{\boldsymbol{\epsilon}}_{\gamma}^{(\mathrm{a})} = -\frac{c'_{\mathrm{e}\gamma}}{c'_{\mathrm{ee}}} \overset{\circ}{\boldsymbol{\zeta}}_{\gamma} \tag{7.59}$$

This means that (7.57) implies that $\overset{\circ}{\boldsymbol{\zeta}}_{\gamma}$ can be expressed in a part of the anelastic deformation. Kluitenberg* showed that the elimination of $\overset{\circ}{\boldsymbol{\zeta}}_{\gamma}$ ends up in an orderly system of equations. Kluitenberg also applies this to the volume deformations. For the volume deformations (see also (7.64))

$$-p^{(\mathrm{eq})} = \rho \frac{\partial f}{\partial I'_{\epsilon}} = k'_{\mathrm{ee}} I'_{\epsilon} + \sum_{\alpha=1}^{\kappa} k'_{\mathrm{e}\alpha} \xi_{\alpha} - b_{\mathrm{e}}(T - T_{\circ})$$

Kluitenberg required

$$p^{(\mathrm{eq})} = 0 \qquad \text{for} \qquad I_{\epsilon}^{(\mathrm{e})} = 0 \qquad \text{and} \qquad T = T_{\circ}$$

which results in

$$k'_{\mathrm{ee}} I_{\epsilon}^{(\mathrm{a})} + \sum_{\alpha=1}^{\kappa} k'_{\mathrm{e}\alpha} \xi_{\alpha} = 0$$

For thermal relaxation in dilute gases it is clear that $I_{\epsilon}^{(\mathrm{a})} = 0$, so that now $\xi_{\alpha} = 0$. No thermal relaxation is found by the method of Kluitenberg, but only structure relaxation. If it is required that

$$p^{(\mathrm{eq})} = 0 \qquad \text{for} \qquad I_{\epsilon}^{(\mathrm{e})} = 0$$

then

$$k'_{\mathrm{ee}} I_{\epsilon}^{(\mathrm{a})} + \sum_{\alpha=1}^{\kappa} k'_{\mathrm{e}\alpha} \xi_{\alpha} - b_{\mathrm{e}}(T - T_{\circ}) = 0$$

The extent of advancement ξ_{α} is now determined by both the volume and the temperature. It therefore seems wise *not* to eliminate ξ_{α} and to express ξ_{α} in terms of $I_{\epsilon}^{(\mathrm{a})}$.

The substitution of (7.59) into (7.52) yields

$$\rho\, du = \rho T\, ds - p^{(\mathrm{eq})} dI'_{\epsilon} + \overset{\circ}{\boldsymbol{\sigma}}^{(\mathrm{eq})} : d\left(\overset{\circ}{\boldsymbol{\epsilon}}^{(\mathrm{e})} + \overset{\circ}{\boldsymbol{\epsilon}}^{(\mathrm{a})}\right)$$
$$- \sum_{\alpha=1}^{\kappa} A_{\alpha}\, d\xi_{\alpha} - \sum_{\beta=1}^{\mu} \vec{H}_{\beta} \cdot d\vec{\eta}_{\beta} - \sum_{\gamma=1}^{\nu} \overset{\circ}{\boldsymbol{P}}_{\gamma} : d\overset{\circ}{\boldsymbol{\epsilon}}_{\gamma}^{(\mathrm{a})} \tag{7.60}$$

---

* Kluitenberg G.A. 1962. Thermodynamic theory of elasticity and plasticity. *Physica*, 28: 217–232; Kluitenberg G.A. 1962. On rheology and thermodynamics of irreversible processes. *Physica*, 28: 1173–1183; Kluitenberg G.A. 1962. On the thermodynamics of viscosity and plasticity. *Physica*, 29: 633–652.

with a modified meaning of $\overset{\circ}{\boldsymbol{P}}_\gamma$ that has now the significance of an *affinity stress*. The Gibbs fundamental equation (7.60) can also be written as

$$\boxed{\begin{aligned}\rho\, du = \rho T\, ds - p^{(\mathrm{eq})} dI'_\epsilon + \overset{\circ}{\boldsymbol{\sigma}}^{(\mathrm{eq})} : d\overset{\circ}{\boldsymbol{\epsilon}}^{(\mathrm{e})} - \sum_{\alpha=1}^{\kappa} A_\alpha\, d\xi_\alpha \\ - \sum_{\beta=1}^{\mu} \vec{H}_\beta \cdot d\vec{\eta}_\beta - \sum_{\gamma=1}^{\nu} \left(\overset{\circ}{\boldsymbol{P}}_\gamma - \overset{\circ}{\boldsymbol{\sigma}}^{(\mathrm{eq})}\right) : d\overset{\circ}{\boldsymbol{\epsilon}}^{(\mathrm{a})}_\gamma\end{aligned}} \tag{7.61}$$

with the application of $(7.59)_1$. For the free energy it can be seen analogously that

$$\begin{aligned}\rho\, df = -\rho s\, dT - p^{(\mathrm{eq})} dI'_\epsilon + \overset{\circ}{\boldsymbol{\sigma}}^{(\mathrm{eq})} : d\overset{\circ}{\boldsymbol{\epsilon}}^{(\mathrm{e})} - \sum_{\alpha=1}^{\kappa} A_\alpha\, d\xi_\alpha \\ - \sum_{\beta=1}^{\mu} \vec{H}_\beta \cdot d\vec{\eta}_\beta - \sum_{\gamma=1}^{\nu} \left(\overset{\circ}{\boldsymbol{P}}_\gamma - \overset{\circ}{\boldsymbol{\sigma}}^{(\mathrm{eq})}\right) : d\overset{\circ}{\boldsymbol{\epsilon}}^{(\mathrm{a})}_\gamma\end{aligned} \tag{7.62}$$

From this equation the independent variables for f can be inferred.

$$f = f(T,\, I'_\epsilon,\, \overset{\circ}{\boldsymbol{\epsilon}}^{(\mathrm{e})},\, \xi_\alpha,\, \vec{\eta}_\beta,\, \overset{\circ}{\boldsymbol{\epsilon}}^{(\mathrm{a})}_\gamma) \tag{7.63}$$

For small deviations from the equilibrium in the reference state, the free energy f can be expanded as follows

$$\begin{aligned}\rho_\circ (f - f_\circ) = \tfrac{1}{2} c_{\mathrm{ee}} \overset{\circ}{\boldsymbol{\epsilon}}^{(\mathrm{e})} : \overset{\circ}{\boldsymbol{\epsilon}}^{(\mathrm{e})} + \tfrac{1}{2} \sum_{\alpha,\beta=1}^{\nu} c_{\alpha\beta} \overset{\circ}{\boldsymbol{\epsilon}}^{(\mathrm{a})}_\alpha : \overset{\circ}{\boldsymbol{\epsilon}}^{(\mathrm{a})}_\beta \\ + \tfrac{1}{2} k_{\mathrm{ee}} I'^{2}_\epsilon + \sum_{\alpha=1}^{\kappa} k_{\mathrm{e}\alpha} I'_\epsilon \xi_\alpha + \tfrac{1}{2} \sum_{\alpha,\beta=1}^{\kappa} k_{\alpha\beta} \xi_\alpha \xi_\beta \\ + \tfrac{1}{2} \sum_{\alpha,\beta=1}^{\mu} l_{\alpha\beta} \left(\vec{\eta}_\alpha \cdot \vec{\eta}_\beta\right) \\ + b_{\mathrm{e}} I'_\epsilon (T - T_\circ) - \sum_{\alpha=1}^{\kappa} b_\alpha \xi_\alpha (T - T_\circ) - \tfrac{1}{2} c_v^{(\circ)} \frac{\rho_\circ}{T_\circ} (T - T_\circ)^2\end{aligned} \tag{7.64}$$

This expansion holds only for isotropic bodies, so that f can depend only on the scalar invariants obtained from the arguments of f given in (7.63). No cross products of $\overset{\circ}{\boldsymbol{\epsilon}}^{(\mathrm{e})}$ and $\overset{\circ}{\boldsymbol{\epsilon}}^{(\mathrm{a})}_\alpha$ occur due to (7.57). From (7.64) the symmetries

$$c_{\alpha\beta} = c_{\beta\alpha} \qquad l_{\alpha\beta} = l_{\beta\alpha} \qquad k_{\alpha\beta} = k_{\beta\alpha} \tag{7.65}$$

are found.

The free energy $f = f_\circ$ is minimal for $T = T_\circ$ if the reference state is a stable equilibrium state, so that the coefficients have to satisfy the conditions

$$\begin{gathered} c_{\mathrm{ee}} \geq 0 \qquad \sum_{\alpha,\beta=1}^{\nu} c_{\alpha\beta} y_\alpha y_\beta \geq 0 \\ k_{\mathrm{ee}} y_e^2 + 2 \sum_{\alpha=1}^{\kappa} k_{\mathrm{e}\alpha} y_e y_\alpha + \sum_{\alpha,\beta=1}^{\kappa} k_{\alpha\beta} y_\alpha y_\beta \geq 0 \\ \sum_{\alpha,\beta=1}^{\mu} l_{\alpha\beta} y_\alpha y_\beta \geq 0 \end{gathered} \tag{7.66}$$

where the y's are arbitrary real variables. The thermostatic equations of state in linearized version results from (7.62) and (7.64)

$$\begin{aligned} s &= -\frac{\partial f}{\partial T} = \frac{1}{\rho_\circ}\left(b_{\mathrm{e}} I'_\epsilon + \sum_{\alpha=1}^{\kappa} b_\alpha \xi_\alpha\right) + \frac{1}{T_\circ} c_v^{(\circ)} (T - T_\circ) \\ -p^{(\mathrm{eq})} &= \rho \frac{\partial f}{\partial I'_\epsilon} = \frac{\rho}{\rho_\circ}\left[k_{\mathrm{ee}} I'_\epsilon + \sum_{\alpha=1}^{\kappa} k_{\mathrm{e}\alpha} \xi_\alpha - b_{\mathrm{e}} (T - T_\circ)\right] \\ -A_\alpha &= \rho \frac{\partial f}{\partial \xi_\alpha} = \frac{\rho}{\rho_\circ}\left[k_{\mathrm{e}\alpha} I'_\epsilon + \sum_{\beta=1}^{\kappa} k_{\alpha\beta} \xi_\beta - b_\alpha (T - T_\circ)\right] \end{aligned} \tag{7.67}$$

$$\begin{aligned} \overset{\circ}{\boldsymbol{\sigma}}{}^{(\mathrm{eq})} &= \rho \frac{\partial f}{\partial \overset{\circ}{\boldsymbol{\epsilon}}{}^{(\mathrm{e})}} = \frac{\rho}{\rho_\circ} c_{\mathrm{ee}} \overset{\circ}{\boldsymbol{\epsilon}}{}^{(\mathrm{e})} \\ -\left(\overset{\circ}{\boldsymbol{P}}_\alpha - \overset{\circ}{\boldsymbol{\sigma}}{}^{(\mathrm{eq})}\right) &= \rho \frac{\partial f}{\partial \overset{\circ}{\boldsymbol{\epsilon}}{}^{(\mathrm{a})}_\alpha} = \frac{\rho}{\rho_\circ} \sum_{\beta=1}^{\nu} c_{\alpha\beta} \overset{\circ}{\boldsymbol{\epsilon}}{}^{(\mathrm{a})}_\beta \\ -\vec{H}_\alpha &= \rho \frac{\partial f}{\partial \vec{\eta}_\alpha} = \frac{\rho}{\rho_\circ} \sum_{\beta=1}^{\mu} l_{\alpha\beta} \vec{\eta}_\beta \end{aligned} \tag{7.68}$$

In the linearized approximation the density is a constant, so that $\rho/\rho_\circ$ can be approximated by unity.

From $(7.67)_1$ it is found with the principle of Le Chatelier that

$$T\left(\frac{\partial s}{\partial T}\right)_{v',\xi} = \frac{T}{T_\circ} c_v^{(\circ)} \approx c_v^{(\circ)} \geq 0 \tag{7.69}$$

in which v' stands for I'_ϵ and ξ for all the ξ_α.

7.4.3. Some historical remarks. The thermodynamic theory of irreversible processes for linear rheological bodies was developed first by Meixner*, who introduced the notions of an *internal process* and of *hidden variables.* His discussions of volume relaxation are very detailed, but the discussion of stress relaxation, deformation, and retardation is treated by Meixner in such a generalized way that it is quite difficult to recognize the various rheological classes. The considerations of Staverman and Schwarzl† contained thermodynamic elements, but did not yet form a completely developed TIP-theory.

Kluitenberg‡ has discussed the theory of thermodynamics of irreversible processes for the fundamental rheological bodies in a series of papers. The elastic deformations need not be restricted to small deformations as long as the irreversible processes are supposed to be linear. This idea is developed by Kluitenberg§ and used by Leonov¶ for the development of his nonlinear models. From the linear models, quasi-linear models can be derived by using objectivity requirements for the formulation of the equations of state. Here,

---

* Meixner J. 1949. Thermodynamik und Relaxationserscheinungen. *Zeitschrift für Naturforschung*, 4a: 594–600; Meixner J. 1953. Die Thermodynamische Theorie der Relaxationserscheinungen und ihre Zusammenhang mit der Nachwirkungstheorie. *Kolloid Zeitschrift*, 134: 3–20; Meixner J. 1954. Thermodynamische Theorie der Elastischen Relaxation. *Zeitschrift für Naturforschung*, 9a: 654–663.

† Staverman A.J. and Schwarzl F. 1952. Thermodynamics of viscoelastic behaviour (Model theory). *Proceedings of the Academy of Sciences, Amsterdam*, B55: 474–485; Staverman A.J. and Schwarzl F. 1952. Non-equilibrium thermodynamics of visco-elastic behaviour. *Proceedings of the Academy of Sciences, Amsterdam*, B55: 486–492.

‡ Kluitenberg G.A. 1962. Thermodynamic theory of elasticity and plasticity. *Physica*, 28: 217–232; Kluitenberg G.A. 1962. A note on the thermodynamics of Maxwell bodies, Kelvin bodies (Voigt bodies), and fluids. *Physica*, 28: 561–568; Kluitenberg G.A. 1962. On the rheology and thermodynamics of irreversible processes. *Physica*, 28: 1173–1183; Kluitenberg G.A. 1962. On the thermodynamics of viscosity and plasticity. *Physica*, 29: 633–652; Kluitenberg G.A. 1966. Application of the thermodynamics of irreversible processes to continuum mechanics. §6 of *Non-equilibrium Thermodynamics, Variational Techniques, and Stability.* University of Chicago Press, Illinois; Kluitenberg G.A. 1967. On heat dissipation due to irreversible mechanical phenomena in continuous media. *Physica*, 35: 177–192; Kluitenberg G.A. 1968. A thermodynamic derivation of the stress-strain relations for Burgers media and related substances. *Physica*, 38: 513–548; Kluitenberg G.A. 1972. On plasticity, elastic relaxation phenomena, and a stress-strain relation which characterizes Schofield–Scott Blair media. *Applied Scientific Research*, 25: 383–398; Kluitenberg G.A. 1977. A thermodynamic discussion of the possibility of singular yield conditions in plasticity theory. *Physica*, 88A: 122–134; Kluitenberg G.A. and Ciancio V. 1978. On linear dynamical equations of state for isotropic media I. *Physica*, 93A: 273–286; Ciancio V. and Kluitenberg G.A. 1979. On linear dynamical equations of state for isotropic media II. *Physica*, 99A: 592–600.

§ Kluitenberg G.A. 1964. A unified thermodynamic theory for large deformations in elastic media and Kelvin (Voigt) media, and for viscous fluid flow. *Physica*, 30: 1945–1972.

¶ Leonov A.I. 1976. Nonequilibrium thermodynamics and rheology of viscoelastic polymer media. *Rheologica Acta*, 15: 85–98.

the pioneering work is done by Oldroyd*. But Oldroyd did not stop at quasi-linear models. He went on to discuss fully nonlinear models for an elastic solid as well in his 1958 paper.

Much of the present discussion is derived from the papers of Kluitenberg, particularly those parts dealing with the deviatoric deformations. Especially, the idea of splitting $\boldsymbol{\epsilon}$ into an elastic and an anelastic part is borrowed from Kluitenberg, where he expressed the internal variables as parts of $\boldsymbol{\epsilon}^{(\mathrm{a})}$. As already remarked, this is correct for deviatoric deformations, but not for volumetric deformations if thermal relaxation is taken into account. Isotropic bodies are postulated here. The discussion of volume relaxations is inspired more by the work of Meixner.

7.5. ENTROPY PRODUCTION

The two points of departure for the calculation of the entropy production are

(1) The thermal energy equation (4.123) without diffusion and electric currents (Note the opposite sign conventions for $\mathbf{P}$ and $\boldsymbol{\sigma}$)

$$\rho\dot{u} = -\mathrm{div}\,\vec{\Phi}_{(\mathrm{u})} + \boldsymbol{\sigma} : \boldsymbol{D} \tag{7.70}$$

in which for small deformations it can be assumed that

$$\boldsymbol{D} = \dot{\boldsymbol{\epsilon}} = \dot{\boldsymbol{\epsilon}}^{(\mathrm{e})} + \dot{\boldsymbol{\epsilon}}^{(\mathrm{a})} + \dot{\boldsymbol{\epsilon}}^{(\mathrm{v})} \tag{7.71}$$

by (7.36) in which (7.42) has been substituted.

(2) The principle of the local equilibrium, according to which (7.61) applies in the form

$$\rho\dot{u} = \rho T\dot{s} - p^{(\mathrm{eq})}\dot{I}'_{\epsilon} + \mathring{\boldsymbol{\sigma}}^{(\mathrm{eq})} : \dot{\mathring{\boldsymbol{\epsilon}}}^{(\mathrm{e})} - \sum_{\alpha=1}^{\kappa} A_{\alpha}\dot{\xi}_{\alpha} - \sum_{\beta=1}^{\mu} \vec{H}_{\beta}\cdot\dot{\vec{\eta}}_{\beta} - \sum_{\gamma=1}^{\nu}\left(\mathring{\boldsymbol{P}}_{\gamma} - \mathring{\boldsymbol{\sigma}}^{(\mathrm{eq})}\right) : \dot{\mathring{\boldsymbol{\epsilon}}}_{\gamma}^{(\mathrm{a})} \tag{7.72}$$

---

* Oldroyd J.G. 1950. On the formulation of rheological equations of state. *Proceedings of the Royal Society of London*, A200: 523–541; Oldroyd J.G. 1950. Finite strains in an anisotropic elastic continuum. *Proceedings of the Royal Society of London*, A202: 345–358; Oldroyd J.G. 1958. Non-Newtonian effects in steady motion of some idealized elastico-viscous liquids. *Proceedings of the Royal Society of London*, A245: 278–297; Oldroyd J.G. 1965. Some steady flows of the general elastico-viscous liquid. *Proceedings of the Royal Society of London*, A283: 115–133; Oldroyd J.G. 1970. Equations of state of continuous matter in general relativity. *Proceedings of the Royal Society of London*, A316: 1–28; Oldroyd J.G. 1984. An approach to non-Newtonian fluid mechanics. *Journal of Non-Newtonian Fluid Mechanics*, 14: 9–46.

Elimination of $\dot{u}$ from (7.70) and (7.72) yields

$$\rho T\dot{s} = -\mathrm{div}\,\vec{\Phi}_{(\mathrm{u})} + \boldsymbol{\sigma} : \left(\dot{\boldsymbol{\epsilon}}^{(\mathrm{e})} + \dot{\boldsymbol{\epsilon}}^{(\mathrm{a})} + \dot{\boldsymbol{\epsilon}}^{(\mathrm{v})}\right) + p^{(\mathrm{eq})}\dot{I}'_{\epsilon} - \mathring{\boldsymbol{\sigma}}^{(\mathrm{eq})} : \dot{\mathring{\boldsymbol{\epsilon}}}^{(\mathrm{e})}$$
$$+ \sum_{\alpha=1}^{\kappa} A_{\alpha}\dot{\xi}_{\alpha} + \sum_{\beta=1}^{\mu} \vec{H}_{\beta}\cdot\dot{\vec{\eta}}_{\beta} + \sum_{\gamma=1}^{\nu}\left(\mathring{\boldsymbol{P}}_{\gamma} - \mathring{\boldsymbol{\sigma}}^{(\mathrm{eq})}\right) : \dot{\mathring{\boldsymbol{\epsilon}}}_{\gamma}^{(\mathrm{a})}$$

or

$$\rho T\dot{s} = -\mathrm{div}\,\vec{\Phi}_{(\mathrm{u})} - \left(p - p^{(\mathrm{eq})}\right)\left(\dot{I}_{\epsilon}^{(\mathrm{e})} + \dot{I}_{\epsilon}^{(\mathrm{a})}\right) - p\dot{I}_{\epsilon}^{(\mathrm{v})} + \left(\mathring{\boldsymbol{\sigma}} - \mathring{\boldsymbol{\sigma}}^{(\mathrm{eq})}\right) : \dot{\mathring{\boldsymbol{\epsilon}}}^{(\mathrm{e})}$$
$$+ \mathring{\boldsymbol{\sigma}} : \dot{\mathring{\boldsymbol{\epsilon}}}^{(\mathrm{v})} + \sum_{\alpha=1}^{\kappa} A_{\alpha}\dot{\xi}_{\alpha} + \sum_{\beta=1}^{\mu} \vec{H}_{\beta}\cdot\dot{\vec{\eta}}_{\beta} + \sum_{\gamma=1}^{\nu}\left(\mathring{\boldsymbol{P}}_{\gamma} + \mathring{\boldsymbol{\sigma}} - \mathring{\boldsymbol{\sigma}}^{(\mathrm{eq})}\right) : \dot{\mathring{\boldsymbol{\epsilon}}}_{\gamma}^{(\mathrm{a})}$$

This equation can be written as a balance equation for the entropy

$$\rho\dot{s} = -\mathrm{div}\left(\vec{\Phi}_{(\mathrm{u})}/T\right) + \pi_{(\mathrm{s})} \tag{7.73}$$

where $\pi_{(\mathrm{s})}$ is the rate of entropy production per unit of volume and per unit of time. The dissipation and the rate of entropy production are given by

$$\mathfrak{D} = T\pi_{(\mathrm{s})} = -\frac{1}{T}\vec{\Phi}_{(\mathrm{u})}\cdot\mathrm{grad}\,T - \left(p - p^{(\mathrm{eq})}\right)\left(\dot{I}_{\epsilon}^{(\mathrm{e})} + \dot{I}_{\epsilon}^{(\mathrm{a})}\right) - p\dot{I}_{\epsilon}^{(\mathrm{v})}$$
$$+ \left(\mathring{\boldsymbol{\sigma}} - \mathring{\boldsymbol{\sigma}}^{(\mathrm{eq})}\right) : \dot{\mathring{\boldsymbol{\epsilon}}}^{(\mathrm{e})} + \mathring{\boldsymbol{\sigma}} : \dot{\mathring{\boldsymbol{\epsilon}}}^{(\mathrm{v})}$$
$$+ \sum_{\alpha=1}^{\kappa} A_{\alpha}\dot{\xi}_{\alpha} + \sum_{\beta=1}^{\mu} \vec{H}_{\beta}\cdot\dot{\vec{\eta}}_{\beta} + \sum_{\gamma=1}^{\nu}\left(\mathring{\boldsymbol{P}}_{\gamma} + \mathring{\boldsymbol{\sigma}} - \mathring{\boldsymbol{\sigma}}^{(\mathrm{eq})}\right) : \dot{\mathring{\boldsymbol{\epsilon}}}_{\gamma}^{(\mathrm{a})} \tag{7.74}$$

Comparison of (7.74) with the dissipation function (4.150) shows that the vector and the tensor affinities are added to this expression and that the term $-\boldsymbol{\Pi} : \boldsymbol{D} = \boldsymbol{\sigma} : \boldsymbol{D}$ has been expanded. Because the medium is supposed to be chemically simple, there is no difference between the energy flux $\vec{\Phi}_{(\mathrm{u})}$ and the reduced heat flux $\vec{\Phi}_{(\mathrm{q})}$. Yet from now on $\vec{\Phi}_{(\mathrm{q})}$ will be used in the entropy production and the dissipation function to be in line with the notation for multicomponent systems given in (4.150), and also since for the single component system it can be stated that

$$\boxed{\vec{\Phi}_{(\mathrm{q})} = \vec{\Phi}_{(\mathrm{u})}}$$

The elastic deformations are not allowed in the entropy production, so that

$$\left(p - p^{(\mathrm{eq})}\right)\dot{I}_{\epsilon}^{(\mathrm{e})} = 0 \qquad \text{and} \qquad \left(\mathring{\boldsymbol{\sigma}} - \mathring{\boldsymbol{\sigma}}^{(\mathrm{eq})}\right) : \dot{\mathring{\boldsymbol{\epsilon}}}^{(\mathrm{e})} = 0 \tag{7.75}$$

Based on Reiner's first rheological axiom it is posed that in principle for the volume deformations $I_\epsilon^{(e)} \neq 0$, so that always

$$p = p^{(eq)} \qquad \text{for} \qquad I_\epsilon^{(e)} \neq 0 \tag{7.76}$$

The conditions $(7.75)_2$ result into two cases

$$\begin{aligned} \mathring{\boldsymbol{\sigma}} &= \mathring{\boldsymbol{\sigma}}^{(eq)} && \text{if} \quad \mathring{\boldsymbol{\epsilon}}^{(e)} \neq 0 \\ \mathring{\boldsymbol{\sigma}} &\neq \mathring{\boldsymbol{\sigma}}^{(eq)} = 0 && \text{if} \quad \mathring{\boldsymbol{\epsilon}}^{(e)} = 0 \end{aligned} \tag{7.77}$$

In the second case it follows that $\mathring{\boldsymbol{\sigma}}^{(eq)} = 0$ from the fact that $\mathring{\boldsymbol{\epsilon}}^{(e)}$ is found in the thermostatic potentials, so that

$$\mathring{\boldsymbol{\sigma}}^{(eq)} = \rho \frac{\partial f}{\partial \mathring{\boldsymbol{\epsilon}}^{(e)}} = 0 \qquad \text{for} \qquad \mathring{\boldsymbol{\epsilon}}^{(e)} = 0 \tag{7.78}$$

From (7.77) it follows that there is a distinction between *solid* bodies with $\mathring{\boldsymbol{\epsilon}}^{(e)} \neq 0$ and *fluid* bodies with $\mathring{\boldsymbol{\epsilon}}^{(e)} = 0$.

The dissipation function (7.74) becomes with the substitution of (7.75)

$$\boxed{\begin{aligned} \mathfrak{D} = T\pi_{(s)} = -\frac{1}{T}\vec{\Phi}_{(q)} \cdot \operatorname{grad} T - p\dot{I}_\epsilon^{(v)} + \sum_{\alpha=1}^{\kappa} A_\alpha \dot{\xi}_\alpha + \sum_{\beta=1}^{\mu} \vec{H}_\beta \cdot \dot{\vec{\eta}}_\beta \\ + \mathring{\boldsymbol{\sigma}} : \dot{\mathring{\boldsymbol{\epsilon}}}^{(v)} + \sum_{\gamma=1}^{\nu} \left(\mathring{\boldsymbol{P}}_\gamma + \mathring{\boldsymbol{\sigma}} - \mathring{\boldsymbol{\sigma}}^{(eq)} \right) : \dot{\mathring{\boldsymbol{\epsilon}}}_\gamma^{(a)} \end{aligned}} \tag{7.79}$$

The entropy principle of the thermodynamics demands that

$$\pi_{(s)} \geq 0 \tag{7.80}$$

The inequality sign dictates the direction in which the irreversible processes proceed. For example, for a single internal scalar process it is necessary that

$$\begin{aligned} A_\alpha \geq 0 \quad &\Longleftrightarrow \quad \dot{\xi}_\alpha \geq 0 \\ A_\alpha \leq 0 \quad &\Longleftrightarrow \quad \dot{\xi}_\alpha \leq 0 \end{aligned} \tag{7.81}$$

so that in equilibrium, for which the equality sign in (7.80) applies, it follows that

$$A_\alpha = 0 \quad \Longleftrightarrow \quad \dot{\xi}_\alpha = 0$$

This remains valid for all terms in (7.79), so that equilibrium is characterized by

$$
\begin{array}{rclrcl}
p = 0 & \Leftrightarrow & \overset{\bullet}{I}{}_{\epsilon}^{(\mathrm{v})} = 0 & A_{\alpha} = 0 & \Leftrightarrow & \dot{\xi}_{\alpha} = 0 \\
\vec{\Phi}_{(\mathrm{q})} = \vec{0} & \Leftrightarrow & \operatorname{grad} T = \vec{0} & \vec{H}_{\beta} = \vec{0} & \Leftrightarrow & \dot{\vec{\eta}}_{\beta} = \vec{0} \\
\mathring{\boldsymbol{\sigma}} = 0 & \Leftrightarrow & \overset{\bullet}{\mathring{\boldsymbol{\epsilon}}}{}^{(\mathrm{v})} = 0 & \mathring{\boldsymbol{P}}_{\gamma} + \mathring{\boldsymbol{\sigma}} - \mathring{\boldsymbol{\sigma}}^{(\mathrm{eq})} = 0 & \Leftrightarrow & \overset{\bullet}{\mathring{\boldsymbol{\epsilon}}}{}_{\gamma}^{(\mathrm{a})} = 0
\end{array}
\tag{7.82}
$$

Remember that $\mathring{\boldsymbol{\sigma}}$ in (7.82) is defined with respect to the reference state, so that p is also defined with respect to the reference state. If p is the absolute pressure, then p has to be replaced everywhere by $p - p_{\mathrm{o}}$. Consider for $(7.82)_3$ that in equilibrium, $\mathring{\boldsymbol{\sigma}} = \mathring{\boldsymbol{\sigma}}^{(\mathrm{eq})}$ always applies, so that only $\mathring{\boldsymbol{P}}_{\gamma}$ is left over. The conditions in (7.82) represent the equilibrium of the reference state. There is also the possibility of a constrained equilibrium, frozen or false equilibrium, which differs from the reference state. Internal processes are constrained if they are unequal to zero, but not proceed, so if

$$
\begin{array}{rclrcl}
A_{\alpha} \neq 0 & \Leftrightarrow & \dot{\xi}_{\alpha} = 0 & \vec{H}_{\beta} \neq \vec{0} & \Leftrightarrow & \dot{\vec{\eta}}_{\beta} = \vec{0} \\
& & \mathring{\boldsymbol{P}}_{\gamma} + \mathring{\boldsymbol{\sigma}} - \mathring{\boldsymbol{\sigma}}^{(\mathrm{eq})} \neq 0 & \Leftrightarrow & \overset{\bullet}{\mathring{\boldsymbol{\epsilon}}}{}_{\gamma}^{(\mathrm{a})} = 0 &
\end{array}
\tag{7.83}
$$

In the constrained state the rate of entropy production $\pi_{(\mathrm{s})} = 0$, and formally there is equilibrium. This type of equilibrium might be observed for very fast variations of the macroscopic conditions that can not be followed by the internal processes due to their retarded return to equilibrium, although the affinities differ from zero because of the deviations from the true equilibrium state.

Clearly, the coefficient $c_{\mathrm{v}}^{(\mathrm{o})}$ as defined by (7.69) represents the specific heat at constant volume and constrained internal processes ξ_{α}.

The dissipation can be calculated as follows. From $(7.67)_1$

$$
\rho \dot{s} = b_{\mathrm{e}} \overset{\bullet}{I}{}'_{\epsilon} + \sum_{\alpha=1}^{\kappa} b_{\alpha} \dot{\xi}_{\alpha} + \frac{\rho}{T} c_v^{(\mathrm{o})} \dot{T}
$$

Eliminate $\dot{s}$ with the help of (7.73)

$$
\frac{\rho}{T} c_v^{(\mathrm{o})} \dot{T} = -\operatorname{div}\left(\vec{\Phi}_{(\mathrm{q})}/T\right) + \pi_{(\mathrm{s})} - b_{\mathrm{e}} \overset{\bullet}{I}{}'_{\epsilon} - \sum_{\alpha=1}^{\kappa} b_{\alpha} \dot{\xi}_{\alpha}
$$

$$
\rho c_v^{(\mathrm{o})} \dot{T} = -\operatorname{div} \vec{\Phi}_{(\mathrm{q})} - T \left[b_{\mathrm{e}} \overset{\bullet}{I}{}'_{\epsilon} + \sum_{\alpha=1}^{\kappa} b_{\alpha} \dot{\xi}_{\alpha} \right] + \mathfrak{D}^{+}
\tag{7.84}
$$

in which the dissipation function $\mathfrak{D}^+$ is given by

$$\mathfrak{D}^+ = T\pi_{(\mathrm{s})} + \frac{1}{T}\vec{\Phi}_{(\mathrm{q})} \cdot \operatorname{grad} T \tag{7.85}$$

or with (7.79)

$$\boxed{\begin{aligned}\mathfrak{D}^+ = -p\dot{I}_\epsilon^{(\mathrm{v})} + \sum_{\alpha=1}^{\kappa} A_\alpha \dot{\xi}_\alpha + \sum_{\beta=1}^{\mu} \vec{H}_\beta \cdot \dot{\vec{\eta}}_\beta + \mathring{\boldsymbol{\sigma}} : \mathring{\dot{\boldsymbol{\epsilon}}}^{(\mathrm{v})} \\ + \sum_{\gamma=1}^{\nu} \left(\mathring{\boldsymbol{P}}_\gamma + \mathring{\boldsymbol{\sigma}} - \mathring{\boldsymbol{\sigma}}^{(\mathrm{eq})} \right) : \mathring{\dot{\boldsymbol{\epsilon}}}_\gamma^{(\mathrm{a})} \geq 0\end{aligned}} \tag{7.86}$$

The dissipation function $\mathfrak{D}^+$ represents that part of the work that is transformed into thermal energy and cannot be regained as useful work. The dissipation is caused by viscous deformations and internal processes according to (7.86). The dissipation by internal processes is sometimes called *internal friction.*

7.6. PHENOMENOLOGICAL EQUATIONS

From the true equilibrium conditions (7.82) it is found that, in the first approximation, the thermodynamic fluxes are homogeneous, linear functions of the thermodynamic forces. Thermodynamic fluxes are process quantities, which can be defined from the divergence terms in the balance equations, so that $\vec{\Phi}_{(\mathrm{q})}$, p, and $\mathring{\boldsymbol{\sigma}}$ can be considered as fluxes. Furthermore, $\dot{\xi}_\alpha$, $\dot{\vec{\eta}}_\beta$, and $\mathring{\dot{\boldsymbol{\epsilon}}}_\gamma^{(\mathrm{a})}$ can be considered as fluxes, since these are time derivatives of the thermodynamic variables; in particular they are the extents of change of the thermodynamic variables.

The scalar quantities depend only on scalar quantities, according to the Curie principle for isotropic media, so that the *scalar phenomenological equations* become

$$\boxed{\begin{aligned}\dot{\xi}_\alpha &= \sum_{\beta=1}^{\kappa} \kappa_{\alpha\beta} A_\beta + \kappa_{\alpha\mathrm{v}} \dot{I}_\epsilon^{(\mathrm{v})} \\ -p &= \sum_{\alpha=1}^{\kappa} \kappa_{\mathrm{v}\alpha} A_\alpha + \kappa_{\mathrm{vv}} \dot{I}_\epsilon^{(\mathrm{v})}\end{aligned}} \tag{7.87}$$

In a time reversal, A_α does dot change sign, while $\dot{I}_\epsilon^{(\mathrm{v})}$—like all velocity gradients—changes its sign, so that $\dot{I}_\epsilon^{(\mathrm{v})}$ is an odd quantity and A_α an even quantity. The Onsager Casimir reciprocal relations are

$$\kappa_{\alpha\mathrm{v}} = -\kappa_{\mathrm{v}\alpha} \qquad \text{and} \qquad \kappa_{\alpha\beta} = \kappa_{\beta\alpha} \tag{7.88}$$

The entropy production, associated with the scalar processes, is with (7.79)

$$\mathfrak{D}^{(s)} = \left(T\pi_{(s)}\right)_{\text{scalar}} = -p\dot{I}_{\epsilon}^{(v)} + \sum_{\alpha=1}^{\kappa} A_\alpha \dot{\xi}_\alpha$$
$$= \kappa_{vv}\dot{I}_{\epsilon}^{(v)2} + \sum_{\alpha,\beta=1}^{\kappa} \kappa_{\alpha\beta} A_\alpha A_\beta \geq 0 \tag{7.89}$$

so that

$$\kappa_{vv} \geq 0 \qquad \text{and} \qquad \sum_{\alpha,\beta=1}^{\kappa} \kappa_{\alpha\beta} y_\alpha y_\beta \geq 0 \tag{7.90}$$

The scalar processes were discussed briefly at the end of chapter four. The coefficients used here are provided with subscripts to indicate the processes to which they are related. If it is taken into account that the stress tensor $\mathbf{P}$ has the sign opposite to that of the stress tensor $\boldsymbol{\sigma}$, then the coefficients $L_{\alpha\beta}$, λ_α and κ used in section 4.10 correspond to $\kappa_{\alpha\beta}$, $\kappa_{v\alpha}$, and κ_{vv} respectively.

The thermodynamic fluxes with a vector character depend in isotropic media only on the vector thermodynamic fluxes. This restriction results in the following *vector phenomenological equations*

$$\boxed{\begin{aligned} \dot{\vec{\eta}}_\beta &= \sum_{\alpha=1}^{\mu} \lambda_{\beta\alpha} \vec{H}_\alpha + \frac{1}{T}\lambda_{\beta q} \operatorname{grad} T \\ -\vec{\Phi}_{(q)} &= \lambda_{qq} \operatorname{grad} T + \sum_{\beta=1}^{\mu} \lambda_{q\beta} \vec{H}_\beta \end{aligned}} \tag{7.91}$$

The vector affinity $\vec{H}$ is an odd* variable, while T and $\operatorname{grad} T$ are even variables for time reversal (see Chapter 5). The Onsager Casimir reciprocal relations then become

$$\lambda_{q\alpha} = -\lambda_{\alpha q} \qquad \text{and} \qquad \lambda_{\alpha\beta} = \lambda_{\beta\alpha} \tag{7.92}$$

According to (7.79) $(\operatorname{grad} T)/T$ can be seen as a force, so that it is appropriate to include the factor $1/T$ in λ_{qq}.

The entropy production associated with the vector processes becomes

$$\mathfrak{D}^{(v)} = \left(T\pi_{(s)}\right)_{\text{vector}} = -\frac{1}{T}\vec{\Phi}_{(q)} \cdot \operatorname{grad} T + \sum_{\beta=1}^{\mu} \vec{H}_\beta \cdot \dot{\vec{\eta}}_\beta$$
$$= -\frac{1}{T}\lambda_{qq} (\operatorname{grad} T)^2 + \sum_{\alpha,\beta=1}^{\mu} \lambda_{\alpha\beta} \vec{H}_\alpha \cdot \vec{H}_\beta \geq 0 \tag{7.93}$$

* Although internal vector processes are considered, an analogy can be drawn with the vector processes in section 6.3.1. Compare $\dot{\vec{\eta}}_\beta$ with $\vec{d}_K$, and $\vec{H}_\beta$ with the odd variable $\vec{V}_L$. Only the coefficient λ_{qq} then corresponds to the coefficient L_{TT}/T, while the coefficients $\lambda_{q\alpha}$, $\lambda_{\alpha q}$ and $\lambda_{\alpha\beta}$ can be compared with L_{TK}, L_{KT} and pL_{KL} respectively.

so that

$$\lambda_{\mathrm{qq}} \geq 0 \qquad \text{and} \qquad \sum_{\alpha,\beta=1}^{\mu} \lambda_{\alpha\beta} y_\alpha y_\beta \geq 0 \tag{7.94}$$

Finally, the *tensor phenomenological equations* for isotropic media are given by

$$\boxed{\begin{aligned} \overset{\circ}{\dot{\boldsymbol{\epsilon}}}{}^{(\mathrm{a})}_\gamma &= \sum_{\beta=1}^{\nu} \phi'_{\gamma\beta} \left(\overset{\circ}{\boldsymbol{P}}_\beta + \overset{\circ}{\boldsymbol{\sigma}} - \overset{\circ}{\boldsymbol{\sigma}}{}^{(\mathrm{eq})} \right) + \phi'_{\gamma\mathrm{v}} \overset{\circ}{\dot{\boldsymbol{\epsilon}}}{}^{(\mathrm{v})} \\ \overset{\circ}{\boldsymbol{\sigma}} &= \phi'_{\mathrm{vv}} \overset{\circ}{\dot{\boldsymbol{\epsilon}}}{}^{(\mathrm{v})} + \sum_{\gamma=1}^{\nu} \phi'_{\mathrm{v}\gamma} \left(\overset{\circ}{\boldsymbol{P}}_\gamma + \overset{\circ}{\boldsymbol{\sigma}} - \overset{\circ}{\boldsymbol{\sigma}}{}^{(\mathrm{eq})} \right) \end{aligned}} \tag{7.95}$$

In this expression the viscous rate of deformation $\overset{\circ}{\dot{\boldsymbol{\epsilon}}}{}^{(\mathrm{v})}$ is an odd variable. In section 4.10 only the viscous part of the deformations is accounted for, in which the coefficient ϕ'_{vv} can be compared with 2η, while the other coefficients $\phi'_{\mathrm{v}\gamma}$ are introduced in this section.

A stress tensor is even, so that also $\overset{\circ}{\boldsymbol{P}}_\alpha$ is an even variable, if $\overset{\circ}{\boldsymbol{P}}_\alpha$ is given the meaning of an affinity stress tensor. The reciprocal relations therefore become

$$\phi'_{\mathrm{v}\alpha} = -\phi'_{\alpha\mathrm{v}} \qquad \text{and} \qquad \phi'_{\alpha\beta} = \phi'_{\beta\alpha} \tag{7.96}$$

From (7.95) it follows that

$$\begin{aligned} \overset{\circ}{\dot{\boldsymbol{\epsilon}}}{}^{(\mathrm{v})} &= \frac{1}{\phi'_{\mathrm{vv}}} \overset{\circ}{\boldsymbol{\sigma}} - \sum_{\gamma=1}^{\nu} \frac{\phi'_{\mathrm{v}\gamma}}{\phi'_{\mathrm{vv}}} \left(\overset{\circ}{\boldsymbol{P}}_\gamma + \overset{\circ}{\boldsymbol{\sigma}} - \overset{\circ}{\boldsymbol{\sigma}}{}^{(\mathrm{eq})} \right) \\ \overset{\circ}{\dot{\boldsymbol{\epsilon}}}{}^{(\mathrm{a})}_\gamma &= \frac{\phi'_{\gamma\mathrm{v}}}{\phi'_{\mathrm{vv}}} \overset{\circ}{\boldsymbol{\sigma}} + \sum_{\beta=1}^{\nu} \left(\phi'_{\gamma\beta} - \frac{\phi'_{\gamma\mathrm{v}} \phi'_{\mathrm{v}\beta}}{\phi'_{\mathrm{vv}}} \right) \left(\overset{\circ}{\boldsymbol{P}}_\beta + \overset{\circ}{\boldsymbol{\sigma}} - \overset{\circ}{\boldsymbol{\sigma}}{}^{(\mathrm{eq})} \right) \end{aligned}$$

By defining the following phenomenological coefficients

$$\begin{gathered} \phi_{\mathrm{vv}} = \frac{1}{\phi'_{\mathrm{vv}}} \qquad \phi_{\mathrm{v}\gamma} = -\frac{\phi'_{\mathrm{v}\gamma}}{\phi'_{\mathrm{vv}}} \qquad \phi_{\gamma\mathrm{v}} = \frac{\phi'_{\gamma\mathrm{v}}}{\phi'_{\mathrm{vv}}} \\ \phi_{\alpha\beta} = \phi'_{\alpha\beta} - \frac{\phi'_{\alpha\mathrm{v}} \phi'_{\mathrm{v}\beta}}{\phi'_{\mathrm{vv}}} \end{gathered} \tag{7.97}$$

the tensor phenomenological equations then become

$$\begin{aligned} \overset{\circ}{\dot{\boldsymbol{\epsilon}}}{}^{(\mathrm{v})} &= \phi_{\mathrm{vv}} \overset{\circ}{\boldsymbol{\sigma}} + \sum_{\gamma=1}^{\nu} \phi_{\mathrm{v}\gamma} \left(\overset{\circ}{\boldsymbol{P}}_\gamma + \overset{\circ}{\boldsymbol{\sigma}} - \overset{\circ}{\boldsymbol{\sigma}}{}^{(\mathrm{eq})} \right) \\ \overset{\circ}{\dot{\boldsymbol{\epsilon}}}{}^{(\mathrm{a})}_\gamma &= \phi_{\gamma\mathrm{v}} \overset{\circ}{\boldsymbol{\sigma}} + \sum_{\beta=1}^{\nu} \phi_{\gamma\beta} \left(\overset{\circ}{\boldsymbol{P}}_\beta + \overset{\circ}{\boldsymbol{\sigma}} - \overset{\circ}{\boldsymbol{\sigma}}{}^{(\mathrm{eq})} \right) \end{aligned} \tag{7.98}$$

in which (7.96) implies that

$$\phi_{\mathrm{v}\gamma} = \phi_{\gamma\mathrm{v}} \qquad \text{and} \qquad \phi_{\alpha\beta} = \phi_{\beta\alpha} \tag{7.99}$$

The following entropy production is associated with the tensor processes

$$\mathfrak{D}^{(\mathrm{t})} = \left(T\pi_{(\mathrm{s})}\right)_{\mathrm{tensor}} = \mathring{\boldsymbol{\sigma}} : \dot{\mathring{\boldsymbol{\epsilon}}}^{(\mathrm{v})} + \sum_{\gamma=1}^{\nu} \left(\mathring{\boldsymbol{P}}_{\gamma} + \mathring{\boldsymbol{\sigma}} - \mathring{\boldsymbol{\sigma}}^{(\mathrm{eq})}\right) : \dot{\mathring{\boldsymbol{\epsilon}}}_{\gamma}^{(\mathrm{a})}$$

or, with the substitution of (7.98),

$$\mathfrak{D}^{(\mathrm{t})} = \left(T\pi_{(\mathrm{s})}\right)_{\mathrm{tensor}} = \phi_{\mathrm{vv}}\mathring{\boldsymbol{\sigma}} : \mathring{\boldsymbol{\sigma}} + 2\sum_{\gamma=1}^{\nu} \phi_{\mathrm{v}\gamma}\mathring{\boldsymbol{\sigma}} : \left(\mathring{\boldsymbol{P}}_{\gamma} + \mathring{\boldsymbol{\sigma}} - \mathring{\boldsymbol{\sigma}}^{(\mathrm{eq})}\right)$$
$$+ \sum_{\alpha,\beta=1}^{\nu} \phi_{\alpha\beta} \left(\mathring{\boldsymbol{P}}_{\alpha} + \mathring{\boldsymbol{\sigma}} - \mathring{\boldsymbol{\sigma}}^{(\mathrm{eq})}\right) : \left(\mathring{\boldsymbol{P}}_{\beta} + \mathring{\boldsymbol{\sigma}} - \mathring{\boldsymbol{\sigma}}^{(\mathrm{eq})}\right) \geq 0 \tag{7.100}$$

so that

$$\phi_{\mathrm{vv}}y_{\mathrm{v}}^2 + 2\sum_{\gamma=1}^{\nu} \phi_{\mathrm{v}\gamma}y_{\mathrm{v}}y_{\gamma} + \sum_{\alpha,\beta=1}^{\nu} \phi_{\alpha\beta}y_{\alpha}y_{\beta} \geq 0 \tag{7.101}$$

By elimination of the internal quantities, the macroscopic constitutive equations can be derived from the thermodynamic equations of state and the phenomenological equations. This procedure will be illustrated first for simple cases.

7.7. SIMPLE BODIES WITH INTERNAL PROCESSES

Basic equations. For the simplest cases, consider one scalar, one vector and one tensor internal process. Then the Gibbs fundamental equation (7.61) becomes

$$\rho\, du = \rho T\, ds - p^{(\mathrm{eq})} dI_{\epsilon}' - A\, d\xi - \vec{H} \cdot d\vec{\eta}$$
$$+ \mathring{\boldsymbol{\sigma}}^{(\mathrm{eq})} : d\mathring{\boldsymbol{\epsilon}}^{(\mathrm{e})} - \left(\mathring{\boldsymbol{P}} - \mathring{\boldsymbol{\sigma}}^{(\mathrm{eq})}\right) : d\mathring{\boldsymbol{\epsilon}}^{(\mathrm{a})} \tag{7.102}$$

(7.62) becomes

$$\rho\, df = -\rho s\, dT - p^{(\mathrm{eq})} dI_{\epsilon}' - A\, d\xi - \vec{H} \cdot d\vec{\eta}$$
$$+ \mathring{\boldsymbol{\sigma}}^{(\mathrm{eq})} : d\mathring{\boldsymbol{\epsilon}}^{(\mathrm{e})} - \left(\mathring{\boldsymbol{P}}_{\gamma} - \mathring{\boldsymbol{\sigma}}^{(\mathrm{eq})}\right) : d\mathring{\boldsymbol{\epsilon}}^{(\mathrm{a})} \tag{7.103}$$

(7.64) becomes

$$\rho_{\circ}\left(f - f_{\circ}\right) = \tfrac{1}{2}c_{\mathrm{ee}}\mathring{\boldsymbol{\epsilon}}^{(\mathrm{e})} : \mathring{\boldsymbol{\epsilon}}^{(\mathrm{e})} + \tfrac{1}{2}c_{\mathrm{aa}}\mathring{\boldsymbol{\epsilon}}^{(\mathrm{a})} : \mathring{\boldsymbol{\epsilon}}^{(\mathrm{a})}$$
$$+ \tfrac{1}{2}k_{\mathrm{ee}}I_{\epsilon}'^{2} + k_{\mathrm{e}\xi}I_{\epsilon}'\xi + \tfrac{1}{2}k_{\xi\xi}\xi^{2}$$
$$+ \tfrac{1}{2}l_{\eta\eta}\eta^{2}$$
$$+ b_{\mathrm{e}}I_{\epsilon}'\left(T - T_{\circ}\right) - b_{\xi}\xi\left(T - T_{\circ}\right) - \tfrac{1}{2}c_{v}^{(\circ)}\frac{\rho_{\circ}}{T_{\circ}}\left(T - T_{\circ}\right)^{2} \tag{7.104}$$

The thermostatic equations of state (7.67) and (7.68) become

$$\left.\begin{aligned} s = -\frac{\partial f}{\partial T} &= \frac{1}{\rho_\circ}\left(b_{\mathrm{e}} I'_\epsilon + b_\xi \xi\right) + \frac{1}{T_\circ} c_v^{(\circ)}\left(T - T_\circ\right) \\ -p^{(\mathrm{eq})} = \rho \frac{\partial f}{\partial I'_\epsilon} &= k_{\mathrm{ee}} I'_\epsilon + k_{\mathrm{e}\xi}\xi - b_{\mathrm{e}}\left(T - T_\circ\right) \\ -A = \rho\frac{\partial f}{\partial \xi} &= k_{\mathrm{e}\xi} I'_\epsilon + k_{\xi\xi}\xi - b_\xi\left(T - T_\circ\right) \end{aligned}\right\} \tag{7.105}$$

$$-\vec{H} = \rho\frac{\partial f}{\partial \vec{\eta}} = l_{\eta\eta}\vec{\eta} \tag{7.106}$$

$$\left.\begin{aligned} \mathring{\boldsymbol{\sigma}}^{(\mathrm{eq})} &= \rho\frac{\partial f}{\partial \mathring{\boldsymbol{\epsilon}}^{(\mathrm{e})}} = c_{\mathrm{ee}}\mathring{\boldsymbol{\epsilon}}^{(\mathrm{e})} \\ -\left(\mathring{\boldsymbol{P}} - \mathring{\boldsymbol{\sigma}}^{(\mathrm{eq})}\right) &= \rho\frac{\partial f}{\partial \mathring{\boldsymbol{\epsilon}}^{(\mathrm{a})}} = c_{\mathrm{aa}}\mathring{\boldsymbol{\epsilon}}^{(\mathrm{a})} \end{aligned}\right\} \tag{7.107}$$

For the thermostatic coefficients

$$\begin{array}{ccc} & c_{\mathrm{ee}} \geq 0 \qquad\qquad c_{\mathrm{aa}} \geq 0 & \\ k_{\mathrm{ee}} \geq 0 & k_{\xi\xi} \geq 0 & k_{\mathrm{ee}}k_{\xi\xi} - k_{\mathrm{e}\xi}^2 \geq 0 \\ & l_{\eta\eta} \geq 0 \qquad\qquad c_v^{(\circ)} \geq 0 & \end{array} \tag{7.108}$$

The dissipation and the rate of entropy production (7.79) become

$$\begin{aligned} \mathfrak{D} = T\pi_{(\mathrm{s})} = &-\frac{1}{T}\vec{\Phi}_{(\mathrm{q})}\cdot \operatorname{grad} T - p\dot{I}_\epsilon^{(\mathrm{v})} + A\dot{\xi} \\ &+ \vec{H}\cdot\dot{\vec{\eta}} + \mathring{\boldsymbol{\sigma}} : \dot{\mathring{\boldsymbol{\epsilon}}}^{(\mathrm{v})} + \left(\mathring{\boldsymbol{P}} + \mathring{\boldsymbol{\sigma}} - \mathring{\boldsymbol{\sigma}}^{(\mathrm{eq})}\right) : \dot{\mathring{\boldsymbol{\epsilon}}}^{(\mathrm{a})} \geq 0 \end{aligned} \tag{7.109}$$

The scalar phenomenological equations (7.87) become

$$\begin{aligned} \dot{\xi} &= \kappa_{\xi\xi}A + \kappa_{\xi\mathrm{v}}\dot{I}_\epsilon^{(\mathrm{v})} \\ -p &= \kappa_{\mathrm{v}\xi}A + \kappa_{\mathrm{vv}}\dot{I}_\epsilon^{(\mathrm{v})} \end{aligned} \tag{7.110}$$

with

$$\kappa_{\mathrm{v}\xi} = -\kappa_{\xi\mathrm{v}} \tag{7.111}$$

and

$$\kappa_{\mathrm{vv}} \geq 0 \qquad\qquad \kappa_{\xi\xi} \geq 0 \tag{7.112}$$

The vector phenomenological equations (7.91) become

$$\begin{aligned} \dot{\vec{\eta}} &= \lambda_{\eta\eta}\vec{H} + \frac{1}{T}\lambda_{\eta\mathrm{q}}\operatorname{grad} T \\ -\vec{\Phi}_{(\mathrm{q})} &= \lambda_{\mathrm{q}\eta}\vec{H} + \lambda_{\mathrm{qq}}\operatorname{grad} T \end{aligned} \tag{7.113}$$

with

$$\lambda_{q\eta} = -\lambda_{\eta q} \tag{7.114}$$

and

$$\lambda_{qq} \geq 0 \qquad\qquad \lambda_{\eta\eta} \geq 0 \tag{7.115}$$

The tensor phenomenological equations (7.98) become

$$\begin{aligned} \overset{\bullet}{\mathring{\boldsymbol{\epsilon}}}{}^{(v)} &= \phi_{vv}\mathring{\boldsymbol{\sigma}} + \phi_{va}\left(\mathring{\boldsymbol{P}} + \mathring{\boldsymbol{\sigma}} - \mathring{\boldsymbol{\sigma}}^{(eq)}\right) \\ \overset{\bullet}{\mathring{\boldsymbol{\epsilon}}}{}^{(a)} &= \phi_{av}\mathring{\boldsymbol{\sigma}} + \phi_{aa}\left(\mathring{\boldsymbol{P}} + \mathring{\boldsymbol{\sigma}} - \mathring{\boldsymbol{\sigma}}^{(eq)}\right) \end{aligned} \tag{7.116}$$

with

$$\phi_{va} = \phi_{av} \tag{7.117}$$

and

$$\phi_{vv} \geq 0 \qquad\qquad \phi_{aa} \geq 0 \qquad\qquad \phi_{vv}\phi_{aa} - \phi_{va}^2 \geq 0 \tag{7.118}$$

It is clear that in isotropic linear bodies the scalar, vector, and tensor transport processes can be discussed independently of each other.

7.8. INTERNAL VECTOR PROCESSES

Heat conduction Of the internal processes, the vector processes are the most simple ones to discuss. The vector affinity $\vec{H}$ can be eliminated by the substitution of (7.106) into (7.113), yielding

$$\begin{aligned} -\vec{\Phi}_{(q)} &= \lambda_{qq}\operatorname{grad} T - \lambda_{q\eta}l_{\eta\eta}\vec{\eta} \\ \dot{\vec{\eta}} &= \frac{1}{T}\lambda_{\eta q}\operatorname{grad} T - \lambda_{\eta\eta}l_{\eta\eta}\vec{\eta} \end{aligned} \tag{7.119}$$

The second equation of (7.119) is the relaxation equation for the internal variable $\vec{\eta}$

$$(\lambda_{\eta\eta}l_{\eta\eta} + \mathcal{D})\,\vec{\eta} = \frac{1}{T}\lambda_{\eta q}\operatorname{grad} T \tag{7.120}$$

where from now on the operator $\mathcal{D}$ in the relaxation equations stands for the operator D/Dt. From (7.120) it is concluded that the internal vector process is affected only by the gradient of the temperature.

Next, $\vec{\eta}$ can be eliminated from $(7.119)_1$ for $\lambda_{q\eta} \neq 0$ and substituted into (7.120)

$$(\lambda_{\eta\eta}l_{\eta\eta} + \mathcal{D})\,\vec{\Phi}_{(q)} = -\left[\lambda_{qq}\lambda_{\eta\eta}l_{\eta\eta} - \frac{1}{T}\lambda_{q\eta}l_{\eta\eta}\lambda_{\eta q} + \lambda_{qq}\mathcal{D}\right]\operatorname{grad} T \tag{7.121}$$

and with (7.114) this can be written as

$$\boxed{(1 + \tau_q\mathcal{D})\,\vec{\Phi}_{(q)} = -\lambda^{(eq)}\,(1 + \bar{\tau}_q\mathcal{D})\operatorname{grad} T} \tag{7.122}$$

in which the relaxation time τ_q and the thermal conductivity $\lambda^{(eq)}$ are defined by

$$\tau_q = \frac{1}{l_{\eta\eta}\lambda_{\eta\eta}} \qquad \text{and} \qquad \lambda^{(eq)} = \lambda_{qq} + \frac{\lambda_{q\eta}^2}{T\lambda_{\eta\eta}} \tag{7.123}$$

respectively, and the retardation time $\bar{\tau}_q$ is defined by

$$\bar{\tau}_q = \frac{\lambda_{qq}\lambda_{\eta\eta}}{\lambda_{qq}\lambda_{\eta\eta} + \lambda_{q\eta}^2/T}\frac{1}{l_{\eta\eta}\lambda_{\eta\eta}} = \lambda_{qq}\frac{\tau_q}{\lambda^{(eq)}} \tag{7.124}$$

For internal vector processes ($l_{\eta\eta} \neq 0$) follows from (7.108) and (7.115) the conditions

$$\tau_q > \bar{\tau}_q \geq 0 \tag{7.125}$$

$$\lambda^{(eq)} \geq 0 \tag{7.126}$$

These important inequalities result from the thermodynamic stability of the material and the entropy principle. The coefficient $\lambda^{(eq)}$ is the heat conduction coefficient for which the internal process is in equilibrium with the macroscopic conditions. ($\dot{\vec{\eta}} = \vec{0}$).

If only relaxation of the heat waves is considered, then the retardation time $\bar{\tau}_q = 0$ for $\lambda_{qq} = 0$, $\lambda_{q\eta} \neq 0$, and $\lambda_{\eta\eta} \neq 0$, so that the differential equation (7.122) reduces to

$$(1 + \tau_q\mathcal{D})\,\vec{\Phi}_{(q)} = -\lambda^{(eq)}\operatorname{grad} T \tag{7.127}$$

This modification of Fourier's law has been proposed by various authors*. Clearly this modified form of the heat conduction law is acceptable by (7.114) (if $\lambda_{q\eta} = \lambda_{\eta q}$ then $\lambda_{qq} = 0$ would imply that $\lambda_{q\eta} = \lambda_{\eta q} = 0$). Substitution of (7.127) into the energy equation (3.66) for a solid leads to a hyperbolic telegraph equation with a finite thermal wave speed $\sqrt{\lambda^{(eq)}/C\tau_q}$. Heat waves in metals travel with a velocity of the order of the speed of light, and Fourier's

---

* Maxwell J.C. 1867. On the dynamical theory of gases. *Philosophical Transactions of the Royal Society of London*, 157: 49–88; Cattaneo C. 1948. Sulla conduzione del calore. *Atti del Seminario Matematico e Fisico della Università di Modena*, 3: 3–21; Cattaneo C. 1958. Sur une forme de l'équation éliminant le paradoxe d'une propagation instantanée. *Comptes Rendus de l'Académie des Sciences, Paris*, 247: 431–433; Chester M. 1963. Second sound in solids. *Physical Review*, 131: 2013–2015; Gurtin M.E. and Pipkin A.C. 1968. A general theory of heat conduction with finite wave speeds. *Archive for Rational Mechanics and Analysis*, 31: 113–126; Vernotte P. 1958. Les paradoxes de la théorie continue de l'équation de la chaleur. *Comptes Rendus Académie des Sciences, Paris*, 247: 3154–3155; Luikov A.V. 1966. Application of irreversible thermodynamics methods for investigation of heat and mass transfer. *International Journal of Heat and Mass Transfer*, 9: 139–152; Lambermont J. and Lebon G. 1973. On a generalization of the Gibbs equation for heat conduction. *Physics Letters* 42A: 499–500.

law is then quite sufficient. Slow speeds of the order of 1000 m/s are measured in dielectric crystals at low temperature. The slowest heat wave speeds are found in liquid helium II where the velocity ranges from zero at 2.2 K to about 100 m/s near absolute zero temperature. Substitution of (7.122) results in a parabolic equation that for small $\bar{\tau}_q \ll \tau_q$ smooths a discontinuity imposed by a sudden jump in the temperature. Not much attention is given in the literature to the retardation of the heat conduction. For an extensive review of the literature on heat waves see Joseph and Preziosi*.

7.9. INTERNAL SCALAR PROCESSES

7.9.1. Thermostatic equations of state. The thermostatic equations of state are given by (7.105). The quantities of state s, p, and $v = 1/\rho$ are defined with respect to the reference state. If these quantities are counted with respect to the reference state, then it can be noted that

$$\delta T = T - T_\circ \qquad \delta p = p - p_\circ \qquad \delta s = s - s_\circ \qquad I_\epsilon = \frac{v - v_\circ}{v_\circ} \approx \frac{\delta v}{v} \tag{7.128}$$

If in the reference state is in true equilibrium it is known that $A = 0$, $\xi = 0$, so that we have $A = \delta A$, and $\xi = \delta\xi$. It will be assumed that the viscous contributions to the volume deformations can be neglected

$$I_\epsilon^{(\mathrm{v})} = 0 \qquad \text{hence} \qquad I'_\epsilon = I_\epsilon^{(\mathrm{e})} + I_\epsilon^{(\mathrm{a})} = I_\epsilon \tag{7.129}$$

so that the volume viscosity will be considered to result from an internal process with a very short natural time. The equations of state (7.105) can thus be written

$$\begin{aligned} -\delta p &= k_{\mathrm{ee}}\frac{\delta v}{v} - b_{\mathrm{e}}\,\delta T + k_{\mathrm{e}\xi}\xi \\ \delta s &= b_{\mathrm{e}}\,\delta v + \frac{1}{T_\circ}c_v^{(\circ)}\,\delta T + v b_\xi \xi \\ -\delta(vA) &= k_{\mathrm{e}\xi}\,\delta v - b_\xi\,\delta T + v k_{\xi\xi}\xi \end{aligned} \tag{7.130}$$

These are the linear approximations of the thermodynamic equations of state, and written in differentials for the f-representation

$$\begin{aligned} -dp &= \frac{1}{v}k_{\mathrm{ee}}\,dv - b_{\mathrm{e}}\,dT + k_{\mathrm{e}\xi}\,d\xi \\ ds &= b_{\mathrm{e}}\,dv + \frac{1}{T_\circ}c_v^{(\circ)}\,dT + v b_\xi\,d\xi \\ -d(vA) &= k_{\mathrm{e}\xi}\,dv - v b_\xi\,dT + v k_{\xi\xi}\,d\xi \end{aligned} \tag{7.131}$$

---

* Joseph D.D. and Preziosi L. 1989. Heat waves. *Reviews of Modern Physics*, 61: 41–73; Joseph D.D. and Preziosi L. 1990. Addendum to the paper "Heat waves". *Reviews of Modern Physics*, 62: 375–391.

In this form the thermodynamic equations of state are exact. The Gibbs fundamental equation for the volume deformation is

$$\rho\, du = \rho T\, ds - p\, dI_\epsilon - A\, d\xi$$

or

$$du = T\, ds - \frac{\rho_\circ}{\rho} p\, dv - vA\, d\xi$$

The occurrence of the ratio $\rho_\circ/\rho$ results from the approximation (7.40) of the work done by the stresses and has to be set equal to unity in the exact equation, so that

$$du = T\, ds - p\, dv - vA\, d\xi = T\, ds - p\, dv - A\, dX' \tag{7.132}$$

where X' is the normalized or specific extent of change. The coefficients in (7.131) are the thermostatic coefficients, which are usually defined as follows

$$\begin{aligned}
&\textit{heat capacities:} && c_p = T\left(\frac{\partial s}{\partial T}\right)_p && \text{and} \quad c_v = T\left(\frac{\partial s}{\partial T}\right)_v \\
&\textit{compressibilities:} && \kappa_T = -\frac{1}{v}\left(\frac{\partial v}{\partial p}\right)_T && \text{and} \quad \kappa_s = -\frac{1}{v}\left(\frac{\partial v}{\partial p}\right)_s \\
&\textit{thermal expansion coefficient:} && \alpha = \frac{1}{v}\left(\frac{\partial v}{\partial T}\right)_p &&
\end{aligned} \tag{7.133}$$

It is clear that the bulk moduli are given by

$$K_T = \frac{1}{\kappa_T} \qquad \text{and} \qquad K_s = \frac{1}{\kappa_s} \tag{7.134}$$

If internal processes are found in the system, then the false equilibrium coefficients or constrained coefficients and the equilibrium coefficients are defined. The *constrained thermostatic coefficients* are defined for ξ is constant or $d\xi = 0$, so that the internal process is constrained—does not proceed—and then (7.131) becomes

$$\left.\begin{aligned}
-dp &= \frac{1}{v} k_{\mathrm{ee}}\, dv - b_{\mathrm{e}}\, dT \\
ds &= b_{\mathrm{e}}\, dv + \frac{1}{T} c_v^{(\circ)}\, dT
\end{aligned}\right\} \quad (d\xi = 0) \tag{7.135}$$

From this it follows immediately that

$$K_T^{(\circ)} = -v\left(\frac{\partial p}{\partial v}\right)_T^{(\circ)} = k_{\mathrm{ee}} \qquad \alpha^{(\circ)} = \frac{1}{v}\left(\frac{\partial v}{\partial T}\right)_p^{(\circ)} = \frac{b_{\mathrm{e}}}{k_{\mathrm{ee}}}$$

$$c_v^{(\circ)} = T\left(\frac{\partial s}{\partial T}\right)_v^{(\circ)}$$

Solving dv for $dp = 0$ in $(7.135)_1$ yields $dv = vb_\mathrm{e}\, dT/k_\mathrm{ee}$, and the substitution into $(7.135)_2$ gives

$$ds = \left(v\frac{b_\mathrm{e}^2}{k_\mathrm{ee}} + \frac{c_v^{(\circ)}}{T} \right) dT = \frac{c_p^{(\circ)}}{T}\, dT$$

Obviously the heat capacity at constant pressure and constrained internal processes is

$$c_p^{(\circ)} = Tv\frac{b_\mathrm{e}^2}{k_\mathrm{ee}} + c_v^{(\circ)}$$

Solving dT for $ds = 0$ in $(7.135)_2$ yields $dT = -Tb_\mathrm{e}\, dv/c_v^{(\circ)}$, and the substitution into $(7.135)_1$ gives

$$-dp = \frac{1}{v}\left(k_\mathrm{ee} + Tv\frac{b_\mathrm{e}^2}{c_v^{(\circ)}} \right) dv = \frac{K_s^{(\circ)}}{v}\, dv$$

The constrained thermostatic coefficients therefore become

$$\left.\begin{array}{c} c_p^{(\circ)} = c_v^{(\circ)} + Tv\dfrac{b_\mathrm{e}^2}{k_\mathrm{ee}} \geq 0 \qquad c_v^{(\circ)} \geq 0 \\ K_T^{(\circ)} = \dfrac{1}{\kappa_T^{(\circ)}} = k_\mathrm{ee} \geq 0 \qquad K_s^{(\circ)} = \dfrac{1}{\kappa_s^{(\circ)}} = k_\mathrm{ee} + Tv\dfrac{b_\mathrm{e}^2}{c_v^{(\circ)}} \geq 0 \\ \alpha^{(\circ)} = \dfrac{b_\mathrm{e}}{k_\mathrm{ee}} \end{array}\right\} \tag{7.136}$$

The inequalities in (7.136) express the principle of Le Chatelier*, by which the intrinsic stability of a phase is guaranteed (the intrinsic stability of a phase is concerned with the possible tendency to turn over into other phases). From (7.136) follows

$$c_p^{(\circ)} - c_v^{(\circ)} = TvK_T^{(\circ)}{\alpha^{(\circ)}}^2 = Tv\frac{{\alpha^{(\circ)}}^2}{\kappa_T^{(\circ)}} \geq 0 \tag{7.137}$$

---

* If the equilibrium of a phase is intrinsically stable, then the spontaneous processes initiated by the deviations from the equilibrium state proceed such that the phase again returns to the equilibrium state. This is in fact the definition of stable equilibrium. From the intrinsic stability and the transformations of Legendre it follows that in each representation $\partial P_i/\partial y_i > 0$ and $\partial y_i/\partial P_i > 0$, where y_i denotes a specific extensive variable and P_i the intensive variable conjugate to the extensive variable y_i. See for instance Prigogine I. and Defay R. 1954. *Chemical Thermodynamics.* Longmans Green & Co, London, p. 262; Le Chatelier H. 1888. *Recherches sur les Équilibres Chimiques.* Paris.

and

$$\kappa_T^{(\circ)} - \kappa_s^{(\circ)} = \frac{1}{k_{\mathrm{ee}}} - \frac{c_v^{(\circ)}}{k_{\mathrm{ee}}\left(c_v^{(\circ)} + Tvb_{\mathrm{e}}^2/k_{\mathrm{ee}}\right)}$$
$$= \frac{1}{k_{\mathrm{ee}}}\left(1 - \frac{c_v^{(\circ)}}{c_p^{(\circ)}}\right) = \frac{\kappa_T^{(\circ)}}{c_p^{(\circ)}}\left(c_p^{(\circ)} - c_v^{(\circ)}\right)$$

or with (7.137)

$$\kappa_T^{(\circ)} - \kappa_s^{(\circ)} = Tv\frac{\alpha^{(\circ)2}}{c_p^{(\circ)}} \geq 0 \tag{7.138}$$

From (7.137) and (7.138) it is also found that

$$\frac{c_p^{(\circ)}}{c_v^{(\circ)}} = \frac{\kappa_T^{(\circ)}}{\kappa_s^{(\circ)}} = \frac{K_s^{(\circ)}}{K_T^{(\circ)}} \geq 1 \tag{7.139}$$

The inequalities (7.137)–(7.139) express the principle of Le Chatelier–Braun*.

The *thermostatic equilibrium coefficients* are defined for true equilibrium, for which $A = 0$ applies, so that $(7.131)_3$ reduces to

$$d\xi = -\frac{1}{v}\frac{k_{\mathrm{e}\xi}}{k_{\xi\xi}}\,dv + \frac{b_\xi}{k_{\xi\xi}}\,dT \qquad (A = 0) \tag{7.140}$$

Substitution into the equations $(7.131)_{1,2}$ gives

$$\begin{aligned} -dp &= \frac{1}{v}\left(k_{\mathrm{ee}} - \frac{k_{\mathrm{e}\xi}^2}{k_{\xi\xi}}\right)dv - \left(b_{\mathrm{e}} - k_{\mathrm{e}\xi}\frac{b_\xi}{k_{\xi\xi}}\right)dT \\ ds &= \left(b_{\mathrm{e}} - k_{\mathrm{e}\xi}\frac{b_\xi}{k_{\xi\xi}}\right)dv + \frac{1}{T}\left(c_v^{(\circ)} + Tv\frac{b_\xi^2}{k_{\xi\xi}}\right)dT \end{aligned} \tag{7.141}$$

From which

$$K_T^{(\mathrm{eq})} = -v\left(\frac{\partial p}{\partial v}\right)_T^{(\mathrm{eq})} = k_{\mathrm{ee}} - \frac{k_{\mathrm{e}\xi}^2}{k_{\xi\xi}} \tag{i}$$

$$\alpha^{(\mathrm{eq})} = \frac{1}{v}\left(\frac{\partial v}{\partial T}\right)_p^{(\mathrm{eq})} = \left(b_{\mathrm{e}} - k_{\mathrm{e}\xi}\frac{b_\xi}{k_{\xi\xi}}\right)\Big/\left(k_{\mathrm{ee}} - \frac{k_{\mathrm{e}\xi}^2}{k_{\xi\xi}}\right) \tag{ii}$$

$$c_v^{(\mathrm{eq})} = T\left(\frac{\partial s}{\partial T}\right)_v^{(\mathrm{eq})} = c_v^{(\circ)} + Tv\frac{b_\xi^2}{k_{\xi\xi}} \tag{iii}$$

---

* The variation of an extensive variable is larger in stable equilibrium systems for a variation of the conjugate intensive variable if instead of an extensive additional variable the conjugate intensive variable is kept constant. The Le Chatelier–Braun theorem is an example of the theorems of moderation. Braun F. 1887 Untersuchungen über die Löslichkeit fester Körper und die den Vorgang der Lösung begleitenden Volum- und Energieänderungen. Einige Bemerkungen zu dem vorstehenden Aufsatze. *Zeitschrift für physikalische Chemie*, 1: 259–272.

For $dp = 0$ and by using (ii), it is found from $(7.141)_1$ that $dv = v\alpha^{(\mathrm{eq})}\,dT$. Substitution into $(7.141)_2$ yields

$$ds = \left[v\alpha^{(\mathrm{eq})}\left(b_{\mathrm{e}} - k_{\mathrm{e}\xi}\frac{b_\xi}{k_{\xi\xi}}\right) + \frac{1}{T}\left(c_v^{(\mathrm{o})} + Tv\frac{b_\xi^2}{k_{\xi\xi}}\right)\right] dT = \frac{c_p^{(\mathrm{eq})}}{T}dT$$

so that

$$c_p^{(\mathrm{eq})} = c_v^{(\mathrm{o})} + Tv\frac{b_\xi^2}{k_{\xi\xi}} + Tv\alpha^{(\mathrm{eq})}\left(b_{\mathrm{e}} - k_{\mathrm{e}\xi}\frac{b_\xi}{k_{\xi\xi}}\right) \tag{iv}$$

For $ds = 0$ and by using (iii), it is found from $(7.141)_2$ that

$$dT = -T\left(b_{\mathrm{e}} - k_{\mathrm{e}\xi}\frac{b_\xi}{k_{\xi\xi}}\right)\Big/ c_v^{(\mathrm{eq})}\,dv$$

and substitution into $(7.141)_1$ results in

$$-dp = \frac{1}{v}\left[k_{\mathrm{ee}} - \frac{k_{\mathrm{e}\xi}^2}{k_{\xi\xi}} + Tv\frac{(b_{\mathrm{e}} - k_{\mathrm{e}\xi}b_\xi/k_{\xi\xi})^2}{c_v^{(\mathrm{eq})}}\right] dv = \frac{1}{v}K_s^{(\mathrm{eq})}\,dv \tag{v}$$

Summarizing, the results (i)–(v) yield the following thermostatic equilibrium coefficients

$$\left.\begin{aligned}
c_p^{(\mathrm{eq})} &= c_v^{(\mathrm{o})} + Tv\frac{b_\xi^2}{k_{\xi\xi}} + TvK_T^{(\mathrm{eq})}{\alpha^{(\mathrm{eq})}}^2 \geq 0 && (a)\\
c_v^{(\mathrm{eq})} &= c_v^{(\mathrm{o})} + Tv\frac{b_\xi^2}{k_{\xi\xi}} \geq 0 && (b)\\
K_T^{(\mathrm{eq})} &= k_{\mathrm{ee}} - \frac{k_{\mathrm{e}\xi}^2}{k_{\xi\xi}} \geq 0 && (c)\\
K_s^{(\mathrm{eq})} &= K_T^{(\mathrm{eq})} + Tv{K_T^{(\mathrm{eq})}}^2\frac{{\alpha^{(\mathrm{eq})}}^2}{c_v^{(\mathrm{eq})}} \geq 0 && (d)\\
\alpha^{(\mathrm{eq})} &= \left(b_{\mathrm{e}} - k_{\mathrm{e}\xi}\frac{b_\xi}{k_{\xi\xi}}\right)\Big/\left(k_{\mathrm{ee}} - \frac{k_{\mathrm{e}\xi}^2}{k_{\xi\xi}}\right) && (e)
\end{aligned}\right\} \tag{7.142}$$

The inequalities express again the principle of Le Chatelier–Braun. From (7.142a) and (7.142b) it follows that

$$c_p^{(\mathrm{eq})} - c_v^{(\mathrm{eq})} = TvK_T^{(\mathrm{eq})}{\alpha^{(\mathrm{eq})}}^2 = Tv\frac{{\alpha^{(\mathrm{eq})}}^2}{\kappa_T^{(\mathrm{eq})}} \geq 0 \tag{7.143}$$

For (7.142d) the following can be written

$$\frac{1}{\kappa_s^{(\mathrm{eq})}} - \frac{1}{\kappa_T^{(\mathrm{eq})}} = Tv\frac{\alpha^{(\mathrm{eq})^2}}{\kappa_T^{(\mathrm{eq})^2}}\frac{1}{c_v^{(\mathrm{eq})}}$$

or

$$\kappa_T^{(\mathrm{eq})} - \kappa_s^{(\mathrm{eq})} = Tv\frac{\alpha^{(\mathrm{eq})^2}}{\kappa_T^{(\mathrm{eq})}}\frac{\kappa_s^{(\mathrm{eq})}}{c_v^{(\mathrm{eq})}}$$

or with (7.143)

$$\kappa_T^{(\mathrm{eq})} - \kappa_s^{(\mathrm{eq})} = \frac{\kappa_s^{(\mathrm{eq})}}{c_v^{(\mathrm{eq})}}\left(c_p^{(\mathrm{eq})} - c_v^{(\mathrm{eq})}\right) \geq 0 \tag{7.144}$$

or

$$\frac{c_p^{(\mathrm{eq})}}{c_v^{(\mathrm{eq})}} = \frac{\kappa_T^{(\mathrm{eq})}}{\kappa_s^{(\mathrm{eq})}} = \frac{K_s^{(\mathrm{eq})}}{K_T^{(\mathrm{eq})}} \geq 1 \tag{7.145}$$

From (7.143) and (7.144)

$$\kappa_T^{(\mathrm{eq})} - \kappa_s^{(\mathrm{eq})} = Tv\frac{\alpha^{(\mathrm{eq})^2}}{c_p^{(\mathrm{eq})}} \tag{7.146}$$

The analogies among (7.143), (7.146), (7.145) and (7.137), (7.138), (7.139) are obvious.

The *thermostatic internal coefficients* are defined as the difference between the equilibrium coefficients and the constrained coefficients. The internal coefficients can be considered to represent the contribution of the internal process to the coefficient concerned.

From (7.136) and (7.142)

$$c_p^{(\mathrm{a})} = c_p^{(\mathrm{eq})} - c_p^{(\mathrm{o})} = Tv\left(\frac{b_\xi^2}{k_{\xi\xi}} - \frac{b_\mathrm{e}^2}{k_\mathrm{ee}}\right) + TvK_T^{(\mathrm{eq})}\alpha^{(\mathrm{eq})^2} =$$

$$= Tv\left[\frac{b_\xi^2}{k_{\xi\xi}} - \frac{b_\mathrm{e}^2}{k_\mathrm{ee}} + \frac{\left(b_\mathrm{e} - k_{\mathrm{e}\xi}b_\xi/k_{\xi\xi}\right)^2}{k_\mathrm{ee} - k_{\mathrm{e}\xi}^2/k_{\xi\xi}}\right]$$

or after some algebra

$$c_p^{(\mathrm{a})} = c_p^{(\mathrm{eq})} - c_p^{(\mathrm{o})} = Tv\frac{k_\mathrm{ee}}{k_{\xi\xi}}\frac{\left(b_\xi - b_\mathrm{e}k_{\mathrm{e}\xi}/k_\mathrm{ee}\right)^2}{k_\mathrm{ee} - k_{\mathrm{e}\xi}^2/k_{\xi\xi}} \geq 0 \tag{7.147a}$$

$$c_v^{(\mathrm{a})} = c_v^{(\mathrm{eq})} - c_v^{(\mathrm{o})} = Tv\frac{b_\xi^2}{k_{\xi\xi}} \geq 0 \tag{7.147b}$$

Furthermore

$$\kappa_T^{(\mathrm{a})} = \kappa_T^{(\mathrm{eq})} - \kappa_T^{(\mathrm{o})} = \frac{1}{k_{\mathrm{ee}} - k_{\mathrm{e}\xi}^2/k_{\xi\xi}} - \frac{1}{k_{\mathrm{ee}}}$$

$$= \frac{k_{\mathrm{e}\xi}^2/k_{\mathrm{ee}}}{k_{\mathrm{ee}}k_{\xi\xi} - k_{\mathrm{e}\xi}^2} \geq 0 \tag{7.147c}$$

$$\alpha^{(\mathrm{a})} = \alpha^{(\mathrm{eq})} - \alpha^{(\mathrm{o})} = \frac{b_{\mathrm{e}} - k_{\mathrm{e}\xi}b_\xi/k_{\xi\xi}}{k_{\mathrm{ee}} - k_{\mathrm{e}\xi}^2/k_{\xi\xi}} - \frac{b_{\mathrm{e}}}{k_{\mathrm{ee}}}$$

$$= -k_{\mathrm{e}\xi}\frac{b_\xi - b_{\mathrm{e}}k_{\mathrm{e}\xi}/k_{\mathrm{ee}}}{k_{\mathrm{ee}}k_{\xi\xi} - k_{\mathrm{e}\xi}^2} \tag{7.147d}$$

Also for

$$\kappa_s^{(\mathrm{a})} = \kappa_s^{(\mathrm{eq})} - \kappa_s^{(\mathrm{o})} \geq 0 \tag{7.147e}$$

a relation can be derived, but this one is rather complicated. The inequalities in (7.147a) express again the principle of Le Chatelier–Braun.

From (7.147a, c, d)

$$c_p^{(\mathrm{a})}\kappa_T^{(\mathrm{a})} = Tv{\alpha^{(\mathrm{a})}}^2 \tag{7.148}$$

This relation is typical for one internal process. If several internal processes are present, then

$$c_p^{(\mathrm{a})}\kappa_T^{(\mathrm{a})} > Tv{\alpha^{(\mathrm{a})}}^2 \tag{7.149}$$

as has been shown by Meixner*. The relation (7.148) can be used to ascertain whether only one internal process is present in the system. Clearly, this is a difficult task due to the limited accuracy of the measurements.

Finally, *reaction coefficients* are introduced, which indicate the extent to which the extensive quantities change during the development of the internal process

$$\psi_{Tp} = \left(\frac{\partial\psi}{\partial X'}\right)_{T,p} \qquad \text{for} \qquad \psi = v, s, u \tag{7.150}$$

With the use of the Gibbs fundamental equation (7.132)

$$u_{Tp} = Ts_{Tp} - pv_{Tp} - A \tag{7.151}$$

From $(7.131)_1$ it follows that

$$v_{Tp} = -k_{\mathrm{e}\xi}/k_{\mathrm{ee}} \tag{7.152}$$

while $(7.131)_2$ leads to

$$s_{Tp} = b_{\mathrm{e}}v_{Tp} + b_\xi = b_\xi - k_{\mathrm{e}\xi}b_{\mathrm{e}}/k_{\mathrm{ee}} \tag{7.153}$$

---

* Meixner J. 1952. *Changements de Phases*. Société Chimie Physique, Paris.

7.9.2. Relaxation equations. For $I_\epsilon^{(v)} = 0$ the phenomenological scalar equation (7.87) reduces for one internal process to

$$\dot{\xi} = \rho_\circ \dot{X}' = \kappa_{\xi\xi} A \tag{7.154}$$

Combined with $(7.130)_3$ this equation (7.154) becomes

$$(k_{\xi\xi}\kappa_{\xi\xi} + \mathcal{D})\, X' = -\frac{\kappa_{\xi\xi}}{\rho_\circ}\left(k_{e\xi}\frac{\delta v}{v} - b_\xi\, \delta T\right) \tag{7.155}$$

or

$$X' + \tau_{Tv}\dot{X}' = -\frac{1}{\rho_\circ k_{\xi\xi}}\left(k_{e\xi}\frac{\delta v}{v} - b_\xi\, \delta T\right) = X'^{(eq)}_{Tv} \tag{7.156}$$

or

$$\dot{X}' = \frac{X'^{(eq)}_{Tv} - X'}{\tau_{Tv}} \tag{7.157}$$

where $X'^{(eq)}_{Tv}$ is the equilibrium value of X', which corresponds to a constant δT and δv, and with a relaxation time τ_{Tv}.

Equation (7.157) is a relaxation equation, which expresses the idea that $\dot{X}'$ is caused by perturbations of X' from the equilibrium value $X'^{(eq)}_{Tv}$. If for example

$$X'^{(eq)}_{Tv} = \hat{X}'^{(eq)}_{Tv} H(t) \tag{7.158a}$$

in which $H(t)$ denotes the unit step function and $\hat{X}'^{(eq)}_{Tv}$ a constant, then the solution of (7.157) becomes

$$X' = \hat{X}'^{(eq)}_{Tv}\big(1 - \exp(-t/\tau_{Tv})\big)H(t) \tag{7.158b}$$

so that X' relaxes exponentially to a new equilibrium that corresponds to $\hat{X}'^{(eq)}_{Tv}$. A measure for the delayed return to equilibrium is the relaxation time τ_{Tv}. The general solution of (7.157)) is

$$X'(t) = \frac{1}{\tau_{Tv}}\int_{-\infty}^{t} X'^{(eq)}_{Tv}(t')\exp\big(-(t-t')/\tau_{Tv}\big)\, dt' \tag{7.159}$$

if $X'^{(eq)}_{Tv}(-\infty) = 0$.

So far relaxation under the influence of δT and δv has been considered. One might also be interested in the relaxation under the influence of other combinations of the thermodynamic quantities, for example δT and δp. From $(7.130)_1$

$$\frac{\delta v}{v} = -\frac{1}{k_{ee}}\delta p + \frac{b_e}{k_{ee}}\delta T - \rho_\circ\frac{k_{e\xi}}{k_{ee}}X' \tag{7.160}$$

Substitution into (7.155) yields

$$\left[\left(k_{\xi\xi} - \frac{k_{e\xi}^2}{k_{ee}}\right)\kappa_{\xi\xi} + \mathfrak{D}\right] X' = \frac{\kappa_{\xi\xi}}{\rho_\circ}\left[\frac{k_{e\xi}}{k_{ee}}\delta p + \left(b_\xi - k_{e\xi}\frac{b_e}{k_{ee}}\right)\delta T\right] \qquad (7.161)$$

This equation now possesses another relaxation time. From (7.155) and (7.156) it is found that

$$\tau_{Tv} = \frac{1}{k_{\xi\xi}\kappa_{\xi\xi}} \qquad (7.162)$$

while from (7.161) it follows that

$$\tau_{Tp} = \frac{1}{\kappa_{\xi\xi}\left(k_{\xi\xi} - k_{e\xi}^2/k_{ee}\right)} = \tau_{Tv}\frac{k_{ee}}{k_{ee} - k_{e\xi}^2/k_{\xi\xi}}$$

or with substitution of the thermostatic coefficients (7.136) and (7.142c)

$$\tau_{Tp} = \frac{K_T^{(\circ)}}{K_T^{(\mathrm{eq})}}\tau_{Tv} \qquad (7.163)$$

For (7.161) it can be seen that

$$(1 + \tau_{Tp}\mathfrak{D})\, X' = \frac{1}{\rho_\circ(k_{\xi\xi} - k_{e\xi}^2/k_{ee})}\left[\frac{k_{e\xi}}{k_{ee}}\,\delta p + \left(b_\xi - k_{e\xi}\frac{b_e}{k_{ee}}\right)\delta T\right]$$

With (7.142c), (7.152) and (7.153) this can be rewritten as

$$(1 + \tau_{Tp}\mathfrak{D})\, X' = \frac{k_{ee}}{k_{\xi\xi}}\kappa_T^{(\mathrm{eq})}\left(-v_{Tp}\,\delta p + s_{Tp}\,\delta T\right) = X'^{(\mathrm{eq})}_{Tp} \qquad (7.164)$$

or

$$\dot{X}' = \frac{X'^{(\mathrm{eq})}_{Tp} - X'}{\tau_{Tp}} \qquad (7.165)$$

This equation has the same form as (7.157), but the relaxation times τ_{Tv} and τ_{Tp} are different in general. For all cases it is easily seen that

$$\dot{X}' = \frac{X'^{(\mathrm{eq})}_{PQ} - X'}{\tau_{PQ}} \qquad (7.166)$$

where P and Q are two independent variables of state, by which $X'^{(\mathrm{eq})}_{PQ}$ is determined. It is to be expected that with each choice of P and Q there is a corresponding relaxation time τ_{PQ}. The most commonly used independent variables of state for the volume relaxation are

$$P, Q: \qquad T \qquad s \qquad p \qquad v$$

This leads to six relaxation times

$$\tau_{Ts} \qquad \tau_{Tp} \qquad \tau_{Tv} \qquad \tau_{sp} \qquad \tau_{sv} \qquad \tau_{pv} \tag{7.167}$$

The existence of various relaxation times leads to the introduction of the following nomenclature. An internal *mechanism* results under specific macroscopic conditions—variations of P and Q—in an internal *process* with a specific relaxation time. Usually, several internal processes correspond to one internal mechanism.

The fact that various relaxation times correspond to one internal mechanism was put forward by Meixner* and later by Davies and Lamb†. If the differences in the various relaxation times are not taken into account, then this omission ends in apparent inconsistencies, as for example illustrated by Manus‡.

The relaxation in equation (7.164) is determined by the reaction coefficients. The relaxation is caused by pressure and temperature, according to this equation, so that two cases can be distinguished, namely thermal and pressure relaxation.

Thermal relaxation: Suppose that for constant temperature and pressure the volume is not influenced by the development of the internal process, so that

$$v_{Tp} = 0 \qquad \text{and} \qquad k_{e\xi} = 0 \tag{7.168}$$

where the second equality follows from (7.152). With the substitution of $v_{Tp} = 0$ into (7.164) it can be shown that for thermal relaxation the relaxation is activated only by the temperature, while the specific volume is not altered during the process. Examples are physical processes in gases, in which the number of molecules remains the same during the process, so that the thermal equation of state is unaltered, like the vibration and rotation of molecules caused by collisions where in particular the vibration of the molecules is retarded with respect to the translation of the molecules. These phenomena are important in gas dynamics, where the thermal relaxation might result in

---

* Meixner J. 1949. Thermodynamik und Relaxationserscheinungen. *Zeitschrift für Naturforschung*, 4a: 594–600.

† Davies R.O. and Lamb J. 1956. On the description of rate processes by means of relaxation time. *Proceedings of the Physical Society*, B69: 293–300; Davies R.O. 1956. The macroscopic theory of irreversibility. *Reports on Progress in Physics*, 19: 326–367.

‡ Manus M. 1953. Relationships between kinetics and acoustic phenomena in equilibrium systems. *Journal of Chemical Physics*, 21: 1791–1796.

a broadening of the shock zone. From (7.147c, d), with the substitution of (7.168), it follows that

$$\begin{aligned} \kappa_T^{(\mathrm{a})} &= 0 \qquad \text{or} \qquad \kappa_T^{(\mathrm{eq})} = \kappa_T^{(\circ)} \qquad (a) \\ \alpha^{(\mathrm{a})} &= 0 \qquad \text{or} \qquad \alpha^{(\mathrm{eq})} = \alpha^{(\circ)} \qquad (b) \end{aligned} \tag{7.169}$$

so that thermal relaxation can be recognized experimentally because the constrained coefficients and the equilibrium coefficients are the same for isothermal compressibility and for thermal expansion.

Pressure relaxation: Assume that for constant temperature and pressure the entropy is not influenced by the development of the internal process

$$s_{Tp} = 0 \qquad \text{and} \qquad k_{\mathrm{e}\xi} = k_{\mathrm{ee}} \frac{b_\xi}{b_{\mathrm{e}}} \tag{7.170}$$

where the second equality follows from (7.153). With the substitution of $s_{Tp} = 0$ into (7.164) it is shown that for pressure relaxation, the relaxation mechanism is activated only by the pressure while the specific entropy does not change. From (7.170) and (7.147a, d) it is found that

$$\begin{aligned} c_p^{(\mathrm{a})} &= 0 \qquad \text{or} \qquad c_p^{(\mathrm{eq})} = c_p^{(\circ)} \qquad (a) \\ \alpha^{(\mathrm{a})} &= 0 \qquad \text{or} \qquad \alpha^{(\mathrm{eq})} = \alpha^{(\circ)} \qquad (b) \end{aligned} \tag{7.171}$$

so that pressure relaxation can experimentally be recognized if the constrained coefficients and the equilibrium coefficients are the same for the specific heat at constant pressure and for thermal expansion.

7.9.3. Dynamic equations of state. After solving the relaxation equations, the rheological or dynamic equations of state for scalar processes can be formulated. Substitution of the constrained thermodynamic coefficients (7.136) into $(7.130)_1$ gives

$$-\delta p = K_T^{(\circ)} \frac{\delta v}{v} - \alpha^{(\circ)} K_T^{(\circ)} \, \delta T + k_{\mathrm{e}\xi} \, \xi \tag{7.172}$$

Elimination of the internal variable $\xi = \rho X'$ with the help of (7.156) yields

$$\begin{aligned} -\left(1 + \tau_{Tv} \mathfrak{D}\right) \delta p = {} & K_T^{(\circ)} \left(1 + \tau_{Tv} \mathfrak{D}\right) \frac{\delta v}{v} \\ & - \alpha^{(\circ)} K_T^{(\circ)} \left(1 + \tau_{Tv} \mathfrak{D}\right) \delta T - \frac{k_{\mathrm{e}\xi}^2}{k_{\xi\xi}} \frac{\delta v}{v} + b_\xi \frac{k_{\mathrm{e}\xi}}{k_{\xi\xi}} \delta T \end{aligned}$$

or

$$\begin{aligned} -\left(1 + \tau_{Tv} \mathfrak{D}\right) \delta p = {} & \left[K_T^{(\circ)} - \frac{k_{\mathrm{e}\xi}^2}{k_{\xi\xi}} + K_T^{(\circ)} \tau_{Tv} \mathfrak{D} \right] \frac{\delta v}{v} \\ & - \left[\alpha^{(\circ)} K_T^{(\circ)} - b_\xi \frac{k_{\mathrm{e}\xi}}{k_{\xi\xi}} + \alpha^{(\circ)} K_T^{(\circ)} \tau_{Tv} \mathfrak{D} \right] \delta T \end{aligned}$$

With the constrained coefficients (7.136) and the equilibrium coefficients (7.142) it is found that

$$K_T^{(\mathrm{eq})} = K_T^{(\circ)} - \frac{k_{\mathrm{e}\xi}^2}{k_{\xi\xi}}$$
$$\alpha^{(\mathrm{eq})} K_T^{(\mathrm{eq})} = \alpha^{(\circ)} K_T^{(\circ)} - b_\xi \frac{k_{\mathrm{e}\xi}}{k_{\xi\xi}} \tag{7.173}$$

so that

$$-\left(1+\tau_{Tv}\mathfrak{D}\right)\delta p = \left[K_T^{(\mathrm{eq})} + K_T^{(\circ)}\tau_{Tv}\mathfrak{D}\right]\frac{\delta v}{v} - \left[\alpha^{(\mathrm{eq})}K_T^{(\mathrm{eq})} + \alpha^{(\circ)}K_T^{(\circ)}\tau_{Tv}\mathfrak{D}\right]\delta T \tag{7.174}$$

This dynamic equation of state can finally be written as

$$\boxed{-\left(1+\tau_{Tv}\mathfrak{D}\right)\delta p = K_T^{(\mathrm{eq})}\left(1+\tau_{Tp}\mathfrak{D}\right)\frac{\delta v}{v} - \alpha^{(\mathrm{eq})}K_T^{(\mathrm{eq})}\left(1+\tau_{pv}\mathfrak{D}\right)\delta T} \tag{7.175}$$

with

$$\tau_{Tp} = \frac{K_T^{(\circ)}}{K_T^{(\mathrm{eq})}}\tau_{Tv} \qquad \text{and} \qquad \tau_{pv} = \frac{\alpha^{(\circ)}K_T^{(\circ)}}{\alpha^{(\mathrm{eq})}K_T^{(\mathrm{eq})}}\tau_{Tv} \tag{7.176}$$

Similarly as above the following dynamic equation of state can be derived by elimination of the internal scalar variable. Substitution of (7.136) gives for $(7.130)_2$

$$-\delta s = \alpha^{(\circ)}K_T^{(\circ)}\delta v + c_v^{(\circ)}\frac{\delta T}{T} + v b_\xi\,\xi \tag{7.177}$$

Eliminate $\xi = \rho X'$ with the help of (7.156)

$$-\left(1+\tau_{Tv}\mathfrak{D}\right)\delta s = \alpha^{(\circ)}K_T^{(\circ)}\left(1+\tau_{Tv}\mathfrak{D}\right)\delta v + c_v^{(\circ)}\left(1+\tau_{Tv}\mathfrak{D}\right)\frac{\delta T}{T} - b_\xi\frac{k_{\mathrm{e}\xi}}{k_{\xi\xi}}\delta v + Tv\frac{b_\xi^2}{k_{\xi\xi}}\frac{\delta T}{T}$$

or

$$-\left(1+\tau_{Tv}\mathfrak{D}\right)\delta s = \left[\alpha^{(\circ)}K_T^{(\circ)} - b_\xi\frac{k_{\mathrm{e}\xi}}{k_{\xi\xi}} + \alpha^{(\circ)}K_T^{(\circ)}\tau_{Tv}\mathfrak{D}\right]\delta v + \left[c_v^{(\circ)} + Tv\frac{b_\xi^2}{k_{\xi\xi}} + c_v^{(\circ)}\tau_{Tv}\mathfrak{D}\right]\frac{\delta T}{T}$$

Substitution of (7.147b) and $(7.173)_2$ gives

$$-\left(1+\tau_{Tv}\mathcal{D}\right)\delta s=\left[\alpha^{(\mathrm{eq})}K_T^{(\mathrm{eq})}+\alpha^{(\circ)}K_T^{(\circ)}\tau_{Tv}\mathcal{D}\right]\delta v+\left[c_v^{(\mathrm{eq})}+c_v^{(\circ)}\tau_{Tv}\mathcal{D}\right]\frac{\delta T}{T}$$

Finally this can be written as

$$\boxed{-\left(1+\tau_{Tv}\mathcal{D}\right)\delta s=\alpha^{(\mathrm{eq})}K_T^{(\mathrm{eq})}\left(1+\tau_{Ts}\mathcal{D}\right)\delta v+c_v^{(\mathrm{eq})}\left(1+\tau_{sv}\mathcal{D}\right)\frac{\delta T}{T}}\tag{7.178}$$

with

$$\tau_{Ts}=\frac{\alpha^{(\circ)}K_T^{(\circ)}}{\alpha^{(\mathrm{eq})}K_T^{(eq)}}\tau_{Tv}=\tau_{pv}\qquad\text{and}\qquad\tau_{sv}=\frac{c_v^{(\circ)}}{c_v^{(eq)}}\tau_{Tv}\tag{7.179}$$

Equations (7.175) and (7.178) are the dynamic equations of state, which were first derived in acoustics by Mandelstam and Leontovitsch*, and derived thermodynamically by Meixner†.

For the interpretation of the equations (7.175) and (7.178) fast and slow variations can be considered. For very slow variations the internal mechanism is always in equilibrium with the macroscopic conditions, so that the equations can be approximated by

$$\begin{aligned}-\delta p&=K_T^{(\mathrm{eq})}\frac{\delta v}{v}-\alpha^{(\mathrm{eq})}K_T^{(\mathrm{eq})}\delta T\\-\delta s&=\alpha^{(\mathrm{eq})}K_T^{(\mathrm{eq})}\delta v+c_v^{(\mathrm{eq})}\frac{\delta T}{T}\end{aligned}\tag{7.180}$$

For very fast variations, the internal mechanisms are not able to follow these very fast variations. This means that the internal mechanism is constrained. The dynamic equations (7.175) and (7.178) reduce for $|\tau\mathcal{D}|\gg 1$ to

$$\begin{aligned}-\tau_{Tv}\dot{p}&=K_T^{(\mathrm{eq})}\tau_{Tp}\frac{\dot{v}}{v}-\alpha^{(\mathrm{eq})}K_T^{(\mathrm{eq})}\tau_{pv}\dot{T}\\-\tau_{Tv}\dot{s}&=\alpha^{(\mathrm{eq})}K_T^{(\mathrm{eq})}\tau_{Ts}\dot{v}+c_v^{(\mathrm{eq})}\tau_{sv}\frac{\dot{T}}{T}\end{aligned}\tag{7.181}$$

* Mandelstam L.J. and Leontovitsch M.A. 1937. To the theory of sound absorption in fluids. *Journal of Experimental and Theoretical Physics, U.S.S.R.*, 7: 438–449.

† Meixner J. 1953. Die Thermodynamische Theorie der Relaxationserscheinungen und ihre Zusammenhang mit der Nachwirkungstheorie. *Kolloid Zeitschrift*, 134: 3–20.

For constrained internal equilibrium on the other hand it follows from (7.172) and (7.177) that

$$
\begin{aligned}
-\dot{p} &= K_T^{(\circ)} \frac{\dot{v}}{v} - \alpha^{(\circ)} K_T^{(\circ)} \dot{T} \\
-\dot{s} &= \alpha^{(\circ)} K_T^{(\circ)} \dot{v} + c_v^{(\circ)} \frac{\dot{T}}{T}
\end{aligned}
\tag{7.182}
$$

Comparison of the coefficients in (7.181) and (7.182) leads to (7.176) and (7.179).

With this interpretation other equations of state can be deduced quickly; these equations would otherwise have been derived quite laboriously from (7.175) and (7.178). Consider for example the dynamic relation between p, v and s

$$(1 + \tau_{sv}\mathcal{D})\ \delta p = a\,(1 + \tau_{sp}\mathcal{D})\ \delta v + b\,(1 + \tau_{pv}\mathcal{D})\ \delta s \tag{7.183}$$

in which

$$
\begin{aligned}
a &= \left(\frac{\partial p}{\partial v}\right)_s^{(\mathrm{eq})} = -\frac{1}{v} K_s^{(\mathrm{eq})} \\
b &= \left(\frac{\partial p}{\partial s}\right)_v^{(\mathrm{eq})}
\end{aligned}
\tag{7.184}
$$

The coefficient b has yet to be formulated in terms of the equilibrium coefficients. According to (7.132)

$$du = T\,ds - p\,dv \tag{7.185}$$

for $A = 0$ (equilibrium) or $d\xi = 0$ (constrained equilibrium). For $A = 0$ or $d\xi = 0$ accordingly

$$\left(\frac{\partial s}{\partial p}\right)_v = \left(\frac{\partial s}{\partial T}\right)_v \left(\frac{\partial T}{\partial p}\right)_v \tag{i}$$

$$dv = \left(\frac{\partial v}{\partial p}\right)_T dp + \left(\frac{\partial v}{\partial T}\right)_p dT$$

or

$$0 = \left(\frac{\partial v}{\partial p}\right)_T + \left(\frac{\partial v}{\partial T}\right)_p \left(\frac{\partial T}{\partial p}\right)_v$$

or

$$\left(\frac{\partial T}{\partial p}\right)_v = -\left(\frac{\partial v}{\partial p}\right)_T \Big/ \left(\frac{\partial v}{\partial T}\right)_p = \frac{\kappa_T}{\alpha} = \frac{1}{\alpha K_T}$$

With this it is found for (i) that

$$\left(\frac{\partial s}{\partial p}\right)_v = \frac{c_v}{\alpha T K_T} = \frac{c_p}{\alpha T K_s} \tag{ii}$$

The last equality follows from (7.139) and (7.145). This therefore yields

$$dp = \left(\frac{\partial p}{\partial v}\right)_s dv + \left(\frac{\partial p}{\partial s}\right)_v ds = -K_s \frac{dv}{v} + \frac{\alpha T K_s}{c_p} ds \tag{7.186}$$

For $A = 0$ it is seen that

$$dp = -K_s^{(\mathrm{eq})} \frac{dv}{v} + \frac{\alpha^{(\mathrm{eq})} T K_s^{(\mathrm{eq})}}{c_p^{(\mathrm{eq})}} ds \tag{7.187}$$

so that b is now also specified and (7.183) becomes

$$(1 + \tau_{sv}\mathcal{D})\, \delta p = -K_s^{(\mathrm{eq})} (1 + \tau_{sp}\mathcal{D}) \frac{\delta v}{v} + \frac{\alpha^{(\mathrm{eq})} T K_s^{(\mathrm{eq})}}{c_p^{(\mathrm{eq})}} (1 + \tau_{pv}\mathcal{D})\, \delta s \tag{7.188}$$

This result illustrates the scalar mechanical relaxation, in which now the relaxation of the entropy is also taken into account. For unconstrained equilibrium or true equilibrium it follows from (7.188) with (7.128) and (7.187) that

$$\tau_{sv}\dot{p} = -K_s^{(\mathrm{eq})} \tau_{sp} \frac{\dot{v}}{v} + \frac{\alpha^{(\mathrm{eq})} T K_s^{(\mathrm{eq})}}{c_p^{(\mathrm{eq})}} \tau_{pv} \dot{s}$$

For constrained equilibrium it follows according to (7.186) that

$$\dot{p} = -K_s^{(\circ)} \frac{\dot{v}}{v} + \frac{\alpha^{(\circ)} T K_s^{(\circ)}}{c_p^{(\circ)}} \dot{s}$$

The comparison of the coefficients leads to

$$\tau_{sp} = \frac{K_s^{(\circ)}}{K_s^{(\mathrm{eq})}} \tau_{sv} \qquad \text{and} \qquad \tau_{pv} = \frac{\alpha^{(\circ)} T K_s^{(\circ)}}{c_p^{(\circ)}} \frac{c_p^{(\mathrm{eq})}}{\alpha^{(\mathrm{eq})} T K_s^{(\mathrm{eq})}} \tau_{sv} \tag{7.189}$$

The relations (7.176), (7.179) and (7.189) are summarized by

$$\left.\begin{aligned} \tau_{Tp} &= \frac{K_T^{(\circ)}}{K_T^{(\mathrm{eq})}} \tau_{Tv} = \frac{K_T^{(\circ)}}{K_T^{(\mathrm{eq})}} \frac{c_v^{(\mathrm{eq})}}{c_v^{(\circ)}} \tau_{sv} = \frac{\alpha^{(\mathrm{eq})}}{\alpha^{(\circ)}} \tau_{pv} \\ &= \frac{c_p^{(\mathrm{eq})}}{c_p^{(\circ)}} \tau_{sp} = \frac{K_s^{(\circ)}}{K_s^{(\mathrm{eq})}} \frac{c_p^{(\mathrm{eq})}}{c_p^{(\circ)}} \tau_{sv} = \frac{\alpha^{(\mathrm{eq})}}{\alpha^{(\circ)}} \tau_{Ts} \\ & \tau_{pv} = \tau_{Ts}, \end{aligned}\right\} \tag{7.190}$$

where (7.140) and (7.146) are used. These relations were derived by Davies*.

---

* Davies R.O. and Lamb J. 1956. On the description of rate processes by means of relaxation time. *Proceedings of the Physical Society*, B69: 293–300; Davies R.O. 1956. The macroscopic theory of irreversibility. *Reports on Progress in Physics*, 19: 326–367.

From (7.147a) the following conditions are found

$$\begin{aligned} c_p^{(\text{eq})}/c_p^{(\text{o})} &\geq 1 \qquad & c_v^{(\text{eq})}/c_v^{(\text{o})} &\geq 1 \\ K_T^{(\text{o})}/K_T^{(\text{eq})} &\geq 1 \qquad & K_s^{(\text{o})}/K_s^{(\text{eq})} &\geq 1 \end{aligned} \tag{7.191}$$

and the inequalities below follow from (7.190) and (7.191)

$$\begin{aligned} \tau_{sv} \leq \tau_{Tv} \leq \tau_{Tp} \\ \tau_{sv} \leq \tau_{sp} \leq \tau_{Tp} \end{aligned} \tag{7.192}$$

These inequalities were derived by Meixner* and are the result of the principle of Le Chatelier–Braun. According to (7.192), the relaxation time for isothermal pressure relaxation (τ_{Tv}) is not smaller than the relaxation time for isentropic pressure relaxation (τ_{sv}), that is $\tau_{Tv} \geq \tau_{sv}$. The inequality is illustrated by considering a gas in which the molecules are translating, rotating and vibrating. Let the temperature that corresponds to the translational energy and the rotational energy be T. Suppose furthermore that only the first vibrational level is occupied and that the occupation of this vibrational level is larger than the one that corresponds to the Boyle temperature† T. The internal temperature $T^{(\text{a})}$ is introduced as the internal variable ξ with $T^{(\text{a})} > T$, where the internal temperature is the temperature of the gas, for which the given occupation of the vibrational level would be in equilibrium with the translational energy and the rotational energy (in the next subsection the notion of the internal temperature is specified further). At constant T and v a lowering of the internal temperature occurs, for which $T^{(\text{a})}$ decreases to T. At constant s and v the internal temperature is also lowered, since energy transfer of the vibrational to the translational and the rotational energies is possible. However, the internal temperature decreases then not so strongly, so that the relaxation at constant s and v will proceed faster than at constant T and v. This means that $\tau_{Tv} \geq \tau_{sv}$.

The dynamic equations of state are sufficient to describe the behavior of the material. However, mechanical models are sometimes useful. As Frenkel‡ already remarked, the relation between δp and δv can be illustrated by means of a mechanical model of Poynting–Thomson. Consider for instance (7.188) for $\delta s = 0$

$$(1 + \tau_{sv}\mathfrak{D})\, \delta p = -K_s^{(\text{eq})}\, (1 + \tau_{sp}\mathfrak{D})\, \frac{\delta v}{v} \qquad (\delta s = 0) \tag{7.193}$$

---

* Meixner J. 1953. Die Thermodynamische Theorie der Relaxationserscheinungen und ihre Zusammenhang mit der Nachwirkungstheorie. *Kolloid Zeitschrift*, 134: 3–20.

† The temperature for which the second virial coefficient is zero. The ideal gas law is then a good approximation.

‡ Frenkel J. 1946. *A Kinetic Theory of Liquids*. Clarendon Press. Oxford. See pp. 208–249.

In the mechanical model applies for a spring that $\delta p = -K\,\delta v/v$ and for a piston in a cylinder ('dashpot') that $\delta p = -\mu\,\delta\dot{v}/v$. Two representations of the equation of state are given in figure 7.5.

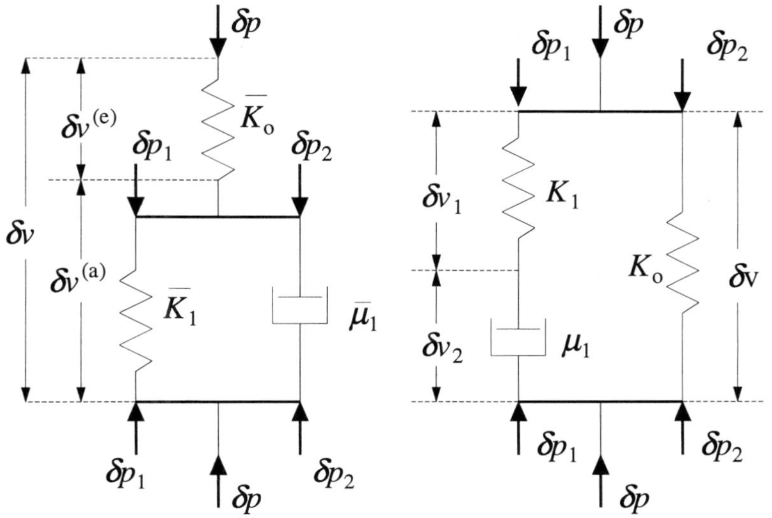

Figure 7.5. Kelvin representation and Maxwell representation of the dynamic equation of state (7.193).

In the Kelvin representation we have

$$\delta p = -\bar{K}_0 \frac{\delta v^{(\mathrm{e})}}{v} \qquad \text{and} \qquad \delta p_1 + \delta p_2 = \delta p = -\bar{K}_1 \frac{\delta v^{(\mathrm{a})}}{v} - \bar{\mu}_1 \frac{\delta \dot{v}^{(\mathrm{a})}}{v}$$

or with $\delta v = \delta v^{(\mathrm{e})} + \delta v^{(\mathrm{a})}$

$$\begin{aligned}\delta p &= -\left(\bar{K}_1 + \bar{\mu}_1 \mathcal{D}\right) \frac{\delta v^{(\mathrm{a})}}{v} \\ &= -\left(\bar{K}_1 + \bar{\mu}_1 \mathcal{D}\right)\left(\frac{\delta v}{v} - \frac{\delta v^{(\mathrm{e})}}{v}\right) = -\left(\bar{K}_1 + \bar{\mu}_1 \mathcal{D}\right)\left(\frac{\delta v}{v} + \frac{\delta p}{\bar{K}_0}\right)\end{aligned}$$

yielding

$$\left(1 + \frac{\bar{\mu}_1}{\bar{K}_0 + \bar{K}_1}\mathcal{D}\right)\delta p = -\frac{\bar{K}_1 \bar{K}_0}{\bar{K}_0 + \bar{K}_1}\left(1 + \frac{\bar{\mu}_1}{\bar{K}_1}\mathcal{D}\right)\frac{\delta v}{v} \tag{7.194}$$

Comparison with (7.193) results for the coefficients in

$$\tau_{sv} = \frac{\bar{\mu}_1}{\bar{K}_0 + \bar{K}_1} \qquad \tau_{sp} = \frac{\bar{\mu}_1}{\bar{K}_1} \qquad K_s^{(\mathrm{eq})} = \frac{\bar{K}_0 \bar{K}_1}{\bar{K}_0 + \bar{K}_1} \tag{7.195}$$

In the Maxwell representation we have

$$\delta p_1 = -K_1 \frac{\delta v_1}{v} = -\mu_1 \mathcal{D} \frac{\delta v_2}{v} \qquad \text{and} \qquad \delta p_2 = -K_0 \frac{\delta v}{v}$$

so that

$$\delta \dot{p}_1 = -K_1 \left(\frac{\delta \dot{v}}{v} - \frac{\delta \dot{v}_2}{v} \right) = -K_1 \left(\frac{\delta \dot{v}}{v} + \frac{\delta p_1}{\mu_1} \right)$$

or

$$\left(\frac{K_1}{\mu_1} + \mathcal{D} \right) \delta p_1 = -K_1 \frac{\delta \dot{v}}{v}$$

or

$$\left(\frac{K_1}{\mu_1} + \mathcal{D} \right) (\delta p - \delta p_2) = \left(\frac{K_1}{\mu_1} + \mathcal{D} \right) \left(\delta p + K_0 \frac{\delta v}{v} \right) = -K_1 \frac{\delta \dot{v}}{v}$$

and finally

$$\left(1 + \frac{\mu_1}{K_1} \mathcal{D} \right) \delta p = -K_0 \left[1 + \mu_1 \left(\frac{1}{K_0} + \frac{1}{K_1} \right) \mathcal{D} \right] \frac{\delta v}{v} \tag{7.196}$$

Comparison with (7.193) yields for the coefficients

$$\tau_{sv} = \frac{\mu_1}{K_1} \qquad \tau_{sp} = \frac{\mu_1}{K_0} + \frac{\mu_1}{K_1} \qquad K_s^{(\text{eq})} = K_0 \tag{7.197}$$

Both representations are equivalent and lead to the conclusion that $\tau_{sp} \geq \tau_{sv}$, as it should be.

According to Meixner and Reik* the volume viscosity can be considered as the effect of an internal process with very short relaxation times. Take for example again (7.188) and assume that

$$|\tau_{sv} \mathcal{D}| \ll 1 \tag{7.198}$$

This means that τ_{sv} is much smaller than the macroscopic time scale. In that case (7.188) becomes

$$\delta p = -K_s^{(\text{eq})} \left[1 + (\tau_{sp} - \tau_{sv}) \mathcal{D} \right] \frac{\delta v}{v} + \frac{T \alpha^{(\text{eq})} K_s^{(\text{eq})}}{c_p^{(\text{eq})}} \left[1 + (\tau_{pv} - \tau_{sv}) \mathcal{D} \right] \delta s \tag{7.199}$$

With the substitution of (7.190) this equation can be written as

$$\delta p = K_s^{(\text{eq})} \frac{\delta \rho}{\rho} + \left(K_s^{(\text{o})} - K_s^{(\text{eq})} \right) \tau_{sv} \frac{\dot{\rho}}{\rho} + T \frac{\alpha^{(\text{eq})} K_s^{(\text{eq})}}{c_p^{(\text{eq})}} \delta s + T \left(\frac{\alpha^{(\text{o})} K_s^{(\text{o})}}{c_p^{(\text{o})}} - \frac{\alpha^{(\text{eq})} K_s^{(\text{eq})}}{c_p^{(\text{eq})}} \right) \tau_{sv} \dot{s} \tag{7.200}$$

---

* Meixner J. and Reik H.G. 1959. Thermodynamik der irreversiblen Prozessen. *Encyclopedia of Physics*, III/2. Springer-Verlag, Berlin. See pp. 413–523.

For isentropic changes of state ($\dot{s} = 0$, $\delta s = 0$), when the continuity equation is used, the equation reduces to

$$\delta p = (\delta p)^{(\mathrm{eq})} - \kappa \operatorname{div} \vec{v} \tag{7.201}$$

with the volume viscosity

$$\kappa = \left(K_s^{(\mathrm{o})} - K_s^{(\mathrm{eq})} \right) \tau_{sv} \geq 0 \tag{7.202}$$

For gases this interpretation of the volume viscosity agrees well with the molecular theory. This interpretation can also be used for liquids.

7.9.4. Thermal relaxation. Thermal relaxation is found in gas dynamics in relation to the excitation of rotation and vibration of the gas molecules. The excitation of these molecules is now considered as an internal process. During such a process the number of molecules is not altered, so that in dilute gases the density is not influenced by the internal process. This leads—see (7.168) and (7.169)—to

$$v_{Tp} = 0 \qquad \Rightarrow \qquad k_{e\xi} = 0 \tag{7.203}$$

and

$$\kappa_T^{(\mathrm{eq})} = \kappa_T^{(\mathrm{o})} = \kappa_T \qquad \alpha^{(\mathrm{eq})} = \alpha^{(\mathrm{o})} = \alpha \tag{7.204}$$

Therefore the thermal equation of state—in particular the relation between p, ρ, and T that reduces for an ideal gas to $p = \rho R T$—is not influenced by the internal process either.

With the substitution of (7.204) the relaxation times (7.190) become

$$\left.\begin{aligned} &\tau_{Tp} = \tau_{Tv} = \tau_{pv} = \tau_{Ts} \\ &\tau_{sv} = \frac{c_v^{(\mathrm{o})}}{c_v^{(\mathrm{eq})}} \tau_{Tp} \qquad \tau_{sp} = \frac{c_p^{(\mathrm{o})}}{c_p^{(\mathrm{eq})}} \tau_{Tp} \end{aligned}\right\} \tag{7.205}$$

With substitution into the dynamical equation of state (7.175) it follows that the relation among p, ρ, and T is not influenced by the internal process. This follows exactly from $(7.131)_1$ and $(7.203)_2$

$$dp = -K_T \frac{dv}{v} + \alpha K_T \, dT \tag{7.206}$$

From (7.147a)a) and $(7.203)_2$ it is furthermore found that

$$c_p^{(\mathrm{eq})} - c_p^{(\mathrm{o})} = c_v^{(\mathrm{eq})} - c_v^{(\mathrm{o})} = c^{(\mathrm{a})} = Tv \frac{b_\xi^2}{k_{\xi\xi}} \geq 0 \tag{7.207}$$

and from (7.138) and (7.146) it follows with (7.204) and (7.207) that

$$\kappa_s^{(\text{eq})} - \kappa_s^{(\circ)} = Tv\alpha^2 \frac{c^{(\text{a})}}{c_p^{(\circ)} c_p^{(\text{eq})}} \geq 0 \tag{7.208}$$

If the gas is ideal, then $u^{(\text{eq})}$ and $u^{(\circ)}$ depend unambiguously only on T. The thermal relaxation in ideal gases is usually described with the notion of *internal temperature.* This temperature can be defined as the temperature that would be found in the gas if it had the same value of the internal energy in thermodynamic equilibrium as it has in the nonequilibrium situation.

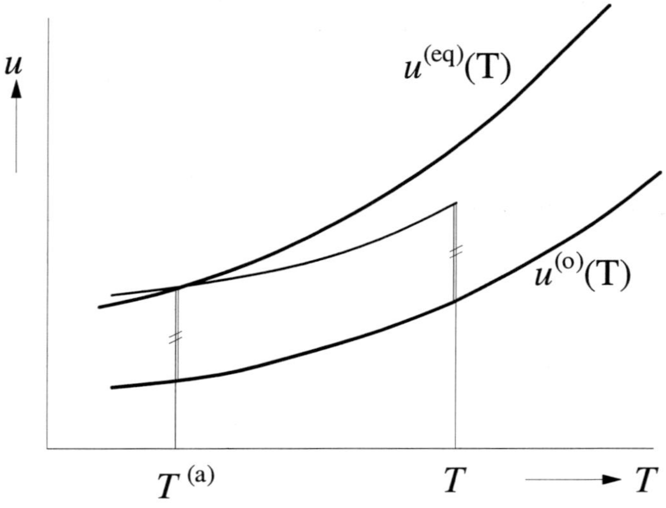

Figure 7.6. Internal energy in equilibrium and constrained equilibrium.

The introduction of the internal temperature can be most easily been discussed based on figure 7.6. The internal energy of an ideal gas in a nonequilibrium situation depends on T and ξ. Now the 'constrained' internal energy can be introduced, which is equal to the internal energy for the temperature T and an internal variable $\xi = 0$. Let this internal energy be denoted by $u^{(\circ)}(T)$

$$u(T, \xi) = u^{(\circ)}(T) + u^{(\text{a})}(T, \xi) \tag{7.209}$$

Suppose that in equilibrium ($A = 0$) the internal energy is $u^{(\text{eq})}(T)$. The internal temperature $T^{(\text{a})}$ is now defined by

$$u(T, \xi) - u^{(\circ)}(T) = u^{(\text{eq})}(T^{(\text{a})}) - u^{(\circ)}(T^{(\text{a})}) \tag{7.210}$$

This definition of the internal temperature is unambiguous if $u^{(\text{eq})}$ and $u^{(\circ)}$ are single-valued functions of T. With

$$u^{(\text{a})}(T^{(\text{a})}) = u^{(\text{eq})}(T^{(\text{a})}) - u^{(\circ)}(T^{(\text{a})}) \tag{7.211}$$

the internal energy (7.209) becomes

$$u(T,\xi) = u^{(\text{o})}(T) + u^{(\text{a})}(T^{(\text{a})}) \tag{7.212}$$

The differential of the internal energy becomes, with (7.207) and (7.211)

$$du = c_v^{(\text{o})}\, dT + c^{(\text{a})}\, dT^{(\text{a})} \tag{7.213}$$

so that

$$du^{(\text{a})}(T^{(\text{a})}) = c^{(\text{a})}\, dT^{(\text{a})} \tag{7.214}$$

In equilibrium $T = T^{(\text{a})}$, so that $T - T^{(\text{a})}$ or $u^{(\text{a})}(T^{(\text{a})})$ are measures for the deviations of the equilibrium. $T^{(\text{a})}$ or $u^{(\text{a})}(T^{(\text{a})})$ can be considered as the internal parameter X'. Since X' is originally an extensive variable, it seems wise to put

$$X' = u^{(\text{a})}\left(T^{(\text{a})}\right) \tag{7.215}$$

since $u^{(\text{a})}$ is also basically extensive, while $T^{(\text{a})}$ is always an intensive variable.

The affinity corresponding to (7.215) has yet to be calculated. For thermal relaxation $k_{e\xi} = 0$, so that $(7.131)_3$ becomes

$$-dA = -b_\xi\, dT + \rho_\text{o} k_{\xi\xi}\, dX' \tag{7.216}$$

With $v_{Tp} = 0$ the reaction coefficients (7.151) and (7.153) become

$$\begin{aligned} u_{Tp} &= T s_{Tp} - A \\ s_{Tp} &= b_\xi \end{aligned} \tag{7.217}$$

According to (7.207)

$$c^{(\text{a})} = Tv\frac{b_\xi^2}{k_{\xi\xi}} \qquad \Rightarrow \qquad \rho_\text{o} k_{\xi\xi} = T\frac{b_\xi^2}{c^{(\text{a})}} = T\frac{s_{Tp}^2}{c^{(\text{a})}} \tag{7.218}$$

Substitution of (7.218) and $(7.217)_2$ into (7.216) gives

$$dA = s_{Tp}\, dT - T\frac{s_{Tp}^2}{c^{(\text{a})}}\, dX' \tag{7.219}$$

From (7.212) and (7.215) it follows that $u_{Tp} = 1$, so that $(7.217)_1$ yields

$$T s_{Tp} = 1 + A \tag{7.220}$$

and the substitution into (7.219)

$$dA = \frac{1+A}{T}\, dT - \frac{(1+A)^2}{Tc^{(\text{a})}}\, du^{(\text{a})} \tag{7.221}$$

dA is an exact differential, hence

$$\frac{\partial A}{\partial T} = \frac{1+A}{T} \qquad \Rightarrow \qquad 1 + A = BT \tag{7.222}$$

Furthermore from (7.221) and (7.214) it follows that

$$\frac{\partial A}{\partial u^{(a)}} = -\frac{(1+A)^2}{T\, du^{(a)}/dT^{(a)}} \qquad \text{for} \qquad T = \text{constant}$$

B is a constant of integration in (7.222) that can depend only on $u^{(a)}$, so that with (7.222)

$$\frac{\partial A}{\partial u^{(a)}} = T \frac{dB}{du^{(a)}} = -\frac{B^2 T}{du^{(a)}/dT^{(a)}}$$

or

$$\frac{dB}{dT^{(a)}} = -B^2 \qquad \Rightarrow \qquad B = \frac{1}{T^{(a)} + \text{constant}}$$

and the substitution into (7.222) gives

$$A = \frac{T}{T^{(a)} + \text{ constant}} - 1$$

The requirement $A = 0$ for $T = T^{(a)}$ (equilibrium) leads to the conclusion that the constant is zero, so that

$$A = \frac{T}{T^{(a)}} - 1 \tag{7.223}$$

The Gibbs fundamental equation (7.132) therefore becomes

$$du = T\, ds - p\, dv - \left(\frac{T}{T^{(a)}} - 1\right) du^{(a)} \left(T^{(a)}\right) \tag{7.224}$$

or with (7.214)

$$du = T\, ds - p\, dv - \left(\frac{T}{T^{(a)}} - 1\right) c^{(a)}\, dT^{(a)} \tag{7.225}$$

The relaxation equation (7.165) becomes, with (7.215)

$$\dot{u}^{(a)} \left(T^{(a)}\right) = \frac{u^{(a)}(T) - u^{(a)}\left(T^{(a)}\right)}{\tau_{Tp}} \tag{7.226}$$

in which due to (7.162) and (7.204)

$$\tau_{Tp} = \tau_{Tv} = \tau_{pv} = \tau_{Ts} = \frac{1}{k_{\xi\xi}\kappa_{\xi\xi}} \tag{7.227}$$

In (7.218) it has been shown that

$$\rho_\circ k_{\xi\xi} = T s_{Tp}^2 / c^{(a)}$$

From (7.220) and (7.223) it follows that

$$T s_{Tp} = 1 + A = \frac{T}{T^{(a)}}$$

so that

$$k_{\xi\xi} = \frac{T}{\rho_\circ c^{(a)} T^{(a)2}} \tag{7.228}$$

The dissipation and rate of the entropy production per unit of volume become, with (7.154)

$$\mathfrak{D}^{(s)} = T\pi_{(s)} = \rho A \dot{X}' = \kappa_{\xi\xi} A^2$$

or with (7.223), (7.227) and (7.228)

$$\mathfrak{D}^{(s)} = T\pi_{(s)} = \frac{\rho_\circ c^{(a)}}{\tau_{Tp}} \frac{\left(T - T^{(a)}\right)^2}{T} \tag{7.229}$$

In the derivation it was not necessary to linearize the relation between u and T. In (7.226) it follows, for small differences between T and $T^{(a)}$, that

$$u^{(a)}(T) - u^{(a)}\left(T^{(a)}\right) \approx c^{(a)}\left(T - T^{(a)}\right) \tag{7.230}$$

Furthermore $\dot{u}^{(a)} = c^{(a)} \dot{T}^{(a)}$, so that (7.226) becomes

$$\dot{T}^{(a)} = \frac{T - T^{(a)}}{\tau_{Tp}} \tag{7.231}$$

This relation is often used in gas dynamics.

The concept of 'internal' temperature has been introduced by Herzfeld and Rice*, Ubbelohde† and Schäfer et al‡. Especially Rutgers§ introduced a relaxation equation of the type (7.231). Statistically the theory of the thermal relaxation is well developed. For certain gases the relaxation time τ_{Tp} can be calculated quantum mechanically.

---

* Herzfeld K. and Rice R.O. 1928. Dispersion and absorption of high frequency sound waves. *Physical Review*, 31: 691–695.

† Ubbelohde A.R. 1935. The thermal conductivity of polyatomic gases. *Journal of Chemical Physics*, 3: 219–223.

‡ Schäfer K., Rating W. and Eucken A. 1942. Über den Einfluß des gehemmten Austauschs der Translations- und Schwingungsenergie auf das Wärmeleitvermögen der Gase. *Annalen der Physik*, 42(5): 176–202.

§ Rutgers A.J. 1933. Zur Dispersionstheorie des Schalles. *Annalen der Physik*, 16(5): 350–359.

7.9.5. Velocity of sound. The velocity of sound a is defined as the velocity of a small disturbance propagating adiabatically in a medium at rest ($v = 1/\rho$)

$$a^2 = -v^2 \left(\frac{\partial p}{\partial v}\right)_{\mathrm{ad}} = \left(\frac{\partial p}{\partial \rho}\right)_{\mathrm{ad}} \tag{7.232}$$

It is assumed in the definition of the speed of sound that the disturbances fluctuate so fast that the time needed for the volume-element to exchange heat with its environment is too slow with respect to the fluctuating disturbances, so that these disturbances propagate adiabatically.

If the disturbances are small of the first order, then the entropy production is small of the second order and can be neglected in a first order theory, so that (7.232) becomes, with the substitution of the adiabatic compressibility $(7.133)_4$

$$a^2 = \left(\frac{\partial p}{\partial \rho}\right)_s = \frac{v}{\kappa_s} = \frac{1}{\rho\kappa_s} \tag{7.233}$$

If only one relaxation mechanism is found, then it follows, according to (7.188) for constant s, that

$$(1 + \tau_{sv}\mathfrak{D})\ \delta p = -K_s^{(\mathrm{eq})}\ (1 + \tau_{sp}\mathfrak{D})\ \frac{\delta v}{v} \tag{7.234}$$

For harmonic disturbances $\Re\{\delta\hat{p}\, e^{i\omega t}\}$, where $\delta\hat{p}$ is a complex amplitude, ω the circular frequency of the harmonic disturbance, and $\Re$ means: 'the real part of', the equation for the complex amplitudes (denoted by the circumflex) becomes

$$(1 + i\omega\tau_{sv})\ \delta\hat{p} = \frac{K_s^{(\mathrm{eq})}}{\rho}\ (1 + i\omega\tau_{sp})\ \delta\hat{\rho} \tag{7.235}$$

With this equation it follows for (7.233) that

$$a^2 = \Re\{\hat{a}^2\} = \frac{K_s^{(\mathrm{eq})}}{\rho}\Re\left\{\frac{1 + i\omega\tau_{sp}}{1 + i\omega\tau_{sv}}\right\} \tag{7.236}$$

For very small frequencies $\omega \to 0$ the internal mechanism is almost in equilibrium with the very slow changing circumstances, so that

$$a^2 \to a_\circ^2 = \frac{K_s^{(\mathrm{eq})}}{\rho} \qquad \text{for} \qquad \omega \to 0 \tag{7.237}$$

For very high frequencies ($\omega \to \infty$) the internal mechanism cannot follow the fast fluctuations, so that the mechanism is now constrained. For a constrained mechanism (7.236) reduces to

$$a^2 \to a_\infty^2 = a_\circ^2 \frac{\tau_{sp}}{\tau_{sv}} = \frac{K_s^{(\circ)}}{\rho} \geq a_\circ^2 \qquad \text{for} \qquad \omega \to \infty \tag{7.238}$$

In which the inequality follows from the principle of Le Chatelier–Braun ($K_s^{(\text{o})} \geq K_s^{(\text{eq})}$). With (7.237) the square of the complex amplitude of the speed of sound (7.236) can be found to be

$$\hat{a}^2 = a_\circ^2 \frac{1 + i\omega\tau_{sp}}{1 + i\omega\tau_{sv}} \tag{7.239}$$

which, with the substitution of (7.238), can be written alternatively as

$$\hat{a}^2 = a_\infty^2 - \frac{a_\infty^2 - a_\circ^2}{1 + i\omega\tau_{sv}} \tag{7.240}$$

The real part of $\hat{a}$ is the phase velocity. For $a_\infty > a_\circ$, the phase velocity is a monotonically increasing function of ω according to (7.240), until the value a_∞ for $\omega = \infty$ is reached. Then dispersion is observed.

For a plane wave, the fluctuating quantities are proportional to

$$\exp[i\omega(t - x/a)]$$

The imaginary part of a determines the attenuation, which is zero for $\omega = 0$ and $\omega = \infty$, and has its maximum value for $\omega\tau_{sv} \approx 1$. For gases the values for τ_{sv} are found in the range from 10^{-3} s to 10^{-6} s.

The group velocity v_g is defined by the variation of the circular frequency ω with the wave numbers k ($k = \omega/a$, the number of wave lengths in 2π) of the constituent waves

$$v_\text{g} = \frac{d\omega}{dk} = a + k\frac{da}{dk} \tag{7.241}$$

If a is a monotonically increasing function of ω, then the group velocity is larger than the phase velocity, and this is called *anomalous dispersion.* For anomalous dispersion the group velocity is not always equal to the signal velocity. The signal velocity is defined as the velocity of the main signal which has the intensity of the order of the input signal*. Internal mechanisms with even variables apparently yield to the anomalous dispersion of sound waves. It can be noted that the numerous kinematic models used to explain the dispersion and absorption of acoustic waves correspond well with (7.240), or can be derived from (7.240) as special cases.

7.10. INTERNAL TENSOR PROCESSES

7.10.1. Basic equations for fundamental linear rheological bodies. A fundamental rheological body shows elastic deformations $\mathring{\boldsymbol{\epsilon}}^{(\text{e})}$, anelastic deformations $\mathring{\boldsymbol{\epsilon}}^{(\text{a})}$ and viscous deformations $\mathring{\boldsymbol{\epsilon}}^{(\text{v})}$

$$\boxed{\mathring{\boldsymbol{\epsilon}} = \mathring{\boldsymbol{\epsilon}}^{(\text{e})} + \mathring{\boldsymbol{\epsilon}}^{(\text{a})} + \mathring{\boldsymbol{\epsilon}}^{(\text{v})}} \tag{7.242}$$

* Brillouin L. 1960. *Wave Propagation and Group Velocity.* Academic Press, New York.

For simplicity it is assumed that only one internal mechanism causes $\overset{\circ}{\boldsymbol{\epsilon}}^{(\mathrm{a})}$.

In isotropic linear bodies the deviatoric deformations can be discussed separately. Then the Gibbs fundamental equation (7.61) reduces to

$$\boxed{\rho\, du = \rho T\, ds + \overset{\circ}{\boldsymbol{\sigma}}^{(\mathrm{eq})} : d\overset{\circ}{\boldsymbol{\epsilon}}^{(\mathrm{e})} - \left(\overset{\circ}{\boldsymbol{P}} - \overset{\circ}{\boldsymbol{\sigma}}^{(\mathrm{eq})}\right) : d\overset{\circ}{\boldsymbol{\epsilon}}^{(\mathrm{a})}} \tag{7.243}$$

The elastic function $\mathfrak{E} = \rho_\circ\,(f - f_\circ)$ (7.64) for small values of $|\overset{\circ}{\boldsymbol{\epsilon}}|$ becomes

$$\boxed{\mathfrak{E} = \rho_\circ\,(f - f_\circ) = \tfrac{1}{2} c_{\mathrm{ee}} \overset{\circ}{\boldsymbol{\epsilon}}^{(\mathrm{e})} : \overset{\circ}{\boldsymbol{\epsilon}}^{(\mathrm{e})} + \tfrac{1}{2} c_{\mathrm{aa}} \overset{\circ}{\boldsymbol{\epsilon}}^{(\mathrm{a})} : \overset{\circ}{\boldsymbol{\epsilon}}^{(\mathrm{a})} \geq 0} \tag{7.244}$$

while the thermostatic equations are given by $(7.68)_{1,2}$

$$\begin{aligned} \overset{\circ}{\boldsymbol{\sigma}}^{(\mathrm{eq})} &= c_{\mathrm{ee}} \overset{\circ}{\boldsymbol{\epsilon}}^{(\mathrm{e})} \\ -\left(\overset{\circ}{\boldsymbol{P}} - \overset{\circ}{\boldsymbol{\sigma}}^{(\mathrm{eq})}\right) &= c_{\mathrm{aa}} \overset{\circ}{\boldsymbol{\epsilon}}^{(\mathrm{a})} \end{aligned} \tag{7.245}$$

According to (7.77)

$$\begin{aligned} &\overset{\circ}{\boldsymbol{\sigma}} = \overset{\circ}{\boldsymbol{\sigma}}^{(\mathrm{eq})} && \text{if} \quad \overset{\circ}{\boldsymbol{\epsilon}}^{(\mathrm{e})} \neq 0 \\ &\overset{\circ}{\boldsymbol{\sigma}} \neq \overset{\circ}{\boldsymbol{\sigma}}^{(\mathrm{eq})} = 0 && \text{if} \quad \overset{\circ}{\boldsymbol{\epsilon}}^{(\mathrm{e})} = 0 \end{aligned} \tag{7.246}$$

The dissipation function $\mathfrak{D}^{+}$ due to internal friction is, according to (7.86)

$$\boxed{\mathfrak{D}^{+} = \overset{\circ}{\boldsymbol{\sigma}} : \dot{\overset{\circ}{\boldsymbol{\epsilon}}}^{(\mathrm{v})} + \left(\overset{\circ}{\boldsymbol{P}} + \overset{\circ}{\boldsymbol{\sigma}} - \overset{\circ}{\boldsymbol{\sigma}}^{(\mathrm{eq})}\right) : \dot{\overset{\circ}{\boldsymbol{\epsilon}}}^{(\mathrm{a})} \geq 0} \tag{7.247}$$

The phenomenological or kinetic equations are, according to (7.98)

$$\begin{aligned} \dot{\overset{\circ}{\boldsymbol{\epsilon}}}^{(\mathrm{v})} &= \phi_{\mathrm{vv}} \overset{\circ}{\boldsymbol{\sigma}} + \phi_{\mathrm{va}} \left(\overset{\circ}{\boldsymbol{P}} + \overset{\circ}{\boldsymbol{\sigma}} - \overset{\circ}{\boldsymbol{\sigma}}^{(\mathrm{eq})}\right) \\ \dot{\overset{\circ}{\boldsymbol{\epsilon}}}^{(\mathrm{a})} &= \phi_{\mathrm{av}} \overset{\circ}{\boldsymbol{\sigma}} + \phi_{\mathrm{aa}} \left(\overset{\circ}{\boldsymbol{P}} + \overset{\circ}{\boldsymbol{\sigma}} - \overset{\circ}{\boldsymbol{\sigma}}^{(\mathrm{eq})}\right) \end{aligned} \tag{7.248}$$

The symmetry relation (7.99) becomes

$$\phi_{\mathrm{av}} = \phi_{\mathrm{va}} \tag{7.249}$$

and the requirement that the minimum of the quadratic function (7.101) is positive is guaranteed if

$$\phi_{\mathrm{vv}} \geq 0 \qquad \phi_{\mathrm{aa}} \geq 0 \qquad \phi_{\mathrm{vv}} \phi_{\mathrm{aa}} - \phi_{\mathrm{va}}^2 \geq 0 \tag{7.250}$$

From the positiveness of the elastic function (7.244) it follows obviously that

$$c_{\mathrm{ee}} \geq 0 \qquad \text{and} \qquad c_{\mathrm{aa}} \geq 0 \tag{7.251}$$

The elimination from $(7.245)_2$ and $(7.248)_2$ of $\mathring{\boldsymbol{P}} - \mathring{\boldsymbol{\sigma}}^{(\mathrm{eq})}$ results in the *dynamic equation*

$$c_{\mathrm{aa}}\phi_{\mathrm{aa}}\mathring{\boldsymbol{\epsilon}}^{(\mathrm{a})} + \dot{\mathring{\boldsymbol{\epsilon}}}^{(\mathrm{a})} = (\phi_{\mathrm{av}} + \phi_{\mathrm{aa}})\,\mathring{\boldsymbol{\sigma}} \tag{7.252}$$

The deviatoric, anelastic deformations are thus excited by the deviatoric stress tensor. The time

$$\bar{\tau} = \frac{1}{c_{\mathrm{aa}}\phi_{\mathrm{aa}}} \tag{7.253}$$

is the *retardation time* of the deformation. One speaks of *retardation* if the cause consists of intensive variables, and the effect consists of extensive or specific variables.

With the substitution of $(7.245)_2$, the phenomenological equation $(7.248)_1$ can be written as

$$\dot{\mathring{\boldsymbol{\epsilon}}}^{(\mathrm{v})} = (\phi_{\mathrm{vv}} + \phi_{\mathrm{va}})\,\mathring{\boldsymbol{\sigma}} - \phi_{\mathrm{va}}c_{\mathrm{aa}}\mathring{\boldsymbol{\epsilon}}^{(\mathrm{a})}$$

From this result and from (7.252) it follows that

$$(\phi_{\mathrm{av}} + \phi_{\mathrm{aa}})\,\dot{\mathring{\boldsymbol{\sigma}}} = (\phi_{\mathrm{aa}}c_{\mathrm{aa}} + \mathfrak{D})\,\dot{\mathring{\boldsymbol{\epsilon}}}^{(\mathrm{a})}$$

and

$$(\phi_{\mathrm{vv}} + \phi_{\mathrm{va}})\,(\phi_{\mathrm{aa}}c_{\mathrm{aa}} + \mathfrak{D})\,\mathring{\boldsymbol{\sigma}} - \phi_{\mathrm{va}}c_{\mathrm{aa}}\,(\phi_{\mathrm{av}} + \phi_{\mathrm{aa}})\,\mathring{\boldsymbol{\sigma}} = (\phi_{\mathrm{aa}}c_{\mathrm{aa}} + \mathfrak{D})\,\dot{\mathring{\boldsymbol{\epsilon}}}^{(\mathrm{v})}$$

Adding these two equations gives

$$\begin{aligned} c_{\mathrm{aa}}\left(\phi_{\mathrm{aa}}\phi_{\mathrm{vv}} - \phi_{\mathrm{va}}^2\right)\mathring{\boldsymbol{\sigma}} + (\phi_{\mathrm{vv}} + 2\phi_{\mathrm{av}} + \phi_{\mathrm{aa}})\,\dot{\mathring{\boldsymbol{\sigma}}} \\ = (\phi_{\mathrm{aa}}c_{\mathrm{aa}} + \mathfrak{D})\left(\dot{\mathring{\boldsymbol{\epsilon}}}^{(\mathrm{v})} + \dot{\mathring{\boldsymbol{\epsilon}}}^{(\mathrm{a})}\right) \end{aligned} \tag{7.254}$$

For $\boldsymbol{\epsilon}^{(\mathrm{e})} = 0$ this leads to

$$c_{\mathrm{aa}}\left(\phi_{\mathrm{aa}}\phi_{\mathrm{vv}} - \phi_{\mathrm{va}}^2\right)\mathring{\boldsymbol{\sigma}} + (\phi_{\mathrm{vv}} + 2\phi_{\mathrm{va}} + \phi_{\mathrm{aa}})\,\dot{\mathring{\boldsymbol{\sigma}}} = \phi_{\mathrm{aa}}c_{\mathrm{aa}}\dot{\mathring{\boldsymbol{\epsilon}}} + \ddot{\mathring{\boldsymbol{\epsilon}}} \tag{7.255}$$

For $\boldsymbol{\epsilon}^{(\mathrm{e})} \neq 0$, then according to (7.246) $\mathring{\boldsymbol{\sigma}} = \mathring{\boldsymbol{\sigma}}^{(\mathrm{eq})}$, so that $(7.245)_1$ becomes

$$\mathring{\boldsymbol{\sigma}} = c_{\mathrm{ee}}\mathring{\boldsymbol{\epsilon}}^{(\mathrm{e})} \tag{7.256}$$

From (7.242), (7.254) and (7.256) it follows that

$$
\boxed{
\begin{aligned}
&c_{\mathrm{ee}}c_{\mathrm{aa}}\left(\phi_{\mathrm{aa}}\phi_{\mathrm{vv}}-\phi_{\mathrm{va}}^{2}\right)\mathring{\boldsymbol{\sigma}} \\
&\qquad+\left[c_{\mathrm{ee}}\left(\phi_{\mathrm{vv}}+2\phi_{\mathrm{va}}+\phi_{\mathrm{aa}}\right)+\phi_{\mathrm{aa}}c_{\mathrm{aa}}\right]\dot{\mathring{\boldsymbol{\sigma}}}+\ddot{\mathring{\boldsymbol{\sigma}}} \\
&\qquad\qquad=c_{\mathrm{ee}}\left(\phi_{\mathrm{aa}}c_{\mathrm{aa}}\dot{\mathring{\boldsymbol{\epsilon}}}+\ddot{\mathring{\boldsymbol{\epsilon}}}\right)
\end{aligned}
}
\tag{7.257}
$$

For $c_{\mathrm{ee}}=\infty$ (7.257) reduces to (7.255).

Equation (7.257) is the simplest rheological equation, in which the three deformations $\mathring{\boldsymbol{\epsilon}}^{(\mathrm{e})}$, $\mathring{\boldsymbol{\epsilon}}^{(\mathrm{a})}$ and $\mathring{\boldsymbol{\epsilon}}^{(\mathrm{v})}$ are taken into account. This rheological equation defines a fundamental body that is named for Burgers*.

7.10.2. The Burgers body. This body is defined by (7.257) and can also be represented by mechanical models. Note that such mechanical models are only illustrative, but the definitions of the models are given by their differential or integral representations. To discuss the mechanical models, consider a simple shear with the shear γ and the shear stress σ.

Figure 7.7 depicts the Kelvin representation of a Burgers body. In this representation we have

$$
\begin{aligned}
\sigma &= \bar{G}_{\mathrm{e}}\gamma^{(\mathrm{e})}=\bar{\eta}_{\mathrm{v}}\dot{\gamma}^{(\mathrm{v})} \\
\sigma &= \left(\bar{G}_{\mathrm{a}}+\bar{\eta}_{\mathrm{a}}\mathcal{D}\right)\gamma^{(\mathrm{a})} \\
\gamma = \gamma^{(\mathrm{e})}+\gamma^{(\mathrm{a})}+\gamma^{(\mathrm{v})} &= \frac{\sigma}{\bar{G}_{\mathrm{e}}}+\frac{\sigma}{\bar{\eta}_{\mathrm{v}}\mathcal{D}}+\frac{\sigma}{\bar{G}_{\mathrm{a}}+\bar{\eta}_{\mathrm{a}}\mathcal{D}}
\end{aligned}
$$

or

$$
\left[\bar{\eta}_{\mathrm{v}}\left(\bar{G}_{\mathrm{a}}+\bar{\eta}_{\mathrm{a}}\mathcal{D}\right)\mathcal{D}+\bar{G}_{\mathrm{e}}\left(\bar{G}_{\mathrm{a}}+\bar{\eta}_{\mathrm{a}}\mathcal{D}\right)+\bar{G}_{\mathrm{e}}\bar{\eta}_{\mathrm{v}}\mathcal{D}\right]\sigma=\bar{G}_{\mathrm{e}}\bar{\eta}_{\mathrm{v}}\left(\bar{G}_{\mathrm{a}}+\bar{\eta}_{\mathrm{a}}\mathcal{D}\right)\dot{\gamma}
$$

or

$$
\frac{\bar{G}_{\mathrm{e}}}{\bar{\eta}_{\mathrm{v}}}\frac{\bar{G}_{\mathrm{a}}}{\bar{\eta}_{\mathrm{a}}}\sigma+\left(\frac{\bar{G}_{\mathrm{e}}}{\bar{\eta}_{\mathrm{v}}}+\frac{\bar{G}_{\mathrm{e}}}{\bar{\eta}_{\mathrm{a}}}+\frac{\bar{G}_{\mathrm{a}}}{\bar{\eta}_{\mathrm{a}}}\right)\dot{\sigma}+\ddot{\sigma}=\bar{G}_{\mathrm{e}}\left(\frac{\bar{G}_{\mathrm{a}}}{\bar{\eta}_{\mathrm{a}}}\dot{\gamma}+\ddot{\gamma}\right)
\tag{7.258}
$$

In the three dimensional generalization the shear γ has to be replaced by $2\mathring{\boldsymbol{\epsilon}}$ and the shear stress σ by $\mathring{\boldsymbol{\sigma}}$, so that from (7.257) and (7.258) it follows that

$$
\begin{aligned}
&2\bar{G}_{\mathrm{e}}=c_{\mathrm{ee}} \qquad \frac{\bar{G}_{\mathrm{e}}}{\bar{\eta}_{\mathrm{v}}}\frac{\bar{G}_{\mathrm{a}}}{\bar{\eta}_{\mathrm{a}}}=c_{\mathrm{ee}}\left(\phi_{\mathrm{aa}}\phi_{\mathrm{vv}}-\phi_{\mathrm{va}}^{2}\right)c_{\mathrm{aa}} \qquad \frac{\bar{G}_{\mathrm{a}}}{\bar{\eta}_{\mathrm{a}}}=\phi_{\mathrm{aa}}c_{\mathrm{aa}} \\
&\frac{\bar{G}_{\mathrm{e}}}{\bar{\eta}_{\mathrm{v}}}+\frac{\bar{G}_{\mathrm{e}}}{\bar{\eta}_{\mathrm{a}}}+\frac{\bar{G}_{\mathrm{a}}}{\bar{\eta}_{\mathrm{a}}}=c_{\mathrm{ee}}\left(\phi_{\mathrm{vv}}+2\phi_{\mathrm{va}}+\phi_{\mathrm{aa}}\right)+\phi_{\mathrm{aa}}c_{\mathrm{aa}}
\end{aligned}
\tag{7.259}
$$

---

* Burgers J.M. 1935. Mechanical considerations—model systems—phenomenological theories of relaxation and of viscosity. Chapter I of the *First Report on Viscosity and Plasticity. Proceedings of the Royal Academy of Sciences Amsterdam*, Physics Section. (First Section), Part XV, No. 3. Noord-Hollandsche Uitg.-My., Amsterdam. See also Reiner M. 1971. *Advanced Rheology*. Lewis & Co. Ltd., London. Chapter M.

From this the parameters are found as

$$\begin{aligned} \bar{G}_e &= \tfrac{1}{2} c_{ee} & \bar{\eta}_v &= \tfrac{1}{2}\frac{\phi_{aa}}{\phi_{aa}\phi_{vv} - \phi_{va}^2} \\ \bar{G}_a &= \tfrac{1}{2} c_{aa}\frac{\phi_{aa}^2}{(\phi_{aa}+\phi_{va})^2} & \bar{\eta}_a &= \tfrac{1}{2}\frac{\phi_{aa}}{(\phi_{aa}+\phi_{va})^2} \end{aligned} \tag{7.260}$$

The model is in fact determined by four parameters. The Kelvin representation in figure 7.7 is also called the standard four-parameter Voigt model*.

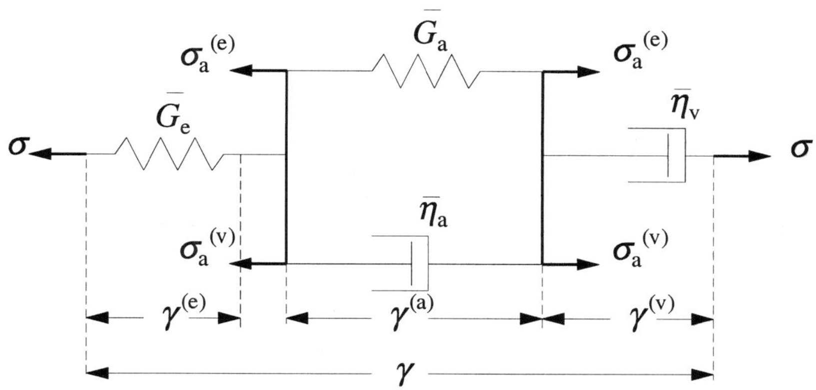

Figure 7.7. Kelvin representation of the Burgers body defined by the differential equation (7.257).

The other classical mechanical model representation of the Burgers body is the Maxwell-representation. This representation is depicted in figure 7.8, and then we have

$$\gamma_1^{(e)} = \frac{\sigma_1}{G_1} \qquad \dot{\gamma}_1^{(v)} = \frac{\sigma_1}{\eta_1} \qquad \Rightarrow \qquad \dot{\gamma} = \frac{1}{\eta_1}\left(1 + \frac{\eta_1}{G_1}\mathcal{D}\right)\sigma_1$$

Likewise

$$\dot{\gamma} = \frac{1}{\eta_2}\left(1 + \frac{\eta_2}{G_2}\mathcal{D}\right)\sigma_2$$

and using $\sigma = \sigma_1 + \sigma_2$ results

$$\sigma = \left(\frac{\eta_1}{1 + \frac{\eta_1}{G_1}\mathcal{D}} + \frac{\eta_2}{1 + \frac{\eta_2}{G_2}\mathcal{D}}\right)\dot{\gamma}$$

* Tschoegl N.W. 1989. *The Phenomenological Theory of Linear Viscoelastic Behavior.* Springer-Verlag, Berlin. See p. 93.

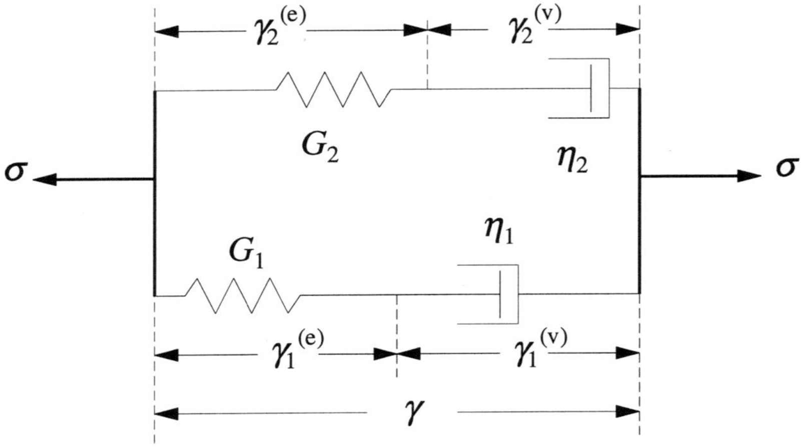

Figure 7.8. Maxwell representation of the Burgers body defined by the differential equation (7.257).

Evaluation leads to the differential equation

$$\frac{G_1}{\eta_1}\frac{G_2}{\eta_2}\sigma + \left(\frac{G_1}{\eta_1} + \frac{G_2}{\eta_2}\right)\dot{\sigma} + \ddot{\sigma} = (G_1 + G_2)\left(\frac{G_1 G_2}{G_1 + G_2}\frac{\eta_1 + \eta_2}{\eta_1 \eta_2}\dot{\gamma} + \ddot{\gamma}\right) \tag{7.261}$$

Comparison with (7.257) gives

$$\left.\begin{aligned} 2(G_1 + G_2) &= c_{ee} \\ \frac{G_1 G_2}{G_1 + G_2}\frac{\eta_1 + \eta_2}{\eta_1 \eta_2} &= \phi_{aa} c_{aa} \\ \frac{G_1}{\eta_1}\frac{G_2}{\eta_2} &= c_{ee}\left(\phi_{aa}\phi_{vv} - \phi_{va}^2\right) c_{aa} \\ \frac{G_1}{\eta_1} + \frac{G_2}{\eta_2} &= c_{ee}\left(\phi_{vv} + 2\phi_{va} + \phi_{aa}\right) + \phi_{aa} c_{aa} \end{aligned}\right\} \tag{7.262}$$

For the calculation of the canonical parameters G_1, G_2, η_1 and η_2, two relaxation times

$$\tau_1 = \eta_1 / G_1 \qquad \text{and} \qquad \tau_2 = \eta_2 / G_2 \tag{7.263}$$

are introduced. In the Maxwell representation the cause is extensive or specific and the effect is intensive. Then one speaks of *relaxation*.

According to $(7.262)_{3,4}$ these relaxation times are determined by the roots of the quadratic equation

$$\tau^{-2} - \left[\phi_{aa} c_{aa} + c_{ee}\left(\phi_{vv} + 2\phi_{va} + \phi_{aa}\right)\right]\tau^{-1} + c_{ee} c_{aa}\left(\phi_{aa}\phi_{vv} - \phi_{va}^2\right) = 0 \tag{7.264}$$

If τ_1, and τ_2 are calculated from this equation, then the other parameters follow from (7.262).

Clearly the Kelvin representation is better suited to represent the starting point than the Maxwell representation. However, the two representations are equivalent and conjugate to each other. Both representations are also called 'canonical', since these representations contain the least number of parameters. All other representations use more parameters. Alfrey* has derived general rules for the construction of the conjugate representation from a given primitive representation.

The Burgers body is the simplest model in the viscoelastic class, and it is also called the *standard linear liquid.* From this body the other classical bodies of the other classes can be derived.

7.10.3. The Poynting–Thomson body. The Poynting–Thomson body does not have viscous deformations and is a fundamental body of the firmoviscous class. Zener† called this body a *standard linear solid.*

Assume in the Burgers body that

$$\phi_{\mathrm{vv}} = 0 \tag{7.265}$$

Inequality (7.250) implies that $\phi_{\mathrm{va}} = 0$, so that (7.257) becomes

$$(c_{\mathrm{ee}} + c_{\mathrm{aa}})\,\phi_{\mathrm{aa}}\dot{\mathring{\boldsymbol{\sigma}}} + \ddot{\mathring{\boldsymbol{\sigma}}} = c_{\mathrm{ee}}\left(\phi_{\mathrm{aa}}c_{\mathrm{aa}}\dot{\mathring{\boldsymbol{\epsilon}}} + \ddot{\mathring{\boldsymbol{\epsilon}}}\right) \tag{7.266}$$

and (7.260) becomes

$$\begin{aligned} \bar{G}_{\mathrm{e}} &= \tfrac{1}{2}c_{\mathrm{ee}} & \qquad \bar{\eta}_{\mathrm{v}} &= \infty \\ \bar{G}_{\mathrm{a}} &= \tfrac{1}{2}c_{\mathrm{aa}} & \qquad \bar{\eta}_{\mathrm{a}} &= \frac{1}{2\phi_{\mathrm{aa}}} \end{aligned} \tag{7.267}$$

For $\bar{\eta}_{\mathrm{v}} = \infty$ the differential equation (7.258) of the Burgers body reduces to

$$\frac{\bar{G}_{\mathrm{e}} + \bar{G}_{\mathrm{a}}}{\bar{\eta}_{\mathrm{a}}}\dot{\sigma} + \ddot{\sigma} = \bar{G}_{\mathrm{e}}\left(\frac{\bar{G}_{\mathrm{a}}}{\bar{\eta}_{\mathrm{a}}}\dot{\gamma} + \ddot{\gamma}\right)$$

* Conjugate models with three or more elements can be derived from each other if: the number of springs and dashpots is the same; the series combination of a spring and a dashpot is replaced by a parallel combination of a spring and a dashpot and vice versa; and if an isolated dashpot or spring is present, then an isolated spring or dashpot is absent in the conjugate model and vice versa. Alfrey T.J. and Doty P. 1945. The methods of specifying the properties of viscoelastic materials. *Journal of Applied Physics*, 16: 700–713. See also Tschoegl N.W. 1989. *The Phenomenological Theory of Linear Viscoelastic Behavior.* Springer-Verlag, Berlin. Chapter 3.

† Zener C. 1952. *Elasticity and Anelasticity of Metals.* University of Chicago Press, Chicago, Illinois. See p. 43.

or

$$\dot{\sigma} + \tau\ddot{\sigma} = \bar{G}^{(\mathrm{eq})}\left(\dot{\gamma} + \bar{\tau}\ddot{\gamma}\right) \tag{7.268}$$

with

$$\begin{aligned}\tau &= \frac{\bar{\eta}_{\mathrm{a}}}{\bar{G}_{\mathrm{e}} + \bar{G}_{\mathrm{a}}} = \frac{1}{\phi_{\mathrm{aa}}\left(c_{\mathrm{ee}} + c_{\mathrm{aa}}\right)}\\ \bar{\tau} &= \frac{\bar{\eta}_{\mathrm{a}}}{\bar{G}_{\mathrm{a}}} = \frac{1}{\phi_{\mathrm{aa}}c_{\mathrm{aa}}}\\ \bar{G}^{(\mathrm{eq})} &= \frac{\bar{G}_{\mathrm{e}}\bar{G}_{\mathrm{a}}}{\bar{G}_{\mathrm{e}} + \bar{G}_{\mathrm{a}}} = \tfrac{1}{2}\frac{c_{\mathrm{ee}}c_{\mathrm{aa}}}{c_{\mathrm{ee}} + c_{\mathrm{aa}}}\end{aligned} \tag{7.269}$$

in which τ denotes a relaxation time and $\bar{\tau}$ a retardation time. From (7.251) and (7.269) it follows that

$$\tau \leq \bar{\tau} \tag{7.270}$$

so that in the class of firmo-viscous bodies the relaxation time is smaller than the retardation time.

In the representation of Maxwell

$$\sigma + \frac{\eta_1}{G_1}\dot{\sigma} = G_2\left[\gamma + \left(\frac{\eta_1}{G_1} + \frac{\eta_1}{G_2}\right)\dot{\gamma}\right] \tag{7.271}$$

so that now

$$\begin{aligned}\tau = \frac{\eta_1}{G_1} \leq \bar{\tau} &= \eta_1\left(\frac{1}{G_1} + \frac{1}{G_2}\right)\\ \bar{G}^{(\mathrm{eq})} &= G_2\end{aligned} \tag{7.272}$$

This model was introduced by Poynting and Thomson* for the rheological description of glass wires and for the creep of metals. These authors were the first to represent the rheological behavior of materials by mechanical models composed of viscous and elastic elements. The Kelvin and Maxwell representation is depicted in figure 7.9. Zener† applied this model to describe the anelasticity of metals. Integration and substitution of (7.269) into (7.266) yields for the differential equation of the Poynting–Thomson

$$\boxed{\overset{\circ}{\boldsymbol{\sigma}} + \tau\dot{\overset{\circ}{\boldsymbol{\sigma}}} = 2\bar{G}^{(\mathrm{eq})}\left(\overset{\circ}{\boldsymbol{\epsilon}} + \bar{\tau}\dot{\overset{\circ}{\boldsymbol{\epsilon}}}\right)} \tag{7.273}$$

* Poynting J.H. and Thomson J.J. 1902. *Properties of Matter.* C. Griffin & Co., London. See p. 57.

† Zener C. 1952. *Elasticity and Anelasticity of Metals.* University of Chicago Press, Chicago, Illinois.

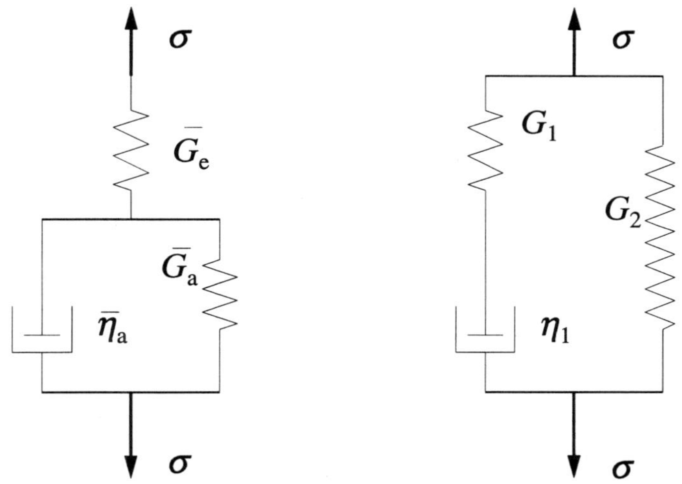

Figure 7.9. Kelvin- and Maxwell-representation of the Poynting–Thomson body defined by the differential equation (7.273).

7.10.4. The Jeffreys body. This body, the mechanical model of which is depicted in figure 7.10, does not possess elastic deformations and is fundamental for the class of the elastic fluids.

Assume now in the Burgers body that

$$c_{\mathrm{ee}} = \infty \tag{7.274}$$

so that (7.257) becomes

$$c_{\mathrm{aa}}\left(\phi_{\mathrm{aa}}\phi_{\mathrm{vv}} - \phi_{\mathrm{va}}^2\right)\mathring{\boldsymbol{\sigma}} + \left(\phi_{\mathrm{vv}} + 2\phi_{\mathrm{va}} + \phi_{\mathrm{aa}}\right)\dot{\mathring{\boldsymbol{\sigma}}} = \phi_{\mathrm{aa}}c_{\mathrm{aa}}\dot{\mathring{\boldsymbol{\epsilon}}} + \ddot{\mathring{\boldsymbol{\epsilon}}} \tag{7.275}$$

and (7.260) reduces to

$$\begin{aligned} \bar{G}_{\mathrm{e}} &= \infty & \bar{\eta}_{\mathrm{v}} &= \tfrac{1}{2}\frac{\phi_{\mathrm{aa}}}{\phi_{\mathrm{aa}}\phi_{\mathrm{vv}} - \phi_{\mathrm{va}}^2} \\ \bar{G}_{\mathrm{a}} &= \tfrac{1}{2}c_{\mathrm{aa}}\frac{\phi_{\mathrm{aa}}^2}{\left(\phi_{\mathrm{aa}} + \phi_{\mathrm{va}}\right)^2} & \bar{\eta}_{\mathrm{a}} &= \tfrac{1}{2}\frac{\phi_{\mathrm{aa}}}{\left(\phi_{\mathrm{aa}} + \phi_{\mathrm{va}}\right)^2} \end{aligned} \tag{7.276}$$

With this (7.258) reduces to

$$\frac{\bar{G}_{\mathrm{a}}}{\bar{\eta}_{\mathrm{v}}\bar{\eta}_{\mathrm{a}}}\sigma + \left(\frac{1}{\bar{\eta}_{\mathrm{v}}} + \frac{1}{\bar{\eta}_{\mathrm{a}}}\right)\dot{\sigma} = \frac{\bar{G}_{\mathrm{a}}}{\bar{\eta}_{\mathrm{a}}}\dot{\gamma} + \ddot{\gamma}$$

or

$$\sigma + \tau\dot{\sigma} = \bar{\eta}_{\mathrm{v}}\left(\dot{\gamma} + \bar{\tau}\ddot{\gamma}\right) \tag{7.277}$$

with

$$\tau = \frac{\bar{\eta}_{\mathrm{v}} + \bar{\eta}_{\mathrm{a}}}{\bar{G}_{\mathrm{a}}} \geq \bar{\tau} = \frac{\bar{\eta}_{\mathrm{a}}}{\bar{G}_{\mathrm{a}}} \tag{7.278}$$

In comparison with (7.270) it turns out now that in the class of the elastic fluids the relaxation time is larger than the retardation time.

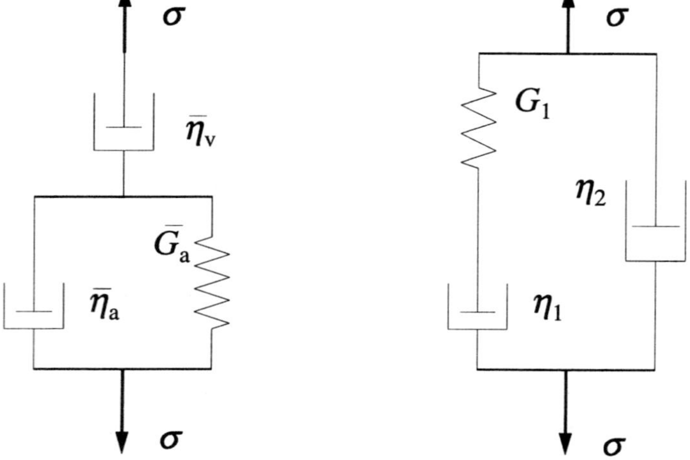

Figure 7.10. Kelvin- and Maxwell-representation of the Jeffreys body defined by the differential equation (7.280).

In the Maxwell representation, again the differential equation (7.277) is found with

$$\tau = \frac{\eta_1}{G_1} \geq \bar{\tau} = \frac{1}{G_1} \frac{\eta_1 \eta_2}{\eta_1 + \eta_2} \qquad \bar{\eta}_{\mathrm{v}} = \eta_1 + \eta_2 \tag{7.279}$$

Jeffreys* introduced this body to model the behavior of the Maxwell and the Kelvin–Voigt bodies mathematically by one equation, while Lethersich† suggested this model for the description of the linear behavior of bitumens. For a dilute suspension of elastic spheres Fröhlich and Sack‡ derived relations for the three constants in the Jeffreys model in terms of the parameters describing the structure of the suspension. Oldroyd§ showed that, besides suspensions, emulsions can also be described by the Jeffreys model.

* Jeffreys H. 1924. *The Earth: its Origin, History and Physical Constitution.* Cambridge University Press, Cambridge. 2nd ed. 1929, p. 265; 6th ed. 1976, p. 10.

† Lethersich W. 1941. See Reiner M. 1960. *Deformation, Strain and Flow.* H.K. Lewis & Co., London. pp. 148, 201.

‡ Fröhlich R. and Sack R. 1946. Theory of the rheological properties of dispersions. *Proceedings of the Royal Society of London*, A185: 415–430.

§ Oldroyd J.G. 1953. The elastic and viscous properties of emulsions and suspensions. *Proceedings of the Royal Society of London*, A218: 122–132.

With the substitution of (7.276) and (7.279) into (7.275) the differential equation for the Jeffreys body becomes

$$\boxed{\overset{\circ}{\sigma} + \tau \dot{\overset{\circ}{\sigma}} = 2\bar{\eta}_{\mathrm{v}}\left(\dot{\overset{\circ}{\epsilon}} + \bar{\tau}\ddot{\overset{\circ}{\epsilon}}\right)} \tag{7.280}$$

7.10.5. The Maxwell body. The Maxwell bodies do not possess anelastic deformations, and are fundamental for the elastico-viscous class

$$\phi_{\mathrm{aa}} = 0 \tag{7.281}$$

so that $(7.250)_2$ implies that $\phi_{\mathrm{av}} = 0$ and (7.257) reduces to

$$c_{\mathrm{ee}}\phi_{\mathrm{vv}}\dot{\overset{\circ}{\sigma}} + \ddot{\overset{\circ}{\sigma}} = c_{\mathrm{ee}}\ddot{\overset{\circ}{\epsilon}} \tag{7.282}$$

or

$$\overset{\circ}{\sigma} + \frac{1}{c_{\mathrm{ee}}\phi_{\mathrm{vv}}}\dot{\overset{\circ}{\sigma}} = \frac{1}{\phi_{\mathrm{vv}}}\dot{\overset{\circ}{\epsilon}} = 2\bar{\eta}_{\mathrm{v}}\dot{\overset{\circ}{\epsilon}} \tag{7.283}$$

In the Maxwell body only relaxation of the stress is found and no retardation of the deformation. The mechanical model is a spring and a dashpot in series as depicted in figure 7.11.

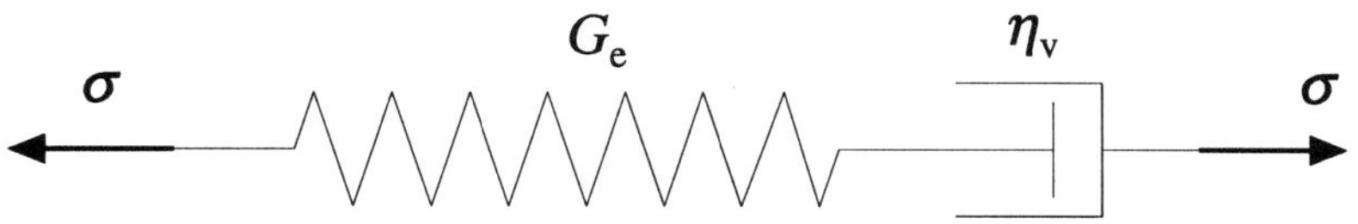

Figure 7.11. The Maxwell body defined by the differential equation (7.285).

The Maxwell body is, in the linear region, the only representative of the elastico-viscous class. The Maxwell model can be deduced from the Jeffreys model by substituting $\phi_{\mathrm{aa}} = 0$ into (7.275), so that the Maxwell body can also be considered as an approximation of the elastic fluids with $\bar{\tau} = 0$.

Maxwell* thought that gases might have a delayed return to the equilibrium state of the translation energy of the gas molecules. The relaxation time is then

$$\tau = \frac{1}{c_{\mathrm{ee}}\phi_{\mathrm{vv}}} = \frac{\bar{\eta}_{\mathrm{v}}}{\bar{G}_{\mathrm{e}}} \approx \frac{\eta}{p} \tag{7.284}$$

in which η is the viscosity of the gas and p the pressure. Under normal conditions η/p is very small. The discussion of Maxwell means that in principle elastic effects are found even in gases. In liquids the relaxation time τ can

---

* Maxwell J.C. 1868. On the dynamical theory of gases. *Philosophical Magazine and Journal of Science*, 35(4): 129–145, 185–217.

have much larger values, although then it is advisable to consider in addition the retardation of the deformation. The differential equation (7.283) for the Maxwell body becomes with the substitution of (7.284)

$$\boxed{\overset{\circ}{\sigma} + \tau \dot{\overset{\circ}{\sigma}} = 2\bar{\eta}_{\mathrm{v}} \dot{\overset{\circ}{\epsilon}}} \tag{7.285}$$

7.10.6. The Kelvin–Voigt body. The Kelvin–Voigt bodies are fundamental for the anelastic class and can be derived from the Poynting–Thomson model by substituting $c_{\mathrm{ee}} = \infty$ into (7.266), or from the Jeffreys model by substituting $\phi_{\mathrm{vv}} = 0$ into (7.275) (note that due to (7.250) $\phi_{\mathrm{vv}} = 0$ implies that $\phi_{\mathrm{va}} = 0$). In both cases it is found that the differential equation for the Kelvin–Voigt body is given by

$$\boxed{\overset{\circ}{\sigma} = \bar{G}_{\mathrm{a}} \big(\overset{\circ}{\epsilon} + \bar{\tau} \dot{\overset{\circ}{\epsilon}} \big)} \tag{7.286}$$

with

$$\bar{\tau} = \frac{1}{c_{\mathrm{aa}} \phi_{\mathrm{aa}}} \qquad \text{and} \qquad \bar{G}_{\mathrm{a}} = \tfrac{1}{2} c_{\mathrm{aa}} \tag{7.287}$$

The Kelvin–Voigt model does not show relaxation of the stresses, but it shows retardation of the deformation. In this respect the Kelvin–Voigt model is complementary to the Maxwell model. The mechanical model is now a spring and dashpot in parallel as depicted in figure 7.12.

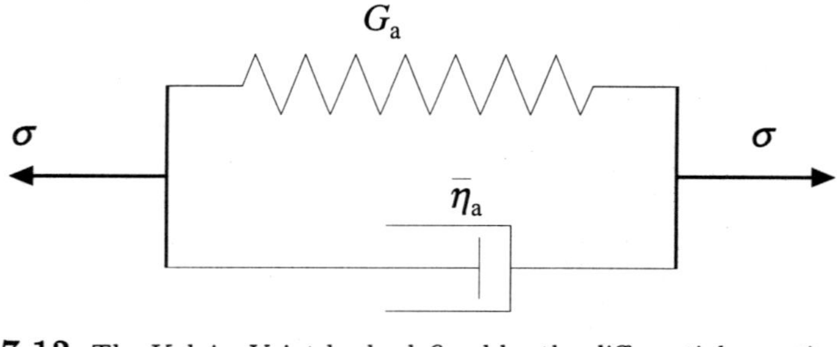

Figure 7.12. The Kelvin–Voigt body defined by the differential equation (7.286).

Kelvin* formulated this model by considering that the elastic deformations which proceed with finite velocity are accompanied by dissipation in principle. Kelvin thought of a model that consists of elastic grains that are bordered on each other by fluid films. The resistance against deformation is combined in parallel. Later on Voigt† introduced the model without any reference to a physical mechanism.

* Thomson W. (Lord Kelvin). 1878. Elasticity. In the *Encyclopedia Britannica*, 9th ed., Vol. VII, p. 803d. Charles Scribner's, New York.

† Voigt W. 1892. Ueber innere Reibung fester Körper, insbesondere der Metalle. *Annalen der Physik und Chemie*, Neue Folge 47: 671–693.

7.10.7. Classical bodies. If $c_{\mathrm{ee}} = \infty$ is substituted into the model for the Maxwell body then a body results that shows only viscous deformation. This result can also be obtained if $\phi_{\mathrm{aa}} = 0$ is substituted into the Jeffreys model (then also $\phi_{\mathrm{va}} = 0$)

$$\phi_{\mathrm{vv}}\mathring{\boldsymbol{\sigma}} = \dot{\mathring{\boldsymbol{\epsilon}}} \qquad \text{or} \qquad \boxed{\mathring{\boldsymbol{\sigma}} = 2\bar{\eta}_{\mathrm{v}}\dot{\mathring{\boldsymbol{\epsilon}}} = 2\bar{\eta}_{\mathrm{v}}\mathring{\boldsymbol{D}}} \tag{7.288}$$

This law was formulated by Newton*, so that these bodies are referred to as *Newtonian bodies* or linear viscous bodies. Fluids consisting of relatively simple molecules usually satisfy the Newtonian law quite well, if the velocity gradients are not too large. A measure for the relative magnitude of the velocity gradient is

$$Tr = \frac{\eta}{p}\sqrt{\mathring{II}_D} \tag{7.289}$$

with $\eta = \bar{\eta}_{\mathrm{v}}$ and $\mathring{II}_D = \mathring{\boldsymbol{D}}\colon \mathring{\boldsymbol{D}}$. The number Tr was introduced by Truesdell† as the *truncation number* and is a special case of the familiar Deborah‡ number De used in rheology. For $Tr \ll 1$, the Newton law for fluids, that the stress depends linearly on the rate of deformation, is quite succesful. For air and water the deformation gradients have to be very large in order that $Tr \approx 1$. For $\eta = \eta_{\mathrm{v}} = 0$, $Tr = 0$ for all p and $\mathring{II}_D$. In that case $\mathring{\boldsymbol{\sigma}}$ is zero for all $\mathring{\boldsymbol{D}}$. The fluid cannot sustain finite deviatoric stresses at finite values of $\mathring{\boldsymbol{D}}$. This may thus happen for $Tr \ll 1$. Then the *rheological ideal Pascal fluids* result that are important in fluid dynamics calculations.

Bodies that show only elastic deformation are obtained by substituting $\phi_{\mathrm{aa}} = 0$ into the Poynting–Thomson model or $\phi_{\mathrm{vv}} = 0$ into the Maxwell model. In both cases it is found that

$$\boxed{\mathring{\boldsymbol{\sigma}} = c_{\mathrm{ee}}\mathring{\boldsymbol{\epsilon}} = 2\bar{G}_{\mathrm{e}}\mathring{\boldsymbol{\epsilon}}} \tag{7.290}$$

These are the linear-elastic bodies that are named for Hooke§, so they are spoken of as Hooke bodies. Particularly for metals the Hooke model (7.290) is a good first approximation to the mechanical behavior.

---

* Newton I. 1687. Book II, *The Motion of Bodies (In Resisting Mediums)*. In *Philosophiae Naturalis Principia Mathematica*. London.

† Truesdell C. and Toupin R. 1960. The classical field theories. In the *Encyclopedia of Physics*, Vol. III/1. Springer-Verlag, Berlin. See p. 721.

‡ Reiner M. 1964. The Deborah number. *Physics Today*, 17: 62.

§ Hooke R. 1678. *Lectures de Potentia Restitutiva*. London.

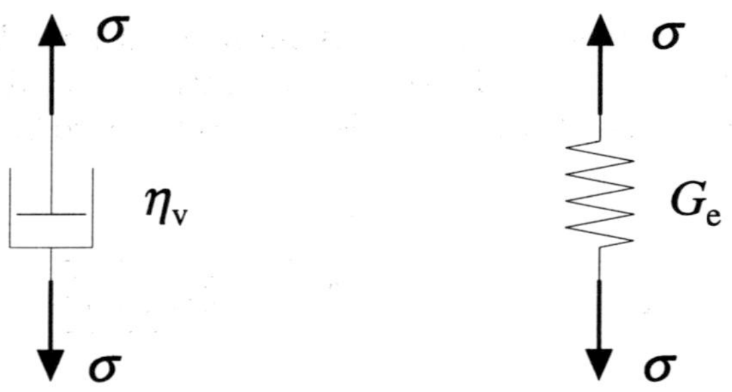

Figure 7.13. Bodies of Newton and Hooke defined by the differential equation (7.288) and the equation (7.290) respectively.

For $c_{ee} = \infty$ the deviatoric deformation is zero for all deviatoric stresses. The 'spring' in the mechanical representation in figure 7.13 has now become a rigid rod. If these reductions apply equally for the volume deformations due to pressures then the rigid *Euclidean bodies* result that can be considered as a kind of zero order approximation of the behavior of 'solid' bodies.

The bodies of Newton, Pascal, Hooke and Euclid are called classical in rheology as these models are not of major interest for rheologists. The models can be derived in various ways from a master model that includes all basic types of deformation, in this case the Burgers model. In the discussion above, following the considerations about the classifications of the non-plastic bodies, successively $\overset{\circ}{\boldsymbol{\epsilon}}^{(e)}$, $\overset{\circ}{\boldsymbol{\epsilon}}^{(a)}$ and $\overset{\circ}{\boldsymbol{\epsilon}}^{(v)}$ were set equal to zero. This systematics is depicted in the diagram of figure 7.14, in which the coefficients are the coefficients that occur in the differential equation (7.257) for the Burgers body.

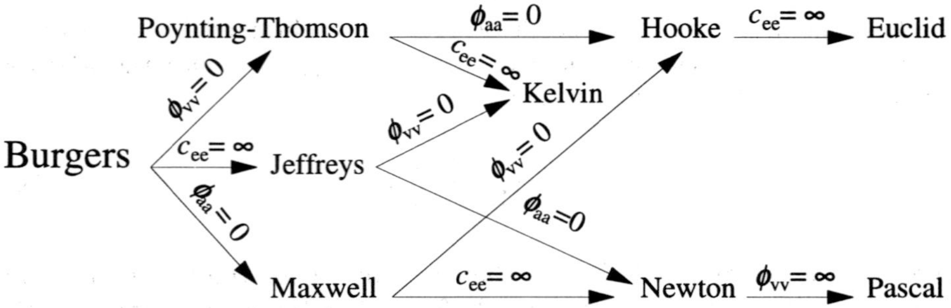

Figure 7.14. Diagram to derive the models from the Burgers model. The coefficients are found in the differential equation (7.257) for the Burgers body.

7.11. BODIES WITH A NUMBER OF INTERNAL PROCESSES

7.11.1. Basic equations. Next, only deviatoric deformations and deviatoric stresses are considered. For volumetric deformations and stresses the results are analogous.

The linearized thermostatic equations of state are given by $(7.68)_{1,2}$

$$\begin{aligned} \overset{\circ}{\boldsymbol{\sigma}}^{(\mathrm{eq})} &= c_{\mathrm{ee}} \overset{\circ}{\boldsymbol{\epsilon}}^{(\mathrm{e})} \\ -\left(\overset{\circ}{\boldsymbol{P}}_{\alpha} - \overset{\circ}{\boldsymbol{\sigma}}^{(\mathrm{eq})}\right) &= \sum_{\beta=1}^{\nu} c_{\alpha\beta} \overset{\circ}{\boldsymbol{\epsilon}}_{\beta}^{(\mathrm{a})} \end{aligned} \tag{7.291}$$

The corresponding elastic function is, according to (7.64)

$$\mathfrak{E} = \rho_{\circ}\left(f - f_{\circ}\right) = \tfrac{1}{2} c_{\mathrm{ee}} \overset{\circ}{\boldsymbol{\epsilon}}^{(\mathrm{e})} : \overset{\circ}{\boldsymbol{\epsilon}}^{(\mathrm{e})} + \tfrac{1}{2} \sum_{\alpha,\beta=1}^{\nu} c_{\alpha\beta} \overset{\circ}{\boldsymbol{\epsilon}}_{\alpha}^{(\mathrm{a})} : \overset{\circ}{\boldsymbol{\epsilon}}_{\beta}^{(\mathrm{a})} \geq 0 \tag{7.292}$$

The phenomenological or kinetic equations are given by (7.98)

$$\begin{aligned} \dot{\overset{\circ}{\boldsymbol{\epsilon}}}^{(\mathrm{v})} &= \phi_{\mathrm{vv}} \overset{\circ}{\boldsymbol{\sigma}} + \sum_{\alpha=1}^{\nu} \phi_{v\alpha} \left(\overset{\circ}{\boldsymbol{P}}_{\alpha} + \overset{\circ}{\boldsymbol{\sigma}} - \overset{\circ}{\boldsymbol{\sigma}}^{(\mathrm{eq})}\right) \\ \dot{\overset{\circ}{\boldsymbol{\epsilon}}}_{\alpha}^{(\mathrm{a})} &= \phi_{\alpha v} \overset{\circ}{\boldsymbol{\sigma}} + \sum_{\beta=1}^{\nu} \phi_{\alpha\beta} \left(\overset{\circ}{\boldsymbol{P}}_{\beta} + \overset{\circ}{\boldsymbol{\sigma}} - \overset{\circ}{\boldsymbol{\sigma}}^{(\mathrm{eq})}\right) \end{aligned} \tag{7.293}$$

The corresponding dissipation function is, according to (7.86)

$$\mathfrak{D}^{+} = \overset{\circ}{\boldsymbol{\sigma}} : \dot{\overset{\circ}{\boldsymbol{\epsilon}}}^{(\mathrm{v})} + \sum_{\alpha=1}^{\nu} \left(\overset{\circ}{\boldsymbol{P}}_{\alpha} + \overset{\circ}{\boldsymbol{\sigma}} - \overset{\circ}{\boldsymbol{\sigma}}^{(\mathrm{eq})}\right) : \dot{\overset{\circ}{\boldsymbol{\epsilon}}}_{\alpha}^{(\mathrm{a})} \geq 0 \tag{7.294}$$

The inequality in (7.292) expresses the thermodynamic stability of the reference state. The inequality (7.294) expresses the entropy principle. Both inequalities are closely related to each other.

Suppose now that $\nu > 1$ internal processes are active in the body causing the anelastic deformations. To facilitate the solution of (7.293) it is assumed that the viscous deformation is caused by $N - \nu$ internal processes. Then (7.293) can be replaced by

$$\dot{\overset{\circ}{\boldsymbol{\epsilon}}}_{\alpha} = \sum_{\beta=1}^{N} \phi_{\alpha\beta} \left(\overset{\circ}{\boldsymbol{P}}_{\beta} + \overset{\circ}{\boldsymbol{\sigma}} - \overset{\circ}{\boldsymbol{\sigma}}^{(\mathrm{eq})}\right) \tag{7.295}$$

and (7.294) can be written as

$$\mathfrak{D}^{+} = \sum_{\alpha=1}^{N} \left(\overset{\circ}{\boldsymbol{P}}_{\alpha} + \overset{\circ}{\boldsymbol{\sigma}} - \overset{\circ}{\boldsymbol{\sigma}}^{(\mathrm{eq})}\right) : \dot{\overset{\circ}{\boldsymbol{\epsilon}}} \geq 0 \tag{7.296}$$

The internal variables $\dot{\mathring{\boldsymbol{\epsilon}}}_\alpha$ are chosen such that $\phi_{\alpha\beta}$ is positive definite, so that the internal variables are not singular. The constitutive equations (7.291) can be replaced by

$$\begin{aligned} \mathring{\boldsymbol{\sigma}}^{(\mathrm{eq})} &= c_{\mathrm{ee}}\mathring{\boldsymbol{\epsilon}}^{(\mathrm{e})} \\ -\left(\mathring{\boldsymbol{P}}_\alpha - \mathring{\boldsymbol{\sigma}}^{(\mathrm{eq})}\right) &= \sum_{\beta=1}^{\nu} c_{\alpha\beta}\mathring{\boldsymbol{\epsilon}}_\beta \end{aligned} \tag{7.297}$$

with

$$\mathfrak{E} = \tfrac{1}{2}c_{\mathrm{ee}}\mathring{\boldsymbol{\epsilon}}^{(\mathrm{e})} : \mathring{\boldsymbol{\epsilon}}^{(\mathrm{e})} + \tfrac{1}{2}\sum_{\alpha,\beta=1}^{\nu} c_{\alpha\beta}\mathring{\boldsymbol{\epsilon}}_\alpha : \mathring{\boldsymbol{\epsilon}}_\beta \geq 0 \tag{7.298}$$

The $N - \nu$ internal variables, associated with the viscous deformations, are not found by definition in (7.297) and (7.298). If ν internal variables describe the anelastic deformation, then the rank of the matrix $c_{\alpha\beta}$ is given by ν. If there are in total N internal variables, then the matrix $c_{\alpha\beta}$ is singular for $\nu < N$. Since $\phi_{\alpha\beta}$ is not singular, (7.295) can be inverted

$$\sum_{\beta=1}^{N} \phi^{-1}_{\alpha\beta}\, \dot{\mathring{\boldsymbol{\epsilon}}}_\beta = \mathring{\boldsymbol{P}}_\alpha + \mathring{\boldsymbol{\sigma}} - \mathring{\boldsymbol{\sigma}}^{(\mathrm{eq})} \tag{7.299}$$

and the substitution of $(7.297)_2$ yields

$$\sum_{\beta=1}^{N} \left(\phi^{-1}_{\alpha\beta}\dot{\mathring{\boldsymbol{\epsilon}}}_\beta + c_{\alpha\beta}\,\mathring{\boldsymbol{\epsilon}}_\beta\right) = \mathring{\boldsymbol{\sigma}} \tag{7.300}$$

This result summarizes the relaxation equations that show that the internal processes are only excited by the deviatoric stresses.

The dissipation function (7.296) can be written with the substitution of (7.299) as

$$\mathfrak{D}^{+} = \sum_{\alpha,\beta=1}^{N} \phi^{-1}_{\alpha\beta}\,\dot{\mathring{\boldsymbol{\epsilon}}}_\alpha : \dot{\mathring{\boldsymbol{\epsilon}}}_\beta \geq 0 \tag{7.301}$$

With (7.298) and (7.301), the deviatoric stress (7.300) can be written in a 'Lagrangian formulation'

$$\frac{\partial \mathfrak{E}}{\partial \mathring{\boldsymbol{\epsilon}}_\alpha} + \frac{\partial \mathfrak{D}^{+}/2}{\partial \dot{\mathring{\boldsymbol{\epsilon}}}_\alpha} = \mathring{\boldsymbol{\sigma}} \tag{7.302}$$

Some authors have a preference for variational formulations, for instance Biot*.

---

* Biot M.A. 1954. Theory of stress-strain relations in anisotropic viscoelasticity and relaxation phenomena. *Journal of Applied Physics*, 25: 1385–1391; Biot M.A. 1955. Variational principles in irreversible thermodynamics with application to viscoelasticity. *Physical Review*, 97: 1463–1469.

7.11.2. Solution of the relaxation equations. Introduce the variables

$$q_\alpha = \overset{\circ}{\epsilon}_\alpha \qquad \text{and} \qquad s_\alpha = \overset{\circ}{\boldsymbol{\sigma}} \tag{7.303}$$

The q_α's ($\alpha = 1, 2, \cdots, N$) constitute a N-dimensional space that is the space of the internal variables. The relaxation equations (7.300) written in the variables (7.303) become

$$\sum_{\beta=1}^{N} \left(\phi_{\alpha\beta}^{-1} \dot{q}_\beta + c_{\alpha\beta} q_\beta \right) = s_\alpha \qquad (\alpha = 1, 2, \cdots, N) \tag{7.304}$$

These equations may be solved by reducing them to diagonal form with a principal-axes transformation. To this purpose, $\phi_{\alpha\beta}^{-1}$ and $c_{\alpha\beta}$ are transformed simultaneously* to their principal-axes with

$$\sum_{\beta=1}^{N} \left(\phi_{\alpha\beta}^{-1} - \bar{\tau} c_{\alpha\beta} \right) q_\beta = 0 \tag{7.305}$$

These N linear equations have nontrivial solutions if

$$\det \left| \phi_{\alpha\beta}^{-1} - \bar{\tau} c_{\alpha\beta} \right| = 0 \tag{7.306}$$

This condition is an algebraic equation for $\bar{\tau}$ of the degree N that yields real roots $\bar{\tau}_K$ ($K = 1, 2 \cdots, N$) if $\phi_{\alpha\beta}$ and $c_{\alpha\beta}$ are real symmetric matrices and $\phi_{\alpha\beta}$ is positive definite. The real roots $\bar{\tau}_K$ are the principal values—eigenvalues—of the principal-axes problem. To each principal value $\bar{\tau}_K$ there corresponds a principal vector $\vec{q}_K$ with components $q_{K,\alpha}$ in the space of the internal variables that are determined by

$$\sum_{\beta=1}^{N} \left(\phi_{\alpha\beta}^{-1} - \bar{\tau}_K c_{\alpha\beta} \right) q_{K,\beta} = 0 \tag{7.307}$$

It may be noted that $c_{\alpha\beta}$ is of the rank $\nu < N$, so that there exist ν linear independent principal vectors $\vec{q}_K$, for which

$$\sum_{\beta=1}^{N} c_{\alpha\beta} q_{K,\beta} \neq 0 \qquad (K = 1, 2, \cdots, \nu)$$

---

* Simultaneous reduction to diagonal forms can be imagined as follows: a real symmetric matrix $c_{\alpha\beta}$ can be transformed in the space of the internal variables at its principal-axes. Rotate the coordinate system such that it coincides with the principal directions and then multiply the coordinate axes so that the ellipsoid representing the matrix $c_{\alpha\beta}$ reduces to a sphere. Subsequently, rotate the principal directions so that they coincide with the principal directions of $\phi_{\alpha\beta}^{-1}$.

The subspace spanned by these ν principal vectors $\vec{q}_K$ is the subspace of the anelastic internal variables.

If $\phi_{\alpha\beta}$ has the rank N and $c_{\alpha\beta}$ has rank $\nu < N$, then there exist $N-\nu$ linear independent principal vectors $\vec{\hat{q}}_P$, for which

$$\sum_{\beta=1}^{N} c_{\alpha\beta}\hat{q}_{P,\beta} = 0 \qquad \sum_{\beta=1}^{N} \phi^{-1}_{\alpha\beta}\,\hat{q}_{P,\beta} \neq 0 \qquad (P = \nu+1, \cdots, N) \tag{7.308}$$

The subspace spanned by these $N-\nu$ principal vectors is the subspace of the viscous internal variables.

Therefore in the subspace of the anelastic internal variables (7.307) applies, and in the subspace of the viscous internal variables (7.308) applies. Formally this means that (7.306) includes roots $\bar{\tau}_P = \infty$, for which (7.307) reduces to (7.308).

Consider first the subspace of the anelastic internal variables. Multiplying (7.307) with $q_{K,\alpha}$ and summing over α yield for the principal value

$$\bar{\tau}_K = \sum_{\alpha,\beta=1}^{\nu} q_{K,\alpha}\phi^{-1}_{\alpha\beta}\,q_{K,\beta} \Bigg/ \sum_{\alpha,\beta=1}^{\nu} q_{K,\alpha}c_{\alpha\beta}q_{K,\beta} \tag{7.309}$$

Since in this subspace both $\phi_{\alpha\beta}$ and $c_{\alpha\beta}$ are positive definite, it follows that

$$0 < \bar{\tau}_K < \infty \qquad (K = 1, 2, \cdots, \nu) \tag{7.310}$$

For $\vec{q}_K = \vec{\hat{q}}_P$ it is clear that it follows from (7.308) and (7.309) that $\bar{\tau}_K = \infty$, so that the subspace of the viscous internal variables corresponds with the roots $\bar{\tau} = \infty$ of (7.306). For each $\bar{\tau}_K$ the equations (7.307) yield at least ∞^1 solutions, so that a normalization condition can also be imposed, which is chosen as

$$\sum_{\alpha,\beta=1}^{N} q_{K,\alpha}c_{\alpha\beta}q_{M,\beta} = \delta_{KM} \qquad (K, M = 1, 2 \cdots, \nu) \tag{7.311}$$

With this extra condition, (7.307) gives

$$\sum_{\alpha,\beta=1}^{N} q_{K,\alpha}\phi^{-1}_{\alpha\beta}q_{M,\beta} = \bar{\tau}_K\delta_{KM} \qquad (K, M = 1, 2 \cdots, \nu) \tag{7.312}$$

In the subspace of the viscous internal variables the normalization can be imposed by

$$\left.\begin{aligned} &\sum_{\beta=1}^{N} c_{\alpha\beta}\hat{q}_{P,\beta} = 0 \\ &\sum_{\alpha,\beta=1}^{N} \hat{q}_{P,\alpha}\phi^{-1}_{\alpha\beta}\hat{q}_{Q,\beta} = \delta_{PQ} \end{aligned}\right\} \qquad (P, Q = \nu+1, \cdots, N) \tag{7.313}$$

Multiplication of (7.307) with $\hat{q}_{P,\alpha}$ and summing over α yield

$$\sum_{\alpha,\beta=1}^{N} \hat{q}_{P,\alpha} \left(\phi_{\alpha\beta}^{-1} - \bar{\tau}_K c_{\alpha\beta}\right) q_{K,\beta} = 0$$

or by using $(7.313)_1$

$$\sum_{\alpha,\beta=1}^{N} \hat{q}_{P,\alpha} \phi_{\alpha\beta}^{-1} q_{K,\beta} = 0 \qquad (K = 1, \cdots, \nu; P = \nu + 1, \cdots, N) \tag{7.314}$$

The principal-axes transformation becomes

$$q_\alpha = \sum_{K=1}^{\nu} q_{K,\alpha} Q_K + \sum_{P=\nu+1}^{N} \hat{q}_{P,\alpha} \hat{Q}_P \tag{7.315}$$

Q_K is thus the component of $\vec{q}$ with respect to the principal vector $\vec{q}_K$, and $\hat{Q}_P$ is the component with respect to the principal vector $\vec{\hat{q}}_P$.

Denote in (7.304)

$$\mathfrak{D} = D/Dt \tag{7.316}$$

and substitute (7.315) into (7.304)

$$\sum_{\beta=1}^{N} \left[\sum_{M=1}^{\nu} c_{\alpha\beta} q_{M,\beta} Q_M + \sum_{Q=\nu+1}^{N} c_{\alpha\beta} \hat{q}_{Q,\beta} \hat{Q}_Q + \sum_{M=1}^{\nu} \phi_{\alpha\beta}^{-1} q_{M,\beta} \mathfrak{D} Q_M + \sum_{Q=\nu+1}^{N} \phi_{\alpha\beta}^{-1} \hat{q}_{Q,\beta} \mathfrak{D} \hat{Q}_Q \right] = s_\alpha$$

or with $(7.313)_1$

$$\sum_{\beta=1}^{N} \left[\sum_{M=1}^{\nu} + c_{\alpha\beta} q_{M,\beta} Q_M + \sum_{M=1}^{\nu} \phi_{\alpha\beta}^{-1} q_{M,\beta} \mathfrak{D} Q_M + \sum_{Q=\nu+1}^{N} \phi_{\alpha\beta}^{-1} \hat{q}_{Q,\beta} \mathfrak{D} \hat{Q}_Q \right] = s_\alpha \tag{7.317}$$

Multiply (7.317) with $q_{K,\alpha}$, sum over α and apply (7.311), (7.312) and (7.314), to get

$$Q_K + \bar{\tau}_K \mathfrak{D} Q_K = \sum_{\alpha=1}^{N} q_{K,\alpha} s_\alpha \tag{7.318}$$

Multiplication of (7.317) with $\hat{q}_{P,\alpha}$, summing over α, and finally applying (7.313) and (7.314) gives

$$\mathfrak{D} \hat{Q}_P = \sum_{\alpha=1}^{N} \hat{q}_{P,\alpha} s_\alpha \tag{7.319}$$

This leads to

$$
\begin{aligned}
Q_K &= \frac{1}{1+\bar{\tau}_K \mathfrak{D}} \sum_{\beta=1}^{N} q_{K,\beta} s_\beta \qquad (K = 1, 2 \cdots, \nu) \\
\hat{Q}_P &= \frac{1}{\mathfrak{D}} \sum_{\beta=1}^{N} \hat{q}_{P,\beta} s_\beta \qquad (P = \nu+1, \cdots, N)
\end{aligned}
\tag{7.320}
$$

The interpretation of the differential operator in the denominator is given by (7.337). Substitution of (7.320) into (7.315) yields

$$
q_\alpha = \sum_{K=1}^{\nu} \sum_{\beta=1}^{N} \frac{q_{K,\alpha} q_{K,\beta} s_\beta}{1+\bar{\tau}_K \mathfrak{D}} + \sum_{P=\nu+1}^{N} \sum_{\beta=1}^{N} \frac{\hat{q}_{P,\alpha} \hat{q}_{P,\beta} s_\beta}{\mathfrak{D}} \tag{7.321}
$$

Written in the strains and stresses this becomes, with the definitions (7.303)

$$
\mathring{\epsilon}_\alpha = \sum_{K=1}^{\nu} \sum_{\beta=1}^{N} \frac{q_{K,\alpha} q_{K,\beta}}{1+\bar{\tau}_K \mathfrak{D}} \mathring{\boldsymbol{\sigma}} + \sum_{P=\nu+1}^{N} \sum_{\beta=1}^{N} \hat{q}_{P,\alpha} \hat{q}_{P,\beta} \frac{1}{\mathfrak{D}} \mathring{\boldsymbol{\sigma}} \tag{7.322}
$$

and finally

$$
\boxed{2\left(\mathring{\boldsymbol{\epsilon}}^{(a)} + \mathring{\boldsymbol{\epsilon}}^{(v)}\right) = 2\sum_{\alpha=1}^{N} \mathring{\epsilon}_\alpha = \sum_{K=1}^{\nu} \frac{\bar{C}_K}{1+\bar{\tau}_K \mathfrak{D}} \mathring{\boldsymbol{\sigma}} + \varphi_\infty \frac{1}{\mathfrak{D}} \mathring{\boldsymbol{\sigma}}} \tag{7.323}
$$

with

$$
\left.
\begin{aligned}
\bar{C}_K &= 2 \sum_{\alpha,\beta=1}^{N} q_{K,\alpha} q_{K,\beta} \\
\hat{C}_P &= 2 \sum_{\alpha,\beta=1}^{N} \hat{q}_{P,\alpha} \hat{q}_{P,\beta} \qquad \varphi_\infty = \sum_{P=\nu+1}^{N} \hat{C}_P
\end{aligned}
\right\}
\tag{7.324}
$$

Now

$$
\sum_{\alpha,\beta=1}^{N} q_{K,\alpha} q_{K,\beta} y_\alpha y_\beta = \left(\sum_{\alpha=1}^{N} q_{K,\alpha}\, y_\alpha \right)^2 \geq 0
$$

This applies also for $y_\alpha = 1$, so that

$$
\bar{C}_K \geq 0 \qquad \text{and} \qquad \varphi_\infty \geq 0 \tag{7.325}
$$

In the thermodynamics of irreversible processes the important inequalities (7.310) and (7.325) follow from

(1) thermodynamic stability of the reference state,
(2) entropy principle,
(3) Onsager Casimir reciprocal relations.

7.11.3. Calculation of the energy functions. Write the elastic function (7.298) as

$$\begin{aligned} \mathfrak{E} &= \tfrac{1}{2} c_{\text{ee}} \mathring{\boldsymbol{\epsilon}}^{(\text{e})} : \mathring{\boldsymbol{\epsilon}}^{(\text{e})} + \mathring{\mathfrak{E}} \\ \mathring{\mathfrak{E}} &= \tfrac{1}{2} \sum_{\alpha,\beta=1}^{N} c_{\alpha\beta} \mathring{\boldsymbol{\epsilon}}_{\alpha} : \mathring{\boldsymbol{\epsilon}}_{\beta} \end{aligned} \tag{7.326}$$

According to (7.77)

$$\begin{aligned} \mathring{\boldsymbol{\sigma}} &= \mathring{\boldsymbol{\sigma}}^{(\text{eq})} & \text{if} \quad & \mathring{\boldsymbol{\epsilon}}^{(\text{e})} \neq 0 \\ \mathring{\boldsymbol{\sigma}} &\neq \mathring{\boldsymbol{\sigma}}^{(\text{eq})} = 0 & \text{if} \quad & \mathring{\boldsymbol{\epsilon}}^{(\text{e})} = 0 \end{aligned} \tag{7.327}$$

With the substitution of $(7.297)_1$ the elastic function $(7.326)_1$ can be written as

$$\mathfrak{E} = \frac{1}{2c_{\text{ee}}} \mathring{\boldsymbol{\sigma}} : \mathring{\boldsymbol{\sigma}} + \mathring{\mathfrak{E}} \tag{7.328}$$

in which $c_{\text{ee}} = \infty$ for $\mathring{\boldsymbol{\epsilon}}^{(\text{e})} = 0$ to satisfy (7.327). With the substitution of (7.322) the anelastic contribution $(7.326)_2$ becomes

$$2\mathring{\mathfrak{E}} = \sum_{K,M=1}^{N} \sum_{\alpha,\beta=1}^{N} \sum_{\rho,\sigma=1}^{N} c_{\alpha\beta} \left[\frac{q_{K,\alpha} q_{K,\rho}}{1 + \bar{\tau}_K \mathfrak{D}} \mathring{\boldsymbol{\sigma}} + \hat{q}_{K,\alpha} \hat{q}_{K,\rho} \frac{1}{\mathfrak{D}} \mathring{\boldsymbol{\sigma}} \right] : \left[\frac{q_{M,\beta} q_{M,\sigma}}{1 + \bar{\tau}_M \mathfrak{D}} \mathring{\boldsymbol{\sigma}} + \hat{q}_{M,\beta} \hat{q}_{M,\sigma} \frac{1}{\mathfrak{D}} \mathring{\boldsymbol{\sigma}} \right]$$

Due to $(7.313)_1$ all terms with $\mathring{\boldsymbol{\sigma}}/\mathfrak{D}$ cancel, while the use of (7.311) results in

$$2\mathring{\mathfrak{E}} = \sum_{K,M=1}^{\nu} \sum_{\rho,\sigma=1}^{N} \delta_{KM} q_{K,\rho} q_{M,\sigma} \left[\frac{\mathring{\boldsymbol{\sigma}}}{1 + \bar{\tau}_K \mathfrak{D}} \right] : \left[\frac{\mathring{\boldsymbol{\sigma}}}{1 + \bar{\tau}_M \mathfrak{D}} \right]$$

or with $(7.324)_1$

$$2\mathring{\mathfrak{E}} = \tfrac{1}{2} \sum_{K=1}^{\nu} \bar{C}_K \left[\frac{\mathring{\boldsymbol{\sigma}}}{1 + \bar{\tau}_K \mathfrak{D}} \right] : \left[\frac{\mathring{\boldsymbol{\sigma}}}{1 + \bar{\tau}_K \mathfrak{D}} \right]$$

With this result, (7.328) becomes

$$\boxed{\mathfrak{E} = \frac{1}{2c_{\text{ee}}} \mathring{\boldsymbol{\sigma}} : \mathring{\boldsymbol{\sigma}} + \tfrac{1}{4} \sum_{K=1}^{\nu} \bar{C}_K \left[\frac{\mathring{\boldsymbol{\sigma}}}{1 + \bar{\tau}_K \mathfrak{D}} \right] : \left[\frac{\mathring{\boldsymbol{\sigma}}}{1 + \bar{\tau}_K \mathfrak{D}} \right]} \tag{7.329}$$

The dissipation function $\mathfrak{D}$ (7.301) becomes, with (7.322)

$$\mathfrak{D}^{+} = \sum_{K,M=1}^{N} \sum_{\alpha,\beta=1}^{N} \sum_{\rho,\sigma=1}^{N} \phi_{\alpha\beta}^{-1} \left[\frac{q_{K,\alpha} q_{K,\rho}}{1+\bar{\tau}_K \mathcal{D}} \dot{\mathring{\boldsymbol{\sigma}}} + \hat{q}_{K,\alpha} \hat{q}_{K,\rho} \dot{\mathring{\boldsymbol{\sigma}}} \right] : \left[\frac{q_{M,\beta} q_{M,\sigma}}{1+\bar{\tau}_M \mathcal{D}} \dot{\mathring{\boldsymbol{\sigma}}} + \hat{q}_{M,\beta} \hat{q}_{M,\sigma} \dot{\mathring{\boldsymbol{\sigma}}} \right]$$

By using (7.312), $(7.313)_2$ and (7.314) this results in

$$\mathfrak{D}^{+} = \sum_{K,M=1}^{\nu} \sum_{\rho,\sigma=1}^{N} \delta_{KM} \bar{\tau}_K q_{K,\rho} q_{M,\sigma} \left[\frac{\dot{\mathring{\boldsymbol{\sigma}}}}{1+\bar{\tau}_K \mathcal{D}} \right] : \left[\frac{\dot{\mathring{\boldsymbol{\sigma}}}}{1+\bar{\tau}_M \mathcal{D}} \right] + \sum_{P,Q=\nu+1}^{N} \sum_{\rho,\sigma=1}^{N} \delta_{PQ} \hat{q}_{P,\rho} \hat{q}_{Q,\sigma} \left(\dot{\mathring{\boldsymbol{\sigma}}} : \dot{\mathring{\boldsymbol{\sigma}}} \right)$$

or with (7.324)

$$\boxed{\mathfrak{D}^{+} = \tfrac{1}{2} \sum_{K=1}^{\nu} \bar{\tau}_K \bar{C}_K \left[\frac{\dot{\mathring{\boldsymbol{\sigma}}}}{1+\bar{\tau}_K \mathcal{D}} \right] : \left[\frac{\dot{\mathring{\boldsymbol{\sigma}}}}{1+\bar{\tau}_K \mathcal{D}} \right] + \tfrac{1}{2} \varphi_{\infty} (\dot{\mathring{\boldsymbol{\sigma}}} : \dot{\mathring{\boldsymbol{\sigma}}})} \qquad (7.330)$$

The dissipation function (7.329) and the energy function (7.329) are now obtained by solving formally the relaxation equations. The same expressions will be obtained in (7.402) using mechanical models, showing that both methods lead to the same results. The energy function and the dissipation function are expressed in the shear compliance in (7.355) and (7.356) and in the shear modulus in (7.415) and (7.416).

7.12. GENERAL LINEAR RHEOLOGICAL EQUATIONS OF STATE FOR ISOTROPIC BODIES

7.12.1. Rheological equation of state for retardation. The total deviatoric strain is the sum of elastic, anelastic and viscous strains

$$\mathring{\boldsymbol{\epsilon}} = \mathring{\boldsymbol{\epsilon}}^{(\mathrm{e})} + \mathring{\boldsymbol{\epsilon}}^{(\mathrm{a})} + \mathring{\boldsymbol{\epsilon}}^{(\mathrm{v})} \qquad (7.331)$$

in which, according to $(7.297)_1$ and (7.327)

$$\mathring{\boldsymbol{\epsilon}}^{(\mathrm{e})} = \mathring{\boldsymbol{\sigma}} / c_{\mathrm{ee}} \qquad (7.332)$$

with the condition that $c_{\mathrm{ee}} = \infty$ for $\mathring{\boldsymbol{\epsilon}}^{(\mathrm{e})} = 0$.

From (7.331), (7.332) and (7.323) it therefore follows that

$$2\mathring{\boldsymbol{\epsilon}} = \left[J_{\circ} + \sum_{K=1}^{\nu} \frac{\bar{C}_K}{1+\bar{\tau}_K \mathcal{D}} + \varphi_{\infty} \frac{1}{\mathcal{D}} \right] \mathring{\boldsymbol{\sigma}} \qquad (7.333)$$

with

$$J_\circ = 2/c_{ee} \tag{7.334}$$

so that $J_\circ = 0$ for $\overset{\circ}{\boldsymbol{\epsilon}}^{(e)} = 0$.

The operators in (7.333) still have to be interpreted. To this end, define an auxiliary function $h(t)$ such that

$$\frac{1}{1+\bar{\tau}_K \mathfrak{D}} f(t) = h(t) \tag{7.335}$$

or

$$f(t) = h(t) + \bar{\tau}_K \frac{\mathrm{D}}{\mathrm{D}t} h(t)$$

or

$$\bar{\tau}_K \frac{\mathrm{D}}{\mathrm{D}t}\left(e^{t/\bar{\tau}_K} h(t)\right) = e^{t/\bar{\tau}_K} f(t)$$

or after integration from $-\infty$ to take into account the whole history of the function $f(t)$

$$\bar{\tau}_K e^{t/\bar{\tau}_K} h(t) = \int_{-\infty}^{t} e^{t'/\bar{\tau}_K} f(t')\, dt'$$

Substitution of the definition of $h(t)$ yields

$$\boxed{\frac{1}{1+\bar{\tau}_K \mathfrak{D}} f(t) = \frac{1}{\bar{\tau}_K} \int_{-\infty}^{t} e^{-(t-t')/\bar{\tau}_K} f(t')\, dt'} \tag{7.336}$$

With this interpretation of the operator, the strain (7.333) becomes

$$2\overset{\circ}{\boldsymbol{\epsilon}} = J_\circ \overset{\circ}{\boldsymbol{\sigma}} + \sum_{K=1}^{\nu} \frac{\bar{C}_K}{\bar{\tau}_K} \int_{-\infty}^{t} e^{-(t-t')/\bar{\tau}_K} \overset{\circ}{\boldsymbol{\sigma}}(t')\, dt' + \varphi_\infty \int_{-\infty}^{t} \overset{\circ}{\boldsymbol{\sigma}}(t')\, dt' \tag{7.337}$$

By integration in parts with $\overset{\circ}{\boldsymbol{\sigma}}(-\infty) = 0$ it is found that

$$\int_{-\infty}^{t} \overset{\circ}{\boldsymbol{\sigma}}(t')\, dt' = t\,\overset{\circ}{\boldsymbol{\sigma}}(t)\Big|_{-\infty}^{t} - \int_{-\infty}^{t} t'\dot{\overset{\circ}{\boldsymbol{\sigma}}}(t')\, dt'$$

or

$$\int_{-\infty}^{t} \overset{\circ}{\boldsymbol{\sigma}}(t')\, dt' = \int_{-\infty}^{t} (t-t')\dot{\overset{\circ}{\boldsymbol{\sigma}}}(t')\, dt' \tag{7.338}$$

and that

$$\int_{-\infty}^{t} e^{t'/\bar{\tau}_K} \overset{\circ}{\boldsymbol{\sigma}}(t')\, dt' = \bar{\tau}_K e^{t/\bar{\tau}_K} \overset{\circ}{\boldsymbol{\sigma}}(t) - \bar{\tau}_K \int_{-\infty}^{t} e^{t'/\bar{\tau}_K} \dot{\overset{\circ}{\boldsymbol{\sigma}}}(t')\, dt'$$

or

$$\int_{-\infty}^{t} e^{t'/\bar{\tau}_K}\overset{\circ}{\boldsymbol{\sigma}}(t')\,dt' = \bar{\tau}_K\int_{-\infty}^{t}\left(e^{t/\bar{\tau}_K} - e^{t'/\bar{\tau}_K}\right)\dot{\overset{\circ}{\boldsymbol{\sigma}}}(t')\,dt' \tag{7.339}$$

Substitution of (7.338) and (7.339) into (7.337) yields

$$2\overset{\circ}{\boldsymbol{\epsilon}}(t) = \int_{-\infty}^{t}\left[J_\circ + \varphi_\infty(t-t') + \sum_{K=1}^{\nu}\bar{C}_K\left(1 - e^{-(t-t')/\bar{\tau}_K}\right)\right]\dot{\overset{\circ}{\boldsymbol{\sigma}}}(t')\,dt'$$

In the phenomenological theory of after effects, this result is written as follows

$$\boxed{2\overset{\circ}{\boldsymbol{\epsilon}}(t) = \int_{-\infty}^{t} J(t-t')\dot{\overset{\circ}{\boldsymbol{\sigma}}}(t')\,dt' = \int_{0}^{\infty} J(t')\dot{\overset{\circ}{\boldsymbol{\sigma}}}(t-t')\,dt'} \tag{7.340}$$

with

$$\boxed{J(t) = \left[J_\circ + \varphi_\infty t + \sum_{K=1}^{\nu}\bar{C}_K\left(1 - \exp(-t/\bar{\tau}_K)\right)\right]H(t)} \tag{7.341}$$

Here the Heaviside unit step function $H(t)$ is used to indicate that $J(t)$ is defined only for $t > 0$, and can be set to zero for $t < 0$ (principle of causality). $J(t)$ is called the *shear creep compliance*, $J_\circ$ the *instantaneous compliance*, φ_∞ the *fluidity for retardation* or *steady-state fluidity* or *steady-flow fluidity* and the function

$$j(t) = \sum_{K=1}^{\nu}\bar{C}_K\left(1 - e^{-t/\bar{\tau}_K}\right)H(t) \tag{7.342}$$

is called the *retarded creep compliance* or the *delayed creep compliance* or in short the *retardation function.* The times $\bar{\tau}_K$ are the *retardation times,* which constitute together the *retardation spectrum.* This spectrum is a line spectrum that can be represented by

$$\bar{F}(\tau) = \sum_{K=1}^{\nu}\bar{C}_K\delta(\tau - \bar{\tau}_K) \tag{7.343}$$

From (7.342) and (7.343) it is found that

$$\frac{dj}{dt} = \int_0^\infty \frac{1}{\tau}\bar{F}(\tau)e^{-t/\tau}\,d\tau \tag{7.344}$$

or

$$\boxed{j(t) = \int_0^\infty \bar{F}(\tau)\left(1 - e^{-t/\tau}\right)d\tau} \tag{7.345}$$

For $\bar{C}_K > 0$ it applies that dj/dt is completely monotonic and (7.344) represents as such the Bernstein theorem* for complete monotonic functions. Complete monotonicity is characteristic for *creep behavior* of the deformation.

For simple shear with shear strain γ and shear stress σ, (7.340) becomes

$$\gamma(t) = \int_{-\infty}^{t} J(t - t')\dot{\sigma}(t')\, dt' \tag{7.346}$$

from which it is found that

$$\gamma = J(t)\,\sigma_\circ \qquad \text{for} \qquad \sigma = \sigma_\circ\, H(t) \tag{7.347}$$

This illustrates the meaning of the creep compliance: the creep compliance is the response to the unit step load in a simple shear experiment. This interpretation opens the possibility of measuring the creep compliance with 'static' retardation experiments. (in general: transition experiments). Starting from the experience with recovery experiments where the total deformation shows elastic, anelastic and viscous deformation, the thermodynamics of irreversible processes then result in (7.340) and (7.341). Now, it can be verified that (7.340) with (7.341) corresponds to these observations.

For a simple recovery experiment as sketched in figure 7.15 a shear stress is applied such that

$$\sigma = [H(t) - H(t - t_\circ)]\,\sigma_\circ$$

According to (7.346) and (7.341)

$$\begin{aligned}
\gamma/\sigma_\circ &= J(t) - J(t - t_\circ) \qquad\qquad \text{for} \qquad t \geq t_\circ \\
&= J_\circ[H(t) - H(t - t_\circ)] + \varphi_\infty t_\circ + \sum_{K=1}^{\nu} \bar{C}_K \left(e^{t_\circ/\bar{\tau}_K} - 1\right) e^{-t/\bar{\tau}_K}
\end{aligned}$$

From this, the elastic, anelastic and viscous deformations are found to be respectively

$$\begin{aligned}
\gamma^{(\mathrm{e})} &= J_\circ \sigma_\circ \\
\gamma^{(\mathrm{a})} &= \sum_{K=1}^{\nu} \bar{C}_K \left(e^{t_\circ/\bar{\tau}_K} - 1\right) e^{-t/\bar{\tau}_K}\, \sigma_\circ \\
\gamma^{(\mathrm{v})} &= \varphi_\infty t_\circ \sigma_\circ
\end{aligned}$$

* For the demonstration that (7.344) is necessary and sufficient to guarantee that dj/dt is completely monotonic, see books on Laplace transformations; for instance Widder D.V. 1946. *The Laplace Transform.* Princeton University Press, Princeton, Princeton, NJ.

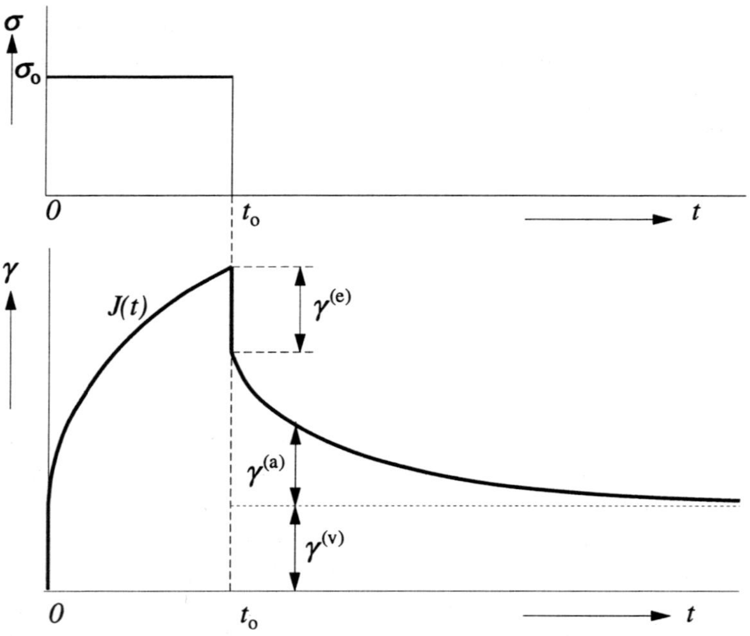

Figure 7.15. Response of the deformation γ in a fundamental recovery experiment.

The anelastic deformation is thus indeed completely monotonic, while $\gamma^{(v)}$ cannot be determined within the instrumental accuracy of the measurement if $t_\circ$ (the time scale of the recovery experiment) is too small.

Small-amplitude sinusoidal retardation experiments are also performed. These type of experiments were called dynamical experiments in the past. The response to the *small-amplitude sinusoidal* stresses follows from (7.340)

$$2\overset{\circ}{\boldsymbol{\epsilon}}(t) = \Re\left\{z\left(\int_0^\infty J(t')e^{z(t-t')}\,dt'\right)\hat{\boldsymbol{\sigma}}\right\} \qquad \text{for} \qquad \overset{\circ}{\boldsymbol{\sigma}} = \Re\left\{\hat{\boldsymbol{\sigma}}\,e^{zt}\right\} \quad (7.348)$$

with $z = \beta + i\omega$, and where the stress $\hat{\boldsymbol{\sigma}}$ is complex ($\Re\{\hat{\boldsymbol{\sigma}}\}$ denotes its real part). For $\beta > 0$ the requirement that $\overset{\circ}{\boldsymbol{\sigma}}(-\infty) = 0$ is satisfied. For (7.348) it can be written that

$$2\overset{\circ}{\boldsymbol{\epsilon}}(t) = \Re\left\{J^\star(z)e^{zt}\,\hat{\boldsymbol{\sigma}}\right\} \qquad \text{for} \qquad \overset{\circ}{\boldsymbol{\sigma}} = \Re\left\{\hat{\boldsymbol{\sigma}}\,e^{zt}\right\} \quad (7.349)$$

with

$$J^\star(z) = z\mathcal{L}[J(t)] \quad (7.350)$$

where the Laplace transform $\mathcal{L}$ is defined by

$$\mathcal{L}[f(t)] \equiv \int_0^\infty e^{-zt} f(t)\,dt \quad (7.351)$$

$J^\star(z)$ is the *complex shear* or *harmonic shear compliance*, which is defined by (7.350) in the complex plane for $\beta > 0$. The complex shear compliance becomes, with (7.341) and (7.350)

$$\boxed{J^\star(z) = J_\circ + \frac{\varphi_\infty}{z} + j^\star(z)} \tag{7.352}$$

with

$$j^\star(z) = z\mathcal{L}[j(t)] = \sum_{K=1}^{\nu} \frac{\bar{C}_K}{1 + \bar{\tau}_K z} \tag{7.353}$$

$j^\star(z)$ is called the *retarded* or *delayed complex shear compliance.* With (7.343) the delayed complex shear compliance (7.353) can be written as

$$\boxed{j^\star(z) = \int_0^\infty \frac{\bar{F}(\tau)}{1 + \tau z}\, d\tau} \tag{7.354}$$

With (7.352) and (353 the creep compliance $J^\star(z)$ is now also defined for $\beta \leq 0$ (analytic continuation of $J^\star(z)$). Obviously, $J^\star(z)$ is a meromorphic function of z, the poles and the zeros of which are found on the nonpositive real axis of the complex z-plane. That there are no poles or zeros found on the positive axis of the z-plane is typical for the behavior of materials exhibiting creep (in these materials no damped or undamped eigenfrequencies are found).

For the retardation of the deformation three functions are therefore important

Shear creep compliance: $J(t)$,
Complex shear compliance: $J^\star(z)$,
Retardation spectrum: $\bar{F}(\tau)$.

The energy function can be expressed in term of the creep compliance. For example, (7.329) becomes, with the substitution of (7.334) and (7.336)

$$\mathfrak{E} = \tfrac{1}{4} J_\circ \mathring{\boldsymbol{\sigma}} : \mathring{\boldsymbol{\sigma}} + \tfrac{1}{4} \sum_{K=1}^{\nu} \frac{\bar{C}_K}{\bar{\tau}_K^2} \int_{-\infty}^{t} dt' \int_{-\infty}^{t} dt'' \\ \exp\big(-(t-t')/\bar{\tau}_K\big) \exp\big(-(t-t'')/\bar{\tau}_K\big) \mathring{\boldsymbol{\sigma}}(t') : \mathring{\boldsymbol{\sigma}}(t'')$$

or with (7.339)

$$\mathfrak{E} = \tfrac{1}{4} J_\circ \mathring{\boldsymbol{\sigma}} : \mathring{\boldsymbol{\sigma}} + \tfrac{1}{4} \sum_{K=1}^{\nu} \bar{C}_K \int_{-\infty}^{t} dt' \int_{-\infty}^{t} dt'' \\ \big[1 - \exp\big(-(t-t')/\bar{\tau}_K\big)\big] \big[1 - \exp\big(-(t-t'')/\bar{\tau}_K\big)\big] \dot{\mathring{\boldsymbol{\sigma}}}(t') : \dot{\mathring{\boldsymbol{\sigma}}}(t'')$$

or with (7.342)

$$\boxed{\begin{aligned}\mathfrak{E} = \tfrac{1}{4} J_\circ \overset{\circ}{\boldsymbol{\sigma}} : \overset{\circ}{\boldsymbol{\sigma}} + \tfrac{1}{4} \int_{-\infty}^{t} dt' \int_{-\infty}^{t} dt'' \\ [j(t-t') + j(t-t'') - j(2t-t'-t'')]\, \dot{\overset{\circ}{\boldsymbol{\sigma}}}(t') : \dot{\overset{\circ}{\boldsymbol{\sigma}}}(t'')\end{aligned}} \tag{7.355}$$

Note that $\overset{\circ}{\boldsymbol{\sigma}} : \overset{\circ}{\boldsymbol{\sigma}} = 2\sigma^2$ if there are only shear shear stresses.

The dissipation function (7.330) becomes, with (7.336)

$$\begin{aligned}\mathfrak{D}^+ = {} & \tfrac{1}{2}\varphi_\infty (\dot{\overset{\circ}{\boldsymbol{\sigma}}} : \dot{\overset{\circ}{\boldsymbol{\sigma}}}) \\ & + \tfrac{1}{2} \sum_{K=1}^{\nu} \frac{\bar{C}_K}{\bar{\tau}_K} \int_{-\infty}^{t} dt' \int_{-\infty}^{t} dt'' e^{-(2t-t'-t'')/\bar{\tau}_K} \dot{\overset{\circ}{\boldsymbol{\sigma}}}(t') : \dot{\overset{\circ}{\boldsymbol{\sigma}}}(t'')\end{aligned}$$

or with (7.342)

$$\boxed{\mathfrak{D}^+ = \tfrac{1}{2}\varphi_\infty (\dot{\overset{\circ}{\boldsymbol{\sigma}}} : \dot{\overset{\circ}{\boldsymbol{\sigma}}}) + \tfrac{1}{2} \int_{-\infty}^{t} dt' \int_{-\infty}^{t} dt'' \dot{j}(2t-t'-t'')\, \dot{\overset{\circ}{\boldsymbol{\sigma}}}(t') : \dot{\overset{\circ}{\boldsymbol{\sigma}}}(t'')} \tag{7.356}$$

Expressions like (7.355) and (7.356) were first derived by Staverman and Schwarzl*.

7.12.2. Rheological equation of state for relaxation. Retardation experiments can be realized more easily than the relaxation experiments. The discussion in the previous section was based on experience with recovery experiments. The result is then a description of the retardation of the deformation due to prescribed stresses.

For solving the balance equations, however, it is more appropriate that the stresses are known as functions of the deformations. This implies that the rheological equation of state obtained for the retardation has to be inverted. The small-amplitude sinusoidal responses can be inverted most easily; (7.349) yields for instance

$$\overset{\circ}{\boldsymbol{\sigma}}(t) = \Re\left\{2G^\star(z)\hat{\boldsymbol{\epsilon}}\, e^{zt}\right\} \qquad \text{for} \qquad \overset{\circ}{\boldsymbol{\epsilon}} = \Re\left\{\hat{\boldsymbol{\epsilon}}\, e^{zt}\right\} \tag{7.357}$$

with

$$\boxed{G^\star(z)J^\star(z) = 1} \tag{7.358}$$

$G^\star(z)$ is the *complex shear* or *complex relaxation modulus.*

* Staverman A.J. and Schwarzl F. 1952. Thermodynamics of viscoelastic behaviour (Model theory). *Proceedings of the Academy of Sciences, Amsterdam*, B55: 474–485; Staverman A.J. and Schwarzl F. 1952. Non-equilibrium thermodynamics of visco-elastic behaviour. *Proceedings of the Academy of Sciences, Amsterdam*, B55: 486–492.

Analogous to (7.350) put

$$G^\star(z) = z\mathcal{L}[G(t)] \tag{7.359}$$

where $G(t)$ is called the *shear relaxation modulus.* Substitution of (7.350) and (7.359) into (7.358) gives

$$z\mathcal{L}[G(t)]\mathcal{L}[J(t)] = \frac{1}{z} \tag{7.360}$$

When the convolution theorem from the theory of Laplace transformations is used, the inversion of (7.360) leads to

$$\frac{d}{dt}\int_{0-}^{t+} G(t')\,J(t-t')\,dt' = 1 \tag{7.361}$$

The relation between $J(t)$ and $G(t)$ is therefore given by a Volterra integral equation. With the Laplace transformation of (7.357), the deviatoric stress can be obtained from

$$\mathcal{L}[\mathring{\boldsymbol{\sigma}}(t)] = 2G^\star(z)\mathcal{L}[\mathring{\boldsymbol{\epsilon}}(t)] = 2\mathcal{L}[G(t)]z\mathcal{L}[\mathring{\boldsymbol{\epsilon}}(t)] = 2\mathcal{L}[G(t)]\mathcal{L}[\dot{\mathring{\boldsymbol{\epsilon}}}(t)]$$

which yields on inversion

$$\boxed{\mathring{\boldsymbol{\sigma}}(t) = 2\int_{-\infty}^{t} G(t-t')\,\dot{\mathring{\boldsymbol{\epsilon}}}(t')\,dt' = 2\int_{0}^{\infty} G(t')\,\dot{\mathring{\boldsymbol{\epsilon}}}(t-t')\,dt'} \tag{7.362}$$

a result analogous to (7.340). However, the shear relaxation modulus $G(t)$ is still unknown. It could be obtained by solving (7.361) or by performing relaxation experiments but, integral equations are difficult to solve and no new information arise from the relaxation experiments, since the linear behavior of the material is already contained in the shear compliance $J(t)$. So, next we want to show how to get $G(t)$ from $J(t)$ by using the simple relation (7.358). We obtain first the complex relaxation modulus $G^\star(z)$ in equation (7.370), and then $G(t)$ in (7.376) by applying an inverse Laplace transformation.

The relation (7.358) can be rewritten by using (7.352) and (7.353) as

$$G^\star(z) = \left[J_\circ + \frac{\varphi_\infty}{z} + \sum_{K=1}^{\nu}\frac{\bar{C}_K}{1+\bar{\tau}_K z}\right]^{-1} \tag{7.363}$$

$G^\star(z)$ is thus also a meromorphic function of z, the poles of which coincide with the zeros of $J^\star(z)$. For the subsequent investigation of $G^\star(z)$, the behavior in $z = 0$ and in $z = \infty$ is considered more closely. For $z \to 0$ (7.363) is approximated by

$$\left.\begin{aligned} G^\star(z) &= G_\infty + \mathcal{O}(z) && \text{if} \quad \varphi_\infty = 0 \\ G^\star(z) &= \frac{z}{\varphi_\infty} + \mathcal{O}(z^2) && \text{if} \quad \varphi_\infty > 0 \end{aligned}\right\} \quad \text{for} \quad z \to 0 \tag{7.364}$$

with

$$G_\infty = \frac{1}{J_\infty} = \frac{1}{J_\circ + \sum_{K=1}^{\nu} \bar{C}_K} \tag{7.365}$$

where the subscript ∞ is used, since $z \to 0$ corresponds to $t \to \infty$. J_∞ is called the *equilibrium shear compliance* or *steady-state compliance**.

From (7.364) it follows that either $G_\infty = 0$ or $\varphi_\infty = 0$, so that

$$G_\infty \varphi_\infty = 0 \tag{7.366}$$

For $z \to \infty$, corresponding to $t \to 0$ (indicated by the subscript $\circ$), it is found from (7.363) that

$$\left.\begin{aligned} G^\star(z) &= G_\circ + \mathcal{O}(\frac{1}{z}) && \text{if} \quad J_\circ > 0 \\ G^\star(z) &= \eta_\circ^- z + \mathcal{O}(1) && \text{if} \quad J_\circ = 0 \end{aligned}\right\} \quad \text{for} \quad z \to \infty \tag{7.367}$$

with

$$\left.\begin{aligned} G_\circ J_\circ &= 1 \\ \frac{1}{\eta_\circ^-} = \varphi_\circ &= \varphi_\infty + \sum_{K=1}^{\nu} \frac{\bar{C}_K}{\bar{\tau}_K} \end{aligned}\right\} \tag{7.368}$$

From (7.367) the conclusion is reached that

$$J_\circ \eta_\circ^- = 0 \tag{7.369}$$

With reference to (7.364) and (7.367) it follows that $G^\star(z)$ can be formulated as

$$\boxed{G^\star(z) = G_\infty + \eta_\circ^- z + \sum_{K=1}^{\nu'} C_K \frac{z\tau_K}{1 + z\tau_K}} \tag{7.370}$$

* In the official nomenclature J_∞ is denoted by J_s. Dealy J.M. 1994. Official nomenclature for material functions describing the response of a viscoelastic fluid to various shearing and extensional deformations. *Journal of Reology*, 38: 179–191.

in which G_∞ and $\eta_\circ^-$ are calculated by (7.365) or by $(7.368)_2$, if $J(t)$ has been given. Inversely it follows from (7.370) that

$$\left.\begin{aligned} \frac{1}{J_\circ} &= G_\circ = G_\infty + \sum_{K=1}^{\nu'} C_K \qquad (z\to\infty \quad \text{for} \quad J_\circ > 0,\ \eta_\circ^- = 0) \\ \frac{1}{\varphi_\infty} &= \eta_\infty^- = \eta_\circ^- + \sum_{K=1}^{\nu'} C_K\tau_K \quad (z\to 0 \quad \text{for} \quad G_\infty = 0,\ \varphi_\infty > 0) \end{aligned}\right\} \tag{7.371}$$

where $G_\circ$ is the *glasslike shear modulus.*

The poles $z = -1/\tau_K$ of $G^\star(z)$ are the same as the zeros of $J^\star(z)$. These zeros can be calculated by putting $z = -1/\tau$ in $J^\star(z)$

$$\bar{K}(\tau) = J^\star(-\frac{1}{\tau}) = J_\circ - \tau\varphi_\infty + \sum_{K=1}^{\nu} \frac{\bar{C}_K}{1 - \bar{\tau}_K/\tau} \tag{7.372}$$

If one is interested only in the relaxation times τ_K, which are neither zero nor infinite, then these relaxation times are determined by the roots of $\bar{K}(\tau) = 0$. Subsequently, C_K can be calculated using the Cauchy theorem.

Suppose that the roots of $\bar{K}(\tau) = 0$ are simple, so that the corresponding pole $z = -1/\tau_K$ of $G^\star(z)$ is also simple. A contour integration of an analytic function over a small circle, not of radius zero, around this simple pole on the negative real axis is equal to $2\pi i$ times the residue at the pole, according to the Cauchy theorem. This yields, using (7.370)

$$\oint G^\star(z)\,dz = -2\pi i \frac{C_K}{\tau_K} = \oint \frac{dz}{J^\star(z)} \tag{7.373}$$

The residue in the pole $z = -1/\tau_K$ can be found by expanding $J^\star(z)$ around the point $z = -1/\tau_K$

$$J^\star(z) \approx \left(\frac{dJ^\star}{dz}\right)_{z=-1/\tau_K} \left(z + \frac{1}{\tau_K}\right) = \tau_K^2 \left[\frac{dJ^\star(-1/\tau)}{d\tau}\right]_{\tau=\tau_K} \left(z + \frac{1}{\tau_K}\right)$$

or with $(7.372)_1$

$$J^\star(z) \approx \tau_K^2 \bar{K}'(\tau_K) \left(z + \frac{1}{\tau_K}\right)$$

with $\bar{K}' \equiv d\bar{K}/d\tau$. With this result it is found that

$$\oint \frac{dz}{J^\star(z)} = 2\pi i \left[\tau_K^2 \bar{K}'(\tau_K)\right]^{-1}$$

And the substitution into (7.373) leads to

$$C_K = [-\tau_K K'(\tau_K)]^{-1} \tag{7.374}$$

For multiple roots the residues of higher order have to be calculated, but the above discussion shows that it is possible to derive $G^\star(z)$ in the form (7.370) from $J^\star(z)$.

The Laplace transformation of the relaxation modulus can readily be get from (7.359) and (7.370)

$$\mathcal{L}[G(t)] = \frac{1}{z}G^\star(z) = \frac{1}{z}G_\infty + \eta_\circ^- + \sum_{K=1}^{\nu'} \frac{C_K\tau_K}{1+z\tau_K} \tag{7.375}$$

Inversion yields

$$\boxed{G(t) = G_\infty H(t) + \eta_\circ^-\delta(t) + \sum_{K=1}^{\nu'} C_K e^{-t/\tau_K} H(t)} \tag{7.376}$$

G_∞ is called the *equilibrium (shear) modulus*, and $\eta_\circ^-$ the *stress relaxation viscosity*. The times τ_K are the *relaxation times*, which together constitute the *relaxation spectrum*

$$F(\tau) = \sum_{K=1}^{\nu'} C_K\delta(\tau - \tau_K) \tag{7.377}$$

The *relaxing (shear) modulus* is

$$\boxed{g(t) = \sum_{K=1}^{\nu'} C_K e^{-t/\tau_K} = \int_0^\infty F(\tau)e^{-t/\tau}\, d\tau} \tag{7.378}$$

$$\boxed{g^\star(z) = z\mathcal{L}[g(t)] = \sum_{K=1}^{\nu'} C_K \frac{\tau_K z}{1+\tau_K z} = \int_0^\infty F(\tau)\frac{\tau z}{1+\tau z}\, d\tau} \tag{7.379}$$

The relaxing modulus $g(t)$ is obviously completely monotonic.

The three important relaxation functions are therefore

Shear relaxation modulus: $G(t)$,

Complex shear modulus: $G^\star(z)$,

Relaxation spectrum: $F(\tau)$.

The deviatoric stress and deformation states in linear isotropic bodies are therefore defined by six characteristic functions that can be transformed into each other. A linear body results from the superposition principle characterized completely by one scalar response function.

7.12.3. Classification of the linear bodies with discrete spectra. The relations (7.366) and (7.369) result in four classes of linear bodies that can be distinguished in their behavior at $t = 0^+$ and $t = \infty$ in a static (transition) experiment. This behavior is either elastic or viscous.

For $J_\circ > 0$ and $\eta_\circ^- = 0$ the material behaves elastically for $t = 0^+$. In other words: for fast deformations the material behaves as an elastic body.

By contrast, the material behaves viscously for $t = 0^+$ for $J_\circ = 0$ and $\eta_\circ^- > 0$. This means that for fast deformations the material behaves as a viscous body.

For $G_\infty > 0$ and $\varphi_\infty = 0$ the material behaves elastically for $t \to \infty$, or: for slow deformations the material behaves as an elastic body.

For $G_\infty = 0$ and $\varphi_\infty > 0$ the material behaves viscously for $t \to \infty$, or: for slow deformations the material behaves as a viscous body.

Clearly, four classes of materials can be distinguished. Each class will be indicated by the letter S with two subscripts, the first indicates elastic or viscous behavior at $t = 0^+$ and the second elastic or viscous behavior at $t \to \infty$. The classes are:

Class $\mathbf{S}_{ee}$: $J_\circ > 0, \eta_\circ^- = 0$, and $G_\infty > 0, \varphi_\infty = 0$.

For $t \to 0^+$ the material is elastic, and this is also the case for $t \to \infty$, so that the retarded compliance $j(t)$ and the relaxing modulus $g(t)$ describe the transition between two elastic states, as sketched in figure 7.16. From (7.365) it follows that

$$J_\circ < J_\infty \tag{7.380}$$

or with $J_\circ G_\circ = 1$ and $J_\infty G_\infty = 1$

$$G_\circ > G_\infty \tag{7.381}$$

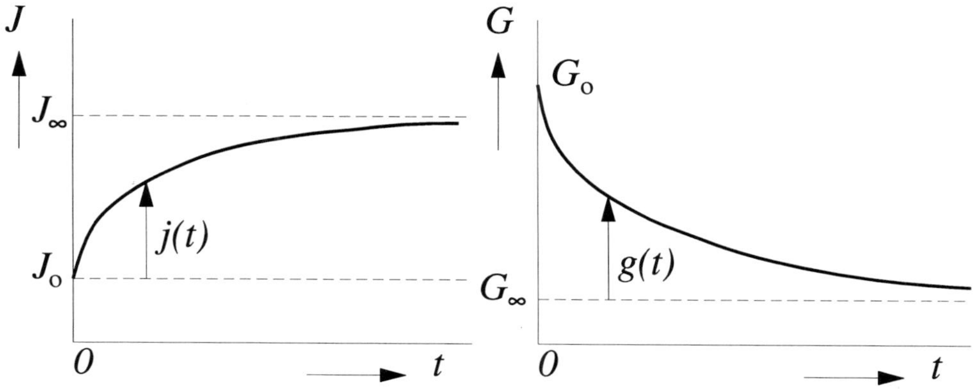

Figure 7.16. Response of the retarded compliance $j(t)$ to a step stress excitation and the response of the relaxing modulus $g(t)$ to a step strain excitation for the material class S_{ee}.

In a static step stress experiment there is found a transition from a state with a high stress modulus ('brittle state', *glassy state*) to a state with a lower modulus (*rubbery state*). These states are connected to each other by the retarded compliance $j(t)$ and the relaxing modulus $g(t)$. The compliance $J_\circ$ is called the *glasslike compliance.*

In the S_{ee}-class it applies that $\overset{\circ}{\boldsymbol{\epsilon}}^{(e)} \neq 0$, $\overset{\circ}{\boldsymbol{\epsilon}}^{(a)} \neq 0$, and $\overset{\circ}{\boldsymbol{\epsilon}}^{(v)} = 0$, so that the S_{ee}-class is the linear subclass of the firmoviscous class, which is described by the differential equation (7.273) of the Poynting–Thomson body. For $j(t) = 0$, $g(t) = 0$ the Hooke bodies (linear bodies in the elastic class) are obtained.

Solid polymers, in which the long polymer chains form networks, are well-known examples of the S_{ee}-class. On a logarithmic time scale the transition from the glassy state to the rubbery state can span several decades. For the glasslike modulus it is found that $G_\circ \approx 10^9$ to 10^{11} Pa and for the equilibrium modulus $G_\infty \approx 10^5$ to 10^7 Pa. An excellent summary of experimental data on polymer responses has been given by Ferry* for the bodies belonging to the various classes.

Class $\mathbf{S_{ev}}$: $J_\circ > 0, \eta_\circ^- = 0$, and $G_\infty = 0, \varphi_\infty > 0$.

For $t \to 0^+$ this material behaves elastically, but for $t \to \infty$ the material behaves viscously. Therefore, for fast deformations the materials of this class behave the same as an elastic body, while for slow deformations it behave the same as a viscous body. These responses are depicted in figure 7.17.

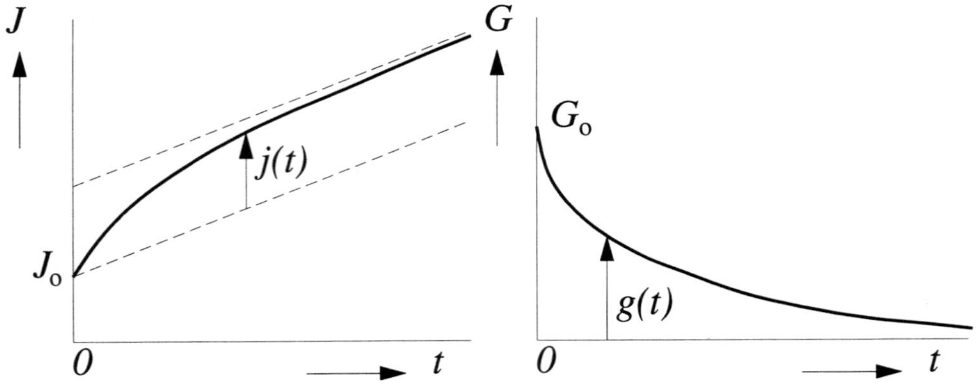

Figure 7.17. Response of the retarded compliance $j(t)$ to a step stress excitation and the response of the relaxing modulus $g(t)$ to a step strain excitation for the material class S_{ev}.

As sketched in the figure, the retarded compliance $j(t)$ and the relaxing modulus $g(t)$ now describe a transition from a glassy state to a fluid state.

* Ferry J.D. 1980. it Viscoelastic Properties of Polymers. John Wiley & Sons, New York, 3rd ed. See Ch. 2.

Since $\overset{\circ}{\epsilon}^{(e)} \neq 0$, $\overset{\circ}{\epsilon}^{(a)} \neq 0$, $\overset{\circ}{\epsilon}^{(v)} \neq 0$, the material class S_{ev} is the linear subclass of the viscoelastic bodies. For $j = 0$, and $g = 0$ the linear Maxwell bodies (elastic-viscous bodies) are found, and these are described by the differential equation (7.285).

An example of the S_{ev} class is the group of solid polymer molecules, in which the polymers do not form permanent networks. For these materials $\eta_\infty^- = 1/\varphi_\infty$ can be very large: $\eta_\infty^- \approx 10^4$ to 10^{14} Pa·s. For 'normal' Newtonian fluids values are found from $\eta \approx 10^{-3}$ to 10^3 Pa·s.

Class $\mathbf{S_{ve}}$: $J_\circ = 0, \eta_\circ^- > 0$, and $G_\infty > 0, \varphi_\infty = 0$.

For fast deformations, that is for $t \to 0^+$, this material behaves in the same way as a viscous body, while for slow deformations, that is for $t \to \infty$, this material behaves in the same way as an elastic body. In figure 7.18 the retarded compliance $j(t)$ and the relaxing modulus $g(t)$ now describe a transition from a viscous state to an elastic equilibrium state. The double arrow indicates the delta function response to the step strain excitation of a viscous fluid with stress relaxation viscosity $\eta_\circ^-$.

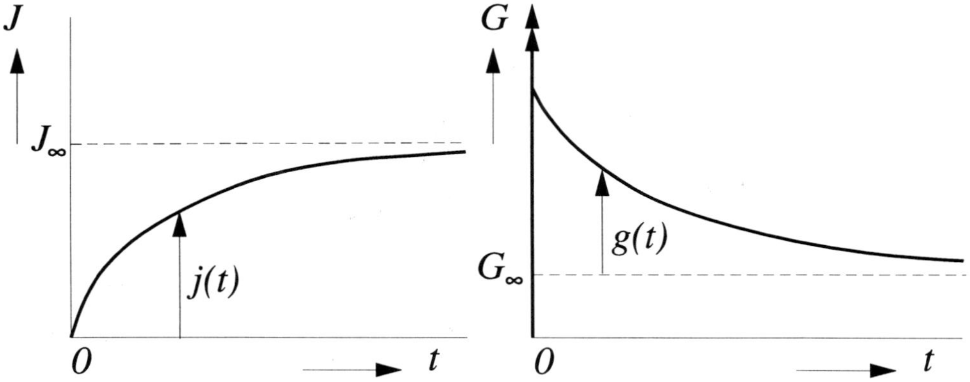

Figure 7.18. Response of the retarded compliance $j(t)$ to a step stress excitation and the response of the relaxing modulus $g(t)$ to a step strain excitation for the material class S_{ve}. The double arrow indicates the delta function response to the step strain excitation of a viscous fluid with stress relaxation viscosity $\eta_\circ^-$.

Since $\overset{\circ}{\epsilon}^{(e)} = 0$, $\overset{\circ}{\epsilon}^{(a)} \neq 0$, $\overset{\circ}{\epsilon}^{(v)} = 0$, the class S_{ve} is the linear subclass of the anelastic bodies, which is described by the differential equation (7.286) for Kelvin–Voigt bodies. According to Kelvin, a granular material with a liquid between the grains is an example of the S_{ve}-class. According to Reiner* poured concrete can be regarded as a material of the S_{ve}-class. It is for this reason

* Reiner M. 1960. *Deformation, Strain and Flow, An Elementary Introduction to Rheology.* H.K. Lewis, London.

that poured concrete is often locally vibrated to ensure that it fills all the spaces in a framing.

Class $\mathbf{S}_{vv}$: $J_\circ = 0, \eta_\circ^- > 0$, and $G_\infty = 0, \varphi_\infty > 0$.

For fast as well as for slow deformations, that is for $t \to 0^+$ and for $t \to \infty$, the material belonging in this class behaves in the same way as a viscous body. According to (7.368)

$$\varphi_\circ > \varphi_\infty \tag{7.382}$$

or with $\eta_\circ^- \varphi_\circ = 1$ and $\eta_\infty^- \varphi_\infty = 1$

$$\eta_\circ^- < \eta_\infty^- \tag{7.383}$$

Therefore the retarded compliance function $j(t)$ and the relaxing modulus function $g(t)$ describe (as sketched in figure 7.19) the transition of a state with a high stress relaxation viscosity $\eta_\circ^-$ to a state with a low stress relaxation viscosity.

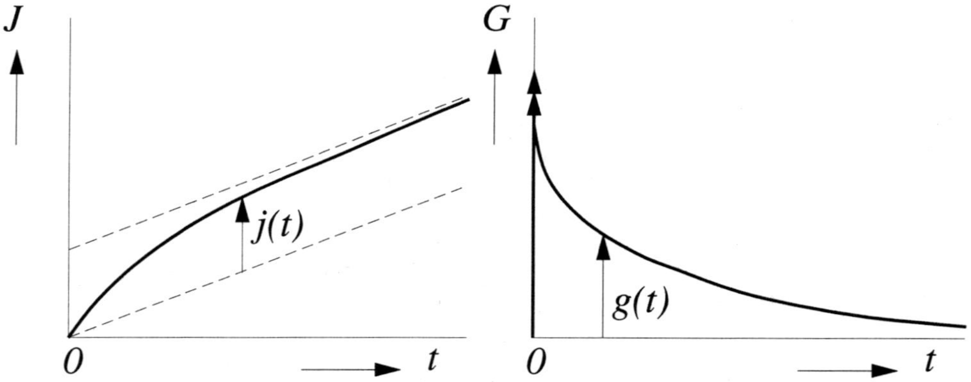

Figure 7.19. Response of the retarded compliance $j(t)$ to a step stress excitation and the response of the relaxing modulus $g(t)$ to a step strain excitation for the material class S_{vv}. The double arrow indicates the delta function response to the step strain excitation of a viscous fluid with stress relaxation viscosity $\eta_\circ^-$.

Since $\overset{\circ}{\epsilon}{}^{(e)} = 0$, $\overset{\circ}{\epsilon}{}^{(a)} \neq 0$, $\overset{\circ}{\epsilon}{}^{(v)} \neq 0$, the class S_{vv} is the linear subclass of the elastic fluids that is described by the differential equation (7.280) of the Jeffreys bodies. Examples of liquids belonging to this class are molten polymers and suspensions or emulsions in Newtonian fluids. Newtonian fluids are defined as fluids with $j(t) = 0$ and $g(0) = 0$.

For linear bodies that satisfy the principle of local action and that can be described by discrete spectra, the possible combinations of elastic and viscous behavior have been exhausted. However, for linear bodies having continuous spectra more possibilities exist.

7.12.4. Spectra and canonical model representations. At first some inequalities for the spectra will be formulated, in which it is assumed that the ν retardation times $\bar{\tau}_K$ are simple* and ordered by increasing magnitude, thus

$$0 < \bar{\tau}_1 < \bar{\tau}_2 < \cdots < \bar{\tau}_{\nu-1} < \bar{\tau}_\nu < \infty \tag{7.384}$$

From (7.372) it follows that

$$\begin{aligned} \bar{K}(\tau) &= J_\circ - \tau\varphi_\infty + \sum_{K=1}^{\nu} \bar{C}_K \frac{\tau}{\tau - \bar{\tau}_K} \\ \bar{K}'(\tau) &= -\varphi_\infty - \sum_{K=1}^{\nu} \bar{C}_K \frac{\bar{\tau}_K}{(\tau - \bar{\tau}_K)^2} \leq 0 \end{aligned} \tag{7.385}$$

so that $\bar{K}(\tau)$ is a monotonic decreasing function of τ. From this it follows that in (7.374) $C_K \geq 0$ if $\tau_K > 0$. Suppose now that there are ν' relaxation times τ_K. The ordering of the ν retardation times and the ν' relaxation times in the various classes is now considered and also if there are more or less retardation times than relaxation times. Consider first

Class $\mathbf{S}_{ee}$: $J_\circ > 0, \varphi_\infty = 0$.

$$\bar{K}(\tau) = J_\circ + \sum_{K=1}^{\nu} \bar{C}_K \frac{\tau}{\tau - \bar{\tau}_K}$$

From this it follows that $\nu' \leq \nu$, since $\bar{K}(\tau) = 0$ is a νth-degree equation in τ. The function jumps from $-\infty$ for values $\bar{\tau}_K-$ just below the retardation times $\bar{\tau}_K$ to ∞ for values $\bar{\tau}_K+$ just above the retardation times $\bar{\tau}_K$ and decreases to $-\infty$ for the next retardation time, so a root (a relaxation time) is found in between. Thus

$$\left.\begin{aligned} \bar{K}(0) &= J_\circ > 0 \\ \bar{K}(\bar{\tau}_1-) &= -\infty \end{aligned}\right\} \Rightarrow \tau_1$$

$$\left.\begin{aligned} \bar{K}(\bar{\tau}_1+) &= +\infty \\ \bar{K}(\bar{\tau}_2-) &= -\infty \end{aligned}\right\} \Rightarrow \tau_2$$

$$\vdots$$

$$\left.\begin{aligned} \bar{K}(\bar{\tau}_\nu-) &= -\infty \end{aligned}\right\} \Rightarrow \tau_\nu$$

$$\bar{K}(\bar{\tau}_\nu+) = +\infty$$

$$\bar{K}(\infty) = J_\circ + \sum_{K=1}^{\nu} \bar{C}_K > 0$$

---

* This means that it is assumed that all $\bar{\tau}_K$ in (7.372) are different. If this is not the case, then $\bar{C}_K$ and ν can be determined such, that all $\bar{\tau}_K$ are different.

From this it is shown that $\nu' = \nu$ and

$$0 < \tau_1 < \bar{\tau}_1 < \tau_2 < \bar{\tau}_2 < \cdots < \tau_\nu < \bar{\tau}_\nu < \infty \tag{7.386}$$

τ_K is therefore positive, and because $\bar{K}'(\tau)$ is negative, relation (7.374) shows that C_K is also positive. In the $\mathrm{S_{ee}}$-class the number of relaxation times equals the number of retardation times. In this class, relaxation proceeds faster than retardation.

Class $\mathbf{S_{ev}}$: $J_\circ > 0$, $\varphi_\infty > 0$.

$$\bar{K}(\tau) = J_\circ - \varphi_\infty \tau + \sum_{K=1}^{\nu} \bar{C}_K \frac{\tau}{\tau - \bar{\tau}_K}$$

From this it follows that $\nu' \leq \nu + 1$, since $\bar{K}(\tau) = 0$ is now a $(\nu+1)$th-degree equation in τ. By roughly calculating the change of sign of this equation it is found that

$$\left.\begin{aligned} \bar{K}(0) &= J_\circ > 0 \\ \bar{K}(\bar{\tau}_1-) &= -\infty \end{aligned}\right\} \Rightarrow \tau_1$$

$$\left.\begin{aligned} \bar{K}(\bar{\tau}_1+) &= +\infty \\ \bar{K}(\bar{\tau}_2-) &= -\infty \end{aligned}\right\} \Rightarrow \tau_2$$

$$\vdots$$

$$\left.\begin{aligned} \bar{K}(\bar{\tau}_\nu-) &= -\infty \end{aligned}\right\} \Rightarrow \tau_\nu$$

$$\left.\begin{aligned} \bar{K}(\bar{\tau}_\nu+) &= +\infty \\ \bar{K}(\infty) &= -\infty \end{aligned}\right\} \Rightarrow \tau_{\nu+1}$$

so that now it is found that $\nu' = \nu + 1$ and

$$0 < \tau_1 < \bar{\tau}_1 < \tau_2 < \bar{\tau}_2 < \cdots < \tau_\nu < \bar{\tau}_\nu < \tau_{\nu+1} < \infty \tag{7.387}$$

In the $\mathrm{S_{ev}}$-class there is one relaxation time more than there are retardation times. Relaxation proceeds faster than retardation, similar as it is found in the $\mathrm{S_{ee}}$-class.

Class $\mathbf{S_{ve}}$: $J_\circ = 0, \varphi_\infty = 0$.

$$\bar{K}(\tau) = \sum_{K=1}^{\nu} \bar{C}_K \frac{\tau}{\tau - \bar{\tau}_K}$$

Since $\tau_1 > 0$, now it is found that $\nu' \leq \nu - 1$ and furthermore that

$$\begin{aligned}
\bar{K}(0) &= J_\circ = 0 \\
\bar{K}(\bar{\tau}_1-) &= -\infty \\
\left.\begin{aligned}\bar{K}(\bar{\tau}_1+) &= +\infty \\ \bar{K}(\bar{\tau}_2-) &= -\infty\end{aligned}\right\} &\Rightarrow \tau_1 \\
&\vdots \\
\left.\begin{aligned}\bar{K}(\bar{\tau}_\nu-) &= -\infty\end{aligned}\right\} &\Rightarrow \tau_{\nu-1} \\
\bar{K}(\bar{\tau}_\nu+) &= +\infty \\
\bar{K}(\infty) &= \sum_{\kappa=1}^{\nu} \bar{C}_\kappa > 0
\end{aligned}$$

For the class S_{ve} it follows that $\nu' = \nu - 1$ and

$$0 < \bar{\tau}_1 < \tau_1 < \bar{\tau}_2 < \tau_2 < \cdots < \tau_{\nu-1} < \bar{\tau}_\nu < \infty \tag{7.388}$$

In the S_{ve}-class there is one relaxation time less than that there are retardation times, and now retardation proceeds faster than relaxation.

Class $\mathbf{S}_{vv}$: $J_\circ = 0$, $\varphi_\infty > 0$.

$$\bar{K}(\tau) = -\varphi_\infty \tau + \sum_{\kappa=1}^{\nu} \bar{C}_\kappa \frac{\tau}{\tau - \bar{\tau}_\kappa}$$

Since $\tau_1 > 0$ it follows that $\nu' \leq \nu$, while it is also found that

$$\begin{aligned}
\bar{K}(0) &= J_\circ = 0 \\
\bar{K}(\bar{\tau}_1-) &= -\infty \\
\left.\begin{aligned}\bar{K}(\bar{\tau}_1+) &= +\infty \\ \bar{K}(\bar{\tau}_2-) &= -\infty\end{aligned}\right\} &\Rightarrow \tau_1 \\
&\vdots \\
\left.\begin{aligned}\bar{K}(\bar{\tau}_\nu-) &= -\infty\end{aligned}\right\} &\Rightarrow \tau_{\nu-1} \\
\left.\begin{aligned}\bar{K}(\bar{\tau}_\nu+) &= +\infty \\ \bar{K}(\infty) &= -\infty\end{aligned}\right\} &\Rightarrow \tau_\nu
\end{aligned}$$

yielding $\nu' = \nu$ and

$$0 < \bar{\tau}_1 < \tau_1 < \bar{\tau}_2 < \tau_2 < \cdots < \tau_{\nu-1} < \bar{\tau}_\nu < \tau_\nu < \infty \tag{7.389}$$

In the S_{vv}-class retardation proceeds faster than relaxation, and the number of relaxation times equals the number of retardation times.

In summarizing it is concluded that the line spectrum of retardation times corresponds to the line spectrum of relaxation times. In the classes S_{ee} and S_{ev} the relaxation times are smaller than the corresponding retardation times, while in the classes S_{ve} and S_{vv} just the opposite is found (compare (7.272) and (7.278)). The class S_{ev} contains one relaxation time more than the class S_{ee}, while the class S_{ve} contains one retardation time less than the class S_{vv}.

To solve the inverse problem, the retardation spectrum can be calculated from the relaxation spectrum*. For this purpose the zeros of $G^\star(z)$—equation (7.370)—have to be determined first from

$$K(\tau) = G^\star(-\frac{1}{\tau}) = G_\infty - \frac{\eta_\circ^-}{\tau} - \sum_{K=1}^{\nu'} C_K \frac{\tau_K}{\tau - \tau_K} \tag{7.390}$$

The function K is defined by (7.390). The roots of $K = 0$ are the retardation times. It is noted that for K it now applies that $K' \equiv dK/d\tau \geq 0$.

With (7.353) it is found that

$$\oint J^\star(z)\, dz = 2\pi i \frac{\bar{C}_K}{\bar{\tau}_K} = \oint \frac{dz}{G^\star(z)} \tag{7.391}$$

with

$$G^\star(z) \approx \bar{\tau}_K^2 K'(\bar{\tau}_K) \left(z + \frac{1}{\bar{\tau}_K} \right)$$

so that

$$\bar{C}_K = \left[\bar{\tau}_K K'(\bar{\tau}_K) \right]^{-1} \tag{7.392}$$

There are canonical model representations connected with the spectra. The retardation spectrum is determined in (7.343) by $\bar{C}_K$ and $\bar{\tau}_K$, in which $\bar{C}_K$

---

* Emri and Tschoegl have reported computer algorithms to determine line spectra from the smoothed experimental data on linear viscoelastic materials in a series of papers. Emri I. and Tschoegl N.W. 1993. Generating line spectra from experimental responses, Part I. Relaxation modulus and creep compliance. *Reologica Acta*, 32: 311–321; Tschoegl N.W. and Emri I. 1993. Generating line spectra from experimental responses, Part II. Storage and loss functions. *Reologica Acta*, 32: 322–327; Tschoegl N.W. and Emri I. 1992. Generating line spectra from experimental responses, Part III. Interconversion between relaxation and retardation behavior. *International Journal of Polymeric Materials*, 18: 117–127; Emri I. and Tschoegl N.W. 1994. Generating line spectra from experimental responses, Part IV. Application to experimental data. *Reologica Acta*, 33: 60–70.

has the dimensions of a compliance, and $\bar{C}_K/\bar{\tau}_K$ the dimensions of a fluidity. It can therefore be expected that the retardation spectrum is directly related to the elastic and dissipated energy of the body.

With (7.333) it is found that

$$2\overset{\circ}{\boldsymbol{\epsilon}} = \left[J_\circ + \sum_{K=1}^{\nu} \frac{\bar{C}_K}{1 + \bar{\tau}_K \mathfrak{D}} + \varphi_\infty \frac{1}{\mathfrak{D}} \right] \overset{\circ}{\boldsymbol{\sigma}}$$

For a simple shear flow the tensor expression reduces to

$$\gamma = \left[J_\circ + \sum_{K=1}^{\nu} \frac{\bar{C}_K}{1 + \bar{\tau}_K \mathfrak{D}} + \varphi_\infty \frac{1}{\mathfrak{D}} \right] \sigma \tag{7.393}$$

This can be interpreted as

$$\gamma = \gamma^{(e)} + \gamma^{(a)} + \gamma^{(v)} \tag{7.394}$$

with

$$\gamma^{(e)} = J_\circ \sigma \qquad \gamma^{(v)} = \varphi_\infty \frac{1}{\mathfrak{D}} \sigma \tag{7.395}$$

$$\gamma^{(a)} = \sum_{K=1}^{\nu} \frac{\bar{C}_K}{1 + \bar{\tau}_K \mathfrak{D}} \sigma \tag{7.396}$$

$(7.395)_1$ represents a 'spring', $(7.395)_2$ a 'dashpot', while (7.396) can be considered as the sum of the deformations of the K elements of a Kelvin representation, as sketched in figure 7.20. For such an element

$$\sigma = \sigma_K^{(e)} + \sigma_K^{(v)} = \bar{G}_K \gamma_K^{(a)} + \bar{\eta}_K \mathfrak{D} \gamma_K^{(a)}$$

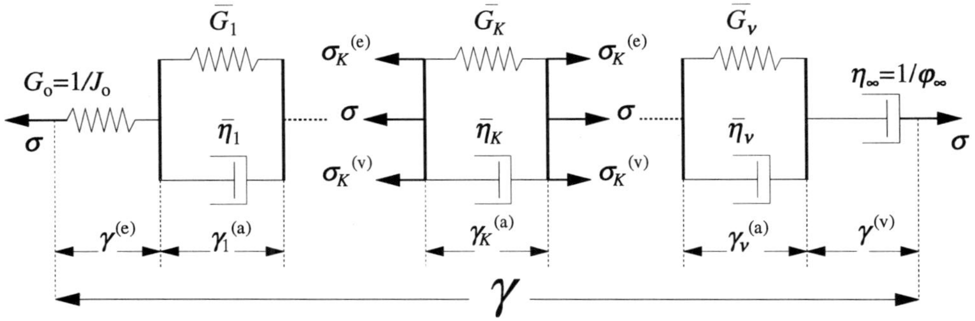

Figure 7.20. Generalized Kelvin representation (retardation spectrum).

or

$$\gamma_K^{(a)} = \frac{1}{\bar{G}_K + \bar{\eta}_K \mathcal{D}} \sigma$$

This corresponds to (7.396) if

$$\bar{G}_K = \frac{1}{\bar{C}_K} \qquad \bar{\eta}_K = \frac{\bar{\tau}_K}{\bar{C}_K} \tag{7.397}$$

The elastic energy of the K-th element is, with the application of (7.396) and (7.397), given by

$$\mathfrak{E}_K^{(a)} = \tfrac{1}{2}\bar{G}_K {\gamma_K^{(a)}}^2 = \tfrac{1}{2}\bar{C}_K \left(\frac{\sigma}{1+\bar{\tau}_K \mathcal{D}}\right)\left(\frac{\sigma}{1+\bar{\tau}_K \mathcal{D}}\right) \tag{7.398}$$

while the dissipation in this element is given by

$$\mathfrak{D}_K^{(a)} = \bar{\eta}_K {\dot{\gamma}_K^{(a)}}^2 = \bar{C}_K \bar{\tau}_K \left(\frac{\dot{\sigma}}{1+\bar{\tau}_K \mathcal{D}}\right)\left(\frac{\dot{\sigma}}{1+\bar{\tau}_K \mathcal{D}}\right) \tag{7.399}$$

The total elastic energy therefore becomes

$$\begin{aligned} \mathfrak{E} &= \mathfrak{E}^{(v)} + \sum_{K=1}^{\nu} \mathfrak{E}_K^{(a)} \\ \mathfrak{E} &= \tfrac{1}{2} J_\circ \sigma^2 + \tfrac{1}{2} \sum_{K=1}^{\nu} \bar{C}_K \left(\frac{\sigma}{1+\bar{\tau}_K \mathcal{D}}\right)\left(\frac{\sigma}{1+\bar{\tau}_K \mathcal{D}}\right) \end{aligned} \tag{7.400}$$

and the total dissipation becomes

$$\begin{aligned} \mathfrak{D}^+ &= \mathfrak{D}^{(v)} + \sum_{K=1}^{\nu} \mathfrak{D}_K^{(a)} \\ \mathfrak{D}^+ &= \varphi_\infty \sigma^2 + \sum_{K=1}^{\nu} \bar{C}_K \bar{\tau}_K \left(\frac{\dot{\sigma}}{1+\bar{\tau}_K \mathcal{D}}\right)\left(\frac{\dot{\sigma}}{1+\bar{\tau}_K \mathcal{D}}\right) \end{aligned} \tag{7.401}$$

Since for simple shear the relation $\sigma^2 = \frac{1}{2}\mathring{\boldsymbol{\sigma}} : \mathring{\boldsymbol{\sigma}}$ holds, the elastic energy (7.400) and the dissipation (7.401) can be written as

$$\boxed{\begin{aligned} \mathfrak{E} &= \tfrac{1}{4} J_\circ \mathring{\boldsymbol{\sigma}} : \mathring{\boldsymbol{\sigma}} + \tfrac{1}{4} \sum_{K=1}^{\nu} \bar{C}_K \left(\frac{\mathring{\boldsymbol{\sigma}}}{1+\bar{\tau}_K \mathcal{D}}\right) : \left(\frac{\mathring{\boldsymbol{\sigma}}}{1+\bar{\tau}_K \mathcal{D}}\right) \\ \mathfrak{D}^+ &= \tfrac{1}{2} \varphi_\infty \mathring{\boldsymbol{\sigma}} : \mathring{\boldsymbol{\sigma}} + \tfrac{1}{2} \sum_{K=1}^{\nu} \bar{C}_K \bar{\tau}_K \left(\frac{\dot{\mathring{\boldsymbol{\sigma}}}}{1+\bar{\tau}_K \mathcal{D}}\right) : \left(\frac{\dot{\mathring{\boldsymbol{\sigma}}}}{1+\bar{\tau}_K \mathcal{D}}\right) \end{aligned}} \tag{7.402}$$

After some calculations these results lead again to (7.355) and (7.356). The correctness of this method for the calculation of the energy functions is thereby demonstrated.

The relaxation spectrum (7.377) can also be represented by a mechanical model. With the substitution of (7.370), the deviatoric stress (7.357) or (7.362) can be written in an operator formulation

$$\overset{\circ}{\boldsymbol{\sigma}} = 2\left[G_\infty + \eta_\circ^- \mathcal{D} + \sum_{K=1}^{\nu'} C_K \frac{\tau_K \mathcal{D}}{1+\tau_K \mathcal{D}}\right] \overset{\circ}{\boldsymbol{\epsilon}} \tag{7.403}$$

For a simple shear this reduces to

$$\sigma = \left[G_\infty + \eta_\circ^- \mathcal{D} + \sum_{K=1}^{\nu'} C_K \frac{\tau_K \mathcal{D}}{1+\tau_K \mathcal{D}}\right] \gamma \tag{7.404}$$

The stress can be expanded as

$$\left.\begin{aligned} \sigma &= \sigma^{(e)} + \sigma^{(a)} + \sigma^{(v)} \\ \sigma^{(a)} &= \sum_{K=1}^{\nu'} \sigma_K^{(a)} \end{aligned}\right\} \tag{7.405}$$

with

$$\left.\begin{aligned} \sigma^{(e)} &= G_\infty \gamma \qquad\qquad \sigma^{(v)} = \eta_\circ^- \dot{\gamma} \\ \sigma_K^{(a)} &= C_K \frac{\tau_K \mathcal{D}}{1+\tau_K \mathcal{D}} \gamma \end{aligned}\right\} \tag{7.406}$$

The stresses (7.405) can also be represented by a Maxwell representation, as sketched in figure 7.21. For element K the following relations apply

$$\gamma_K^{(e)} = \frac{1}{G_K} \sigma_K^{(a)} \qquad \text{and} \qquad \gamma_K^{(v)} = \frac{1}{\eta_K \mathcal{D}} \sigma_K^{(a)}$$

thus

$$\gamma = \gamma_K^{(e)} + \gamma_K^{(v)} = \left(\frac{1}{G_K} + \frac{1}{\eta_K \mathcal{D}}\right) \sigma_K^{(a)}$$

$$\sigma_K^{(a)} = G_K \frac{\tau_K \mathcal{D}}{1+\tau_K \mathcal{D}} \gamma$$

This corresponds to $(7.406)_2$ if

$$G_K = C_K \qquad \text{and} \qquad \tau_K = \eta_K / G_K = \eta_K / C_K \tag{7.407}$$

The elastic energy in element K is

$$\mathfrak{E}_K^{(a)} = \tfrac{1}{2} G_K \gamma_K^{(e)^2} = \tfrac{1}{2} C_K \tau_K^2 \left(\frac{\dot{\gamma}}{1 + \tau_K \mathcal{D}} \right)^2 \tag{7.408}$$

and the dissipation is

$$\mathfrak{D}_K^{(a)} = \eta_K \dot{\gamma}_K^{(v)^2} = \frac{1}{\eta_K} \sigma_K^{(a)^2} = C_K \tau_K \left(\frac{\dot{\gamma}}{1 + \tau_K \mathcal{D}} \right)^2 \tag{7.409}$$

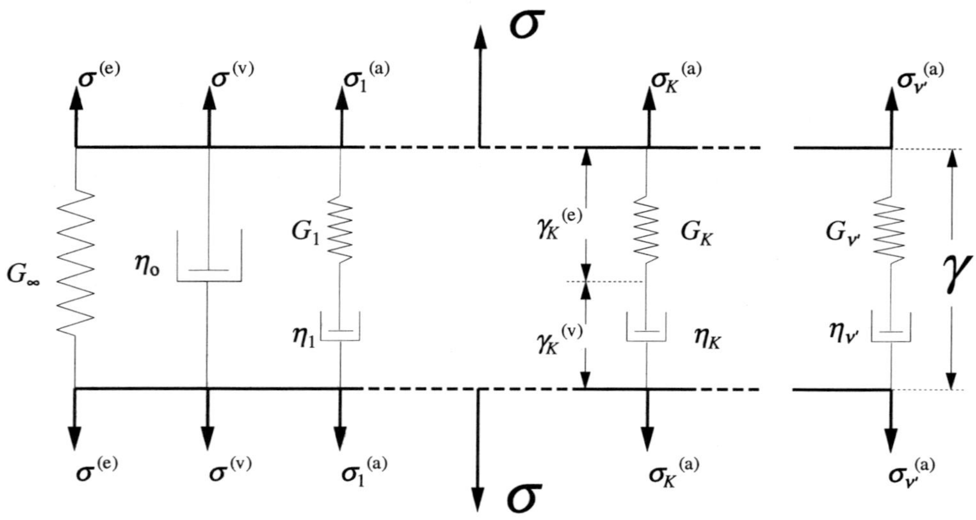

Figure 7.21. Generalized Maxwell representation (relaxation spectrum).

The total elastic energy thus becomes

$$\mathfrak{E} = \mathfrak{E}^{(e)} + \sum_{K=1}^{\nu'} \mathfrak{E}_K^{(a)} = \tfrac{1}{2} G_\infty \gamma^2 + \tfrac{1}{2} \sum_{K=1}^{\nu'} C_K \tau_K^2 \left(\frac{\dot{\gamma}}{1 + \tau_K \mathcal{D}} \right)^2 \tag{7.410}$$

and the total dissipation becomes

$$\mathfrak{D}^+ = \mathfrak{D}^{(v)} + \sum_{K=1}^{\nu'} \mathfrak{D}_K^{(a)} = \eta_\circ^- \dot{\gamma}^2 + \sum_{K=1}^{\nu'} C_K \tau_K \left(\frac{\dot{\gamma}}{1 + \tau_K \mathcal{D}} \right)^2 \tag{7.411}$$

With $\gamma^2 = 2\mathring{\epsilon} : \mathring{\epsilon}$ this yields, for three-dimensional deviatoric deformations

$$\left.\begin{aligned} \mathfrak{E} &= G_\infty \mathring{\epsilon} : \mathring{\epsilon} + \sum_{K=1}^{\nu'} C_K \tau_K^2 \left(\frac{\dot{\mathring{\epsilon}}}{1 + \tau_K \mathcal{D}} \right) : \left(\frac{\dot{\mathring{\epsilon}}}{1 + \tau_K \mathcal{D}} \right) \\ \mathfrak{D}^+ &= 2\eta_\circ^- \dot{\mathring{\epsilon}} : \dot{\mathring{\epsilon}} + 2 \sum_{K=1}^{\nu'} C_K \tau_K \left(\frac{\dot{\mathring{\epsilon}}}{1 + \tau_K \mathcal{D}} \right) : \left(\frac{\dot{\mathring{\epsilon}}}{1 + \tau_K \mathcal{D}} \right) \end{aligned}\right\} \tag{7.412}$$

The energy functions are now found in terms of the relaxation spectrum. With the substitution of the interpretation (7.336) of the differential operator, (7.412) becomes

$$\mathfrak{E} = G_\infty \mathring{\epsilon} : \mathring{\epsilon} + \sum_{K=1}^{\nu'} G_K \int_{-\infty}^{t} dt' \int_{-\infty}^{t} dt'' \, e^{-(2t-t'-t'')/\tau_K} \dot{\mathring{\epsilon}}(t') : \dot{\mathring{\epsilon}}(t'') \quad (7.413)$$

$$\mathfrak{D}^+ = 2\eta_\circ^- (\dot{\mathring{\epsilon}} : \dot{\mathring{\epsilon}}) + 2 \sum_{K=1}^{\nu'} \frac{G_K}{\tau_K} \int_{-\infty}^{t} dt' \int_{-\infty}^{t} dt'' \, e^{-(2t-t'-t'')/\tau_K} \dot{\mathring{\epsilon}}(t') : \dot{\mathring{\epsilon}}(t'') \quad (7.414)$$

and finally with the substitution of (7.378) this can be written as

$$\boxed{\mathfrak{E} = G_\infty (\mathring{\epsilon} : \mathring{\epsilon}) + \int_{-\infty}^{t} \int_{-\infty}^{t} g(2t - t' - t'') \, \dot{\mathring{\epsilon}}(t') : \dot{\mathring{\epsilon}}(t'') \, dt'' \, dt'} \quad (7.415)$$

$$\boxed{\mathfrak{D}^+ = 2\eta_\circ^- (\dot{\mathring{\epsilon}} : \dot{\mathring{\epsilon}}) - 2 \int_{-\infty}^{t} \int_{-\infty}^{t} \dot{g}\,(2t - t' - t'') \, \dot{\mathring{\epsilon}}(t') : \dot{\mathring{\epsilon}}(t'') \, dt'' \, dt'} \quad (7.416)$$

The representations in the figures 7.20 and 7.21 of the Kelvin and Maxwell bodies are called canonical inasmuch these representations contain the least possible number of parameters. The physical meaning of these types of representations is debatable if more than one internal mechanism is present. This was illustrated, among others, by Meixner* as follows.

The introduction of the canonical models is based on a transformation of principal-axes, similar to those used in mechanics for the introduction of 'normal' coordinates, where the kinetic and the potential energies are expressed with respect to the principal-axes. The normal coordinates and the eigenfrequencies associated with these normal coordinates usually have no simple geometric significance, but usually depend in a complicated way on the geometric structure of the mechanical system.

These considerations apply equally for the internal mechanisms. Suppose that the real microscopic or molecular mechanisms are known, so that the

---

* Meixner J. 1949. Thermodynamik und Relaxationserscheinungen. *Zeitschrift für Naturforschung*, 4a: 594–600; Meixner J. 1953. Die Thermodynamische Theorie der Relaxationserscheinungen und ihre Zusammenhang mit der Nachwirkungstheorie. *Kolloid Zeitschrift*, 134: 3–20; Meixner J. 1954. Thermodynamische Theorie der Elastischen Relaxation. *Zeitschrift für Naturforschung*, 9a: 654–663.

originally introduced hidden variables have a known physical meaning. The 'normal' variables—principal vectors in the space of the hidden variables—are in general complicated linear combinations of the primary variables, so that the physical meaning of the normal variables is no longer clear.

The relation between the primary variables and the normal variables is not unique. The primary variables determine unambiguously the normal variables, but with one set of normal variables correspond numerous sets of primary variables. From measurements the spectra can be calculated (this is usually not an easy task either), so that only the normal variables are known. This is however not sufficient to determine which primary mechanisms have been active. This conclusion is not surprising. In general it can not be expected that details of the microscopic processes can be figured out completely with macroscopic measurements. More information is certainly needed for this purpose.

7.13. LINEAR BODIES WITH CONTINUOUS SPECTRA

7.13.1. Transition to continuous spectra. Up to now the number of internal mechanisms has been assumed to be finite. The assumption results in spectra with a finite number of lines and discrete model representations ('lumped circuits'). It is possible that the number of spectral lines may increase indefinitely. This can happen in different ways. It is possible that an accumulation point occurs at infinity, so that the summation over a finite number of terms changes into a summation over an infinite number of terms. In this procedure it is assumed that these summations converge. As Meixner and Reik* argue, it is also possible that the statistical fluctuations in a system result in the introduction of infinitely many internal variables.

Another possibility is that an infinitely denumerable number of spectral lines have an infinitely denumerable number of accumulation points that are dense everywhere. The spectrum is then continuous, and the summations change into integrals. A line spectrum might still be assumed to be found in the continuous spectra (line spectra with 'continuous background'). The physical cause for the occurrence of continuous spectra can be explained according to Fröhlich† by microscopic inhomogeneities that influence the internal processes. The inhomgeneities cause a spread in the free energy of the activation of the internal processes, and that results in a broadening of the spectral lines.

Clearly, in the transition from a summation to an integral the essential properties of retardation and relaxation are conserved. Because of the linearity

* Meixner J. and Reik H.G. 1959. Thermodynamik der irreversiblen Prozessen. *Encyclopedia of Physics*, III/2. Springer-Verlag, Berlin. See pp. 413–523.

† Fröhlich H. 1949. *Theory of Dielectrics*. Oxford University Press, London.

the integral expressions (7.340) and (7.362) remain valid

$$
\begin{aligned}
2\mathring{\boldsymbol{\epsilon}}(t) &= \int_{-\infty}^{t+} J(t-t')\,\dot{\mathring{\boldsymbol{\sigma}}}(t')\,dt' = \int_{0}^{\infty} J(t')\dot{\mathring{\boldsymbol{\sigma}}}(t-t')\,dt' \\
\mathring{\boldsymbol{\sigma}}(t) &= 2\int_{-\infty}^{t+} G(t-t')\,\dot{\mathring{\boldsymbol{\epsilon}}}\,(t')\,dt' = 2\int_{0}^{\infty} G(t')\,\dot{\mathring{\boldsymbol{\epsilon}}}\,(t-t')\,dt'
\end{aligned}
\tag{7.417}
$$

In these equations $J(t)$ and $G(t)$ are, according to (7.341) and (7.376), given in the form

$$
\begin{aligned}
&\boxed{J(t) = [J_\circ + \varphi_\infty t]\,H(t) + j(t)} \\
&\boxed{G(t) = G_\infty H(t) + \eta_\circ^- \delta(t) + g(t)}
\end{aligned}
\tag{7.418}
$$

with

$$
\boxed{J_\circ \geq 0 \qquad \varphi_\infty \geq 0} \qquad \text{and} \qquad \boxed{G_\infty \geq 0 \qquad \eta_\circ^- \geq 0} \tag{7.419}
$$

Causality requires that

$$
j(t) = 0 \qquad \text{and} \qquad g(t) = 0 \qquad \text{for} \qquad t < 0 \tag{7.420}
$$

Furthermore it was established that

$$
\begin{aligned}
&j(0^+) = 0 \qquad j(t) \geq 0 \qquad (-1)^n \frac{d^n}{dt^n}\frac{dj}{dt} \geq 0 \qquad \text{for} \quad n = 0, 1, \cdots \\
&g(\infty) = 0 \qquad (-1)^n \frac{d^n}{dt^n}\, g(t) \geq 0 \qquad \text{for} \quad n = 0, 1, \cdots
\end{aligned}
\tag{7.421}
$$

From (7.366) and (7.369) we have

$$
J_\circ \eta_\circ^- = 0 \qquad \text{and} \qquad G_\infty \varphi_\infty = 0 \tag{7.422}
$$

Consistency of (7.418) with (7.417) requires that

$$
\frac{d}{dt}\int_{0-}^{t+} G(t')J(t-t')\,dt' = 1 \tag{7.423}
$$

With reference to (7.420) and (7.421) the Bernstein theorem can be applied, according to which the functions $\bar{F}$ and F exist such that

$$
\begin{aligned}
j(t) &= \int_0^\infty \bar{F}(\tau)(1 - e^{-t/\tau})\,d\tau \\
g(t) &= \int_0^\infty F(\tau)e^{-t/\tau}\,d\tau
\end{aligned}
\tag{7.424}
$$

where $\bar{F}$ and F for $0 < \tau < \infty$ are non-negative and may possess a denumerable number of delta functions. The dynamical functions are introduced again via (7.350) and (7.359) in the form

$$\begin{aligned} J^\star(z) &= z\mathcal{L}[J(t)] = J_\circ + \frac{\varphi_\infty}{z} + j^\star(z) \\ G^\star(z) &= z\mathcal{L}[G(t)] = G_\infty + \eta_\circ^- z + g^\star(z) \end{aligned} \tag{7.425}$$

Herein

$$\begin{aligned} j^\star(z) &= z\mathcal{L}[j(t)] = \int_0^\infty \bar{F}(\tau)\frac{d\tau}{1+\tau z} \\ g^\star(z) &= z\mathcal{L}[g(t)] = \int_0^\infty F(\tau)\frac{\tau z}{1+\tau z}\,d\tau \end{aligned} \tag{7.426}$$

The relation (7.358) applies, so that

$$G^\star(z)J^\star(z) = 1 \tag{7.427}$$

With use of (7.425) and (7.426) the strain and stress (7.417) can be written again operationally

$$\boxed{\begin{aligned} 2\mathring{\epsilon} &= \left[J_\circ + \varphi_\infty\frac{1}{\mathfrak{D}} + \int_0^\infty \bar{F}(\tau)\frac{d\tau}{1+\tau\mathfrak{D}}\right]\mathring{\sigma} \\ \mathring{\sigma} &= 2\left[G_\infty + \eta_\circ^-\mathfrak{D} + \int_0^\infty F(\tau)\frac{\tau\mathfrak{D}}{1+\tau\mathfrak{D}}\right]\mathring{\epsilon} \end{aligned}} \tag{7.428}$$

which result can be checked with (7.336) and (7.424).

Clearly, the six viscoelastic functions are related to one another by linear operators. The mathematical structure was put forward in particular by Gross*, who also developed a direct relation between the spectra. The mathematical structure is summarized in table 7.2. The diagram gives an overview and shows the various possibilities of transforming the functions into one another. Details of the transformations are not worked out further.

Since the linear isotropic bodies are determined by one scalar characteristic function, a relation has to exist between the real and imaginary parts of $J^\star(z)$ (this applies of course equally for $G^\star(z)$). These relations are known as the *Kronig–Kramers reciprocal relations.*

7.13.2. Supplementary classification with degeneracies. The classification in section 7.12.3 is based on the idea that for very slow and very fast excitations the materials behave either as an elastic body or as a viscous body. This means that due to (7.425) the creep compliance $J(t)$ and the relaxation

* Gross B. 1953. *Mathematical Structure of the Theories of Viscoelasticity.* Hermann et Cie, Paris.

Table 7.2. Mathematical structure according to Gross. The Laplace transformations are indicated by $\mathcal{L}$. The Stieltjes transformations are indicated by $\mathcal{S}$.

	STEP RESPONSE FUNCTIONS	
Shear creep compliance: $J(t) = [J_\circ + \varphi_\infty t]H(t) + j(t)$ $j(0+) = 0 \quad j(t) \geq 0$ $(-1)^n \frac{d^n}{dt^n}\frac{dj}{dt} \geq 0$ $J_\circ \geq 0 \quad \varphi_\infty \geq 0$	← Volterra →	Shear relaxation modulus: $G(t) = G_\infty H(t) + \eta_\circ^- \delta(t) + g(t)$ $g(\infty) = 0 \quad g(t) \geq 0$ $(-1)^n \frac{d^n}{dt^n} g \geq 0$ $G_\infty \geq 0 \quad \eta_\circ^- \geq 0$
$\mathcal{L}$		$\mathcal{L}$
	SPECTRA	
Retardation spectrum: $j(t) = \int_0^\infty \bar{F}(\tau)(1 - e^{-t/\tau})\, d\tau$ $\bar{F}(\tau) \geq 0$	← Gross →	Relaxation spectrum: $g(t) = \int_0^\infty F(\tau) e^{-t/\tau}\, d\tau$ $F(\tau) \geq 0$
$\mathcal{L}$ (step response → complex) ; $\mathcal{S}$ (spectrum → complex)	COMPLEX FUNCTIONS	$\mathcal{S}$ (spectrum → complex) ; $\mathcal{L}$ (step response → complex)
Complex compliance: $J^\star = J_\circ + \frac{\varphi_\infty}{z} + j^\star(z)$ $j^\star = z\mathcal{L}[j] = \int_0^\infty \bar{F}(\tau) \frac{d\tau}{1+\tau z}$	← $J^\star G^\star = 1$ →	Complex modulus: $G^\star = G_\infty + \eta_\circ^- z + g^\star(z)$ $g^\star = z\mathcal{L}[g] = \int_0^\infty F(\tau) \frac{\tau z}{1+\tau z}\, d\tau$
RETARDATION		RELAXATION

modulus $G(t)$ behave as $t \to 0^+$ and $t \to \infty$ the same as found for bodies that are either elastic or viscous. It is obvious that the material at $t = 0^+$ or for $t = \infty$ cannot be elastic as well as viscous, so that the relations (7.422) are more or less trivial.

It might after all be possible that the behavior of the material in a step excitation for $t = 0^+$ and $t = \infty$ is neither that of an elastic body nor that of a viscous body, but that it shows a behavior that can be characterized as being 'between' elastic and viscous. The conditions (7.422) are still satisfied if at $t = 0^+$ it is true that $J_\circ = 0$ and $\eta_\circ^- = 0$, and that at $t = \infty$ it is true that $G_\infty = 0$ and $\varphi_\infty = 0$. For materials with discrete line spectra the foregoing conditions are impossible, but for materials with continuous spectra this can certainly occur. This implies that the classification in section 7.12.3 has to be extended.

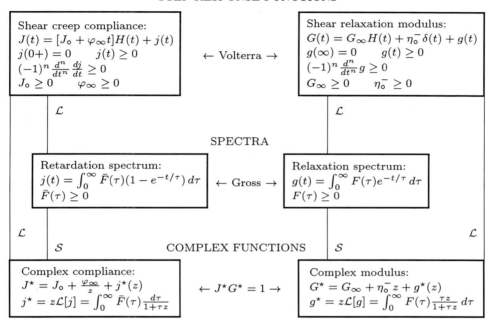

If the material in a step excitation behaves neither the same as an elastic body nor as a viscous body at $t = 0^+$ or $t = \infty$, then the material has a degeneracy in $t = 0^+$ or $t = \infty$. The possibility of degeneracies was pointed out by Schlosser and Kärstner*. For a degeneracy it is necessary that both the delayed compliance $j(t)$ and the relaxing modulus $g(t)$ are present.

At first glance it is to be expected for a degeneracy at $t = 0^+$ that $J_\circ = 0$ and $\eta_\circ^- = 0$. The behavior of $j(t)$ at $t = 0^+$ can now be estimated by $j(t) = \mathcal{O}(t^{\bar{m}})$. Since $j(0^+) = 0$, it is found that $\bar{m} > 0$. Furthermore, since $j(t) \geq 0, dj/dt \geq 0, d^2j/dt^2 \leq 0, \cdots$ (see $(7.421)_1$), it follows that $\bar{m} \leq 1$. For $\bar{m} = 1$ the behavior of the retarded compliance $j(t)$ is again 'normal', and for $0 < \bar{m} < 1$ 'degenerate'. Similarly suppose that as $t \to \infty$ the delayed compliance $j(t) = \mathcal{O}(t^{\bar{m}})$, then complete monotonicity of dj/dt requires that $\bar{m} \leq 1$ holds. In 'normal' cases $j(\infty)$ is finite, so $\bar{m} = 0$. Since dj/dt is always ≥ 0, it follows that $0 \leq \bar{m} \leq 1$. However $\bar{m} = 1$ can be dropped, since in $J(t)$ a term with t is found. For the retarded compliance it can therefore be stated that

$$\begin{aligned} j(t) &= \mathcal{O}(t^{\bar{m}}) \quad \text{with} \quad 0 < \bar{m} \leq 1 \quad \text{as} \quad t \to 0^+ \\ j(t) &= \mathcal{O}(t^{\bar{m}}) \quad \text{with} \quad 0 \leq \bar{m} < 1 \quad \text{as} \quad t \to \infty \end{aligned} \tag{7.429}$$

where the equal signs refer to the normal behavior, and the less-than signs to degenerate behavior. From the sketches in figures 7.22 and 7.23 it is obvious that the degenerate behavior lies between the viscous and elastic behavior.

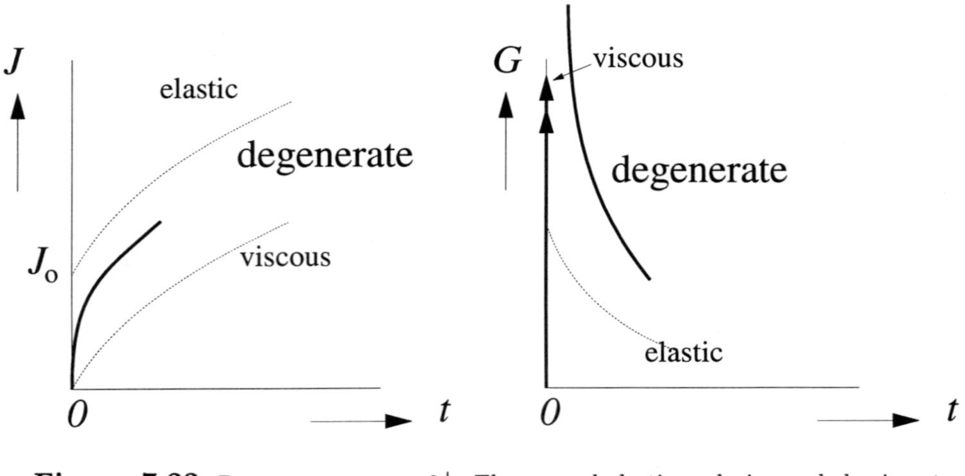

Figure 7.22. Degeneracy at $t = 0^+$. The normal elastic and viscous behavior at $t = 0^+$ is sketched by the dotted lines.

For $g(t)$ it can be assumed that $g(t) = \mathcal{O}(t^{-m})$ as $t \to 0^+$ and $t \to \infty$. Since $g(t)$ is completely monotonic, it follows that $m \geq 0$. In normal cases, $m = 0$

* Schlosser E. and Kärstner S. 1957. Zur phänomenologische Theorie der Visko-elasticität II. Die Stoffklassen der linearen skalaren Relaxationstheorie. *Kolloid Zeitschrift*, 155: 97–106.

as $t \to 0^+$. Furthermore it is found from $(7.418)_2$ that

$$\int_{0^-}^{t} G(t)\,dt = G_\infty t + \eta_\circ^- + \int_0^t g(t)\,dt$$

If as $t \to 0^+$ the relaxing modulus $g(t)$ may represent no greater effect than $\eta_\circ^-$, then $g(t)$ has to be integrable as $t \to 0^+$, so that $m < 1$. As $g(\infty) = 0$ it follows that $m > 0$ as $t \to \infty$. In summary this yields

$$\begin{array}{llll} g(t) = \mathcal{O}(t^{-m}) & \text{with} \quad 0 \le m < 1 & \text{as} & t \to 0^+ \\ g(t) = \mathcal{O}(t^{-m}) & \text{with} \quad 0 < m & \text{as} & t \to \infty \end{array} \tag{7.430}$$

The relaxing modulus $g(t)$ decreases exponentially as $t \to \infty$ for 'normal' standard behavior, so that then $g(t)$ is a rapidly decreasing function as $t \to \infty$. In degenerate cases the relaxing modulus $g(t)$ can decrease as a negative power of t as $t \to \infty$. For the degenerate behavior, the relaxing modulus $g(t)$ is a slowly decreasing function as $t \to \infty$.

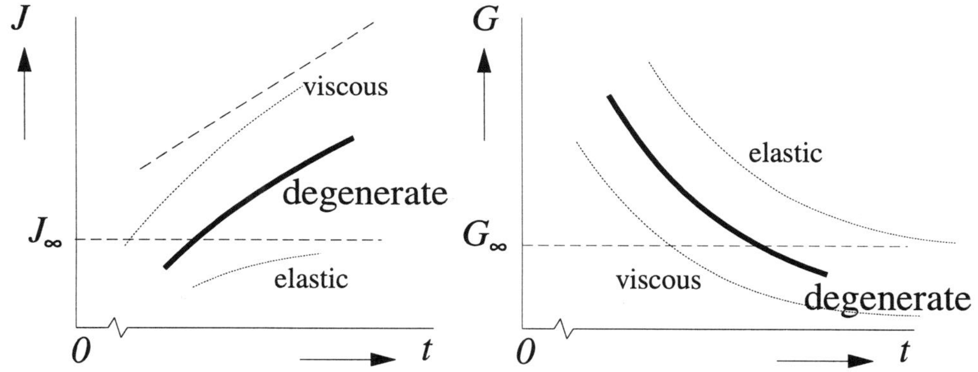

Figure 7.23. Degeneracy as $t \to \infty$. The normal elastic and viscous behavior as $t \to \infty$ is sketched by the dotted lines. The dashed lines are the asymptotes in case of normal elastic and viscous behavior.

It is still an open question as to how far the degeneracies are observed experimentally. The careful determination of $J(t)$ and $G(t)$ as $t \to 0^+$ requires measurements at a small time scale, while as $t \to 0^+$ the measurements can be obscured by inertia effects of the apparatus and the instruments. All these effects make it difficult to examine if, for example, the tangent to $J(t)$ as $t \to 0^+$ is vertical or if there is an instantaneous jump from zero to $J_\circ$. Similarly, the measurements as $t \to \infty$ are not always reliable because the finite time of the measurements. In practice, the behavior of $J(t)$ and of $G(t)$ as $t \to 0^+$ and as $t \to \infty$ has to be obtained by extrapolation of the experimental results. Similarly, sometimes degeneracies are found, so that it make sense to discuss these. It therefore seems unjustified to reject degeneracies in advance, as has been done in the past by some theorists based on the discrete model representations of Maxwell and Kelvin–Voigt.

Schlosser and Kärstner have based their discussions on the integral equation (7.423), but it turns out to be easier to start from the algebraic equation (7.427) that is equivalent to (7.423). To this purpose the Laplace transforms of (7.429) and (7.430) have to be calculated. It is well known that

$$z\mathcal{L}[t^{\bar{m}}] = \frac{\Gamma(\bar{m}+1)}{z^{\bar{m}}} \tag{7.431}$$

In accordance with the asymptotic properties of the Laplace transforms* (7.429) and (7.430) with (7.431) are of the form

$$\left.\begin{array}{lll} j^\star(z) = z\mathcal{L}[j(t)] = \mathcal{O}(z^{-\bar{m}}) & \text{with} & 0 < \bar{m} \leq 1 \\ g^\star(z) = z\mathcal{L}[g(t)] = \mathcal{O}(z^{m}) & \text{with} & 0 \leq m < 1 \end{array}\right\} \quad (z \to \infty) \tag{7.432}$$

and

$$\left.\begin{array}{lll} j^\star(z) = z\mathcal{L}[j(t)] = \mathcal{O}(z^{-\bar{m}}) & \text{with} & 0 \leq \bar{m} < 1 \\ g^\star(z) = z\mathcal{L}[g(t)] = \mathcal{O}(z^{m}) & \text{with} & 0 < m \end{array}\right\} \quad (z \to 0) \tag{7.433}$$

Herein $z \to \infty$ $(z \to 0)$ corresponds to $t \to 0^+$ $(t \to \infty)$.

With the substitution of (7.425) relation (7.427) becomes

$$\Big[J_\circ + \frac{\varphi_\infty}{z} + \underset{z^{-\bar{m}}}{j^\star(z)}\Big]\Big[G_\infty + \eta_\circ^- z + \underset{z^{m}}{g^\star(z)}\Big] = 1 \tag{7.434}$$

Consider first $z \to \infty$ $(t \to 0^+)$. In the normal standard models $\bar{m} = 1$, $m = 0$, so that as $z \to \infty$ the dominant term is given by $J_\circ \eta_\circ^- z$, so that

$$J_\circ \eta_\circ^- = 0 \tag{7.435}$$

which was already established earlier (see (7.422)). This condition follows mathematically from the requirement that the rheological equation of state has to be invertible. Physically, the condition (7.435) means that a material at $t = 0^+$ cannot be at the same time both elastic and viscous.

Suppose now that $J_\circ = 0$, $\eta_\circ^- = 0$, so that (7.434) becomes

$$\underset{z^{-1}}{G_\infty \varphi_\infty \frac{1}{z}} + \underset{z^{m-1}}{\varphi_\infty \frac{1}{z} g^\star(z)} + \underset{z^{-\bar{m}}}{G_\infty j^\star(z)} + \underset{z^{m-\bar{m}}}{j^\star(z) g^\star(z)} = 1 \tag{7.436}$$

* See for example Doetsch G. 1971. *Handbuch der Laplace-Transformation.* Band 1. *Theorie der Laplace-Transformation.* Verlag Birkhäuser, Basel.

For standard behavior $\bar{m} = 1$, $m = 0$ are applicable, so that (7.436) as $z \to \infty$ implies the result $0 = 1$. The requirement $J_\circ = 0$, $\eta_\circ^- = 0$ therefore requires a degeneracy, so that this can be called a *forced degeneracy.* If $0 < \bar{m} < 1$, $0 < m < 1$, then (7.436) as $z \to \infty$ is possible if

$$\lim_{z \to \infty} j^\star(z) g^\star(z) = 1 \tag{7.437}$$

This condition is fulfilled if $\bar{m} = m$, so that

$$j^\star(z) \approx A z^{-\bar{m}} \qquad \text{and} \qquad g^\star(z) \approx \frac{1}{A} z^{\bar{m}} \qquad \text{as} \qquad z \to \infty \tag{7.438}$$

or with (7.431)

$$j^\star(z) \approx \frac{A}{\Gamma(1+\bar{m})} t^{\bar{m}} \qquad g^\star(z) \approx \frac{1}{A\Gamma(1-\bar{m})} t^{-\bar{m}} \quad \text{as} \quad t \to 0^+ \tag{7.439}$$

with $0 < \bar{m} < 1$, and where A is some constant.

Consider now $z \to 0$ ($t \to \infty$). In the normal cases $\bar{m} = 0$, $m > 0$, so that for $z \to 0$ the leading term in (7.434) is given by $G_\infty J_\infty / z$, so

$$G_\infty J_\infty = 0 \tag{7.440}$$

For $G_\infty = 0$ and $\varphi_\infty = 0$ the condition (7.434) becomes, with the substitution of (7.435) and (7.440)

$$\underset{z^m}{J_\circ g^\star(z)} + \underset{z^{1-\bar{m}}}{\eta_\circ^- z j^\star(z)} + \underset{z^{m-\bar{m}}}{j^\star(z) g^\star(z)} = 1$$

In normal standard behavior with $\bar{m} = 0$, $m > 0$ this implies $0 = 1$ as $z \to 0$. There must be a forced degeneracy with

$$\lim_{z \to 0} j^\star(z) g^\star(z) = 1 \tag{7.441}$$

so that $0 < \bar{m} = m < 1$, which again results in (7.438) and (7.439), but now as $z \to 0$ ($t \to \infty$).

Substitution of (7.435) and (7.440) into (7.434) yields

$$\underset{1}{J_\circ G_\infty} + \underset{1}{\varphi_\infty \eta_\circ^-} + \underset{z^m}{J_\circ g^\star(z)} + \underset{z^{m-1}}{\varphi_\infty \frac{1}{z} g^\star(z)} + \underset{z^{-\bar{m}}}{G_\infty j^\star(z)} + \underset{z^{1-\bar{m}}}{\eta_\circ^- z j^\star(z)} + \underset{z^{m-\bar{m}}}{j^\star(z) g^\star(z)} = 1 \tag{7.442}$$

On the contrary, it follows from this that $J_\circ = 0$, $\eta_\circ^- = 0$ if $j(t)$ and $g(t)$ at $t = 0^+$ are simultaneously degenerate. $G_\infty = 0$, $\varphi_\infty = 0$ is applicable if $j(t)$ and $g(t)$ as $t \to \infty$ are simultaneously degenerate.

Table 7.3. The nine material classes of the linear theory. The heavy lined box contains the four standard classes. Bodies with continuous spectra can be a member of one of the five degenerate classes. The abbreviation fr.d. indicates free degeneracies.

$t=\infty$ / $t=0+$	$G_\infty > 0$ $\varphi_\infty = 0$ fr.d.:$g(\infty)$	$G_\infty = 0$ $\varphi_\infty > 0$	$G_\infty = 0$ $\varphi_\infty = 0$
$J_\circ > 0$ $\eta_\circ^- = 0$ fr.d.: $g(0+)$	S_{ee}	S_{ev}	S_{ed}
$J_\circ = 0$ $\eta_\circ^- > 0$ fr.d.: $g(0+)$	S_{ve}	S_{vv}	S_{vd}
$J_\circ = 0$ $\eta_\circ^- = 0$	S_{de}	S_{dv}	S_{dd}

Four standard and five degenerate material classes can be distinguished, and these nine classes are symbolized in table 7.3. For bodies with discrete spectra only the classes S_{ee}, S_{ev}, S_{ve}, and S_{vv} are possible with 'normal' behavior at $t = 0^+$ and $t \to \infty$.

For bodies with continuous spectra *free degeneracies* can be found in addition to the forced degeneracies in the material classes S_{ee}, S_{ev}, S_{ve}, and S_{vv}. In those situations the retarded compliance $j(t)$ and the relaxing modulus $g(t)$ can never be simultaneously degenerate at the same point, since then the degeneracy would be forced. With the help of (7.442) the investigation of the free degeneracies can be done quickly

(1) $t = 0^+$: $j(t)$ degenerate, $g(t)$ normal: $0 < \bar{m} < 1$, $m = 0$.

As $z \to \infty$ in (7.442) $\eta_\circ^- z j^\star(z) \approx \eta_\circ^- z^{1-\bar{m}}$ is the leading term, so that $\eta_\circ^- = 0$ is found. This implies that $J_\circ > 0$, since otherwise the degeneracy is forced.

(2) $t = 0^+$: $j(t)$ normal, $g(t)$ degenerate: $\bar{m} = 1$, $0 < m < 1$.

As $z \to \infty$ in (7.442) $J_\circ g^\star(z) \approx J_\circ z^m$ is the leading term, so that $J_\circ = 0$. This implies for a free degeneracy that $\eta_\circ^- > 0$, since otherwise the degeneracy is forced.

(3) $t = \infty$: $j(t)$ degenerate, $g(t)$ normal: $0 < \bar{m} < 1$, $m \geq 1$ (from the discussion above it follows that $0 < m < 1$) for a degenerate $g(t)$.

As $z \to 0$ in (7.442) the term $G_\infty j^\star(z) \approx G_\infty z^{-\bar{m}}$ is dominant, hence $G_\infty = 0$. This implies that $\varphi_\infty > 0$. The condition is meaningless, since for a degenerate $j(t)$ the creep function cannot have an asymptote. Conclusion: free degeneracies of $j(t)$ are now impossible.

(4) $t = \infty$: $j(t)$ normal, $g(t)$ degenerate: $\bar{m} = 0$, $0 < m < 1$.

As $z \to 0$ in (7.442) the term $\varphi_\infty g^\star(z)/z \approx \varphi_\infty z^{m-1}$ is the dominant term, hence $\varphi_\infty = 0$, $G_\infty > 0$.

The analysis of the degeneracies with the Volterra integral equation is complicated. To obtain the starting point of that analysis write for (7.442) with application of (7.426)

$$J_\circ G_\infty \frac{1}{z} + \varphi_\infty \eta_\circ^- \frac{1}{z} + J_\circ \mathcal{L}[g(t)] + \varphi_\infty \frac{1}{z} \mathcal{L}[g(t)] \\ + G_\infty \mathcal{L}[j(t)] + \eta_\circ^- z \mathcal{L}[j(t)] + z\mathcal{L}[j(t)]\mathcal{L}[g(t)] = \frac{1}{z}$$

Inversion leads to

$$J_\circ G_\infty + \varphi_\infty \eta_\circ^- + J_\circ g(t) + \varphi_\infty \int_0^t g(t')\,dt' \\ + G_\infty j(t) + \eta_\circ^- \frac{dj}{dt} + \frac{d}{dt}\int_0^t j(t')g(t-t')\,dt' = 1 \qquad (7.443)$$

The analysis using the Volterra integral equation is not so easy to follow as the analysis using the algebraic relation between the two complex functions.

Finally, examples of degeneracies have to be considered. In the first place the well-known creep relation of Andrade can be mentioned

$$J(t) = \left(J_\circ + kt^{\bar{m}}\right) H(t) \qquad \text{with} \qquad 0 < \bar{m} < 1 \qquad (7.444)$$

This creep law was suggested for $\bar{m} = \frac{1}{3}$ by Andrade*. With this creep relation, for instance, Henderson† was able to represent well the creep of cadmium wires, and he also succeeded, using this creep equation, in correlating the measurements of Trouton and Rankine‡ for lead and the measurements of Schofield and Scott Blair§. It is clear that (7.444) has a free degeneracy at

* Andrade C.N. da C. 1910. On the viscous flow in metals, and allied phenomena. *Proceedings of The Royal Society of London*, A84: 1–12.

† Henderson C. 1951. The application of the Boltzmann superposition theory to materials exhibiting reversible β-flow. *Proceedings of the Royal Society of London*, A206: 72–86.

‡ Trouton F.T. and Rankine A.O. 1904. On the stretching and torsion of lead wire beyond the elastic limit. *Philosophical Magazine and Journal of Science*, 8: 538–556.

§ Schofield R.K. and Scott Blair G.W. 1933. The relationship between viscosity, elasticity and plastic strength of a soft material as illustrated by some mechanical properties of flour dough.—III. *Proceedings of The Royal Society of London*, A141: 72–85.

$t = 0^+$ and a forced degeneracy at $t = \infty$ (S_{ed} class). Plazek* showed that the creep of, for instance, soft rubbers is very accurately described by a creep relation that is obtained by adding to the Andrade equation the viscous term $\varphi_\infty t$. For $J_\circ = 0$ the Plazek-modified relation of Andrade belongs to the class S_{dv}. For $J_\circ = 0$ relation (7.444) reduces to the empirical Nutting† equation, which is a member of the S_{dd} class. Winter‡ and Larson§ have shown that some crosslinking polymeric liquids close to the gel point can be accurately described by the Nutting equation.

For weakly cross-linking polymers and natural rubbers the relaxation experiments are, according to Chasset and Thirion¶, excellently described by

$$G(t) = G_\infty \left[1 + \left(\frac{t}{t_\circ}\right)^{-m}\right] \tag{7.445}$$

It is clear that (7.445) has a free degeneracy as $t \to \infty$ and a forced degeneracy at $t = 0^+$, so that (7.445) belongs to the class S_{de}.

For polycrystalline materials, a logarithmic creep relation

$$j(t) = k\left[\log\left(\frac{t}{t_\circ} + 1\right)\right]^{\bar{m}} H(t) \tag{7.446}$$

was suggested by Mott and Nabarro‖ with $\bar{m} = \frac{2}{3}$ for alloys. For $\bar{m} = 1$

---

* Plazek D.J. 1960. Dynamic mechanical and creep properties of a 23% cellulose nitrate solution; Andrade creep in polymeric systems. *Journal of Colloid Science*, 15: 50–75; Plazek D.J. 1966. Effect of crosslink density on the creep behavior of natural rubber vulcanizates. *Journal of Polymer Science. Part A-2. Polymer Physics*, 4: 745–763; Plazek D.J., Tan V. and O'Rourke V.M. 1974. The creep behavior of ideally atactic and commercial polymethylmethacrylate. *Rheologica Acta*, 13: 367–376; Ferry J.D. 1980. it Viscoelastic Properties of Polymers. John Wiley & Sons, New York, 3rd ed. See p. 392.

† Nutting P.G. 1921. A study of elastic viscous deformation. *Proceedings American Society for Testing Materials*, 21: 1162–1171.

‡ Winter H.H. 1989. Gel point. In: Kroshwitz J. and Bikales N. (eds.) *Encyclopedia of Polymeric Science and Engineering Supplement*, John Wiley & Sons, New York. See p. 343; Baumgaertel M. and Winter H.H. 1989. Determination of discrete relaxation and retardation time spectra from dynamic mechanical data. *Rheologica Acta*, 28: 511–519; Chambon F. and Winter H.H. 1985. Stopping of crosslinking reaction in a PDMS polymer at the gel point. *Polymer Bulletin*, 13: 499–503; Winter H.H. 1987. Evolution of rheology during chemical gelation. *Progress in Colloid and Polymer Science*, 75: 104–110.

§ Larson R.G. 1985. Constitutive relationships for polymeric materials with power-law distributions of relaxation times. *Reologica Acta*, 24: 327–334; Larson R.G. 1988. *Constitutive Equations for Polymer Melts and Solutions.* Butterworths, Boston. See p. 289.

¶ Chasset R. and Thirion P. 1965. *Proceedings of the Conference on Physics of Non-Crystalline Solids*, Prins J.A. (ed.). North Holland Publ., Amsterdam. See p. 345; Ferry J.D. 1980. it Viscoelastic Properties of Polymers. John Wiley & Sons, New York, 3rd ed. See p. 414.

‖ Mott N.F. and Nabarro F.R.N. 1948. *Reports of a Conference on the Strength of Solids, University of Bristol.* Physical Society of London, pp. 1–20.

relation (7.446) has been derived for glass wires by Bennewitz*, also $J_\circ = 0$, so that this relation belongs to the S_{vd} class. See Meredith† for a discussion of (7.446) with $\bar{m} = 1$ for fibers. Lomnitz‡ showed that the logarithmic relation applies for rocks tested in the laboratory. Jeffreys§ modified this relation such that it fits the data very well for the earth as a whole.

According to Wyatt¶ for polycrystalline copper and aluminum the Andrade creep relation with $\bar{m} = \frac{1}{3}$ applies for high temperatures and the logarithmic creep relation with $\bar{m} = 1$ for low temperatures.

For $0 < \bar{m} \leq 1$ the relation (7.446) is completely monotonic; for $J_\circ > 0$ the relation (7.446) is again a member of the class S_{ed}. For $J_\circ = 0$ the relation (7.446) belongs to the class S_{vd}.

Another creep relation, which behaves logarithmically as $t \to \infty$, reads

$$j(t) = k \left[\ln\left(\frac{t}{t_\circ}\right) - \mathrm{Ei}\left(-\frac{t}{t_\circ}\right) + C \right] H(t) \tag{7.447}$$

in which $C = 0.5772\cdots$ is the Euler constant and

$$\mathrm{Ei}(-x) = -\int_x^\infty \frac{e^{-u}}{u}\, du \tag{7.448}$$

an exponential integral function, so that

$$\frac{dj}{dt} = k \frac{1 - e^{-t/t_\circ}}{t} \tag{7.449}$$

The creep relation (7.447) has been derived by Kuhn, Kunzle and Preismann‖ from the measurements of Brenschede# for hard rubbers. For dielectric after-effects in polymers equation (7.447) has been formulated by Lethersich** and

---

* Bennewitz K. 1920. Über die elastischen Nachwirkung. *Physikalisches Zeitschrift*, 21: 703–705.

† Meredith R. 1958. The rheology of fibers. In *Rheology*, Vol. 2, pp. 261–312, Eirich F.R. (ed.). Academic Press, New York.

‡ Lomnitz C. 1957. Linear dissipation in solids. *Journal of Applied Physics*, 28: 201–205.

§ Jeffreys H. 1976. *The Earth: its Origin, History and Physical Constitution.* 6th ed. Cambridge University Press, Cambridge, See p. 12.

¶ Wyatt O. 1953. Transient creep in pure metals. *Proceedings of the Physical Society of London*, B66: 459–480.

‖ Kuhn W., Künzle O. and Preissmann H. 1947. Relaxationszeitspektrum, Elastizität und Viskosität von Kautschuk I. *Helvetica Chimica Acta*, 30: 307–328; Kuhn W., Künzle O. and Preissmann H. 1947. Relaxationszeitspektrum, Elastizität und Viskosität von Kautschuk II. *Helvetica Chimica Acta*, 30: 464–486.

Brenschede W. 1943. Zur Deformationsmechanik kautschukartiger Körper. Erweiterte Vorstellungen über die Kinetik der Kautschukelastizität. *Kolloid-Zeitschrift*, 104: 1–14.

** Lethersich W. 1950. The rheological properties of dielectric polymers. *Britisch Journal of Applied Physics*, 1: 294–301.

Gross*. Starting from a box spectrum Becker† derived (7.447) for the elastic creep in metals. Also starting from box spectra the relation (7.447) has been formulated by Buchthal and Kaiser‡. For magnetic after-effects in iron this relation was advocated by Richter§. Gross¶ modified (7.447) to eliminate the logarithmic degeneration as $t \to \infty$. Larson‖ multiplied the Nutting equation with an exponential relaxation to eliminate the degeneration as $t \to \infty$. The remark of Larson—that with this modified response functions the constitutive equation of state does not represent a simple fluid—is incorrect, since the fluid is still determined dimensionally only by elasticity and viscosity.

7.13.3. Example: Power law creep equation. The viscoelastic functions corresponding to the creep relation (7.444) will be calculated as an example of a material with continuous spectra.

The complex compliance becomes, for the response function (7.444)

$$J^\star(z) = z\mathcal{L}[J(t)] = z\int_0^\infty (J_\circ + kt^{\bar{m}})e^{-zt}\,dt$$

or

$$J^\star(z) = J_\circ + k\Gamma(1+\bar{m})\,z^{-\bar{m}}$$

or

$$J^\star(z) = J_\circ[1+\lambda z^{-\bar{m}}] \tag{7.450}$$

with

$$\lambda = kG_\circ\Gamma(1+\bar{m}) \qquad \text{and} \qquad G_\circ = \frac{1}{J_\circ} \tag{7.451}$$

From (7.450) it follows that

$$G^\star(z) = \frac{G_\circ}{1+\lambda z^{-\bar{m}}} = z\mathcal{L}[G(t)] \tag{7.452}$$

---

* Gross B. 1953. *Mathematical Structure of the Theories of Viscoelasticity.* Hermann et Cie, Paris. See Ch. XI.

† Becker R. 1925. Elastische Nachwirkung und Plastizität. *Zeitschrift für Physik*, 33: 185–213.

‡ Buchthal F. and Kaiser E. 1951. The rheology of the cross striated muscle fibre with particular reference to isotonic conditions. *Danske Biologiske Meddelelser*, 21: no. 7: 1–318.

§ Richter G. 1937. Über die magnetische Nachwirkung am Carbonyleisen. *Annalen der Physik*, 5.Folge: 29: 605–635.

¶ Gross B. 1953. *Mathematical Structure of the Theories of Viscoelasticity.* Hermann et Cie, Paris.

‖ Larson R.G. 1985. Constitutive relationships for polymeric materials with power-law distributions of relaxation times. *Reologica Acta*, 24: 327–334; Larson R.G. 1988. *Constitutive Equations for Polymer Melts and Solutions.* Butterworths, Boston. See p. 292.

The inverse transformation needed to calculate $G(t)$ will be done through a series expansion for large z (small t)

$$\mathcal{L}[G(t)] = G_\circ \sum_{n=0}^{\infty} (-\lambda)^n z^{-n\bar{m}-1}$$

Now

$$\mathcal{L}^{-1}[z^{-p}] = \frac{t^{p-1}}{\Gamma(p)} \tag{7.453}$$

so that

$$G(t) = G_\circ \sum_{n=0}^{\infty} (-\lambda)^n \frac{t^{\bar{m}n}}{\Gamma(\bar{m}n+1)} \tag{7.454}$$

This result can be expressed in a Mittag-Leffler* function that is defined by

$$E_{\bar{m}}(x) = \sum_{n=0}^{\infty} \frac{x^n}{\Gamma(\bar{m}n+1)} \qquad \text{with} \qquad E_{\bar{m}}(0) = 1 \tag{7.455}$$

For $\bar{m} > 0$ the series converges for all x, the Mittag-Leffler function is defined in the whole complex plane by (7.455).

When the definition (7.455) is used the relaxation modulus (7.454) can be written as

$$G(t) = g(t) = G_\circ E_{\bar{m}}(-\lambda t^{\bar{m}}) \tag{7.456}$$

The result obtained means that

$$\mathcal{L}[E_{\bar{m}}(-\lambda t^{\bar{m}})] = \frac{1}{z}\frac{1}{1+\lambda z^{-\bar{m}}} \tag{7.457}$$

If this expression is used the behavior of the Mittag-Leffler function as $t \to \infty$ $(z \to 0)$ can be investigated. For $z \to 0$ the Laplace transform (7.457) can be expanded as follows

$$\begin{aligned}\mathcal{L}[E_{\bar{m}}(-\lambda t^{\bar{m}})] &= \frac{1}{z}\frac{z^{\bar{m}}}{\lambda}\frac{1}{1+z^{\bar{m}}/\lambda} \\ &= \frac{z^{\bar{m}-1}}{\lambda}\sum_{n=0}^{\infty}(-1)^n \frac{z^{\bar{m}n}}{\lambda^n} = \sum_{n=0}^{\infty}(-1)^n \frac{z^{\bar{m}(n+1)-1}}{\lambda^{n+1}}\end{aligned}$$

Using (7.453) gives

$$E_{\bar{m}}(-\lambda t^{\bar{m}}) \approx \sum_{n=0}^{\infty}(-1)^n \frac{t^{-\bar{m}(n+1)}}{\lambda^{n+1}\Gamma(1-\bar{m}-\bar{m}n)}$$

* See: Erdélyi A. (ed.) 1955. *Higher Transcendental Functions*, Vol. III. McGraw-Hill, New York. See p. 206.

or

$$E_{\bar{m}}(-\lambda t^{\bar{m}}) \approx -\sum_{n=1}^{\infty} \frac{(-\lambda t^{\bar{m}})^{-n}}{\Gamma(1-\bar{m}n)} \qquad \text{for} \qquad t \to \infty \tag{7.458}$$

showing that $G(t) = g(t)$ if $t^{-\bar{m}}$ approaches zero. $G(t)$ then behaves as a slowly decreasing function (the Nutting equation), while from (7.454) it follows that $G(t)$ for $t \to 0^+$ has a vertical tangent ($0 < \bar{m} < 1$).

The retardation spectrum follows from

$$\frac{dj}{dt} = \int_0^{\infty} \frac{1}{\tau}\bar{F}(\tau)e^{-t/\tau}\,d\tau \tag{7.459}$$

With $\tau = 1/s$ this can be written as a Laplace transform

$$\frac{dj}{dt} = \int_0^{\infty} s\bar{N}(s)e^{-st}\,ds \tag{7.460}$$

with

$$\bar{N}(s) = \frac{1}{s^2}\bar{F}(\frac{1}{s}) \qquad \text{or} \qquad \bar{F}(\tau) = \frac{1}{\tau^2}\bar{N}(\frac{1}{\tau}) \tag{7.461}$$

Now $j(t) = kt^{\bar{m}}$ for the Andrade creep law, so that

$$\mathcal{L}[s\bar{N}(s)] = k\bar{m}t^{\bar{m}-1}$$

or

$$s\bar{N}(s) = k\bar{m}\mathcal{L}^{-1}[t^{\bar{m}-1}] = k\bar{m}\frac{s^{-\bar{m}}}{\Gamma(1-\bar{m})}$$

or

$$\bar{N}(s) = \frac{k\bar{m}}{\Gamma(1-\bar{m})}s^{-\bar{m}-1}$$

or

$$\bar{F}(\tau) = \frac{k\bar{m}}{\Gamma(1-\bar{m})}\tau^{\bar{m}-1}$$

or with (7.451)

$$\bar{F}(\tau) = \frac{J_\circ\lambda\bar{m}}{\Gamma(1+\bar{m})\Gamma(1-\bar{m})}\tau^{\bar{m}-1} = J_\circ\lambda\frac{\sin\bar{m}\pi}{\pi}\tau^{\bar{m}-1} \tag{7.462}$$

The retardation spectrum is therefore a power law that decreases from ∞ at $\tau = 0$ to zero as $\tau \to \infty$ ($0 < \bar{m} < 1$).

The relaxation spectrum is defined by

$$g(t) = \int_0^{\infty} F(\tau)e^{-t/\tau}\,d\tau \tag{7.463}$$

or with $\tau = 1/s$

$$g(t) = \int_0^\infty N(s)e^{-st}\,ds \tag{7.464}$$

with

$$N(s) = \frac{1}{s^2}F(\frac{1}{s}) \qquad \text{or} \qquad F(\tau) = \frac{1}{\tau^2}N(\frac{1}{\tau}) \tag{7.465}$$

From (7.464) it follows that

$$\frac{dG}{dt} = \frac{dg}{dt} = -\int_0^\infty sN(s)e^{-st}\,ds$$

therefore

$$\mathcal{L}[\frac{dG}{dt}] = -\int_0^\infty e^{-zt}\,dt \int_0^\infty sN(s)e^{-st}\,ds = -\int_0^\infty N(s)\frac{s}{s+z}\,ds$$

or applying integration by parts

$$\mathcal{L}[\frac{dG}{dt}] = -G_\circ + z\mathcal{L}[G(t)] = -\int_0^\infty N(s)\frac{s}{s+z}\,ds$$

or with (7.456) and (7.457)

$$\int_0^\infty N(s)\frac{s}{s+z}\,ds = G_\circ \frac{1}{1+z^{\bar{m}}/\lambda} \tag{7.466}$$

The left-hand side—Stieltjes integral transformation—is inverted by means of $z = -\beta - i\epsilon$, $\epsilon \to 0^+$

$$\frac{s}{s+z} = \frac{s}{s-\beta-i\epsilon} = \frac{s(s-\beta+i\epsilon)}{(s-\beta)^2+\epsilon^2}$$

or*

$$\frac{s}{s-\beta} + i\pi s \lim_{\epsilon\to 0^+} \frac{1}{\pi}\frac{\epsilon}{(s-\beta)^2+\epsilon^2} = \frac{s}{s-\beta} + i\pi s\,\delta(s-\beta)$$

* The function

$$\delta(t;\epsilon) = \frac{1}{\pi}\int_{-\infty}^\infty \exp(-\epsilon x)\cos xt\,dx = \frac{\epsilon}{\pi(\epsilon^2+t^2)}$$

approaches a δ-function for $\epsilon \to 0$, while for the integral of the function

$$\int_{-\infty}^\infty \delta(t;\epsilon)\,dt = \frac{\epsilon}{\pi}\int_{-\infty}^\infty \frac{dt}{\epsilon^2+t^2} = \frac{1}{\pi}\tan^{-1}\frac{t}{\epsilon}\Big|_{-\infty}^{\infty} = 1$$

Substitution into (7.466) yields

$$\int_0^\infty N(s)\frac{s}{s-\beta}\,ds + i\pi\beta N(\beta) = G_\circ \frac{1}{1+(\beta e^{-i\pi})^{\bar m}/\lambda} \tag{7.467}$$

Obviously it is true that

$$\pi\beta N(\beta) = G_\circ \Im\left(\frac{1}{1+(\beta e^{-i\pi})^{\bar m}/\lambda}\right) = G_\circ \Im\left(\frac{1}{1+\frac{\beta^{\bar m}}{\lambda}(\cos\bar m\pi - i\sin\bar m\pi)}\right)$$

in which $\Im(f)$ is the imaginary part of the function f, so that

$$\pi\beta N(\beta) = G_\circ \frac{\frac{\beta^{\bar m}}{\lambda}\sin\bar m\pi}{(1+\frac{\beta^{\bar m}}{\lambda}\cos\bar m\pi)^2 + (\frac{\beta^{\bar m}}{\lambda}\sin\bar m\pi)^2}$$

or

$$N(\beta) = \frac{G_\circ}{\pi\beta}\frac{\sin\bar m\pi}{\frac{\lambda}{\beta^{\bar m}} + \frac{\beta^{\bar m}}{\lambda} + 2\cos\bar m\pi} \tag{7.468}$$

With this the relaxation spectrum becomes

$$F(\tau) = \frac{G_\circ}{\pi\tau}\frac{\sin\bar m\pi}{\lambda\tau^{\bar m} + \frac{1}{\lambda\tau^{\bar m}} + 2\cos\bar m\pi} \tag{7.469}$$

from which it follows that

$$\begin{aligned} F(\tau) &\approx G_\circ\lambda\frac{\sin\bar m\pi}{\pi}\tau^{\bar m-1} \qquad \text{for} \qquad \tau\to 0^+ \\ F(\tau) &\approx \frac{G_\circ}{\lambda}\frac{\sin\bar m\pi}{\pi}\tau^{-\bar m-1} \qquad \text{for} \qquad \tau\to\infty \end{aligned} \tag{7.470}$$

As $\tau \to 0$ the relaxation spectrum F behaves in a similar way to the retardation spectrum $\bar F$ (see (7.462)), but as $\tau\to\infty$ the relaxation spectrum F decreases faster than the retardation spectrum $\bar F$.

The denominator of (7.469) has extrema if

$$(1+\bar m)\lambda\tau^{\bar m} + \frac{1-\bar m}{\lambda}\tau^{-\bar m} + 2\cos\bar m\pi = 0$$

or

$$\tau^{\bar m} = \frac{-\cos\bar m\pi \pm \sqrt{\cos^2\bar m\pi - (1-\bar m^2)}}{(1+\bar m)\lambda} \tag{7.471}$$

For $\cos^2\bar m\pi = 1-\sin^2\bar m\pi \geq 1-\bar m^2$ the root is real, so a real root follows for $\sin\bar m\pi \leq \bar m$ or $\bar m \geq 0.7365$. For these values of $\bar m$, $\cos\bar m\pi$ is negative, so that τ is real $(0<\bar m<1)$. For $0<\bar m<0.7365$ the relaxation spectrum $F(\tau)$ is therefore monotonic and for $0.7365 < \bar{m} < 1$ it has two extrema (a minimum and a maximum), as sketched in figure 7.24.

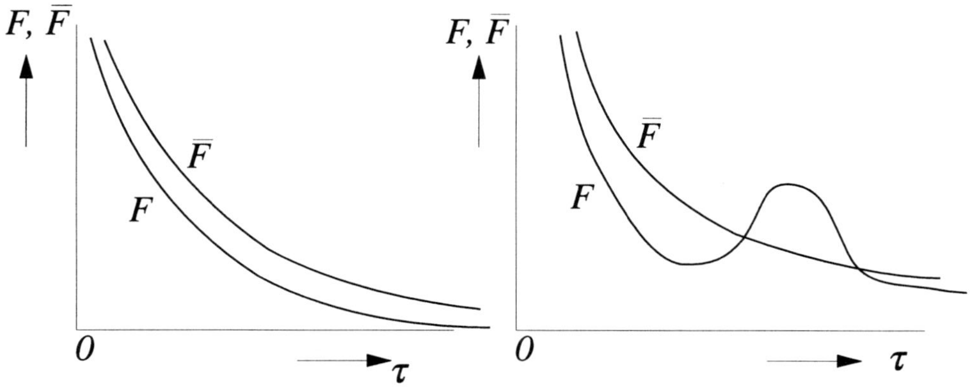

Figure 7.24. The relaxation spectrum F and the retardation spectrum $\bar{F}$ of $J = J_\circ + kt^{\bar{m}}$. For $0 < \bar{m} < 0.7365$ both spectra behave monotonically. For $0.7365 < \bar{m} < 1$ the relaxation spectrum has two local extrema.

In the limit $\bar{m} \to 1^-$ the behavior of the relaxation spectrum F is

$$F = \lim_{\bar{m}\to 1^-} \frac{G_\circ}{\pi\tau} \frac{\sin \bar{m}\pi}{\lambda\tau^{\bar{m}} + \frac{1}{\lambda\tau^{\bar{m}}} + 2\cos\bar{m}\pi}$$

or

$$F = \lim_{\bar{m}\to 1^-} \frac{G_\circ}{\pi\tau} \frac{2\sin\frac{\bar{m}\pi}{2}\cos\frac{\bar{m}\pi}{2}}{\lambda\tau^{\bar{m}} + \frac{1}{\lambda\tau^{\bar{m}}} + 2(\cos^2\frac{\bar{m}\pi}{2} - 1)}$$

or

$$F = \lim_{\bar{m}\to 1^-} \frac{G_\circ}{\pi\tau} \sin\frac{\bar{m}\pi}{2} \frac{2\cos\frac{\bar{m}\pi}{2}}{[\sqrt{\lambda\tau^{\bar{m}}} - \frac{1}{\sqrt{\lambda\tau^{\bar{m}}}}]^2 + 2\cos^2\frac{\bar{m}\pi}{2}}$$

or

$$F = \frac{G_\circ}{\pi\tau}\delta\left(\sqrt{\lambda\tau} - \frac{1}{\sqrt{\lambda\tau}}\right)$$

Now if $\tau_\circ$ is a simple root of $f(\tau) = 0$ it is true* that

$$\delta[f(\tau)] = \frac{\delta(\tau - \tau_\circ)}{|f'(\tau_\circ)|}$$

---

* If $x_\circ$ is a simple root of $\phi(x) = 0$, then in the neighborhood of $x_\circ$

$$\phi(x) = \phi'(x_\circ)(x - x_\circ) \qquad \text{for} \qquad x_\circ - \delta \le x \le x_\circ + \delta$$

$$\int_{x_\circ-\delta}^{x_\circ+\delta} f(x)\delta[\phi(x)]\,dx = \int_{-\phi'\delta}^{+\phi'\delta} f(x_\circ + \frac{\phi}{\phi'})\frac{\delta(\phi)}{\phi'}\,d\phi = \begin{cases} \dfrac{f(x_\circ)}{\phi'(x_\circ)} & \text{for} \quad \phi' > 0 \\ -\dfrac{f(x_\circ)}{\phi'(x_\circ)} & \text{for} \quad \phi' < 0 \end{cases}$$

Then

$$f(\tau) = \sqrt{\lambda\tau} - \frac{1}{\sqrt{\lambda\tau}}$$

from which

$$\tau_\circ = \frac{1}{\lambda} \qquad f' = \tfrac{1}{2}\sqrt{\frac{\lambda}{\tau}} + \tfrac{1}{2}\sqrt{\frac{1}{\lambda\tau^3}} \qquad \Rightarrow \qquad f'(\tau_\circ) = \lambda$$

so that

$$F = \frac{G_\circ}{\pi\tau\lambda}\delta\big(\tau - \frac{1}{\lambda}\big) = G_\circ\delta\big(\tau - \frac{1}{\lambda}\big) \qquad \text{for} \qquad \bar{m} \to 1^- \tag{7.472}$$

As $\bar{m} \to 1^-$ the Andrade relation (7.444) reduces to a Maxwell body with $\bar{F} = 0$ and one relaxation time $1/\lambda$. This can also be deduced directly from (7.455) and (7.456). As $\bar{m} \to 1$ the Mittag-Leffler function (7.455) becomes

$$E_1(x) = \sum_{n=0}^{\infty} \frac{x^n}{\Gamma(n+1)} = e^x \tag{7.473}$$

so that for the relaxation modulus (7.456)

$$G(t) = G_\circ e^{-\lambda t} \qquad \text{for} \qquad \bar{m} \to 1^- \tag{7.474}$$

In terms of a mechanical model, the creep equation of Andrade can thus be represented by a linear spring ($J_\circ = 1/G_\circ$) and a dashpot, in which the creep is formulated by $kt^{\bar{m}}$. In terms of differential equations the Andrade creep relation for $0 < \bar{m} < 1$ turns out to be a fractional Maxwell type of differential equation, for which the ordinary differentials are replaced by fractional differentials ${}_0D_t^{\bar{m}}$

$$\sigma(t) + \frac{1}{\lambda}\,{}_0D_t^{\bar{m}}\sigma(t) = \frac{G_\circ}{\lambda}\,{}_0D_t^{\bar{m}}\gamma(t) \tag{7.475}$$

which for $\gamma(t) = \gamma_\circ H(t)$ reduces to

$$G(t) + \frac{1}{\lambda}\,{}_0D_t^{\bar{m}}G(t) = \frac{G_\circ}{\lambda}\frac{t^{-\bar{m}}}{\Gamma(1-\bar{m})} \tag{7.476}$$

---

so that for a simple root $x_\circ$

$$\delta[\phi(x)] = \frac{\delta(x - x_\circ)}{|\phi'(x_\circ)|}$$

If $\phi(x)$ has several simple roots x_k then

$$\delta[\phi(x)] = \sum_k \frac{\delta(x - x_k)}{|\phi'(x_k)|}$$

with $G(t) = \sigma/\gamma_\circ$. The solution of this fractional* differential equation is given by (7.454).

Gross† suggested eliminating the degeneracy in $t = \infty$ by representing the creep function by a Mittag-Leffler function

$$J(t) = J_\circ + (J_\infty - J_\circ)[1 - E_{\bar{m}}(-\lambda t^{\bar{m}})] \tag{7.477}$$

At $t = 0^+$ there is a free degeneracy, however $j(\infty) < \infty$, so that (7.477) belongs to the class $\mathrm{S_{ee}}$. According to Gross, the experimental results can sometimes be described very well by (7.477). For instance dielectric absorption was described by Cole and Cole‡ with functions of the type (7.477).

---

* For $f(x) = x^m$

$$\frac{d^n x^m}{dx^n} = \frac{m!}{(m-n)!} x^{m-n} = \frac{\Gamma(m+1)}{\Gamma(m-n+1)} x^{m-n}$$

which can be generalized for arbitrary ν by

$$\frac{d^\nu x^m}{dx^\nu} = \frac{\Gamma(m+1)}{\Gamma(m-\nu+1)} x^{m-\nu}$$

Riemann has generalized his classical integral operator into the Liouville-Riemann fractional integral operator by

$${}_0D_t^{-q} F(t) = \frac{1}{\Gamma(q)} \int_0^t (t-\tau)^{q-1} F(\tau)\, d\tau \qquad (q > 0)$$

which is the solution of the equation $d^\nu y/dt^\nu = F(t)$, where $F(t)$ is continuous in the interval $[0, t]$.

The fractional differential operator ${}_0D_t^\nu$ for $\nu > 0$ is defined by

$${}_0D_t^\nu F(t) = \frac{d^n}{dt^n}\left({}_0D_t^{\nu-n} F(t)\right) \qquad (\nu - n < 0)$$

so that 'fractional differentiation' ${}_0D_t^\nu$ can be factorized into a 'fractional integration' ${}_0D_t^{-(n-\nu)}$ followed by an ordinary differentiation d^n/dx^n, whereby n is the smallest possible whole number greater than ν.

See Nonnenmacher T.F 1991. Fractional relaxation equations for viscoelasticity and related phenomena. In Casas-Vázquez J. and Jou D. (Eds.) *Lecture Notes in Physics 381, Rheological Modeling: Thermodynamical and Statistical Approaches.* Springer-Verlag, Berlin, pp. 309–320; Friedrich C. 1991. Relaxation functions of rheological constitutive equations with fractional derivatives: thermodynamical constraints. In Casas-Vázquez J. and Jou D. (Eds.) *Lecture Notes in Physics 381, Rheological Modeling: Thermodynamical and Statistical Approaches.* Springer-Verlag, Berlin, pp. 321–330; Oldham K.B. and Spanier J. 1974. The fractional Calculus. Academic Press, New York, 1974; Ross B. 1975. *Lecture Notes in Mathematics 457.* Springer-Verlag, Berlin, pp. 1–36.

† Gross B. 1947. On creep and relaxation I. *Journal of Applied Physics*, 18: 212–221.

‡ Cole S. and Cole R.H. 1942. Dispersion and absorption in dielectrics II. Direct current characteristics. *Journal of Chemical Physics*, 10: 98–105.

The corresponding relaxation modulus is quickly derived. With (7.457) follows from (7.477)

$$J^\star(z) = z\mathcal{L}[J(t)] = J_\circ + (J_\infty - J_\circ)\left(1 - \frac{1}{1+\lambda z^{-\bar{m}}}\right)$$

or

$$J^\star(z) = J_\infty - \frac{J_\infty - J_\circ}{1+\lambda z^{-\bar{m}}} \tag{7.478}$$

or

$$J^\star(z) = \frac{J_\circ + J_\infty \lambda z^{-\bar{m}}}{1+\lambda z^{-\bar{m}}} \tag{7.479}$$

From this is found that

$$\begin{aligned} G^\star(z) = \frac{1+\lambda z^{-\bar{m}}}{J_\circ + J_\infty \lambda z^{-\bar{m}}} &= G_\infty \frac{J_\infty + J_\infty \lambda z^{-\bar{m}}}{J_\circ + J_\infty \lambda z^{-\bar{m}}} \\ &= G_\infty \frac{J_\circ + J_\infty \lambda z^{-\bar{m}} + J_\infty - J_\circ}{J_\circ + J_\infty \lambda z^{-\bar{m}}} \end{aligned}$$

or

$$G^\star(z) = G_\infty \left(1 + \frac{J_\infty - J_\circ}{J_\circ} \frac{1}{1 + \frac{J_\infty}{J_\circ}\lambda z^{-\bar{m}}}\right)$$

or

$$G^\star(z) = z\mathcal{L}[G(t)] = G_\infty + (G_\circ - G_\infty)\frac{1}{1 + \frac{G_\circ}{G_\infty}\lambda z^{-\bar{m}}} \tag{7.480}$$

Invert with (7.457)

$$G(t) = G_\infty + (G_\circ - G_\infty) E_{\bar{m}}\left(-\frac{G_\circ}{G_\infty}\lambda t^{\bar{m}}\right) \tag{7.481}$$

which relation is also mentioned by Friedrich* as a fractional model belonging to the class S_{ee}. For $\bar{m} \to 1^-$ a Poynting–Thomson body is found.

7.14. EXERCISES

Exercise 7.1. Show that for the creep relation

$$J(t) = \left[J_\circ + k \ln\left(1 + \frac{t}{t_0}\right)\right] H(t)$$

---

* Friedrich C. 1991. Relaxation functions of rheological constitutive equations with fractional derivatives: thermodynamical constraints. In Casas-Vázquez J. and Jou D. (Eds.) *Lecture Notes in Physics 381, Rheological Modeling: Thermodynamical and Statistical Approaches.* Springer-Verlag, Berlin. See p. 324.

the function $\bar{N}(s)$ defined in (7.461) becomes

$$\bar{N}(s) = \frac{k}{s} e^{-k_o s}$$

and the retardation spectrum

$$\bar{F}(\tau) = \frac{k}{\tau} e^{-k_o/\tau}$$

This retardation spectrum has a maximum at $\tau = t_o$. Compare the nonlinear creep function with the linear creep function for a Poynting–Thomson body with a line spectrum. Discuss whether or not the nonlinear effects broaden the lines and make the spectra continuous.

Exercise 7.2. Discuss whether or not large nonlinear elastic deformations are allowed in the thermodynamics of irreversible processes.

Exercise 7.3. Show that the linear differential equations for the various bodies are not objective*. Replace the material time derivatives by the Oldroyd time derivatives and show that for small deformation rates the linear differential equations apply.

Exercise 7.4. Formulate the fractional differential equation (7.475) for three dimensions.

Exercise 7.5. Formulate an objective fractional differential equation.

Exercise 7.6. Substitute the differential equation for the Jeffreys body into the momentum balance. Solve the hyperbolic equation for a half-space subjected to oscillatory shear at the boundary. Discuss the propagation of the shear waves in the half-space.

---

* Bird R.B., Armstrong R.C. and Hassager O. 1987. *Dynamics of Polymeric Liquids*. 2nd ed. John Wiley & Sons, New York.

APPENDIX A
Derivation of the Maxwell electromagnetic field equations

Summary: The Maxwell electromagnetic field equations for moving media are derived from the fact that charged particles produce an electric field if they are at rest in a frame of reference, and that they produce a magnetic field if they are in motion. The Maxwell equations specify the curl and divergence of both fields. The divergence of the electric field expresses the conservation of charges, and the divergence of the magnetic field expresses the fact that there are no magnetic poles. The curl of the magnetic field is proportional to the currents. The curl of the electric field accounts for the induced electric fields, if there are time-dependent magnetic fields.

A.1. THE PHENOMENON OF 'ELECTRICITY'

Thales of Miletus discovered about 26 centuries ago that amber when rubbed attracts small pieces of papyrus. Analogous properties were later found for other substances. For example two glass bars that have been rubbed with silk coth repel one another. The same phenomenon is observed with two resin bars rubbed with cat's fur or a woolen cloth, while a rubbed resin bar and a glass bar rubbed with slik cloth attract each other when brought into close proximity. From experience it is thus found that by close contact (rubbing) a fundamental property can be transferred to some substances. This property is called *electric charge*, and it can increase or decrease in the material, as is established from the different kinds of behavior of glass and resin bars. Historically, a positive charge was assigned to the glass bar rubbed with a silk cloth, and the charged glass bar produces a force field, in which other bodies, charged by rubbing, are attracted or repelled. Experiments show that similar charges repel one another and dissimilar charges attract one another. An electric charge is involved in two actions: (i) the electric charge produces a force field, and (ii) the electric charge is subject to a force because of the force field produced by another charge. Bodies that do not exert electrical forces are called neutral; they either do not possess charges at all, or they possess equal portions of dissimilar positive and negative charges.

Transfering electrical charges to a body is called *electrifying* the body. Some substances can be electrified by being rubbed. Not all materials can be electrified by rubbing. Materials that can be electrified by rubbing are called *insulators* or *dielectrics*. If it is assumed that the electric charge is carried by particles of the solid body, then the charge-carrying particles are fairly well fixed to the surface of the insulator. This is not so for materials that cannot be electrified by rubbing. Such materials are called *conductors.* When an insulated metal plate (conductor) is brought in contact with a rubbed glass rod it is found that the charge distributes itself uniformly over the metal plate; this suggests that the charge carriers in a conductor are mobile.

The observations obtained with electrified bodies can therefore be described by assigning to these bodies a definite charge q; within these bodies this charge is the source of a force field $\vec{E}$. The experiments of Millikan* with oil droplets in an electrical field confirm this view and further show that if a single particle with an electrical charge q is placed in a force field $\vec{E}$, the force on the charged particle is

$$\boxed{\vec{K} = q\,\vec{E}} \tag{A.1}$$

where $\vec{E}$ is called the *electric field strength.* From this definition it follows that $\vec{E}$ at some point $\vec{r}$ in space is the force exerted at this point on a positive unit of charge. This definition of $\vec{E}$ does not make sense if the charge q cannot be measured. To this purpose a second law must also be known.

The measurement of electric forces can be done with a torsion balance, by means of which the forces between two small metal spheres are measured. One small sphere can be charged by bringing it into contact with a rubbed insulator. Next, this small sphere can be brought into contact with the other small sphere, so that the charge is distributed uniformly over both spheres. The force of repulsion can be measured with the torsion balance when the small spheres are a certain distance apart from each other. By measuring the forces while varying the distance, the magnitude and the charge, it is found that the force between the small spheres is along their line of centers and is determined by the distance between the centers of the spheres. With respect to the force, the spheres behave the same as *point charges* located at the centers of the spheres. Finally, from the measurements it is found

$$\vec{K}_{ij} = -\vec{K}_{ji} = \frac{1}{4\pi\epsilon_\circ}\frac{q_i q_j}{|\vec{r}_i - \vec{r}_j|^{3+\epsilon}}(\vec{r}_i - \vec{r}_j) \tag{A.2}$$

with $\vec{K}_{ij}$ the force on sphere i, exerted by sphere j; $\vec{r}_i$ is the position vector of the center of sphere i; q_j the charge of sphere j, and $\epsilon_\circ$ a constant. The result (A.2) was found experimentally by Cavendish (1773) and Coulomb (1785) with ϵ of the order of 10^{-2}. Subsequent investigators increased the accuracy enormously (Williams 1971; $\epsilon = \mathcal{O}(10^{-16})$). Hardly any other

---

* Millikan R.A. 1917. *Electrons (+and-), Protons, Photons, Neutrons, Mesotrons, and Cosmic Rays.* 2nd ed., 2nd impr. University of Chigago Press, Chicago, 1947.

material property has been measured with such accuracy, so that there is much evidence that $\epsilon = 0$; this choice for ϵ leads to the *Coulomb law*

$$\boxed{\vec{K}_{ij} = -\vec{K}_{ji} = \frac{1}{4\pi\epsilon_\circ} \frac{q_i q_j}{|\vec{r}_i - \vec{r}_j|^3}(\vec{r}_i - \vec{r}_j)} \tag{A.3}$$

Equally important is the experimental proof of the *superposition principle* that the forces $\vec{K}_{ij}$ and $\vec{K}_{ik}$ exerted by two charged small spheres j and k on sphere i are pairwise additive

$$\boxed{\vec{K}_i = \vec{K}_{ij} + \vec{K}_{ik}} \tag{A.4}$$

where $\vec{K}_i$ is the total force exerted on sphere i. The SI units cannot yet be fixed, but with (A.3) the electric charge can be measured in 'absolute' units. For this purpose $4\pi\epsilon_\circ$ is set equal to unity, and the *electrostatic unit of charge* (E.S.U.) is defined by: 'an electric charge has the magnitude of 1 E.S.U. of charge if it is repelled with a force of 1 dyne (1 dyne= 10^{-5} N) by a second charge of the same magnitude placed at a distance of 1 cm (10^{-2} m)'.

By measuring the settling speed of droplets in an electric field $\vec{E}$, Millikan showed that the charge q of tiny charged oil droplets is always a multiple of a certain charge. The smallest charge q_e, called the *elementary charge* $\boxed{e}$, is the charge of an electron, and it has the value of $-(4.80286 \pm 0.00009) \times 10^{-10}$ E.S.U.

A.2. ELECTROSTATIC LAWS

From the Coulomb law and the superposition principle equations can be derived, with which the electric field can be calculated for a given distribution of charges. The force exerted by N small charged spheres or charge carriers on a charge carrier i is, according to (A.3) and (A.4), given by

$$\vec{K}_i = \frac{1}{4\pi\epsilon_\circ} \sum_{j \neq i} \frac{q_i q_j}{|\vec{r}_i - \vec{r}_j|^3}(\vec{r}_i - \vec{r}_j) = q_i \vec{E}(\vec{r}_i)$$

The field produced by N charge carriers is given by

$$\vec{E}(\vec{r}) = \frac{1}{4\pi\epsilon_\circ} \sum_j \frac{q_j \delta(\vec{r}_i - \vec{r}_j)}{|\vec{r} - \vec{r}_j|^3}(\vec{r} - \vec{r}_j) \tag{A.5}$$

For a continuous distribution of charge carriers, the *charge density* $\rho_f^{(e)}$ can be introduced, where $\rho_f^{(e)}$ is the density of the charge in free space. Then the electric field (A.5): becomes

$$\vec{E}(\vec{r}) = \frac{1}{4\pi\epsilon_\circ} \int_V \frac{\rho_f^{(e)}(\vec{r}\,')}{|\vec{r} - \vec{r}\,'|^3}(\vec{r} - \vec{r}\,')\, dV' \tag{A.6}$$

which can also be written as

$$\left.\begin{aligned} \vec{E} &= -\operatorname{grad}\mathcal{V} \\ \text{with} \qquad \mathcal{V} &= \frac{1}{4\pi\epsilon_{\rm o}} \fint_V \frac{\rho_{\rm f}^{(\rm e)}(\vec{r}\,')}{|\vec{r}-\vec{r}\,'|}\, dV' \end{aligned}\right\} \tag{A.7}$$

where V denotes the volume that contains the charges, and $\fint_V$ represents the principal value of the volume integral*. $\mathcal{V}$ is the *electrostatic potential.*

From (A.7) it is found that the electrostatic field is irrotational or lamellar. The electrostatic field therefore also satisfies $\operatorname{curl}\vec{E} = \vec{0}$. This equation does not determine $\vec{E}$, since for each scalar function $\psi(\vec{r})$ the combination $\vec{E} + \operatorname{grad}\psi$ is also irrotational. A second equation for $\vec{E}$ can be derived if the flux $\Psi^{(\rm e)}$ of the electrostatic field through the surface A is considered

$$\Psi^{(\rm e)} = \int_A \vec{E}\cdot d\vec{A} \tag{A.8}$$

Suppose that A is a closed surface that encloses a volume V. With the substitution of (A.5) into (A.8) the flux (A.8) of the electrostatic potential becomes

$$\oint_A \vec{E}(\vec{r})\cdot d\vec{A} = \frac{1}{4\pi\epsilon_{\rm o}} \oint_A \sum_j \frac{q_j\delta(\vec{r}-\vec{r}_j)}{|\vec{r}-\vec{r}_j|^3}(\vec{r}-\vec{r}_j)\cdot d\vec{A} \tag{A.9}$$

This integral can be calculated by applying the Gauss divergence theorem and then determining its principal value. The singularities at the points $\vec{r} = \vec{r}_j$ are accounted for by enclosing around each point $\vec{r}_j$ a tiny sphere with the volume b_j and the surface a_j. Now

$$\begin{aligned} \oint_A \vec{E}(\vec{r})\cdot d\vec{A} &= \lim_{b_k\to 0}\int_{V-\sum_k b_k}\left[\operatorname{div}\sum_j \frac{q_j}{4\pi\epsilon_{\rm o}|\vec{r}-\vec{r}_j|^3}(\vec{r}-\vec{r}_j)\right] d\vec{r} \\ &= \lim_{a_k\to 0}\oint_{A+\sum_k a_k}\left[\sum_j \frac{q_j}{4\pi\epsilon_{\rm o}|\vec{r}-\vec{r}_j|^3}(\vec{r}-\vec{r}_j)\right] d\vec{A} \end{aligned} \tag{A.10}$$

---

* If the integrand of an integral becomes unbounded during the integration, then the principal value of such an integral can exist, which is obtained by first enclosing the singular points in tiny spheres and integrating over the remainder, and then reducing the radii of the spheres to zero.

In the region without singularities $\operatorname{div}[(\vec{r}-\vec{r}_j)/(|\vec{r}-\vec{r}_j|^3) = 0$, so that

$$\begin{aligned}\oint_A \vec{E}(\vec{r})\cdot d\vec{A} &= \oint_A \sum_j \frac{q_j}{4\pi\epsilon_\circ|\vec{r}-\vec{r}_j|^3}(\vec{r}-\vec{r}_j)\cdot d\vec{A} \\ &= -\sum_j \lim_{a_j\to 0}\oint_{a_j} \frac{q_j}{4\pi\epsilon_\circ|\vec{r}-\vec{r}_j|^3}(\vec{r}-\vec{r}_j)\cdot d\vec{A} \\ &= \frac{1}{\epsilon_\circ}\sum_j q_j\end{aligned}$$

Obviously for the charges q_j at the points $\vec{r}_j$

$$\epsilon_\circ \oint_A \vec{E}(\vec{r})\cdot d\vec{A} = \sum_j q_j \tag{A.11}$$

For a continuous distribution of the point charges in space this result becomes

$$\epsilon_\circ \oint_A \vec{E}(\vec{r})\cdot d\vec{A} = \int_V \rho_{\mathrm{f}}^{(\mathrm{e})}(\vec{r})\,dV \tag{A.12}$$

or

$$\epsilon_\circ \operatorname{div}\vec{E} = \rho_{\mathrm{f}}^{(\mathrm{e})} \tag{A.13}$$

This is the second equation. The *two electrostatic laws* are therefore

$$\boxed{\epsilon_\circ \operatorname{div}\vec{E} = \rho_{\mathrm{f}}^{(\mathrm{e})}} \quad \text{and} \quad \boxed{\operatorname{curl}\vec{E} = \vec{0}} \tag{A.14}$$

These laws are not only a result of the Coulomb law and the superposition principle, but via the transition of (A.11) to (A.12) the two electrostatic laws also generalize the Coulomb law and the superposition principle so that they apply to continuous distributions of charge carriers.

From the solution of $(\text{A.14})_2$ it follows conversely again that $\vec{E} = -\operatorname{grad}\mathcal{V}$, and the substitution of this into $(\text{A.14})_1$ leads to

$$\nabla^2\mathcal{V} = -\frac{1}{\epsilon_\circ}\rho_{\mathrm{f}}^{(\mathrm{e})} \tag{A.15}$$

The electrostatic potential is thus determined by a *Poisson differential equation.*

To summarize: the electrostatic equations are a result of the observation that charges exist, and that a charge produces an electric field that in turn results in a force on the charge carrier. From measurements of this force the Coulomb law was found and also the superposition principle. With some mathematics and the assumption of a continuous distribution of charges the final electrostatic laws (A.14) are found.

In contrast to insulators, the charge carriers are mobile in conductors, so that the charge carriers are set in motion by applying an electric field. In a stationary electric field, a charge q_j attains a drift velocity $\vec{v}_j$, which depends on the field $\vec{E}$. If the surface of the conductor is A, and there are n charges per unit volume in the conductor, then the total electric current $\vec{I}$ through the conductor under the influence of the stationary field $\vec{E}$ perpendicular to the surface A is

$$\vec{I} = \sum_{j=1}^{n} q_j \vec{v}_j(\vec{E}) A \tag{A.16}$$

The electric current density $\vec{J}$ is

$$\vec{J} = \frac{\vec{I}}{A} = \sum_{j=1}^{n} q_j \vec{v}_j(\vec{E}) \tag{A.17}$$

With several conductors a network can be made. Consider a junction in such a network where k different conductors meet. According to experiments by Kirchhoff at a junction

$$\sum_k I_k = 0 \tag{A.18}$$

if a current that is directed towards the junction is counted positively and a current directed away from the junction is counted negatively or vice versa. The result expresses the conservation of electric charge, so that the flow of electrically charged particles behaves the same as an incompressible fluid. If A_k is the cross-section of the k-th conductor, then (A.18) can be written as

$$\sum_k J_k \, A_k = 0 \tag{A.19}$$

If the junction is surrounded by an arbitrary surface A, then (A.19) can be generalized to

$$\oint_A \vec{J} \cdot d\vec{A} = 0 \tag{A.20}$$

A continuous distribution of electric current result by this generalization to a continuous distribution of conductors, for which from (A.20) the conservation law

$$\boxed{\operatorname{div} \vec{J} = 0} \tag{A.21}$$

for a stationary current of electric charges follows. With (A.14) and (A.21) the time-independent equations are formulated from the two aspects of a material property 'charge', namely, experiments with insulators in which a charge can be held in place and experiments with conductors in which charges are mobile.

A.3. THE PHENOMENON OF 'MAGNETISM'

Just like electricity, magnetism was discovered in antiquity. A kind of magnetic oxide of iron, called lodestone, found near the town Magnesia in Thessaly, Greece, turns out to attract iron. The phenomenon 'magnetism' may be described as follows: if a bar or needle of magnetic material is mounted on a horizontal surface and can rotate freely, then experiment shows that the bar or needle takes a stable position, such, that one end of the bar or needle points roughly to the north. This end is called the north pole of the bar magnet, while the other end is the south pole. If a north pole of a freely rotating magnet is approached by the north pole of another magnet, then the two north poles repel each other. If a south pole approaaches the north pole of a freely rotating magnet, then attraction occurs. It is concluded, therefore, that similar poles repel each other, while dissimilar poles attract each other. Obviously, besides the electric field $\vec{E}$ there exists another field that will be called a magnetic field $\vec{B}$. That the magnetic field is another kind of field is shown by the experimental fact that an electric field $\vec{E}$, produced by fixed charged particles, exerts no forces on a fixed magnet, and conversely the magnetic field $\vec{B}$, produced by a fixed magnet, exerts no forces on the fixed charged particles. Bodies that produce a magnetic field are called magnetized. Just as not all bodies can be electrified, not all bodies can be magnetized, so that magnetism can be considered as a new property of substance. The question arises whether magnetism can be understood in terms of the behavior of charge carriers, just as insulators and conductors were understood in terms of the mobility of the charges.

Oersted in 1820 made the important discovery that an electric current exerts a torque on a magnet. A needle magnet, set up in the neighborhood of a straight conductor, tends to become oriented perpendicular to the conductor. Since action =reaction, the magnet will exert a torque on the conductor; in this way Oersted discovered the interaction between a magnet and an electric current. If torques can be exerted on a magnet only by another magnet through its magnetic field $\vec{B}$, then it follows from the Oersted experiment that a constant electric current is accompanied by a magnetic field.

Biot and Savart discovered experimentally that the magnetic field strength is inversely proportional to the distance R from the straight wire conductor and is proportional to the electric current I in the long wire. In 1820 they published their famous law that an infinitely long straight wire conductor induces a magnetic field that satisfies

$$\vec{B} = \frac{\mu_\circ}{2\pi} \frac{I}{R} \vec{e}_\theta \tag{A.22}$$

where $\mu_\circ$ is a proportionality factor, which—like $\epsilon_\circ$—is determined by the choice of the system of units. In the cylindrical coordinate system (r, θ, z), the straight long wire is situated along the z-axis, and $\vec{e}_\theta$ denotes a unit vector in θ-direction. The Biot–Savart law is an experimental law in an integral formulation, and therefore less appropriate to serve for theoretical calculations. Laplace succeeded in 1821 in putting this law in a differential formulation

$$d\vec{B}(\vec{r}) = \frac{\mu_\circ I}{4\pi} \frac{d\vec{l} \times (\vec{r} - \vec{r}\,')}{|\vec{r} - \vec{r}\,'|^3} \tag{A.23}$$

where $d\vec{B}$ is the differential of the magnetic field at the point $\vec{r}$ that is induced by the current I through element $d\vec{l}$ at $\vec{r}\,'$ in the thin wire conductor. From (A.23) the Biot–Savart law (A.22) follows by integration along an infinitely long straight wire conductor. Integration involves superposition, so that (A.23) has to be supplemented by an assumed superposition principle.

A.4. MAGNETOSTATIC LAWS

The magnetostatic field equations can be derived from the Laplace law (A.23) and the superposition principle If A is the cross-section of the conductor, then

$$I\, d\vec{l} = \vec{J} A\, dl = \vec{J}\, dV \tag{A.24}$$

For a continuous distribution of the electric current in a volume V the differential of the magnetic field (A.23) becomes, with the substitution of (A.24) and use of the superposition principle

$$d\vec{B}(\vec{r}) = \frac{\mu_\circ}{4\pi} \int_V \frac{\vec{J}(\vec{r}\,') \times (\vec{r} - \vec{r}\,')}{|\vec{r} - \vec{r}\,'|^3}\, dV' \tag{A.25}$$

if dV' denotes a volume-element at $\vec{r}\,'$. The differential of the magnetic field can also be written as

$$\begin{aligned} \vec{B} &= -\text{curl}\ \vec{\mathcal{A}} \\ \text{with} \qquad \vec{\mathcal{A}} &= \frac{\mu_\circ}{4\pi} \int_V \frac{\vec{J}(\vec{r}\,')}{|\vec{r} - \vec{r}\,'|}\, dV' \end{aligned} \tag{A.26}$$

From $(A.26)_1$ the first magnetostatic law follows, namely that $\text{div}\,\vec{B} = 0$. If this result is compared with the first electrostatic law $(A.14)_1$, the statement $\text{div}\,\vec{B} = 0$ corresponds to the statement that there exist no free magnetic charges (or poles). Free magnetic charges have not been found in experiments.

Furthermore

$$\text{curl}\ \vec{B} = \text{curl curl}\ \vec{\mathcal{A}} = \text{grad}\,\text{div}\,\vec{\mathcal{A}} - \nabla^2\vec{\mathcal{A}}$$

No objection can be given to setting $\text{div}\,\vec{\mathcal{A}} = 0$, since $\vec{\mathcal{A}}$ is determined within an additive gradient of a scalar function, from which it follows that

$$\text{curl}\ \vec{B} = -\nabla^2\vec{\mathcal{A}} \tag{A.27}$$

The Laplacian of $(A.26)_2$ yields $\nabla^2 \vec{A} = -\mu_\circ \vec{J}$, so that curl $\vec{B} = \mu_\circ \vec{J}$ becomes the second magnetostatic law. The *two magnetostatic laws* are therefore

$$\boxed{\operatorname{div} \vec{B} = 0} \qquad \text{and} \qquad \boxed{\operatorname{curl} \vec{B} = \mu_\circ \vec{J}} \tag{A.28}$$

It can be conjectured from the work of Oersted and of Biot and Savart that the origin of the field $\vec{B}$ may also be electrical, and is caused by charge carriers in motion. This idea was also advocated by Ampère (1820), who shortly after the discovery of Oersted discovered that conductors exert forces on each other. For example, a torque is exerted by a straight current-carrying wire on a current loop that is mounted such that it can rotate freely. completely similarly to the torque exerted by a straight current conductor on a magnet. A north pole and a south pole can, as it were, be assigned to the current loop. Ampère discovered in this way the interaction between electric currents. Apart from this discovery, he made two conjectures, first, that a tiny current loop is equivalent to a magnetic dipole, analogous to an electric dipole, and second, that magnetism in matter is caused by microscopic current loops in the substances.

The electric and magnetic fields are related to one another. From the electrostatic and the magneto-static laws it follows that far away from the electric sources, it is not possible to distinguish between the fields $\vec{E}$ and $\vec{B}$, since in vacuum both $\vec{E}$ and $\vec{B}$ are curl-free and divergence-free fields.

A.5. THE CURRENT LOOP MODEL OF MAGNETISM

To prove the conjecture of Ampère, that magnetism can be described by current loops, the magnetic field is calculated for a closed current loop C by using (A.23), yielding

$$\vec{B}(\vec{r}) = \frac{\mu_\circ I}{4\pi} \oint_C \frac{d\vec{l}' \times (\vec{r} - \vec{r}')}{|\vec{r} - \vec{r}'|^3} \tag{A.29}$$

or

$$\begin{aligned} \vec{B} &= \operatorname{curl} \vec{A} \\ \text{with} \quad \vec{A} &= \frac{\mu_\circ I}{4\pi} \oint_C \frac{d\vec{l}'}{|\vec{r} - \vec{r}'|} \end{aligned} \tag{A.30}$$

Applying the Stokes theorem gives

$$\oint_C \vec{b} \cdot d\vec{l} = \int_A (\operatorname{curl} \vec{b}) \cdot d\vec{A}$$

where $\vec{b}$ is an arbitrary vector field. Assume $\vec{b} = \vec{\pi}$, where $\vec{\pi}$ is an arbitrary constant vector

$$\operatorname{curl} \vec{b} = -\vec{\pi} \times \operatorname{grad} a$$

so that

$$\vec{\pi} \cdot \oint_C a\, d\vec{l} = -\int_A (\vec{\pi} \times \text{grad}\, a) \cdot d\vec{A} = -\vec{\pi} \cdot \int_A (\text{grad}\, a) \times d\vec{A}$$

Therefore

$$\oint_C a\, d\vec{l} = -\int_A (\text{grad}\, a) \times d\vec{A} \tag{A.31}$$

Now with the help of (A.31), with $a = |\vec{r} - \vec{r}\,'|^{-1}$, $(\text{A.30})_2$ may be written as

$$\begin{aligned} \vec{\mathcal{A}} = \frac{\mu_\circ I}{4\pi} \oint_C \frac{d\vec{l}\,'}{|\vec{r} - \vec{r}\,'|} &= -\frac{\mu_\circ I}{4\pi} \int_A \left\{ \text{grad}' \frac{1}{|\vec{r} - \vec{r}\,'|} \right\} \times d\vec{A}\,' \\ = -\frac{\mu_\circ I}{4\pi} \int_A \frac{(\vec{r} - \vec{r}\,')}{|\vec{r} - \vec{r}\,'|^3} \times d\vec{A}\,' &= \frac{\mu_\circ I}{4\pi} \int_A \frac{\vec{n}\,' \times (\vec{r} - \vec{r}\,')}{|\vec{r} - \vec{r}\,'|^3} dA' \end{aligned}$$

Then (A.29) becomes

$$\begin{aligned} \vec{B}(\vec{r}) &= \frac{\mu_\circ I}{4\pi} \int_A \text{curl} \left\{ \frac{\vec{n}\,' \times (\vec{r} - \vec{r}\,')}{|\vec{r} - \vec{r}\,'|} \right\} d\vec{A}\,' \\ &= -\frac{\mu_\circ I}{4\pi} \int_A \text{curl} \left\{ \vec{n}\,' \times \text{grad} \frac{1}{|\vec{r} - \vec{r}\,'|} \right\} d\vec{A}\,' \\ &= -\frac{\mu_\circ I}{4\pi} \int_A \left\{ \vec{n}\,' \nabla^2 \frac{1}{|\vec{r} - \vec{r}\,'|} - (\vec{n}\,' \cdot \vec{\nabla}) \text{grad} \frac{1}{|\vec{r} - \vec{r}\,'|} \right\} d\vec{A}\,' \\ &= \frac{\mu_\circ I}{4\pi} \int_A (\vec{n}\,' \cdot \vec{\nabla}) \left\{ \text{grad} \frac{1}{|\vec{r} - \vec{r}\,'|} \right\} d\vec{A}\,' \\ &= \frac{\mu_\circ I}{4\pi} \int_A \text{grad} \left\{ (\vec{n}\,' \cdot \vec{\nabla}) \frac{1}{|\vec{r} - \vec{r}\,'|} \right\} d\vec{A}\,' \end{aligned}$$

so that $\vec{B}$ can be written as

$$\left. \begin{aligned} \vec{B} &= -\text{grad}\, \mathcal{V}_\text{m} \\ \text{with} \qquad \mathcal{V}_\text{m} &= \frac{\mu_\circ I}{4\pi} \int_A \frac{\vec{n}\,' \cdot (\vec{r} - \vec{r}\,')}{|\vec{r} - \vec{r}\,'|^3} dA' \end{aligned} \right\} \tag{A.32}$$

The expression for the magnetic potential $\mathcal{V}_\text{m}$ corresponds to the potential of an electric dipole. An electric dipole is defined as two equal dissimilar charges $+q$ and $-q$, at a distance $\delta\vec{l}$, such that

$$\lim_{\delta\vec{l} \to \vec{0}} q\, \delta\vec{l} = \vec{p} \tag{A.33}$$

is finite and unequal to zero. $\vec{p}$ is called the *electric dipole moment.* The electric potential of a dipole located at $\vec{r}$ is

$$\mathcal{V}(\vec{r}) = \frac{1}{4\pi\epsilon_\circ}\left(\frac{-q}{R_-} + \frac{q}{R_+}\right)$$

with $\vec{R}_+ = \vec{r} - \vec{r}_+$, and $\vec{R}_- = \vec{r} - \vec{r}_-$, so that it also follows that $\vec{R}_+ = \vec{R}_- - \delta\vec{l}$. Expand R_+^{-1} about R_-. In doing so the dipole has to be kept at position $\vec{r}$ and the differentiations must be performed at $\vec{r}$. This yields

$$\frac{1}{R_+} = \frac{1}{R_-} - (d\vec{l}\cdot\vec{\nabla})\frac{1}{R_-}$$

By means of this the electric potential becomes

$$\mathcal{V}(\vec{r}) = -\frac{q}{4\pi\epsilon_\circ}(\delta\vec{l}\cdot\vec{\nabla})\frac{1}{|\vec{r} - \vec{r}_-|}$$

or in the limit that $\delta\vec{l} \to \vec{0}$, in which also $\vec{r}_- \to \vec{r}\,'$

$$\mathcal{V}(\vec{r}) = -\frac{1}{4\pi\epsilon_\circ}(\vec{p}\cdot\vec{\nabla})\frac{1}{|\vec{r} - \vec{r}\,'|} = \frac{\vec{p}\cdot(\vec{r} - \vec{r}\,')}{4\pi\epsilon_\circ|\vec{r} - \vec{r}\,'|^3} \tag{A.34}$$

Comparison of $\mathcal{V}_{\mathrm{m}}$ in (A.32) with $\mathcal{V}$ in (A.34) confirms Ampère's conjecture that a current loop can be considered as a magnetic double layer, in which each surface element is equivalent to a 'magnetic dipole moment' $d\vec{m}$

$$d\vec{m} = I\,d\vec{A} \tag{A.35}$$

Furthermore it is concluded from these results that magnetism is of electric origin. This view is known as the current loop model* of magnetism. Far

* The atomistic theory confirms this model. In atoms the electrons orbit around the nucleus. Suppose that a particle charged with an electrical charge q orbits around the atomic nucleus. If τ is the period of this motion, then the path of the particle can be considered as a current loop with an electric current q/τ. If μ is the atomic magnetic moment then, according to (A.35)

$$\mu = \frac{q}{\tau}A$$

if A is the surface enclosed by the particle path. The angular momentum $\vec{\sigma}$ of the particle in orbiting motion is

$$\vec{\sigma} = m\vec{r}\times\dot{\vec{r}}$$

if m is the mass of the orbiting particle. Now since $d\vec{A} = \frac{1}{2}\vec{r}\times d\vec{r}$, it can be concluded that

$$\vec{\sigma} = 2m\frac{d\vec{A}}{dt} \approx 2m\frac{\vec{A}}{\tau}$$

away from the source, the potential of a current loop behaves equally as the potential of an electric dipole with dipole $\epsilon_\circ \mu_\circ I\, d\vec{A}$. Although no objection can be raised against the model of a magnetic dipole—the field is measured only outside of the matter and macroscopically it cannot be decided whether the field produced by magnetic material is caused by magnetic dipoles or by current loops with electrons—it is conceptually far better not to introduce an unnecessary physical concept of magnetic dipoles, even more so because a free magnetic charge is not experimentally found. The introduction of the magnetic dipole concept implies two different sets of laws for magnetic fields, one due to poles and one due to currents, while in the current loop model only one concept is needed, that of an electric charged particle (an electron) that produces two types of fields depending on whether it is motion or not. However, the mathematical analogy of the current loop model with the magnetic dipole model may be used in macroscopic theories to obtain expressions for the magnetic field from the results found for electric dipoles.

The potential energy associated with an electric field can be calculated as follows: suppose that a point charge q is shifted over a distance $\delta\vec{r}$. The field

---

Obviously

$$\frac{\mu}{\sigma} = \frac{q}{2m}$$

If the magnetization is caused by the orbiting electrons, then a relation exists between the magnetization and the angular momentum of the electrons the (*magneto-mechanical parallel*). Phenomenologically the parallel is confirmed by the *Richardson-Einstein-de Haas effect.* This effect is as follows: the atomic magnetization is macroscopically observable only if the atomic magnetizations are somewhat aligned. Above a certain temperature (Curie temperature) the thermal motions of the atomic particles are so intense that the ordering of the atomic magnetizations is broken up, resulting in demagnetization. Now mount a magnetized body so that this body can rotate freely around an axis parallel to the magnetization vector. Heat the body above the Curie temperature. In the demagnetized state the body rotates. This can be explained by the magneto-mechanical parallel and the conservation of angular momentum. The inverse effect (magnetization by spinning the body) has been shown by S.J. Barnett. The ratio μ/σ is called the *gyromagnetic ratio.*

An electron carries the smallest quantity of charge, so that it can be expected that an electron represents the smallest quantity of magnetization if this electron is orbiting around the atomic nucleus. According to quantum mechanics the angular momentum of the electron orbit around a nucleus is a multiple of $\hbar$, where $\hbar$ is the Planck constant. The smallest quantity of magnetization would therefore be

$$\mu_B = \frac{e\hbar}{2m_e}$$

with e the charge of the electron, m_e the mass of the electron and μ_B the *Bohr magneton.* Uhlenbeck later discovered that electrons are also spinning with an angular momentum of $\frac{1}{2}\hbar$. The electron is therefore the carrier of the smallest quantity of electric charge and the quantity of magnetization, namely $\frac{1}{2}\mu_B$. Although the picture must not be taken too literally, it shows that magnetization is coupled with the motion of electric charges and emphasizes the great importance of the conjecture of Ampère.

accompanied with q then does the work

$$\vec{K} \cdot \delta \vec{r} = q\vec{E} \cdot \delta \vec{r} = -q(\text{grad}\, \mathcal{V} \cdot \delta \vec{r}) = -q\, \delta \mathcal{V}$$

Thus the potential energy of the electric field decreases by

$$\delta U = -\vec{K} \cdot \delta \vec{r} = q\, \delta \mathcal{V} \qquad \Rightarrow \qquad U = q\mathcal{V}$$

For a dipole this becomes

$$U = -q\mathcal{V}(\vec{r}_-) + q\mathcal{V}(\vec{r}_+)$$

or

$$U = q(d\vec{l} \cdot \vec{\nabla})\mathcal{V}$$

or

$$U = (\vec{p} \cdot \vec{\nabla})\mathcal{V} = -\vec{p} \cdot \vec{E} \tag{A.36}$$

By analogy with the potential energy of an electric dipole (A.36), the potential energy corresponding to the $\vec{B}$ field of a 'magnetic dipole moment', which represents the field produced by a current loop, is given by

$$U_{\mathrm{m}} = -\vec{m} \cdot \vec{B} \tag{A.37}$$

where U_{m} is the potential energy of the magnetic field. With this result the forces between the electric currents can be calculated. Consider a narrow current loop, consisting of two long thin parallel wire conductors at a distance $d\vec{l}$ from each other. The current loop is closed by two short wire conductors of the length dl. The magnetic dipole moment produced by the current in the loop is

$$\vec{m} = \int_A I\, d\vec{A} = \int_A I\vec{n}\, dA \tag{A.38}$$

with A the surface confined by the loop, and $\vec{n}$ the unit normal to the surface. The potential energy of the loop is, according to (A.37)

$$U_{\mathrm{m}} = -\int_A I(\vec{n} \cdot \vec{B})\, dA \tag{A.39}$$

Shift the conductor $d\vec{l}$ over the distance $\delta\vec{s}$. In this process the potential energy changes by an amount

$$\delta U_{\mathrm{m}} = -I\vec{B} \cdot \delta\vec{A} = -I\vec{B} \cdot (\delta\vec{s} \times d\vec{l}) = -I(d\vec{l} \times \vec{B}) \cdot \delta\vec{s}$$

For the infinitesimal displacement $\delta\vec{s}$ of the current conductor $d\vec{l}$ in the constant magnetic field $\vec{B}$, the magnetic field exerts a force $d\vec{K}$ on $d\vec{l}$, by which the potential energy changes by an amount $-\vec{K}\cdot d\vec{s}$, so that

$$\delta U_{\mathrm{m}} = -I(d\vec{l}\times\vec{B}\,)\cdot\delta\vec{s} = -d\vec{K}\cdot d\vec{s}$$

or

$$d\vec{K} = I d\vec{l}\times\vec{B} \tag{A.40}$$

With the substitution of (A.16), (A.40) can be interpreted as follows: if a particle with charge q moves with a velocity $\vec{v}$ in a magnetic field $\vec{B}$, then on this particle a force

$$\boxed{\vec{K} = q\,\vec{v}\times\vec{B}} \tag{A.41}$$

is exerted. This is the *magnetic Lorentz force.* From the study of the behavior of charged particles in magnetic fields, the Lorentz force (A.41) can also be found from experiments, but here it is derived using the change of the potential energy in the field far from the current loop source due to an infinitesimal change in the current loop.

The interaction between two elements $d\vec{l}_1$ and $d\vec{l}_2$ of current conductors can now easily be derived. According to (A.40) the element $d\vec{l}_2$ exerts on the element $d\vec{l}_1$ the following force

$$d^2\vec{K}_{12} = I_1 d\vec{l}_1\times d\vec{B}_{12}$$

where, according to (A.7)

$$d\vec{B}_{12} = \frac{\mu_\circ}{4\pi}I_2\frac{d\vec{l}_2\times(\vec{r}_1-\vec{r}_2)}{|\vec{r}_1-\vec{r}_2|^3}$$

so that

$$d^2\vec{K}_{12} = \frac{\mu_\circ}{4\pi}I_1 I_2\frac{d\vec{l}_1\times[d\vec{l}_2\times(\vec{r}_1-\vec{r}_2)]}{|\vec{r}_1-\vec{r}_2|^3} \tag{A.42}$$

This result is sometimes attributed to Ampère, who in any case predicted that d^2K_{12} is proportional to I_1I_2.

Magnetization can now be understood by means of the current loop model, in terms of the state of motion of the charge carriers. Gauss studied the oscillations of a horizontally mounted rotatable needle magnet that was located near a fixed placed magnet. If the needle magnet is touched, then the oscillations turn out to be damped out; from this it can be deduced that the fixed magnet apparently produces a force field $\vec{B}$ that exerts on the needle magnet a restoring torque. This restoring torque proves to be proportional to the magnetization or the magnetic moment $\vec{m}$ of the needle magnet and the strength of the field $\vec{B}$ produced by the fixed magnet.

The angular torque $\vec{\mathcal{M}}$ on a current loop C is found, by using (A.40), to be

$$\vec{\mathcal{M}} = I \oint_C \vec{r} \times (d\vec{l} \times \vec{B})$$

or

$$\vec{\mathcal{M}} = I \oint_C [(\vec{r} \cdot \vec{B}) d\vec{l} - (\vec{r} \cdot d\vec{l})\vec{B}] \tag{A.43}$$

For a sufficiently small current loop the magnetic field $\vec{B}$ can be assumed to be constant, so that with

$$\oint_C \vec{r} \cdot d\vec{l} = \int_A (\text{curl } \vec{r}) \cdot d\vec{A} = 0$$

(A.43) can be written as

$$\vec{\mathcal{M}} = I \oint_C (\vec{r} \cdot \vec{B})\, d\vec{l}$$

Application of (A.31) with $a = \vec{r} \cdot \vec{B}$ yields

$$\vec{\mathcal{M}} = -I \int_A [\text{grad}\,(\vec{r} \cdot \vec{B})] \times d\vec{A} = -\int_A I\vec{B} \times d\vec{A} = \left(\int_A I\, d\vec{A} \right) \times \vec{B}$$

or with (A.35)

$$\vec{\mathcal{M}} = \vec{m} \times \vec{B} \tag{A.44}$$

This corresponds to the results obtained with the experimental set up of Gauss.

With the discovery of Ampère that there exists an interaction between electric currents, the unit of an electric current can be fixed by measuring the force between two conductors. Consider therefore two straight infinitely long parallel conductors. According to the Biot–Savart law, the first conductor induces a magnetic field given by

$$\vec{B} = \frac{\mu_o}{2\pi} \frac{I_1}{R} \vec{e}_\theta$$

with R the distance between the two conductors. This magnetic field produces on an element $d\vec{l}_2$ of the second conductor a force, which is given by

$$d\vec{K}_{21} = I_2 d\vec{l}_2 \times \vec{B} = \frac{\mu_o}{2\pi} \frac{I_1 I_2}{R} d\vec{l}_2 \times \vec{e}_\theta$$

Per unit of length this force is accordingly

$$K_{21} = \frac{\mu_o}{2\pi}\frac{I_1 I_2}{R} = K_{12} \tag{A.45}$$

If the currents I_1 and I_2 run in the same direction, then both conductors attract each other, and if the currents flow in opposite directions, then the conductors repel each other. The unit of current, the *ampere* is based on (A.45):

'The ampere is that constant current which, if maintained in two straight parallel conductors of infinite length and of negligible circular cross-section, placed 1 metre apart in vacuum, would produce between these conductors a force equal to 2×10^{-7} Newton per metre of length.'

The value of the *permeability in vacuum* follows from this definition

$$\boxed{\mu_o = 4\pi \times 10^{-7}\,\mathrm{N\ A^{-2}}} \tag{A.46}$$

A.6. ELECTROMAGNETIC EQUATIONS IN VACUUM

Consider a deformable loop moving in a stationary inhomogeneous magnetic field with a velocity $\vec{v}$. On each charge carrier of an element $d\vec{l}$ of the loop a magnetic Lorentz force $q\vec{v} \times \vec{B}$ acts. By these forces a current is produced in the closed-loop conductor. This force per unit of charge can be considered as an electric field $\vec{E}'_{\mathrm{ind}}$, effective over the conductor $d\vec{l}$. This field is called the *induced electric field strength of the first kind*, and causes a potential difference $-\vec{E}'_{\mathrm{ind}} \cdot d\vec{l}$. For the complete loop C_c, moving in the field $\vec{B}$ with velocity $\vec{v}$

$$\oint_{C_c} \vec{E}'_{\mathrm{ind}} \cdot d\vec{l} = \oint_{C_c} (\vec{v} \times \vec{B}) \cdot d\vec{l} \tag{A.47}$$

Application of the Stokes theorem gives

$$\begin{aligned}\oint_{C_c} (\vec{v} \times \vec{B}) \cdot d\vec{l} &= \int_{A_c} \mathrm{curl}\ (\vec{v} \times \vec{B}) \cdot d\vec{A} \\ &= \int_{A_c} [(\vec{\nabla} \cdot \vec{B})\vec{v} + (\vec{B} \cdot \vec{\nabla})\vec{v} - (\vec{\nabla} \cdot \vec{v})\vec{B} - (\vec{u} \cdot \vec{\nabla})\vec{B}] \cdot d\vec{A}\end{aligned} \tag{A.48}$$

The surface integral can be rewritten using*

$$\frac{\mathrm{D}}{\mathrm{D}t}\vec{A} = [(\vec{\nabla} \cdot \vec{v})\mathbf{1} - (\vec{\nabla}\vec{v})] \cdot d\vec{A} \tag{A.49}$$

---

* Suppose that $\vec{A}_c(t)$ is the surface enclosed by the deformable current loop, that $\vec{A}_c(t+\Delta t)$ is the deformed surface of the current loop after the displacement $\vec{v}\Delta t$, and that $\Delta\vec{A}_c$ is the surface described by the loop when it moves in the field $\vec{B}$, then it follows from the Gauss

and using the fact that for a stationary magnetic field

$$\frac{\mathrm{D}}{\mathrm{D}t}\vec{B} = (\vec{v} \cdot \vec{\nabla})\vec{B} \qquad \text{for} \qquad \frac{\partial \vec{B}}{\partial t} = \vec{0} \tag{A.50}$$

Since $\operatorname{div} \vec{B} = 0$ the surface integral in (A.48) reduces to

$$\oint_{C_c} (\vec{v} \times \vec{B}) \cdot d\vec{l} = -\int_{A_c} \left\{ \vec{B} \cdot [(\vec{\nabla} \cdot \vec{v})\mathbf{1} - \vec{\nabla}\vec{v})] + (\vec{v} \cdot \vec{\nabla})\vec{B} \right\} \cdot d\vec{A}$$

The substitution of (A.49) and (A.50) yields

$$\oint_{C_c} (\vec{v} \times \vec{B}) \cdot d\vec{l} = -\int_{A_c} \left[\vec{B} \cdot (\frac{\mathrm{D}}{\mathrm{D}t} d\vec{A}) + (\frac{\mathrm{D}}{\mathrm{D}t}\vec{B}) \cdot d\vec{A} \right] = -\frac{\mathrm{D}}{\mathrm{D}t} \int_{A_c} \vec{B} \cdot d\vec{A}$$

so that, for stationary magnetic fields, (A.47) can be written as

$$\oint_{C_c} \vec{E}'_{\text{ind}} \cdot d\vec{l} = -\frac{\mathrm{D}}{\mathrm{D}t} \int_{A_c} \vec{B} \cdot d\vec{A} \qquad \text{for} \qquad \frac{\partial \vec{B}}{\partial t} = \vec{0} \tag{A.51}$$

This is a general relation for induction of the first kind, derived from the magnetic Lorentz force already defined. Until now only stationary fields are considered and induction by time-dependent fields have to be considered. The

---

theorem applied to the three surfaces that $\Delta\vec{A}_c + \Delta\vec{A}_c(t + \Delta t) - \Delta\vec{A}_c(t) = \vec{0}$, so that

$$\frac{\mathrm{D}}{\mathrm{D}t}\vec{A}_c = \lim_{\Delta t \to 0} \frac{1}{\Delta t} \left[\vec{A}_c(t + \Delta t) - \vec{A}_c(t) \right] = \lim_{\Delta t \to 0} \frac{\Delta \vec{A}_c}{\Delta t} = \oint_{C_c} \vec{v} \times d\vec{l}$$

Application of the Stokes theorem: $\oint \vec{b} \cdot d\vec{l} = \int \operatorname{curl} \vec{b} \cdot d\vec{A}$, with $\vec{b} = \vec{\pi} \times \vec{v}$, in which $\vec{\pi}$ is a constant arbitrary vector field, yields

$$\begin{aligned} \vec{\pi} \cdot \oint \vec{v} \times d\vec{l} &= \int \operatorname{curl} (\vec{\pi} \times \vec{v}) \cdot d\vec{A} = \int [\vec{\pi}(\vec{\nabla} \cdot \vec{v}) - (\vec{\pi} \cdot \vec{\nabla})\vec{v}] \cdot d\vec{A} \\ &= \vec{\pi} \cdot \int [(\vec{\nabla} \cdot \vec{v})\mathbf{1} - (\vec{\nabla}\vec{v})] \cdot d\vec{A} \end{aligned}$$

so that

$$\frac{\mathrm{D}}{\mathrm{D}t}\vec{A}_c = \oint_{C_c} \vec{v} \times d\vec{l} = \int_{A_c} [(\vec{\nabla} \cdot \vec{v})\mathbf{1} - (\vec{\nabla}\vec{v})] \cdot d\vec{A}$$

For small $\vec{A}_c$ this becomes

$$\frac{\mathrm{D}}{\mathrm{D}t}\vec{A}_c = [(\vec{\nabla} \cdot \vec{v})\mathbf{1} - (\vec{\nabla}\vec{v})] \cdot d\vec{A}_c$$

induction which is produced in the reverse situation—the loop conductor is fixed and does not change its shape, while the magnetic field is time-dependent—is called *induction of the second kind.* Faraday (1831) investigated this kind of induction experimentally, and his results can be formulated by

$$\oint \vec{E}''_{\text{ind}} \cdot d\vec{l} = -\frac{\partial}{\partial t}\int_A \vec{B} \cdot d\vec{A} = -\int_A \frac{\partial \vec{B}}{\partial t} \cdot d\vec{A} \tag{A.52}$$

in which the surface A enclosed by the loop conductor is now fixed with respect to the observer. Mathematically (A.52) cannot be derived directly from (A.51). An experiment with time-dependent fields is necessary for the further foundation of the electromagnetic field equations, although a relation between the induction of the first kind and the induction of the second kind can be conjectured.

The conjecture can be illustrated by assuming that the magnetic field is produced by a current coil with a current I. A current loop is moved in this constant magnetic field (induced by a constant current I in the coil that is kept in a fixed position). Now induction of the first kind is produced. On the other hand, if the current loop is kept at a fixed position, the coil with the constant current I can be moved. In this situation induction of the second kind is produced. However, it seems obvious to conjecture that for induction only the relative motion of the current loop with respect to the coil is important and vice versa. From this point of view, induction of the first kind implies induction of the second kind. Finally, the current loop and the coil can be fixed in space with respect to each other and the current I in the coil can be varied, so that $\vec{B}$ becomes time-dependent. Induction of the second kind is also then produced. From the point of view of field theory it is conjectured that for induction only the variation of the magnetic flux enclosed by the current loop is important. As a consequence, but not as a matter of course, the Faraday law of induction (A.52) is plausible.

Is a current loop moved in a stationary magnetic field, the total induction is

$$\vec{E}_{\text{ind}} = \vec{E}'_{\text{ind}} + \vec{E}''_{\text{ind}} \tag{A.53}$$

Substitution of (A.51) and (A.52) gives

$$\oint_{C_c} \vec{E}_{\text{ind}} \cdot d\vec{l} = -\frac{\mathrm{D}}{\mathrm{D}t}\int_{A_c} \vec{B} \cdot d\vec{A} \tag{A.54}$$

in which it is now allowed that $\partial\vec{B}/\partial t \neq \vec{0}$.

Furthermore, it can be supposed that in a current loop the total field strength consists of a time-dependent part $\vec{E}_{\text{ind}}$ and a quasi-stationary part $\vec{E}_{\text{s}}$

$$\boxed{\vec{E} = \vec{E}_{\text{ind}} + \vec{E}_{\text{s}}} \qquad \text{with} \qquad \oint_C \vec{E}_{\text{s}} \cdot d\vec{l} = 0 \tag{A.55}$$

being the second Kirchhoff law for the quasi-stationary field that follows from curl $\vec{E}_\mathrm{s} = \vec{0}$. With the substitution of $(\mathrm{A.55})_1$ into (A.54) and using $(\mathrm{A.55})_2$, this equation can be written as

$$\oint_{C_\mathrm{c}} \vec{E} \cdot d\vec{l} = -\frac{\mathrm{D}}{\mathrm{D}t} \int_{A_\mathrm{c}} \vec{B} \cdot d\vec{A} \tag{A.56}$$

By letting the arbitrary fixed loop C_c coincide with the loop C in space the Faraday law is found

$$\oint_{C} \vec{E} \cdot d\vec{l} = -\frac{\partial}{\partial t} \int_{A} \vec{B} \cdot d\vec{A} \tag{A.57}$$

Maxwell assumed that (A.57) applies for each stationary current loop in a magnetic field, so that $\vec{E}$ and $\vec{B}$ are supposed to be continuously differentiable and (A.57) can be written in differential form

$$\boxed{\mathrm{curl}\ \vec{E} = -\frac{\partial \vec{B}}{\partial t}} \tag{A.58}$$

We have obtained the *second Maxwell electromagnetic law.*

The first Kirchhoff law (A.21) applies for a continuous distribution of time-independent electric currents. For time-dependent currents this law of conservation of electric charge also has to be modified. For time-dependent situations the law of the conservation of charges becomes, for a continuous distribution of charge carriers

$$\mathrm{div}\ \vec{J} = -\frac{\partial \rho_\mathrm{f}^{(\mathrm{e})}}{\partial t} \tag{A.59}$$

From the magnetostatic law $(\mathrm{A.28})_2$ it follows that the curl of $\vec{B}$ is determined by the current $\vec{J}$ that satisfies div $\vec{J} = 0$. Consistency of the equations demands that for the time-dependent case, curl $\vec{B} = \mu_\mathrm{o}\vec{J}$ also has to be modified. By taking the partial time derivative of $(\mathrm{A.14})_1$, and by application of (A.59), it follows that

$$\frac{\partial \rho_\mathrm{f}^{(\mathrm{e})}}{\partial t} = \frac{\partial}{\partial t}\mathrm{div}\, \epsilon_\mathrm{o}\vec{E} = \mathrm{div}\, \epsilon_\mathrm{o}\frac{\partial \vec{E}}{\partial t} = -\mathrm{div}\ \vec{J}$$

or

$$\mathrm{div}\,(\vec{J} + \epsilon_\mathrm{o}\frac{\partial \vec{E}}{\partial t}) = 0$$

From this is concluded that in a time-dependent electric field a displacement current

$$\vec{J}_\mathrm{d} = \epsilon_\mathrm{o}\frac{\partial \vec{E}}{\partial t} \tag{A.60}$$

is present. According to Maxwell, the Ampère law for time-dependent fields has to be written as

$$\boxed{\text{curl } \vec{B} = \mu_\circ \left(\vec{J} + \epsilon_\circ \frac{\partial \vec{E}}{\partial t} \right)} \tag{A.61}$$

which gives the *first Maxwell electromagnetic law.* A current can be used to charge, in an open circuit, a condenser producing a magnetic field. The line integral of $\vec{B}$ around a loop is determined by the current through the loop if the arbitrary surface bounded by the loop is crossed by the wire of the circuit or by the displacement current—for a time-dependent electric flux—if the arbitrary surface is thought to pass between the condenser plates. The modification of the Ampère law is needed to deal with these type of situations.

A.7. THE MAXWELL ELECTROMAGNETIC FIELD EQUATIONS IN FREE SPACE

The free space is a space in which only charge carriers are in motion, but in which no polarization effects occur. For free space it has previously been derived that

$$\left.\begin{array}{ll} \epsilon_\circ \text{div } \vec{E} = \rho_{\text{f}}^{(\text{e})} & \text{div } \vec{B} = 0 \\ \text{curl } \vec{E} = -\dfrac{\partial \vec{B}}{\partial t} & \text{curl } \vec{B} = \mu_\circ \left(\vec{J} + \epsilon_\circ \dfrac{\partial \vec{E}}{\partial t} \right) \end{array}\right\} \tag{A.62}$$

In vacuum $\rho_{\text{f}}^{(\text{e})} = 0$ and $\vec{J} = \vec{0}$. For stationary fields in vacuum both the electric field and the magnetic field are curl-free and divergence-free, and there is no distinction between the fields $\vec{E}$ and $\vec{B}$. This is not so for time-dependent electromagnetic fields. In vacuum (A.62) reduces to

$$\left.\begin{array}{ll} \epsilon_\circ \text{div } \vec{E} = 0 & \text{div } \vec{B} = 0 \\ \text{curl } \vec{E} = -\dfrac{\partial \vec{B}}{\partial t} & \text{curl } \vec{B} = \mu_\circ \epsilon_\circ \dfrac{\partial \vec{E}}{\partial t} \end{array}\right\} \tag{A.63}$$

Combining these equations gives

$$\text{curl curl } \vec{E} = \text{grad div } \vec{E} - \nabla \vec{E} = -\frac{\partial}{\partial t} \text{curl } \vec{B} = -\mu_\circ \epsilon_\circ \frac{\partial^2 \vec{E}}{\partial^2 t}$$

or

$$\left.\begin{array}{rl} & \nabla \vec{E} = \mu_\circ \epsilon_\circ \dfrac{\partial^2 \vec{E}}{\partial^2 t} \\ \text{and analogously:} & \nabla \vec{B} = \mu_\circ \epsilon_\circ \dfrac{\partial^2 \vec{B}}{\partial^2 t} \end{array}\right\} \tag{A.64}$$

The $\vec{E}$ and $\vec{B}$ fields therefore both satisfy the Helmholtz wave equation, and as a consequence the electromagnetic fields in vacuum are related to a wave phenomenon, which can be identified under certain conditions (frequency range) with the phenomenon of light. The speed of light in vacuum is therefore

$$\boxed{c_\circ = \frac{1}{\sqrt{\epsilon_\circ \mu_\circ}}} \tag{A.65}$$

Since the permeability in vacuum $\mu_\circ = 4\pi \times 10^{-7}$ SI units, it follows from the measurement of the speed of light, which is currently defined exactly with the definition of the meter, the value of the permittivity in vacuum $\epsilon_\circ$, by which the charge of the electron is also given in SI units. The speed of light in vacuum is defined by

$$\boxed{c_\circ = 2.99792458 \times 10^8 \text{ m/s}} \tag{A.66}$$

The wave equations (A.64) can be solved for plane waves

$$\vec{E} = \Re\{\hat{\vec{E}} \exp[i(\omega t - \vec{k} \cdot \vec{r})]\} \qquad \text{and} \qquad \vec{B} = \Re\{\hat{\vec{B}} \exp[i(\omega t - \vec{k} \cdot \vec{r})]\} \tag{A.67}$$

with

$$k^2 = \frac{\omega^2}{c_\circ^2} \tag{A.68}$$

in which ω is the frequency of the electromagnetic wave and $\vec{k}$ the angular wave vector, which determines the direction of the propagation. Substitution of (A.67) yields, for the divergence equations in (A.63)

$$\vec{k} \cdot \vec{E} = 0 \qquad \text{and} \qquad \vec{k} \cdot \vec{B} = 0 \tag{A.69}$$

so that the electric vector field $\vec{E}$ and the magnetic vector field $\vec{B}$ are perpendicular to the direction of the propagation. These types of waves are called transverse waves. The substitution of (A.67) into the curl equations in (A.63) gives

$$\vec{k} \times \vec{E} = \omega \vec{B} \qquad \text{and} \qquad \vec{k} \times \vec{B} = -\frac{\omega}{c_\circ^2} \vec{E} \tag{A.70}$$

The vectors $\vec{E}$ and $\vec{B}$ are therefore perpendicular to each other, so that $\vec{k}$, $\vec{E}$, and $\vec{B}$ form a right-handed orthogonal set of three of vectors. Finally from (A.70) is found that the ratio of E and B is given by

$$B = \frac{k}{\omega} E = \frac{E}{c_\circ} = \sqrt{\epsilon_\circ \mu_\circ} E \tag{A.71}$$

A.8. POLARIZABLE MATERIALS

A.8.1. Electric polarization. For the study of the electromagnetic fields in matter it is a disadvantage that in matter the field strengths are not measurable directly. Only in free space, outside the block of matter to be studied, can the measurements be performed.

The existence of electric polarization in an insulator can be made plausible phenomenologically as follows. Consider a plate condenser, the plates of which are charged. The charges on the plates and the potential difference between the plates can be measured, but aside from that the plates are isolated. The potential difference between the plates is

$$\int_a^b \vec{E} \cdot d\vec{r} = -\int_a^b (\text{grad}\,\mathcal{V}) \cdot d\vec{r} = -\int_a^b d\mathcal{V} = \mathcal{V}_a - \mathcal{V}_b = \Delta\mathcal{V}$$

Measure first the charge on the plates and $\Delta\mathcal{V}$ in vacuum. Next, insert an insulator between the condenser plates. In measuring the potential difference $\Delta\mathcal{V}$ again, it turns out that it is decreased, while the charge on the plates remains the same.

This measurement can be explained by assuming that the matter of the insulator produces a counter potential difference that acts in opposition to the applied potential difference. The production of the counter potential difference by the insulator substance can be measured only if a cavity is made in the insulator. In a cylindrical cavity, both end surfaces of which are perpendicular to the electric field, a surface charge has to be produced at the end surfaces and such that the field and the surface charges oppose the applied electric field. If a is the cross-section of the small cylinder, and $\sigma^{(\mathrm{e})}$ is the surface charge per unit of area, then

$$\sigma^{(\mathrm{e})} a\vec{d} = \vec{P} ad = \vec{P}\, dV$$

is the *polarization* of the volume element dV. An insulator is polarized by an electric field, and therefore a polarization density $\vec{P}$ per unit volume can be introduced. Electric polarization is usually represented by a dipole. An electric dipole is illustrated by an electric charge of magnitude q and a small vector distance $\vec{d}$, pointing from the negative to the positive charge. If there are n electric dipoles per unit volume, then according to the *dipole model of the electric polarization* the polarization density is given by

$$\vec{P} = nq\vec{d} \tag{A.72}$$

In a dipole a positive charge is always associated with a negative charge. This might be called a bounded charge in opposition to the free charges that represent the positive or negative charge of the more or less freely moving charged particles. The charges associated with the dipoles yield, in a continuous distribution of the dipoles over the media, a contribution to the charge density. Suppose that $\rho_{\mathrm{f}}^{(\mathrm{e})}$ represents the density of the free charged particles. The total charge density in a volume V enclosed by a surface A is now

$$\int_V \rho_{\mathrm{f}}^{(\mathrm{e})}\, dV - \oint_A (\vec{n}\cdot\vec{P})\, dA = \int_V (\rho_{\mathrm{f}}^{(\mathrm{e})} - \mathrm{div}\,\vec{P})\, dV$$

Each time a dipole crosses the surface A in positive direction it contributes to the charges in volume V the charge $-q$. The total number of dipoles cut by the surface element dA is $n\vec{d}\cdot d\vec{A}$. The dipoles cutting dA contribute $-nq\vec{d}\cdot d\vec{A} = -\vec{P}\cdot d\vec{A}$ to the charge density in volume V. The 'true' electric charge density is obviously

$$\boxed{\rho^{(\mathrm{e})} = \rho_{\mathrm{f}}^{(\mathrm{e})} - \mathrm{div}\,\vec{P}} \tag{A.73}$$

The electric dipoles in a moving medium can also contribute to the electric current through a surface enclosed by a loop. Consider an open surface A, bounded by the closed contour C. The number of dipoles that cut the surface A at time t is

$$\int_A n\vec{d}\cdot d\vec{A}$$

The number of dipoles that cut the surface A per unit of time is

$$\frac{\partial}{\partial t}\int_A n\vec{d}\cdot d\vec{A}$$

However, not all these dipoles contribute to the electric current. The dipoles that do not transport charges through the arbitrary surface A slip over the surface into position and are cut by the contour C without crossing the surface A. These dipoles contribute to the net charge of dipoles cut, but not to the current through A. If $\vec{v}$ denotes the velocity of the dipoles, then $n\vec{v}\cdot(\vec{d}\times d\vec{l})$ yields the number of dipoles that do not cut the surface A and slip along the contour element $d\vec{l}$. The total number of these dipoles is

$$\oint_C n\vec{v}\cdot(\vec{d}\times d\vec{l}) = \oint_C n(\vec{v}\times\vec{d})\cdot d\vec{l}$$

The dipoles that pass through the surface A produce a current through A, given by

$$\int_A \frac{\partial}{\partial t}(nq\vec{d})\cdot d\vec{A} - \oint_C n(\vec{v}\times\vec{d})\cdot d\vec{l} = \int_A \left[\frac{\partial\vec{P}}{\partial t} - \mathrm{curl}\;(\vec{v}\times\vec{P})\right]\cdot d\vec{A}$$

The electric polarization $\vec{P}$ therefore adds to the current density a contribution

$$\vec{J}_{\mathrm{p}} = \frac{\partial \vec{P}}{\partial t} - \mathrm{curl}\ (\vec{v} \times \vec{P}) \tag{A.74}$$

The polarization current density contains besides $\partial \vec{P}/\partial t$ the term $-\mathrm{curl}\ (\vec{v} \times \vec{P})$. Since Röntgen discovered the existence of this current experimentally, the contribution $-\mathrm{curl}\ (\vec{v} \times \vec{P})$ is sometimes called the *Röntgen current.*

A.8.2. Magnetic polarization. Magnetic polarization occurs in magnetized media. The magnetization of a continuous body is represented according to the current loop model of Ampère by a continuous distribution of microscopic current loops. Suppose that n^* is the number of current loops per unit volume and that a is the tiny surface enclosed by a tiny current loop. Then the magnetization is

$$\vec{M}_{\mathrm{A}} = n^* I\, \vec{a} \tag{A.75}$$

Current loops do not contribute towards the total electric charge, but they contribute towards the total electric current. Consider a surface A bounded by a curve C. Only those current loops contribute towards the total current for which the surface a of the tiny current loop has a component perpendicular to the curve C and is crossed by the curve C. Then one half of the loop crosses the surface A. The number of loops over a distance $d\vec{l}$ along the curve C that contributes to the current through A is $n^* \vec{a} \cdot d\vec{l}$, so that the current through A becomes

$$\oint_C n^* I \vec{a} \cdot d\vec{l} = \oint_C \vec{M}_{\mathrm{A}} \cdot d\vec{l} = \int_A (\mathrm{curl}\ \vec{M}_{\mathrm{A}}) \cdot d\vec{A}$$

The magnetization therefore yields the following contribution to the current density

$$\vec{J}_{\mathrm{m}} = \mathrm{curl}\ \vec{M}_{\mathrm{A}} \tag{A.76}$$

The total current density becomes, with the contributions (A.74) and (A.76):

$$\boxed{\vec{J} = \vec{J}_{\mathrm{f}} + \frac{\partial \vec{P}}{\partial t} + \mathrm{curl}\ (\vec{M}_{\mathrm{A}} - \vec{v} \times \vec{P})} \tag{A.77}$$

A.9. ELECTRODYNAMIC EQUATIONS FOR MOVING MEDIA

For polarized and magnetized media the electromagnetic field equations (A.62) for free space are now written as follows

$$\left.\begin{array}{ll} \epsilon_{\circ}\mathrm{div}\, \vec{E} = \rho^{(\mathrm{e})} & \mathrm{div}\, \vec{B} = 0 \\[2ex] \mathrm{curl}\ \vec{E} = -\dfrac{\partial \vec{B}}{\partial t} & \mathrm{curl}\ \vec{B} = \mu_{\circ} \left(\vec{J} + \epsilon_{\circ} \dfrac{\partial \vec{E}}{\partial t} \right) \end{array} \right\} \tag{A.78}$$

in which $\rho^{(e)}$ and $\vec{J}$ are now considered to represent the total charge density (A.73) and the total electric current density (A.77). In (A.77) M_A represents the magnetization due to the microscopic current loops, so that $-\vec{v} \times \vec{P}$ can also be considered to represent magnetization. The magnetization by a dipole in motion can be understood in terms of the current loop model, since a dipole moving with the velocity $\vec{v}$ produces a 'loop' with two long sides. It can be imagined that the currents from the short sides cancel each other. The current in the loop is $I = q/\Delta t$. The surface of the loop is $\vec{a} = -(\vec{v} \times \vec{d})\, \Delta t$, so that the equivalent magnetization is

$$\vec{M}_{eq} = -nq(\vec{v} \times \vec{d}) = -\vec{v} \times \vec{P}$$

This magnetization is thus added to the magnetization caused by the microscopic current loops. The total magnetization is therefore

$$\boxed{\vec{M} = \vec{M}_A - \vec{v} \times \vec{P}} \qquad \text{(A.79)}$$

Analogously, magnetization can be accompanied by the current loops causing the equivalent electric polarization $\vec{P}_{eq}$ that amounts to $(\vec{M} \times \vec{v})/c_o^2$, and can be neglected in a nonrelativistic approximation.

The Ampère formulation of the electromagnetic field equations is obtained with the substitution of (A.73) and (A.77) into (A.78). Since inside the media the electromagnetic field is not measurable, the equivalent polarization $\vec{P}_{eq}$ may be also substituted directly into the polarization $\vec{P}$, and the equivalent magnetization $\vec{M}_{eq}$ directly in the magnetization $\vec{M}$. This point of view is called the Boffi formulation of the electromagnetic field equations. In the Boffi formulation, the field equations for the two fields $\vec{E}$ and $\vec{B}$ have the most simple form and (A.78) become with (A.79)

$$\boxed{\begin{array}{ll} \epsilon_o \operatorname{div} \vec{E} = \rho_f^{(e)} - \operatorname{div} \vec{P} & \operatorname{div} \vec{B} = 0 \\ \operatorname{curl} \vec{E} = -\dfrac{\partial \vec{B}}{\partial t} & \operatorname{curl} \vec{B} = \mu_o \left(\vec{J}_f + \dfrac{\partial \vec{P}}{\partial t} + \operatorname{curl} \vec{M} + \epsilon_o \dfrac{\partial \vec{E}}{\partial t} \right) \end{array}} \qquad \text{(A.80)}$$

The Minkowski formulation of the field equations is the best-known, in which the *electric displacement* $\vec{D}$ is introduced

$$\vec{D} = \epsilon_o \vec{E} + \vec{P} \qquad \text{(A.81)}$$

and the *magnetic field strength* $\vec{H}$

$$\vec{H} = \frac{1}{\mu_o} \vec{B} - \vec{M} \qquad \text{(A.82)}$$

Substitution of these fields into (A.78) yields the Minkowski formulation of the electromagnetic field equations

$$\left.\begin{aligned} \epsilon_\circ \operatorname{div} \vec{D} &= \rho_{\mathrm{f}}^{(\mathrm{e})} & \operatorname{div} \vec{B} &= 0 \\ \operatorname{curl} \vec{E} &= -\frac{\partial \vec{B}}{\partial t} & \operatorname{curl} \vec{H} &= \vec{J}_{\mathrm{f}} + \frac{\partial \vec{D}}{\partial t} \end{aligned}\right\} \tag{A.83}$$

The Boffi formulation of the electromagnetic field equations consists of a scalar and a vector equation for the two electromagnetic fields—the electric field $\vec{E}$ and the magnetic induction field $\vec{B}$. The equations have to be supplemented with a vector constitutive equation for $\vec{P}$ and for $\vec{M}$. In contrast, the Minkowski formulation of the electromagnetic field equations consists of four equations for the four fields $\vec{E}$, $\vec{D}$, $\vec{H}$, $\vec{B}$, which have to be supplemented by constitutive equations.

No controversies exist over the formulations of the electromagnetic field equations in vacuum, in which unpolarized charges are moving. There are two field strengths that can be measured in principle with test particles of known mass and electric charge. At the most there may be a difference in the choice for the magnetic field, $\vec{B}$ or $\vec{H}$. In free space, however, $\vec{B} = \mu_\circ \vec{H}$ applies, so that the difference is unimportant, except for the introduction of a new concept of electric field strength $\vec{H}$.

The situation differs for fields inside a material, but inside the materials the electromagnetic fields cannot be measured. Those fields have to be derived from the measurable fields in the free space on the outside of the material. From (A.83) it follows that at the dividing surface between two media the normal component of $\vec{B}$ is continuous, while the normal component of $\vec{H}$ is discontinuous and the tangential component of $\vec{H}$ is continuous. From (A.80) it is noticed that the tangential component of $\vec{B}$ is discontinuous and the normal component of $\vec{B}$ continuous.

If $\vec{H}$ is chosen as the magnetic field, then the magnetization $\vec{M}$ has to be represented by the magnetic dipole model. A continuous distribution of hypothetical magnetic dipoles produces a magnetic surface charge on the dividing surface, at which the normal component of $\vec{H}$ becomes discontinuous, while no surface currents are produced, so that the tangential component of $\vec{H}$ is continuous.

If $\vec{B}$ is chosen as the magnetic field, then the magnetization $\vec{M}$ has to be modeled by the more realistic Ampère current loop model. For a continuous distribution of microscopic current loops a surface current is produced on the dividing surface, at which the tangential component of $\vec{B}$ becomes discontinuous, while no hypothetical magnetic surface charge results, so that the normal component of $\vec{B}$ is continuous.

Penfield and Haus* have shown that, besides some physical interpretations, the various formulations of the electromagnetic field equations for moving media are compatible with each other, since all differences between the various formulations are unmeasurable.

* Penfield P. Jr and Haus H.A. 1967. *Electrodynamics of Moving Media*, Research monograph No. **40**. M.I.T. Press, Cambridge, Massachusetts. See Chapter 7.

APPENDIX B
Summary of vector and tensor notation

Summary: The notation and definitions of vectors and tensors in three-dimensional Euclidean space are summarized. A Cartesian coordinate system is used.

B.1. EUCLIDEAN SPACE AND CARTESIAN COORDINATES

In Chapter 1 it is pointed out that the physical space used in the formulation of the nonrelativistic macroscopic theories may be assumed to be Euclidean. In Euclidean space the theorem of Pythagoras applies, and a Cartesian coordinate system can be used. To establish this coordinate system a fixed point O (fixed with respect to the reference system) has to be chosen as the origin of the Cartesian coordinate system. From this point three mutually perpendicular axes are drawn. The directions of these axes are given by vectors $\vec{e}_i$ $(i = 1, 2, 3)$ of magnitude unity (a vector is defined as a quantity of a given magnitude and direction). These three unit vectors satisfy

$$\vec{e}_i \cdot \vec{e}_j = \delta_{ij} \tag{B.1}$$

where δ_{ij} the Kronecker delta ($\delta_{ij} = 1$ for $i = j$, and $\delta_{ij} = 0$ for $i \neq j$), and the dot between the two unit vectors denotes the scalar product. The scalar product of two vectors gives the projection of $\vec{e}_i$ on $\vec{e}_j$ or vice versa. From (B.1) it follows that the unit vectors $\{\vec{e}_i\}$ are dimensionless. These unit vectors are the basis vectors in the three-dimensional Cartesian space being spanned by these vectors.

Consider two points very close to one another; a vector $d\vec{x}$ be defined by the directed line segment between the two points. The projection of $d\vec{x}$ on a coordinate direction is the component of $d\vec{x}$ in this direction

$$d\vec{x}_i = \vec{e}_i \cdot d\vec{x} \tag{B.2}$$

From the rules for the addition of vectors it follows that

$$d\vec{x} = dx_1\vec{e}_1 + dx_2\vec{e}_2 + dx_3\vec{e}_3 = \sum_{i=1}^{3} dx_i\vec{e}_i \equiv dx_i\vec{e}_i \equiv dx_j\vec{e}_j \tag{B.3}$$

where the identity signs show the meaning of the Cartesian summation convention: if a (dummy) Cartesian coordinate subscript occurs twice in a quantity or in products of quantities, then a summation over the range of values of the subscript is implied.
From (B.1) and (B.3) it follows that

$$(dx)^2 \equiv (d\vec{x}) \cdot (d\vec{x}) = (\vec{e}_i \cdot \vec{e}_j) dx_i\, dx_j = \delta_{ij} dx_i\, dx_j$$
$$\equiv dx_i\, dx_i = (dx_1)^2 + (dx_2)^2 + (dx_3)^2 \quad \text{(B.4)}$$

In (B.4), the magnitude of $d\vec{x}$ squared is given by the Pythagoras theorem. The rules for calculating the magnitude of a line element defines the *metric* of a space, For a Cartesian coordinate system the metric is given by the *metric coefficients* δ_{ij}. The vector $d\vec{x}$ does not change if another triple of unit vectors is chosen to define a Cartesian coordinate system. The vector $d\vec{x}$ is a coordinate invariant quantity. The components of the vector are not coordinate invariant, but transform in a coordinate transformation such that the vector $d\vec{x}$ remains invariant.

The magnitude of a vector need not to be defined by the length of a line segment, but it can also be defined by the area of a surface. The direction of a vector can be indicated by a unit vector. For a vector this unit vector coincides with the line segment, but if the magnitude of a vector is given by a surface it is pointed in a direction perpendicular to that surface. Such a vector is called an axial vector. In an Euclidean space there are no essential differences between vectors and axial vectors. In a Cartesian coordinate system the unit axial vectors point also in the direction of the coordinate axes. The magnitude of the axial vector is now not given by a line segment of unit length, but by the unit area of the square bordered by the two unit vectors perpendicular to the axial vector, so that

$$\vec{e}_i = \vec{e}_j \times \vec{e}_k \qquad \text{for} \qquad i, j, k = 1, 2, 3 \quad \text{or} \quad 3, 1, 2 \quad \text{or} \quad 2, 3, 1 \qquad \text{(B.5)}$$

where the cross between the two unit vectors denotes the vector product of two vectors. For example, the magnitude of $\vec{e}_2 \times \vec{e}_3$ is a unit square, and in the right-handed coordinate system the direction is given by $\vec{e}_1$. This defines also the orientation of the Euclidean space by a right-handed screw.

From (B.1) and (B.5) it follows that the unit cube in the Cartesian coordinate system is

$$\vec{e}_1 \cdot (\vec{e}_2 \times \vec{e}_3) = 1 \qquad \text{(B.6)}$$

and since $\vec{e}_2 \times \vec{e}_3 = -\vec{e}_3 \times \vec{e}_2$, and $\vec{e}_1 \cdot (\vec{e}_1 \times \vec{e}_3) = 0$ and so on for the cyclic permutatations of 1, 2, and 3, it follows that all these results can be summarized in the *permutation symbol* e_{ijk} defined by

$$e_{ijk} \equiv \vec{e}_i \cdot (\vec{e}_j \times \vec{e}_k) = \begin{cases} +1 & \text{for} \quad ijk = 123,\ 231,\ \text{or } 312 \\ -1 & \text{for} \quad ijk = 321,\ 132,\ \text{or } 213 \\ 0 & \text{for any two indices alike} \end{cases} \qquad \text{(B.7)}$$

Bird* noted that also $e_{ijk} = \frac{1}{2}(i-j)(j-k)(k-i)$.

By using (B.7), the unit axial vectors (B.5) can compactly be written as

$$\vec{e}_i \times \vec{e}_j = e_{ijk}\vec{e}_k \tag{B.8}$$

B.2. VECTORS IN EUCLIDEAN SPACE

The scalar product of two vectors $\vec{a}$ and $\vec{b}$ results from a *dot multiplication* between $\vec{a}$ and $\vec{b}$

$$\vec{a} \cdot \vec{b} = \vec{b} \cdot \vec{a} \tag{B.9}$$

The scalar (B.9) is equal to the magnitude of $\vec{a}$ multiplied by the projection of $\vec{b}$ on $\vec{a}$, or vice versa.

The projection of a vector $\vec{a}$ on a coordinate direction is the component of $\vec{a}$ in that direction

$$a_i = \vec{a} \cdot \vec{e}_i \tag{B.10}$$

so that

$$\vec{a} = a_i\,\vec{e}_i \tag{B.11}$$

Substitution of (B.11) into (B.9), and using (B.1) gives

$$\vec{a} \cdot \vec{b} = a_i b_j\,\vec{e}_i \cdot \vec{e}_j = a_i b_j\,\delta_{ij} = a_i b_i \tag{B.12}$$

In a dot multiplication the components on both sides of the dot become the same, so that, by the summation convention, (B.12) is the sum of the products of the corresponding components of $\vec{a}$ and $\vec{b}$.

The vector product of two vectors $\vec{a}$ and $\vec{b}$ is an (axial) vector with a magnitude equal to the area of the parallelogram defined by the two vectors $\vec{a}$ and $\vec{b}$ and perpendicular to the parallelogram in the direction a right-handed screw has if turned from $\vec{a}$ towards $\vec{b}$ over the smallest angle

$$\vec{a} \times \vec{b} = a_i b_j\,\vec{e}_i \times \vec{e}_j = e_{ijk} a_i b_j\,\vec{e}_k \tag{B.13}$$

where use has been made of (B.8). From (B.13) it follows that

$$\vec{a} \times \vec{b} = -\vec{b} \times \vec{a} \tag{B.14}$$

The scalar

$$\vec{a} \cdot (\vec{b} \times \vec{c}) = e_{ijk} a_i b_j c_k = \begin{vmatrix} a_1 & b_1 & c_1 \\ a_2 & b_2 & c_2 \\ a_3 & b_3 & c_3 \end{vmatrix} \tag{B.15}$$

is the volume of the parallelepiped defined by the three vectors $\vec{a}$, $\vec{b}$, and $\vec{c}$.

---

* Bird R.B., Armstrong R.C. and Hassager O. 1987. *Dynamics of Polymeric Liquids*. 2nd ed. John Wiley & Sons, New York. See p. 559.

B.3. TENSORS IN EUCLIDEAN SPACE

A second-order tensor **A** can be defined as a linear operator between two vectors

$$\vec{b} = \mathbf{A} \cdot \vec{a} \tag{B.16}$$

or using (B.10)

$$b_i = \vec{e}_i \cdot \vec{b} = \vec{e}_i \cdot \mathbf{A} \cdot \vec{a} = \vec{e}_i \cdot \mathbf{A} \cdot \vec{e}_j \, a_j = A_{ij} a_j \tag{B.17}$$

where $A_{ij} = \vec{e}_i \cdot \mathbf{A} \cdot \vec{e}_j$ are the nine components of the tensor **A** with respect to the unit vectors $\vec{e}_i$. The dot between **A** and $\vec{a}$ denote the dot multiplication. The effect of a dot multiplication is that the subscripts on both sides of the dot become the same (as in (B.12)) and hence the summation convention applies

$$\mathbf{A} \cdot \vec{a} = A_{ij} a_j \, \vec{e}_i \equiv \sum_{j=1}^{3} A_{ij} a_j \, \vec{e}_i \tag{B.18}$$

The nine components of the tensor are determined by (B.17) only if this relation holds for three non-coplanar vectors $\vec{a}^{\,(i)}$, so that

$$\vec{b}^{\,(p)} = \mathbf{A} \cdot \vec{a}^{\,(p)} \quad \text{for} \quad p = 1, 2, 3 \quad \text{and} \quad \vec{a}^{\,(1)} \cdot (\vec{a}^{\,(2)} \times \vec{a}^{\,(3)}) \neq 0 \tag{B.19}$$

Since a three-dimensional vector-space can be spanned by any three non-coplanar vectors, a second-order tensor can also be defined as an operator that maps two vector-spaces onto each other.

A *dyadic multiplication* of two vectors $\vec{a}$ and $\vec{b}$ is a second-order tensor $\vec{a}\,\vec{b}$, with Cartesian components $a_i b_j$. With (B.18) it can be seen that

$$(\vec{a}\,\vec{b}\,) \cdot \vec{c} = \vec{a}\,(\vec{b} \cdot \vec{c}) \qquad \text{and} \qquad \vec{c} \cdot (\vec{a}\vec{b}\,) = (\vec{c} \cdot \vec{a})\,\vec{c} \tag{B.20}$$

From the definition of the dyadic product and from (B.11) it follows that

$$\vec{a}\vec{b} = a_i b_j \, \vec{e}_i \vec{e}_j \tag{B.21}$$

where the dyadic product $\vec{e}_i\vec{e}_j$ defines a *unit dyad.* and a second-order tensor tensor can be written as

$$\mathbf{A} = A_{ij} \, \vec{e}_i \vec{e}_j \tag{B.22}$$

From (B.22) and (B.20) it follows that

$$A_{ij} = \vec{e}_i \cdot \mathbf{A} \cdot \vec{e}_j \qquad \text{and} \qquad \mathbf{A} = (\vec{e}_i \cdot \mathbf{A} \cdot \vec{e}_j)\, \vec{e}_i \vec{e}_j \tag{B.23}$$

The second-order tensor with components given by the Kronecker delta is the *unit tensor*

$$\mathbf{I} = \delta_{ij} (\vec{e}_i \vec{e}_j) \tag{B.24}$$

for which

$$\vec{a} \cdot \mathbf{I} = \mathbf{I} \cdot \vec{a} = \vec{a} \tag{B.25}$$

The dot multiplication between two second-order tensors yields a second-order tensor

$$\begin{aligned}\mathbf{A} \cdot \mathbf{B} &= A_{ij} B_{kl} \vec{e}_i \vec{e}_j \cdot \vec{e}_k \vec{e}_l = A_{ij} B_{kl} (\vec{e}_j \cdot \vec{e}_k)\, \vec{e}_i \vec{e}_l \\ &= A_{ij} B_{kl} \delta_{jk}\, \vec{e}_i \vec{e}_l = A_{ij} B_{jl}\, \vec{e}_i \vec{e}_l \end{aligned} \tag{B.26}$$

For (B.26) a further contraction of the coordinate subscripts is possible, and this leads to the definition of *double dot multiplication*

$$\mathbf{A} : \mathbf{B} = A_{ij} B_{ji} \tag{B.27}$$

Note the order of the coordinate subscripts in a double dot multiplication!

The *trace* of a second-order tensor $\mathbf{A}$ is

$$\text{trace } \mathbf{A} = \mathbf{A} : \mathbf{I} = \mathbf{I} : \mathbf{A} = A_{ii} = A_{11} + A_{22} + A_{33} \tag{B.28}$$

The *deviator* of a second-order tensor $\mathbf{A}$ is its traceless part

$$\mathring{\mathbf{A}} = \mathbf{A} - \tfrac{1}{3}(\mathbf{I} : \mathbf{A})\, \mathbf{I} \tag{B.29}$$

In the three-dimensional space $\mathbf{I} : \mathbf{I} = 3$, so that from (B.29) it follows that

$$\mathbf{I} : \mathring{\mathbf{A}} = 0 \tag{B.30}$$

The *transpose* of a tensor $\mathbf{A}$ is defined by

$$\mathbf{A}^{\mathrm{T}} = A_{ij}\, \vec{e}_j \vec{e}_i = A_{ji}\, \vec{e}_i \vec{e}_j \tag{B.31}$$

For the transpose of a tensor product obtained by the dot multiplication of two second-order tensors $\mathbf{A}$ and $\mathbf{B}$, it follows that

$$\left(\mathbf{A} \cdot \mathbf{B}\right)^{\mathrm{T}} = A_{jk} B_{ki}\, \vec{e}_i \vec{e}_j = B^{\mathrm{T}}_{ik} A^{\mathrm{T}}_{kj}\, \vec{e}_i \vec{e}_j = \mathbf{B}^{\mathrm{T}} \cdot \mathbf{A}^{\mathrm{T}} \tag{B.32}$$

The *symmetric part* of a tensor is

$$\mathbf{A}^{\mathrm{S}} = \tfrac{1}{2}\left(\mathbf{A} + \mathbf{A}^{\mathrm{T}}\right) = \left(\mathbf{A}^{\mathrm{S}}\right)^{\mathrm{T}} \tag{B.33}$$

and the *anti-symmetric part* of a tensor is

$$\mathbf{A}^{\mathrm{A}} = \tfrac{1}{2}\left(\mathbf{A} - \mathbf{A}^{\mathrm{T}}\right) = -\left(\mathbf{A}^{\mathrm{A}}\right)^{\mathrm{T}} \tag{B.34}$$

Clearly, a tensor **A** can always be split in a symmetric and an antisymmetric part.

With the definition of the dyadic product, higher order tensors can be introduced analogously to (B.22). The third-order tensor with the components given by the permutaion symbol (B.7) is the *permutation tensor* or *alternating unit tensor*

$$\mathbf{e} = e_{ijk}\,\vec{e}_i\vec{e}_j\vec{e}_k \tag{B.35}$$

and the vector product (B.13) can now be written as

$$\vec{a} \times \vec{b} = a_i b_j\,\vec{e}_i \times \vec{e}_j = e_{ijk} a_i b_j\,\vec{e}_k = -a_i e_{ikj} b_j\,\vec{e}_k = -\vec{a} \cdot \mathbf{e} \cdot \vec{b} \tag{B.36}$$

The cross multiplication amounts to the replacement of cross symbol $\times$ by $\cdot\,\mathbf{e}\,\cdot$ and adding a minus sign. Cross multiplications between between tensors or tensors and vectors are now readily written down. For example the cross multiplication between a tensor **A** and a vector $\vec{a}$ yields the tensor product

$$\mathbf{A} \times \vec{a} = -\mathbf{A} \cdot \mathbf{e} \cdot \vec{a} = -A_{ij} e_{jkl} a_l\,\vec{e}_i\vec{e}_k \tag{B.37}$$

The *determinant* of a tensor **A** is (see also (B.15)) defined by

$$\det \mathbf{A} = \begin{vmatrix} A_{11} & A_{12} & A_{13} \\ A_{21} & A_{22} & A_{33} \\ A_{31} & A_{32} & A_{33} \end{vmatrix} = e_{pqr} A_{p1} A_{q2} A_{r3} \tag{B.38}$$

From (B.38) and (B.7) it follows that

$$e_{ijk} \det \mathbf{A} = e_{pqr} A_{pi} A_{qj} A_{rk} \tag{B.39}$$

Multiplication of (B.39) by $a_i b_j c_k$ and use of the definition for the triple scalar product (B.15), the definition of the determinant of a tensor **A** becomes

$$\det \mathbf{A} = \frac{(\mathbf{A} \cdot \vec{a}) \cdot [(\mathbf{A} \cdot \vec{b}) \times (\mathbf{A} \cdot \vec{c})]}{\vec{a} \cdot (\vec{b} \times \vec{c})} \tag{B.40}$$

which is the ratio of the volumes of the parallellepiped defined by the vectors $\mathbf{A} \cdot \vec{a}$, $\mathbf{A} \cdot \vec{b}$, $\mathbf{A} \cdot \vec{c}$, and the parallellepiped defined by the vectors $\vec{a}$, $\vec{b}$, $\vec{c}$.

B.4. DIFFERENTIATION

Differentiation with respect to the spatial coordinate $\vec{x}$ is denoted by

$$\vec{\nabla} \equiv \frac{\partial}{\partial \vec{x}} = \vec{e}_i \frac{\partial}{\partial x_i} \tag{B.41}$$

This operator $\vec{\nabla}$ is a vector differential operator, which can operate on scalars, vectors and tensors. The *gradient* is defined by a dyadic multiplication as follows

$$\operatorname{grad} a = \vec{\nabla} a = \frac{\partial a}{\partial \vec{x}} = \frac{\partial a}{\partial x_i}\,\vec{e}_i \qquad \operatorname{grad} \vec{a} = \vec{\nabla}\vec{a} = \frac{\partial \vec{a}}{\partial \vec{x}} = \frac{\partial a_j}{\partial x_i}\,\vec{e}_i\vec{e}_j$$

$$\operatorname{grad} \mathbf{A} = \vec{\nabla}\mathbf{A} = \frac{\partial \mathbf{A}}{\partial \vec{x}} = \frac{\partial A_{jk}}{\partial x_i}\,\vec{e}_i\vec{e}_j\vec{e}_k$$

The *divergence* is defined by a dot multiplication as follows

$$\operatorname{div} \vec{a} = \vec{\nabla}\cdot\vec{a} = \frac{\partial}{\partial \vec{x}}\cdot\vec{a} = \frac{\partial a_i}{\partial x_i} \qquad \operatorname{div} \mathbf{A} = \vec{\nabla}\cdot\mathbf{A} = \frac{\partial}{\partial \vec{x}}\cdot\mathbf{A} = \frac{\partial A_{ji}}{\partial x_j}\,\vec{e}_i \tag{B.42}$$

The *curl* is defined by a cross multiplication as follows

$$\operatorname{curl} \vec{a} = \vec{\nabla}\times\vec{a} = e_{ijk}\frac{\partial a_k}{\partial x_j}\,\vec{e}_i \qquad \operatorname{curl} \mathbf{A} = \vec{\nabla}\times\mathbf{A} = e_{ipq}\frac{\partial A_{qj}}{\partial x_p}\,\vec{e}_i\vec{e}_j \tag{B.43}$$

The *Laplacian* operator is defined as the scalar differential operator that results from a dot multiplication between two nabla operators, or

$$\vec{\nabla}\cdot\vec{\nabla} = \nabla^2 = \frac{\partial^2}{\partial x_1^2} + \frac{\partial^2}{\partial x_2^2} + \frac{\partial^2}{\partial x_3^2} \tag{B.44}$$

AUTHOR INDEX

SUBJECT INDEX

E

M

P

S